Springer Texts in Statistics

Springer Texts in Statistics

Alfred: Elements of Statistics for the Life and Social Sciences
Berger: Introduction to Probability and Stochastic Processes, Second Edition
Bilodeau and Brenner: Theory of Multivariate Statistics
Blom: Probability and Statistics: Theory and Applications
Brockwell and Davis: An Introduction to Time Series and Forecasting
Carmona: Statistical Analysis of Financial Data in S-Plus
Chow and Teicher: Probability Theory: Independence, Interchangeability, Martingales, Third Edition
Christensen: Advanced Linear Modeling: Multivariate, Times Series, and Spatial Data; Nonparametric Regression and Response Surface Maximization, Second Edition
Christensen: Log-Linear Models and Logistic Regression, Second Edition
Christensen: Plane Answers to Complex Questions: The Theory of Linear Models, Second Edition
Creighton: A First Course in Probability Models and Statistical Inference
Davis: Statistical Methods for the Analysis of Repeated Measurements
Dean and Voss: Design and Analysis of Experiments
Dekking, Kraaikamp, Lopuhaä, and Meester: A Modern Introduction to Probability and Statistics
du Toit, Steyn, and Stumpf: Graphical Exploratory Data Analysis
Durrett: Essential of Stochastic Processes
Edwards: Introduction to Graphical Modeling, Second Edition
Everitt: An R and S-PLUS® Companion to Multivariate Analysis
Finkelstein and Levin: Statistics for Lawyers
Flury: A First Course in Multivariate Statistics
Gut: Probability: A Graduate Course
Heiberger and Holland: Statistical Analysis and Data Display: An Intermediate Course with Examples in S-PLUS, R, and SAS
Jobson: Applied Multivariate Data Analysis, Volume I: Regression and Experimental Design
Jobson: Applied Multivariate Data Analysis, Volume II: Categorical and Multivariate Methods
Kalbfleisch: Probability and Statistical Inference, Volume I: Probability, Second Edition
Kalbfleisch: Probability and Statistical Inference, Volume II: Statistical Interference, Second Edition
Karr: Probability
Keyfitz: Applied Mathematical Demography, Second Edition
Kiefer: Introduction to Statistical Inference
Kokoska and Nevison: Statistical Tables and Formulae
Kulkarni: Modeling, Analysis, Design, and Control of Stochastic Systems
Lange: Applied Probability
Lange: Optimization

(continued after index)

David Ruppert

Statistics and Finance

An Introduction

David Ruppert
School of Operations Research and
Industrial Engineering
Cornell University
Ithaca, NY 14853-3801
USA
dr24@cornell.edu

Library of Congress Cataloging-in-Publication Data
Ruppert, David.
Statistics and finance: an introduction / David Ruppert.
p. cm.—(Springer texts in statistics)
Includes bibliographical references and index.
ISBN 0-387-20270-6 (alk. paper)
1. Finance—Statistical methods. 2. Statistics. I. Title. II. Series.
HG176.5.R87 2004
332`.01`5195—dc22 2003063814

ISBN 0-387-20270-6 Printed on acid-free paper.

Printed in the United States of America.

9 8 7 6 5 4 3 (Corrected as of the 2nd printing, 2006)

springer.com

To Ray,

a true friend and a wonderful colleague.

Preface

This book is an introduction to both statistical modeling and finance, with special attention to the interaction between the two. The reader is assumed to have some background in probability and statistics, but perhaps only an introductory course, and all of the statistical theory needed for the applications in finance is introduced as required. No background in finance is needed.

This textbook grew out of lecture notes for the third and fourth year undergraduate and masters level course, OR&IE 473, at Cornell. Most of the students are in my own School of Operations Research & Industrial Engineering, though students in a wide variety of other majors find the course interesting. Many of the OR&IE majors go on to careers in banking and finance and our Master of Engineering (MENG) program has a popular financial engineering option. However, the course is also popular with students planning careers outside finance. For these students I see finance mainly as a means for learning how to apply the basic tools of operations research: probability, optimization, simulation, and, especially, my own area of statistics.

The prerequisites for the course are two years of engineering mathematics including matrix algebra and multivariate calculus, probability, and statistics, which are all required courses in the OR&IE major. Many of the students have no prior knowledge of finance and economics.

The emphasis in OR&IE 473 is on empirical research methods, that is, data analysis and statistical inference. I have been teaching and conducting research in statistics for over 25 years and I have worked on a variety of applications, but my study of finance began only recently in response to student interest and my own curiosity. I have found finance to be a fascinating subject, both in itself and for its ties with statistics.

This textbook and the course it is based upon have several goals:

- To reinforce the material taught in earlier courses by illustrating the main concepts of probability and statistics with concrete examples from finance. I have found that even after a year of probability and statistics, many students still have hazy notions of the underlying concepts and how they

can be applied to the real world. After seeing statistics and probability applied to problems in finance, their understanding is much deeper.
- To serve as a capstone course integrating statistics, probability and, to a lesser extent, optimization and simulation.
- To introduce undergraduates to finance and financial engineering.
- To help students appreciate the role of empirical research in finance and in operations research.
- To expose students to new statistical methods: time series, GARCH models, resampling (bootstrap), and nonparametric regression.
- To provide examples of using the MATLAB and SAS software packages. This book contains many examples of programming in SAS. MATLAB is used occasionally, for example, for interpolation, for finding efficient portfolios in Chapter 5, and for resampling in Chapter 10, where calculation cannot be done with standard SAS statistical procedures.[1] This book is not an instruction manual for MATLAB or SAS, but students in my course have learned to use these languages by following the examples in the book. Ideally, students will already have some familiarity with these languages or will receive additional instruction in a course. For the independent reader, I have provided references to instruction manuals for SAS and MATLAB at the end of Chapters 2 and 4.

It was difficult to decide upon the statistical software to use in this book. Other statistical packages such as Splus and R are also very powerful and have many useful features, and many analysts prefer them to SAS. I chose SAS since many of the students in my department wish to learn it because SAS is so widely used in industry.

The graphical software in MATLAB is excellent because of the high quality figures it produces, its many features, and the small size of the Postscript files it exports. For this reason, all figures in this book have been produced by MATLAB.

I have intentionally omitted material on stochastic calculus, since that topic deserves (and has) courses of its own and is already well covered in other texts, such as Baxter and Rennie (1996), Neftci (1996), and Mikosch (1998).

My colleague at Cornell, Marty Wells, first taught OR&IE 473. It was his insight that financial engineers would benefit from a better appreciation of statistical methods and data analysis and that perhaps a statistician could best design a course to serve this need. Marty taught the course for two years, and then I took over. As I prepared to teach OR&IE 473 for the first time, I realized that there were no textbooks at the undergraduate level that I could use, so I began typing rough notes to put on a course website. The notes

[1] SAS contains a programming language called IML and optimization routines in its OR package that greatly extend SAS's range of applications and, in particular, could be used for portfolio optimization or resampling, but this book does not discuss these features of SAS and instead uses MATLAB.

have been expanded and refined over the years, and a number of colleagues suggested that I turn them into a textbook.

I have found teaching this material very rewarding. Students like to see "real data." Finance is an excellent source of examples for a statistics course, and financial data are readily available from many websites and sources such as Datastream, Bloomberg, and Reuters. Most students have some familiarity with the stock market and are eager to learn more. Everyone is interested in money.

Although written at an introductory level, this book is able to cover some relatively recent topics such as the bootstrap, penalized splines, and behavioral finance. I especially enjoyed writing Chapter 10 on the bootstrap. The bootstrap has been applied before in finance and, in particular, to the portfolio optimization by Michaud (1998), but I felt that more could be done. Markowitz's theory of optimal portfolios is found in most textbooks on finance, but rarely is an appreciation given of just how sensitive the optimal portfolios are to small, and inevitable, estimation errors in the inputs. The bootstrap shows how serious this problem can be and where the difficulty lies. I particularly like the bootstrap because bootstrap results can be plotted to illustrate the underlying ideas. Throughout the book, I have tried as much as possible to use visualization rather than mathematical calculations to explain concepts, and the bootstrap fits well into this pedagogy.

This book could be used as a text in a course taught to mathematically prepared undergraduates, e.g., majoring in mathematics, statistics, or engineering. I believe that it will also be useful to practicing financial engineers who find that they need more knowledge of probability and statistics. I have been the advisor on many Master of Engineering projects sponsored by clients in business and industry, many in the finance industry. I often find that the students have not mastered the statistical tools needed for the data analysis required by these projects, so I frequently play the role of statistical consultant. A number of the topics in the textbook were developed from consulting sessions. Many of the chapters, especially 2, 4, 6, 10, and 12, will be useful to someone interested in learning applied statistics and data analysis but without a particular interest in finance, provided that such a reader is not bothered that all the examples come from finance.

At the end of each chapter, I have provided a section of bibliographic notes to indicate some of the papers and books that an interested reader might wish to explore. These notes should by no means be considered as complete literature reviews but rather as a sparse sampling of the literature. I apologize to authors whom I have omitted. An attempt was made to include for each subject a few recent references that can be consulted for more complete bibliographies.

This book contains a brief glossary. Boldface is used in the text to indicate concepts explained in the glossary. The glossary does not give precise mathematical definitions; these are in the text.

The book has grown to the point where it now contains too much material to cover at a leisurely pace in one semester. This has the advantage that an instructor has some choice in what to cover. I feel that most of Chapters 1 to 9 are core material, but starred sections either in these chapters or later in the book contain specialized or more advanced material and can be omitted. Risk management is a hot topic, so Chapter 11 on VaR probably should be included. Section 11.2.5 requires Chapter 10 on the bootstrap, but if that section is skipped then Chapter 10 is not necessary. Chapter 12 on GARCH models is not essential but was added because of student interest. The last time I taught from this book, I did not cover Chapters 11 on VaR (because it had not yet been written) or 14 on behavioral finance and I omitted several sections of other chapters. The book could serve as a textbook for a course in statistics aimed at students in computational finance. In that case, much of the material in Chapters 3, 5, 7, 8, and 9 might already be familiar to the students and then the course could easily cover all of the remaining chapters.

I thank my colleagues Bob Jarrow, Phil Protter, Jeremy Staum, Marty Wells, and Yan Yu for guidance as I was learning finance. I am especially grateful to Antje Berndt, Ciprian Crainiceanu, Kay Giesecke, and Anne Shapiro for their careful reading of earlier drafts and many helpful comments. This book has benefited from the comments of anonymous reviewers to whom I am very grateful. Of course, I am solely responsible for any errors.

While this book was being written, I had the good fortune to be given the Andrew Schultz, Jr. Professorship of Engineering. Funds from the endowment of this chair supported my work on the book in the summer of 2003. I am very appreciative for this support which has allowed me to finish this project in a timely manner.

I would much appreciate feedback and suggestions. They can be emailed to me at dr24@cornell.edu.

References

Baxter, M., and Rennie, A. (1998) *Financial Calculus: An Introduction to Derivative Pricing*, Cambridge University Press, Cambridge.

Michaud, R. O. (1998) *Efficient Asset Management*, Harvard Business School Press, Boston.

Mikosch, T. (1998) *Elementary Stochastic Calculus with Finance in View*, World Scientific, Singapore.

Neftci, S. N. (1996) *An Introduction To The Mathematics Of Financial Derivatives*, Academic Press, San Diego.

Ithaca, New York
January 2004

David Ruppert

Contents

Notation

The following conventions are observed as much as possible:

- Lower-case letters, e.g., a and b, are used for non random scalars.
- Lower-case boldface letters, e.g., $\boldsymbol{a}$, $\boldsymbol{b}$, and $\boldsymbol{\theta}$, are used for non random vectors.
- Upper-case letters, e.g., X and Y, are used for random variables.
- Upper-case bold letters either early in the Roman alphabet or Greek, e.g., $\boldsymbol{A}$, $\boldsymbol{B}$, and $\boldsymbol{\Omega}$, are used for non random matrices.
- Upper-case bold letters either later in the Roman alphabet or in the Greek alphabet with a "hat," e.g., $\boldsymbol{X}$, $\boldsymbol{Y}$, and $\widehat{\boldsymbol{\theta}}$, will be used for random vectors.
- A Greek letter denotes a parameter, e.g., θ.
- A boldface Greek letter, e.g., $\boldsymbol{\theta}$, denotes a vector of parameters.
- A hat over a parameter or parameter vector, e.g., $\widehat{\theta}$ and $\widehat{\boldsymbol{\theta}}$, denotes an estimator of the corresponding parameter or parameter vector.
- $A \cap B$ and $A \cup B$ are, respectively, the intersection and union of the sets A and B.
- If A is some statement, then $I\{A\}$ is called the indicator of A and is equal to 1 if A is true and equal to 0 if A is false.

Vectors are column vectors and transposed vectors are rows, e.g.,

$$\boldsymbol{x} = \begin{pmatrix} x_1 \\ \vdots \\ x_n \end{pmatrix}$$

and

$$\boldsymbol{x}^{\mathsf{T}} = \begin{pmatrix} x_1 & \cdots & x_n \end{pmatrix}.$$

1

Introduction

The book grew out of a course first called "Empirical Research Methods in Financial Engineering." *Empirical* means derived from experience, observation, or experiment, so the book is about working with data and doing statistical analysis. **Financial engineering** is the construction of financial products such as stock options, interest rate derivatives, and credit derivatives. The course has been renamed "Operations Research Tools for Financial Engineering," because it also covers applications of probability, simulation, and optimization to financial engineering.

Much of finance is concerned with measuring and managing financial risk. The **return** on an investment is its revenue as a fraction of the initial investment. If one invests at time t_1 in an asset with price P_{t_1} and the price later at time t_2 is P_{t_2}, then the net return for the holding period from t_1 to t_2 is $(P_{t_2} - P_{t_1})/P_{t_1}$. For most assets, future returns cannot be known exactly and therefore are random variables. **Risk** means uncertainty in future returns from an investment, in particular, that the investment could earn less than the expected return and even result in a loss, that is, a negative return. Risk is often measured by the standard deviation of the return, which we also call the volatility. Recently there has been a trend toward measuring risk by Value-at-Risk, often denoted by VaR. VaR focuses on the maximum loss and is a more direct indication of financial risk than the standard deviation of the return; see Chapter 11. Because risk depends upon the probability distribution of a return, probability and statistics are fundamental tools for finance. Probability is needed for risk calculations, and statistics is needed to estimate parameters such as the standard deviation of a return or to test hypotheses such as the so-called "random walk hypothesis" which states that future returns are independent of the past.

Finance makes extensive use of probability models, for example, those used to derive the famous Black-Scholes formula. Use of these models raises important questions of a statistical nature such as: Are these models supported by financial markets data? How are the parameters in these models estimated? Can the models be simplified or, conversely, should they be elaborated?

This chapter provides glimpses into the rest of the book. Do not worry about the details, since they are given later. First, let us look ahead to Chapter 8 and the famous Black-Scholes formula for the price of a European call option. A **European call option** gives one the right, but *not* the obligation, to purchase a specified number of shares, usually 100, of a stock at a specified price called the **exercise or strike price** at a specified future date called the maturity.[1] If one expects a stock's price to rise, then one might consider investing in the stock, but this entails risk because the price may decline despite one's expectation. The risk is potentially large since the price could decline all the way to 0. Instead of purchasing the stock, one might buy a call option on the stock. Purchasing a call is a means to earn a profit if the stock's price rises while managing the risk that the stock's price will decline. If the stock's price is above the exercise price K at the maturity date, then one can exercise the option, buy the stock at price K, and then sell it at the market price and earn an immediate profit. If the price is below K, then one does not exercise the option and the only loss is the initial cost of purchasing the option, which is called the **premium**.

Now let us look at the problem of pricing a European call option. For notation, "now" is called time 0 and the maturity date of the option is denoted by T. For simplicity, assume the option is for one share. Let S_t be the price of the stock at time t for $0 \leq t \leq T$. At time 0, S_0, T, and K are known but S_T is unknown. At time T, S_T will become known. If at time T we learn that $S_T > K$ then we will exercise the option and purchase one share. We can immediately sell the share for S_T dollars and earn a profit of $S_T - K$ dollars. If at time T, $S_T < K$ then we do not exercise the option. The option expires and we lose the amount of the premium, but no more. The value of the option at time T is, therefore, $\max\{0, S_T - K\}$ where "max" denotes the maximum of a set.[2] But right now at time 0 with S_T unknown, what is the value of the option, i.e., the premium for which it should sell on the market? This is a very challenging question whose solution won a Nobel Prize in Economic Sciences.

Prior to the 1970s, options were priced by the "seat of the pants." Then Black, Scholes, and Merton deduced the price of a call option that makes the market free from arbitrage opportunities. One of the fundamental concepts of finance is arbitrage. **Arbitrage** means making a risk-free profit with no invested capital. "Risk-free profit" means that no matter what happens in the market you will at least break even and there is a nonzero probability that you will make money. For example, if you could borrow for one year from a bank at 2% and use the money to purchase one-year Treasury bills paying 3% you would be guaranteed a profit with none of your own capital invested.

[1] European options can only be exercised at the maturity date, but American options can be exercised at any time before or at the maturity date.

[2] The maximum of a set of numbers is the largest number in the set. Therefore, $\max\{0, S_T - K\}$ equals 0 if $S_T - K$ is negative and equals $S_T - K$ if $S_T - K$ is positive.

There is no limit to the amount you could make. Arbitrage opportunites are expected to be rare and to disappear quickly when they do exist. Black and Scholes assumed that the market has no arbitrage possibilities and then from a mathematical model (and much hard thinking) they arrived at a formula for the price of a call. They assumed that one can lend and borrow at a risk-free rate r.[3] Thus, if F_t is the price at time t of a risk-free bond purchased for \$1 at time 0, then $F_0 = 1$ and, assuming continuous compounding as explained shortly,

$$F_t = F_0 \exp(rt) = \exp(rt).$$

Let S_t be the price of the underlying stock. Black, Scholes, and Merton also assumed that

$$S_t = S_0 \exp\{(\mu - \sigma^2/2)t + \sigma B_t\},$$

where μ is a drift or growth rate or mean rate of return, B_t is a Brownian motion stochastic process, and σ is a standard deviation that measures the volatility of the stock. In Chapter 8, you will learn more precisely what this model means. Right now, the "take home" message is that there are mathematical models of stock price movements that lead to widely used formulas and that can be checked against the data. Also, there are important parameters such as μ and σ that must be estimated from data.[4]

The *Black-Scholes formula* is

$$C = \Phi(d_1)S_0 - \Phi(d_2)K \exp(-rT),$$

where C is the price of the option at time 0, Φ is the standard normal cumulative distribution function (CDF),

$$d_1 = \frac{\log(S_0/K) + (r + \sigma^2/2)T}{\sigma\sqrt{T}}, \quad \text{and} \quad d_2 = d_1 - \sigma\sqrt{T}.$$

Here, as elsewhere in the book, log *always* means natural logarithm. The formula is, quite obviously, complicated and is not easy to derive, but it is easy to compute and was hardwired into calculators almost immediately after it was discovered; the Black-Scholes formula and handheld calculators both emerged in the early 1970s.

We are interested in the underlying assumptions behind the formula. Remember: GI — GO (garbage in, garbage out)! If the assumptions do not hold, then there is no reason to trust the Black-Scholes formula, despite the impressive mathematics behind it. The assumptions can be checked by analyzing the behavior of actual stock prices using the techniques discussed in this book.

[3] Risk-free borrowing and lending means that there is no risk to the lender of default by the borrower, as is true for Treasury securities or deposits guaranteed by the U.S. government.

[4] For options pricing, the value of μ is not needed, but this parameter would need to be estimated in other contexts such as finding an efficient portfolio by the methods explained in Chapter 5.

The equation $F_t = \exp(rt)$ of continuous compounding is the solution to the differential equation

$$\frac{dF_t}{dt} = rF_t.$$

The general solution is $F_t = F_0 \exp(rt)$ and $F_0 = 1$ since we have assumed that the bond can be purchased for \$1 at time 0.

Where does

$$S_t = S_0 \exp\{(\mu - \sigma^2/2)t + \sigma B_t\},$$

come from? If σ were 0, then this would be exponential growth, $S_t = S_0 \exp(\mu t)$, just like the bond price F_t. The term σB_t comes from the random behavior of stock prices. The parameter σ is a standard deviation of the changes in the log of the stock price and is a measure of volatility and risk. The random process B_t is something we will learn much more about.

In this book we study statistical models of financial markets, learn to test the models (do they fit financial markets data adequately?), and estimate parameters in the models such as σ that are essential for correct pricing of financial products. One key question that we address is "how do the prices of stocks and other financial assets behave?" Chapter 3 introduces the now classic **efficient market hypothesis** while Chapter 14 discusses more recent studies suggesting that the market often does not price assets nearly as efficiently as economists have liked to believe.

Another question is how to allocate capital to a variety of possible investments. Risks can be reduced by diversification, that is, by spreading one's money around rather than concentrating on a few assets. If one holds a collection of stocks and other assets then this collection is called a *portfolio*. The return on a portfolio is a weighted average of the returns on the individual assets in the portfolio and a weighted average of random variables is generally less variable than the individual random variables, which is why diversification can reduce risk. Portfolio analysis discusses how one should diversify and addresses questions such as what portfolio has the smallest standard deviation of its return subject to the expected return being at least a certain specified value. Portfolio theory leads to one of the cornerstones of modern finance, the capital asset pricing model (CAPM). When I first studied the CAPM a few years ago, as a statistician I had a strong sense of being in familiar territory since much of the CAPM theory is identical to that of linear regression. For example, R^2, a well-known statistical concept, has an interesting financial interpretation in CAPM as the proportion of squared risk (variance) of an asset due to price fluctuations of the entire financial market.

After Chapter 2 introduces probability and statistics, our study of finance begins in Chapter 3 with the behavior of returns on assets. Chapter 4 looks at "ARIMA models" for "time series." Time series are sequences of data sampled over time, so much of the data from financial markets are time series. ARIMA models are stochastic processes, that is, probability models for sequences of random variables. After looking at returns and time series models of return behavior, in Chapter 5 we study optimal portfolios of risky assets (e.g., stocks) and of risky assets and risk-free assets (e.g., short-term U.S. Treasury bills). Chapter 6 covers one of the most important areas of applied statistics, regression. In Chapter 7 portfolio theory and regression are applied to the CAPM. Looking even farther ahead, in Chapter 8 we return to the pricing of stock options by the Black-Scholes formula and Chapters 9 to 13 cover other areas of statistics and finance such as fixed-income securities, GARCH models of non-constant volatility, resampling, risk management, splines, and the term structure of interest rates. In Chapter 14 we examine empirical tests of the efficient market hypothesis and the relatively new field of behavioral finance dealing with the psychology of financial markets.

The main purpose of an introduction to finance should be to acquaint students with the foundations of finance such as portfolio optimization, the CAPM, the efficient market hypothesis, and the geometric random walk and Brownian motion models. Many of these theories and models are under intense challenge by academicians and new ones are being developed. Factor models and arbitrage pricing theory appears to be replacing the CAPM, at least to some extent. The efficient market hypothesis has been subject to heated debate since the development of behavioral finance. In the area of risk management, VaR has become the industry standard but academics have shown that VaR has serious shortcomings and they have labelled it "incoherent."[5] Nonetheless, I believe that the classic material covered here is essential for students to learn, both because it is still widely used and, more importantly, because newer theories of finance will be built upon it. I try to convey to the reader the understanding that the classic models should not be accepted as "the truth" but rather regarded as tools that have proven useful in the past and now are, at the very least, useful hypotheses to test with financial markets data. Improvements to them will be based on empirical research, which is the theme of this book.

1.1 References

Baxter, M., and Rennie, A. (1998) *Financial Calculus: An Introduction to Derivative Pricing, Corrected Reprinting*, Cambridge University Press, Cambridge.

[5] See Section 11.5.

2

Probability and Statistical Models

2.1 Introduction

It is assumed that the reader is already at least somewhat familiar with the basics of probability and statistics. The goals of this chapter are to

1. review these basics;
2. discuss more advanced; topics needed in our empirical study of financial markets data such as random vectors, covariance matrices, best linear prediction, heavy-tailed distributions, maximum likelihood estimation, and likelihood ratio tests;
3. provide glimpses of how probability and statistics are applied to finance problems in this book; and
4. introduce notation that is used throughout the book.

Readers without a solid background in statistics should read this chapter carefully before proceeding. Readers well prepared in statistics should at least skim this chapter before proceeding with the main topics of the book, and then refer back to relevant material as necessary.

2.2 Axioms of Probability

Probability theory is a mathematical model for chance phenomena. We suppose there is a set S that contains all possible outcomes from a *random experiment* which could be any process where exactly one of many possible outcomes will occur. For each subset $A \in S$, which is called an *event*, there is a probability $P(A)$. The following axioms are assumed.

1. $P(S) = 1$;
2. If $A \cap B = \emptyset$, then $P(A \cup B) = P(A) + P(B)$.

A large number of important mathematical results follow from these two basic assumptions. For example, since $A \cup A^c = S$ and $A \cap A^c = \emptyset$,[1] it follows from Assumptions 1 and 2 that

$$P(A^c) = 1 - P(A).$$

The *conditional probability* of A given B is defined whenever $P(B) > 0$ and is

$$P(A|B) = \frac{P(A \cap B)}{P(B)}. \tag{2.1}$$

Similarly, if $P(A) > 0$, then

$$P(B|A) = \frac{P(A \cap B)}{P(A)}. \tag{2.2}$$

Rearranging (2.1) and (2.2) gives the very useful multiplicative laws:

$$P(A \cap B) = P(A|B)P(B) = P(B|A)P(A). \tag{2.3}$$

2.2.1 Independence

The events $A_1, \ldots, A_n$ are **independent** if for any $1 \le i_1 < \cdots < i_k \le n$

$$P\{A_{i_1} \cap \cdots \cap A_{i_k}\} = P\{A_{i_1}\} \cdots P\{A_{i_k}\}.$$

2.2.2 Bayes' law

Suppose that $B_1, \ldots, B_K$ is a partition of S meaning that $B_i \cap B_j = \emptyset$ if $i \neq j$ and $B_1 \cup B_2 \cup \cdots \cup B_K = S$. Then for any set A, we have that

$$A = (A \cap B_1) \cup \cdots \cup (A \cap B_K),$$

and therefore

$$P(A) = P(A \cap B_1) + \cdots + P(A \cap B_K). \tag{2.4}$$

It follows from (2.2) through (2.4) that

$$P(B_j|A) = \frac{P(A|B_j)P(B_j)}{P(A)} = \frac{P(A|B_j)P(B_j)}{P(A|B_1)P(B_1) + \cdots + P(A|B_K)P(B_K)}. \tag{2.5}$$

Equation (2.5) is called **Bayes' law**, also known as Bayes' rule or Bayes' theorem. Bayes' law is a simple, almost trivial, mathematical result, but its implications are profound. In fact, there is an entire branch of statistics, called **Bayesian statistics**, that is based upon Bayes' law and is now playing a very wide role in applied statistics. The importance of Bayes' law comes from its usefulness when updating probabilities. Here is an example, one that is too simple to be realistic but that illustrates the basic idea behind applying Bayes' law.

---

[1] Here A^c is the complement of A, i.e., the set of all elements of S that are not in A. Also, $\emptyset$ is the empty set, the set with no elements.

Example 2.1. Suppose that our prior knowledge about a stock indicates that the probability the price will rise on any given day, which we denote by θ, is either 0.4 or 0.6. Based upon past data, say from similar stocks, we believe that θ is equally likely to be 0.4 or 0.6. Thus, we have the *prior* probabilities

$$P(\theta = 0.4) = 0.5 \quad \text{and} \quad P(\theta = 0.6) = 0.5.$$

We observe the stock for three consecutive days and its price rises on all three days. Assume that the price changes are independent across days so that the probability that the price rises on each of three consecutive days is θ^3. Given this further information, we may suspect that $\theta = 0.6$, not 0.4. Therefore the probability that θ is 0.6, given three consecutive price increases, should be greater than the prior probability of 0.5, but how much greater? As notation, let A be the event that the prices rises on three consecutive days. Then, using Bayes' law we have

$$P(\theta = 0.6|A) = \frac{P(A|\theta = 0.6)P(\theta = 0.6)}{P(A|\theta = 0.6)P(\theta = 0.6) + P(A|\theta = 0.4)P(\theta = 0.4)}$$

$$= \frac{(0.6)^3(0.5)}{(0.6)^3(0.5) + (0.4)^3(0.5)} = \frac{(0.6)^3}{(0.6)^3 + (0.4)^3} = \frac{0.2160}{0.2160 + 0.0640} = 0.7714.$$

Thus, our probability that θ is 0.6 was 0.5 before we observed three consecutive price increases but is 0.7714 after observing this event. Probabilities before observing data are called the *prior probabilities* and the probabilities conditional on observed data are called the *posterior probabilities*, so the prior probability that θ equals 0.6 is 0.5 and the posterior probability is 0.7714.

Bayes' law is so important because it tells us exactly how to update our beliefs in light of new information. Revising beliefs after receiving additional information is something that humans do poorly without the help of mathematics.[2] There is a human tendency to put either too little or too much emphasis on new information, but this problem can be mitigated by using Bayes' law for guidance.

2.3 Probability Distributions

2.3.1 Random variables

A quantity such as the change in a stock price is random when it can take on many possible values, but only one of those values will actually occur. We call such quantities **random variables**. The set of all possible values of a random variable and the probabilities of each of these values are together called the *probability distribution* of the random variable.

[2] See Edwards (1982).

A simple example occurs in the random walk model for a stock price. Assume that at each time step the price can either increase or decrease by a fixed amount, $\Delta > 0$. Suppose that P_1, where $0 < P_1 < 1$, is the probability of an increase and $P_2 = 1 - P_1$ is the probability of a decrease. If X is the change in a single step, then the set of possible values of X is $\{x_1 = \Delta, x_2 = -\Delta\}$ and their probabilities are $\{P_1, P_2\}$. A random variable such as this one with only a finite number of values is called *discrete*. Other random variables that can take on any value in an infinite sequence of numbers are also called discrete. For example, let N be the number of steps in a random walk until a down-step. Assuming that the steps are independent, the probability that $N = n$ is $P_1^{n-1}P_2$, which is the probability of $n-1$ up-steps and then a down-step. Clearly, N can take on any value in the sequence $\{1, 2, 3, \ldots\}$, so N is discrete. This random walk model is called *binomial* because a step can have only two possible values. A more general random walk model allows the steps to have any probability distribution.

The random variable X taking on any value in some interval is called *continuous*. A continuous random variable has a **probability density function (PDF)**, often called simply a density, f_X such that

$$P\{X \in A\} = \int_A f_X(x)dx$$

for all[3] sets A. The famous normal density, or bell curve, is the best known example of a continuous distribution.

2.3.2 Independence

Independence of random variables is much like independence of sets. The random variables $X_1, \ldots, X_n$ are *mutually independent* if

$$P[\{X_1 \in A_1\} \cap \cdots \cap \{X_n \in A_n\}] = P\{X_1 \in A_1\} \cdots P\{X_n \in A_n\}$$

for all sets $A_1, \ldots, A_n$.

2.3.3 Cumulative distribution functions

The **cumulative distribution function (CDF)** of X is defined as

$$F_X(x) = P\{X \leq x\}.$$

If X has a PDF f_X then

[3] For those with a background in measure theory, here and elsewhere in this book "all sets" really means "all *measurable* sets." Those readers without knowledge of measure theory need not worry about this mathematical subtlety since all sets we encounter here are measurable. Nonmeasurable sets are truly bizarre and cannot be described without advanced mathematics.

$$F_X(x) = \int_{-\infty}^{x} f_X(u)\,du.$$

The CDF is often more useful for finding probabilities than the PDF. For example, suppose that we want the probability that X is between a and b for some pair of numbers $a < b$. For simplicity, assume that X is continuous so that the probability of X equaling either exactly a or exactly b is zero. Then $P\{a \le X \le b\} = F_X(b) - F_X(a)$ which is easily calculated assuming that F_X can be evaluated.[4]

2.3.4 Quantiles and percentiles

If the CDF of X is continuous and strictly increasing then it has an inverse[5] function F^{-1}. For each q between 0 and 1, $F^{-1}(q)$ is called the q-**quantile** or $100q$th percentile. The probability that a continuous X is below its q-quantile is precisely q, but as we show this is not exactly true for discrete random variables. The median is the 50% percentile or 0.5-quantile. The 25% and 75% percentiles (0.25- and 0.75-quantiles) are called the first and third quartiles and the median is the second quartile. The three quartiles divide the range of a continuous random variable into four groups of equal probability. Similarly, the 20%, 40%, 60%, and 80% percentiles are called quintiles and the 10%, 20%, ..., 90% percentiles are called deciles.

Quantiles are a bit more complicated for discrete random variables or for continuous random variables whose CDF is not strictly increasing, as Figure 2.1 illustrates. The top left plot has a continuous, strictly increasing CDF. For all q, $0 < q < 1$, and in particular the $q = 0.5$ illustrated here, there is a unique q-quantile $F^{-1}(q)$ and $P\{X < F^{-1}(q)\} = P\{X \le F^{-1}(q)\} = q$. The top right plot has a discrete CDF. For the q illustrated in that plot, there is a unique q-quantile denoted by $F^{-1}(q)$[6] but $P\{X < F^{-1}(q)\} < q < P\{X \le F^{-1}(q)\}$. The bottom left plot has a continuous, but not strictly increasing CDF. For the q illustrated here, there is an interval of q-quantiles between the vertical dashed lines and each quantile $F^{-1}(q)$[7] in this interval satisfies $P\{X < F^{-1}(q)\} = P\{X \le F^{-1}(q)\} = q$.

---

[4] CDFs of distributions that are important in applications are available in tables and can be calculated by many computer software packages.

[5] One says that h is the *inverse* of the function g if $h\{g(x)\} = x$ for all x in the *domain* of g, that is, all x where $g(x)$ is defined.

[6] $F^{-1}(q)$ is not a true inverse since F merely jumps from below q to above q, but does not actually equal q, at $F^{-1}(q)$.

[7] Here again $F^{-1}(q)$ is not a true inverse function, in this case because it is not unique.

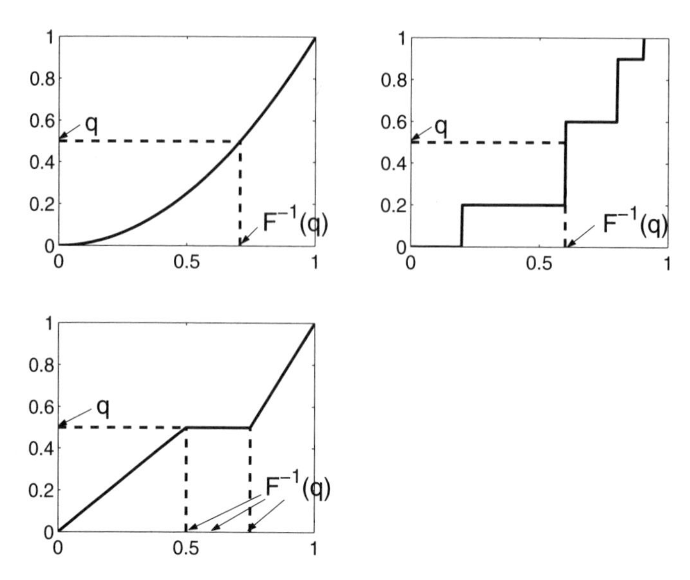

Fig. 2.1. *Quantiles. In each plot the thick solid curve is the CDF.* ***Top left:*** *The CDF is continuous and increasing, so the q-quantile is unique and the probability below the q-quantile is exactly q.* ***Top right:*** *The CDF is discrete, the q-quantile is unique, but the probability below the q-quantile is less than q; that probability is* 0.2 *but* $q = 0.5$. ***Bottom left:*** *The CDF is continuous but not strictly increasing and the q-quantile is not unique. Rather, there is an entire interval,* [0.5, 0.75], *of q-quantiles.*

2.3.5 Expectations and variances

The **expectation** or expected value of a continuous random variable X is

$$E(X) = \int_{-\infty}^{+\infty} x f_X(x)\, dx.$$

If X is discrete then

$$E(X) = \sum_{i=1}^{n} x_i P_i,$$

where n is the number of possible values of X and could be infinity, $x_1, \ldots, x_n$ are the possible values, and $P_1, \ldots, P_n$ are their probabilities.

The **variance** of X is

$$\sigma_X^2 = E\{X - E(X)\}^2.$$

If X is continuous, then

$$\sigma_X^2 = \int \{x - E(X)\}^2 f_X(x)\, dx,$$

and if X is discrete then

$$\sigma_X^2 = \sum_{i=1}^{n} \{x_i - E(X)\}^2 P_i.$$

The following formula is often useful

$$\sigma_X^2 = E(X^2) - \{E(X)\}^2. \tag{2.6}$$

Example 2.2. Suppose that X is the move on a single step of the binomial random walk described earlier. Then the expectation of X is

$$E(X) = \Delta P_1 + (-\Delta)(1 - P_1) = \Delta(2P_1 - 1).$$

Not surprisingly, the expected step is 0 if $P_1 = 1/2$. Also,

$$E(X^2) = \Delta^2 P_1 + (-\Delta)^2(1 - P_1) = \Delta^2,$$

so that

$$\sigma_X^2 = \Delta^2 - \{\Delta(2P_1 - 1)\}^2 = (2\Delta)^2 P_1(1 - P_1).$$

The **standard deviation**, denoted by σ_X, is the square root of the variance:

$$\sigma_X = \sqrt{E\{X - E(X)\}^2}.$$

2.3.6 Does the expected value exist?

The expected value of a random variable could be infinite or not exist at all. Also, a random variable need not have a well-defined and finite variance. To appreciate these facts, let X be a random variable with density f_X. The expectation of X is

$$\int_{-\infty}^{\infty} x f_X(x) dx$$

provided that this integral is defined. If

$$\int_{-\infty}^{0} x f_X(x) dx = -\infty \tag{2.7}$$

and

$$\int_{0}^{\infty} x f_X(x) dx = \infty \tag{2.8}$$

then the expectation is, formally, $-\infty + \infty$ which is not defined so the expectation does not exist. If integrals on the left-hand sides of (2.7) and (2.8) are both finite, then $E(X)$ exists and equals the sum of these two integrals. The expectation can exist but be infinite, because if

$$\int_{-\infty}^{0} x f_X(x) dx = -\infty \text{ and } \int_{0}^{\infty} x f_X(x) dx < \infty$$

then $E(X) = -\infty$ and if

$$\int_{-\infty}^{0} x f_X(x) dx > -\infty \text{ and } \int_{0}^{\infty} x f_X(x) dx = \infty$$

then $E(X) = \infty$.

If $E(X)$ is not defined or is infinite, then the variance that involves $E(X)$ cannot be defined either. If $E(X)$ is defined and finite, then the variance is also defined and is ∞ if and only if $E(X^2) = \infty$.

The nonexistence of finite expected values and variances is of importance for modeling financial markets data, because, for example, the popular GARCH models discussed in Chapter 12 need not have finite expected values and variances, and the implications of these problems are described in Section 12.10.1.

One could argue that any variable X derived from financial markets will be bounded, that is, that there is a constant $M < \infty$ such that $P(|X| \leq M) = 1$. In this case, the integrals in (2.7) and (2.8) are both finite, in fact at most M, and $E(X)$ exists and is finite. Also, $E(X^2) \leq M^2$ so the variance of X is finite. So should we worry at all about the mathematically niceties of whether expected values and variance exist and are finite? I assert that we should. A random variable might be bounded in absolute value by a very large constant M and yet, if M is large enough, behave much like a random variable that does not have an expected value or has an expected value that is infinite or has a finite expected value but an infinite variance. This can be seen in the simulations of GARCH processes in Section 12.10.1. Results from computer simulations are bounded by the maximum size of a number in the computer. Yet these simulations behave as if the variance were infinite.

2.4 Functions of Random Variables

Suppose that X is a random variable with PDF $f_X(x)$ and $Y = g(X)$ for g a strictly increasing function.[8] Since g is strictly increasing it has an inverse which we denote by h. Then Y is also a random variable and its CDF is

$$F_Y(y) = P(Y \leq y) = P\{g(X) \leq y\} = P\{X \leq h(y)\} = F_X\{h(y)\}. \quad (2.9)$$

Differentiating (2.9) we find the PDF of Y:

[8] The function g is increasing if $g(x_1) \leq g(x_2)$ whenever $x_1 < x_2$ and strictly increasing if $g(x_1) < g(x_2)$ whenever $x_1 < x_2$. Decreasing and strictly decreasing are defined similarly and g is (strictly) monotonic if it is either (strictly) increasing or (strictly) decreasing.

$$f_Y(y) = f_X\{h(y)\}h'(y).$$

Applying a similar argument to the case where g is strictly decreasing, one can show that whenever g is strictly monotonic then

$$f_Y(y) = f_X\{h(y)\}|h'(y)|. \tag{2.10}$$

Also from (2.9), when g is strictly increasing then

$$F_Y^{-1}(p) = g\{F_X^{-1}(p)\} \tag{2.11}$$

so that the pth quantile of Y is found by applying g to the pth quantile of X. When g is strictly decreasing then it maps the pth quantile of X to the $(1-p)$th quantile of Y; see Problem 15 for an example.

Result 2.4.1 *Suppose that $Y = a + bX$ for some constants a and $b \neq 0$. Then let $g(x) = a + bx$ so that $h(y) = (y-a)/b$ and $h'(y) = 1/b$. Therefore,*

$$\begin{aligned} F_Y(y) &= F_X\{b^{-1}(y-a)\},\ b > 0, \\ &= 1 - F_X\{b^{-1}(y-a)\},\ b < 0, \end{aligned}$$

$$f_Y(y) = |b|^{-1} f_X\{b^{-1}(y-a)\},$$

and

$$\begin{aligned} F_Y^{-1}(p) &= a + bF_X^{-1}(p),\ b > 0 \\ &= a + bF_X^{-1}(1-p),\ b < 0. \end{aligned}$$

2.5 Random Samples

We say that $X_1, \ldots, X_n$ are a **random sample** from a probability distribution if they each have that probability distribution and if they are independent. In this case, we also say that they are *independent and identically distributed* or simply **i.i.d.** The probability distribution is often called the population and its expected value, variance, CDF, and quantiles are called the *population mean, population variance, population CDF*, and *population quantiles.*

If $X_1, \ldots, X_n$ is a sample from an unknown probability distribution, then the population mean can be estimated by the **sample mean**

$$\overline{X} = n^{-1}\sum_{i=1}^{n} X_i \tag{2.12}$$

and the population variance can be estimated by the *sample variance*

$$s_X^2 = \frac{\sum_{i=1}^{n}(X_i - \overline{X})^2}{n-1}. \tag{2.13}$$

The reason for the denominator of $n-1$ rather than n is discussed in Section 2.18.1. The **sample standard deviation** is s_X, the square root of s_X^2.

2.6 The Binomial Distribution

Suppose that we conduct n experiments for some fixed (nonrandom) integer n. On each experiment there are two possible outcomes called "success" and "failure;" the probability of a success is p, and the probability of a failure is $q = 1 - p$. It is assumed that p and q are the same for all n experiments. Let X be the total number of successes, so that X will equal 0, 1, 2, ..., or n. If the experiments are independent, then

$$P(X = k) = \binom{n}{k} p^k q^{n-k} \text{ for } k = 0, 1, 2, \ldots, n,$$

where

$$\binom{n}{k} = \frac{n!}{k!(n-k)!}.$$

The distribution of X is called the **binomial distribution** and denoted Binomial(n, p). The expected value of X is np and its variance is npq. The Binomial$(1, p)$ distribution is also called the Bernoulli distribution.

The number of up-steps after n steps of a binomial random walk is an example of a binomial random variable.

Example 2.3. Let X be the number of up-steps after 10 steps of a random walk with probability 0.6 of an up-step. Then the expected value of X is $E(X) = (10)(0.6) = 6$ and $\text{Var}(X) = (10)(0.6)(0.4) = 2.4$. The probability that $X = x$ is

$$\frac{10!}{x!\,(10-x)!}(0.6)^x(0.4)^{10-x}.$$

These probabilities are shown in Figure 2.2.

MATLAB functions

The MATLAB functions `binocdf`, `binopdf`, and `binoinv` in the Statistics Toolbox compute the binomial PDF, CDF, and inverse CDF, respectively. Since these functions are in MATLAB's Statistics Toolbox, one can use them only if this toolbox is installed. The arguments of these functions are x (the argument of the function itself), n, and p, so, for example, `binopdf(3,10,.5)` is $P(X = 3)$ if X is Binomial$(10, 0.5)$. Figure 2.2 is produced by the MATLAB program:

```
x=0:10 ;
y=binopdf(x,10,.6)
bar(x,y) ;
set(gca,'fontsize',16) ;
set(gca, 'ylim',[0,.26]) ;
set(gca,'xlim',[-.5 10.5])
xlabel('x','fontsize',16) ;
ylabel('P(X=x)','fontsize',16) ;
```

The statement `bar(x,y)` produces a bar graph. The label "n=10, p=0.6" on the figure was added interactively within MATLAB.

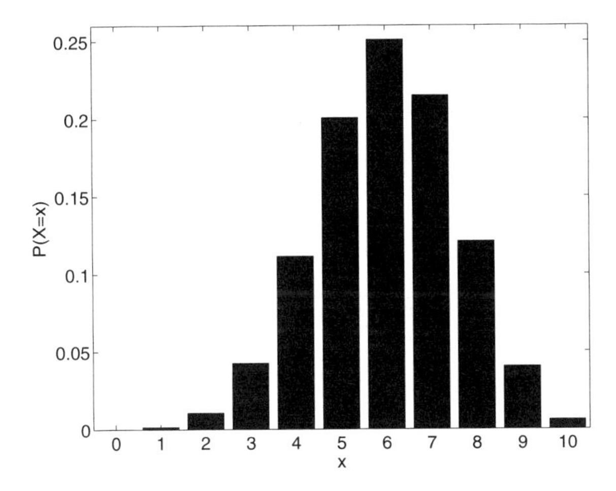

Fig. 2.2. *Binomial probability distribution with $n = 10$ and $p = 0.6$.*

2.7 Location, Scale, and Shape Parameters

Parameters are sometimes classified as location, scale, or shape parameters depending upon which properties of a distribution they determine.

A *location parameter* is a parameter that shifts a distribution to the right or left without changing the distribution's shape or standard deviation. A parameter is a *scale parameter* if it is a constant multiple of the standard deviation. Many examples of location and scale parameters are shown in the following sections.

If $f(x)$ is any fixed density, then $f(x-\mu)$ is a family of distributions with location parameter μ, $\theta^{-1}f(x/\theta)$, $\theta > 0$, is a family of distributions with a scale parameter θ, and $\theta^{-1}f\{\theta^{-1}(x-\mu)\}$ is a family of distributions with location parameter μ and scale parameter θ. If X has density $f(x)$ and $\theta > 0$, then by Result 2.4.1 $X + \mu$ has density $f(x-\mu)$, θX has density $\theta^{-1}f(\theta^{-1}x)$, and $\theta X + \mu$ has density $\theta^{-1}f\{\theta^{-1}(x-\mu)\}$.

A *shape* parameter generally refers to any parameter that is not a location or scale parameter. The parameter p of the Binomial(n,p) distribution is a shape parameter, and other examples come later in this chapter.

2.8 Some Common Continuous Distributions

2.8.1 Uniform distributions

The uniform distribution on the interval $[a,b]$ is denoted by Uniform$[a,b]$ and has PDF equal to $1/(b-a)$ on $[a,b]$ and equal to 0 outside this interval. It is easy to check that if X is Uniform$[a,b]$ then its expectation is

$$E(X) = \frac{1}{b-a}\int_a^b x\,dx = \frac{a+b}{2},$$

which is the midpoint of the interval. Also,

$$E(X^2) = \frac{1}{b-a}\int_a^b x^2\,dx = \frac{x^3|_a^b}{3(b-a)} = \frac{b^2+ab+a^2}{3}.$$

Therefore, using (2.6)

$$\sigma_X^2 = \frac{b^2+ab+a^2}{3} - \left(\frac{a+b}{2}\right)^2 = \frac{(b-a)^2}{12}.$$

Reparameterization means replacing the parameters of a distribution by an equivalent set. The uniform distribution can be reparameterized by using $\mu = (a+b)/2$ and $\sigma = (b-a)/\sqrt{12}$ as the parameters. Then μ is a location parameter and σ is the scale parameter. Which parameterization of a distribution is used depends upon which aspects of the distribution one wishes to emphasize. The parameterization (a, b) of the uniform specifies its endpoints while the parameterization (μ, σ) gives the mean and standard deviation. One is free to move back and forth between two or more parameterizations, using whichever is most useful in a given context. The uniform distribution does not have a shape parameter since the shape of its density is always rectangular.

2.8.2 Normal distributions

The **standard normal distribution** has density

$$\phi(x) = \frac{1}{\sqrt{2\pi}}\exp\left(-x^2/2\right), \quad -\infty < x < \infty.$$

The standard normal has mean 0 and variance 1. If Z is standard normal, then the distribution of $\mu + \sigma Z$ is called the *normal distribution with mean μ and variance σ^2* and denoted by $N(\mu, \sigma^2)$. By Result 2.4.1, the $N(\mu, \sigma^2)$ density is

$$\frac{1}{\sigma}\phi\left(\frac{x-\mu}{\sigma}\right). \tag{2.14}$$

The parameter μ is a location parameter and σ is a scale parameter. The normal distribution does not have a shape parameter since its density is always a bell-shaped curve. Several normal densities are shown in Figure 2.3. The standard normal CDF is

$$\Phi(x) = \int_{-\infty}^x \phi(u)du.$$

Φ can be evaluated using tables or more easily using software such as MATLAB or SAS. If X is $N(\mu, \sigma^2)$, then since $X = \mu + \sigma Z$, where Z is standard normal, by Result 2.4.1

$$F_X(x) = \Phi\{(x-\mu)/\sigma\}. \tag{2.15}$$

Example 2.4. If X is $N(5,4)$, then what is $P\{X \leq 7\}$.

Answer: Using $x = 7$, $\mu = 5$, and $\sigma^2 = 4$, we have $(x - \mu)/\sigma = (7-5)/2 = 1$ and then $\Phi(1) = 0.8413$. In MATLAB 6, "`cdfn`" is the standard normal CDF and "`cdfn(1)`" gives "`ans = 0.8413`".

Normal quantiles

The q-quantile of the $N(0,1)$ distribution is $\Phi^{-1}(q)$ and, more generally, the q-quantile of an $N(\mu, \sigma^2)$ distribution is $\mu + \sigma\Phi^{-1}(q)$. The $(1-\alpha)$-quantile of Φ, that is, $\Phi^{-1}(1-\alpha)$, is denoted by z_α. As shown later, z_α is widely used for confidence intervals.

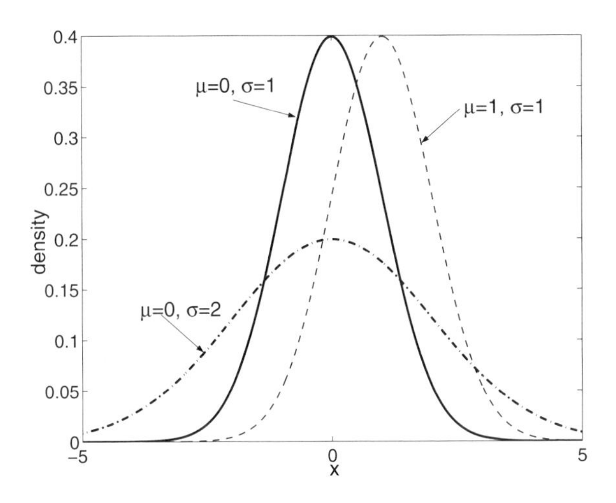

Fig. 2.3. *Examples of normal probability densities.*

MATLAB functions

The MATLAB 6 functions `normcdf`, `normpdf`, and `norminv` in the Statistics Toolbox compute the normal PDF, CDF, and inverse CDF, respectively. The function `cdfn` also computes the standard normal CDF and is available in the MATLAB base product. Each function has three arguments: the argument of the CDF, PDF, or inverse CDF; the mean; and the standard deviation. The default values of the mean and standard deviation are 0 and 1, respectively. For example, `normcdf(2,1,5)` computes the value at 2 of the $N(1,5^2)$ CDF and `norminv(.95,0,1)` and `norminv(.95)` compute the 0.95-quantile of the standard normal distribution.

2.8.3 The lognormal distribution

If Z is distributed $N(\mu, \sigma^2)$, then $X = \exp(Z)$ is said to have a Lognormal(μ, σ^2) distribution. In other words, X is *lognormal* if its logarithm is normally distributed.

The median of X is $\exp(\mu)$ and the expected value of X is $\exp(\mu + \sigma^2/2)$.[9] The expectation is larger than the median because the lognormal distribution is right skewed, and the skewness is more extreme with larger values of σ. Right skewness means that for an x such that $0 < x < 0.5$, the $(0.5 + x)$-quantile is farther above the median than the $(0.5 - x)$quantile is below the median. Skewness is discussed further in Section 2.11. The probability density functions of several lognormal distributions are shown in Figure 2.4.

The parameter μ is a scale parameter and σ is a shape parameter. The lognormal distribution does not have a location parameter since its location is fixed to start at 0.

MATLAB functions

The MATLAB functions `logncdf`, `lognpdf`, and `logninv` in the Statistics Toolbox compute the lognormal PDF, CDF, and inverse CDF, respectively. The arguments are the argument of the PDF, CDF, and inverse CDF, μ, and σ.

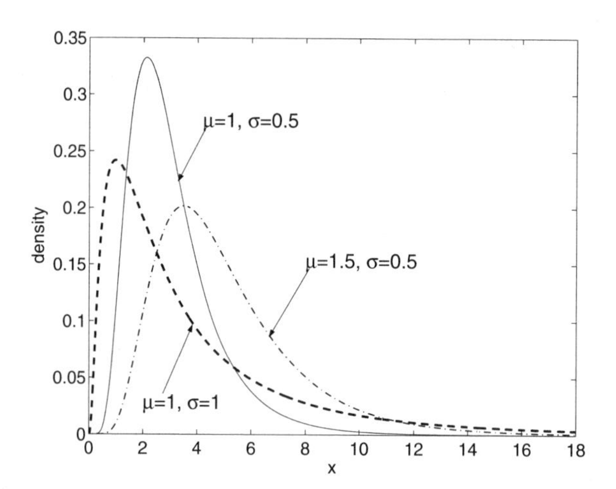

Fig. 2.4. *Examples of lognormal probability densities. Here μ and σ are the mean and standard deviation of the logarithm of a random variable with the indicated density.*

[9] It is important to remember that if X is lognormal(μ, σ), then μ is the expected value of $\log(X)$, not of X.

2.8.4 Exponential and double exponential distributions

The *exponential distribution* with scale parameter $\theta > 0$, which we denote by Exponential(θ), has CDF

$$F(x) = 1 - e^{-x/\theta}, \quad x > 0.$$

The Exponential(θ) distribution has PDF

$$f(x) = \frac{e^{-x/\theta}}{\theta},$$

expected value θ, and standard deviation θ. The inverse CDF is

$$F^{-1}(y) = -\theta \log(1 - y), \quad 0 < y < 1.$$

The *double exponential* or *Laplace distribution* with mean μ and scale parameter θ has PDF

$$f(x) = \frac{e^{-|x-\mu|/\theta}}{2\theta}, \tag{2.16}$$

so if X has a double exponential distribution with mean 0, then $|X|$ has an exponential distribution. A double exponential distribution has a standard deviation of $\sqrt{2}\theta$. The mean μ is a location parameter and θ is a scale parameter.

2.9 Sampling a Normal Distribution

A common situation is that we have a random sample from a normal distribution and we wish to have confidence intervals for the mean and variance or test hypotheses about these parameters. Then the following distributions are very useful. Also, they are the basis for confidence intervals and tests about regression parameters needed in Chapter 6 and elsewhere.

2.9.1 Chi-squared distributions

Suppose that $Z_1, \ldots, Z_n$ are i.i.d. $N(0,1)$. Then the distribution of $Z_1^2 + \cdots + Z_n^2$ is called the *chi-squared distribution* with n *degrees of freedom*. This distribution has an expected value of n and a variance of $2n$. The $(1-\alpha)$-quantile of this distribution is denoted by $\chi^2_{\alpha,n}$ and is used in tests and confidence intervals about variances; see Section 2.9.1 for the latter. Also, as discussed in Section 2.20.6, $\chi^2_{\alpha,n}$ is used in likelihood ratio testing.

The functions `chi2cdf`, `chi2pdf`, and `chi2inv`, which compute the CDF, PDF, and inverse CDF, are available in MATLAB's Statistics Toolbox.

2.9.2 t-distributions

If Z is $N(0,1)$, W is chi-squared with ν degrees of freedom, and Z and W are independent, then the distribution of $Z/\sqrt{W/\nu}$ is called the *t-distribution* with ν *degrees of freedom* and denoted t_ν. The $(1-\alpha)$-quantile of the t_ν-distribution is denoted by $t_{\alpha,\nu}$ and is used in tests and confidence intervals about population means, regression coefficients, and parameters in time series models. See Section 2.19.1 for confidence intervals for the mean.

The variance of a t_ν is finite and equals $\nu/(\nu-2)$ if $\nu > 2$.[10] If X has a t_ν-distribution with $\nu > 2$, then

$$\mu + \frac{\sigma X}{\sqrt{\nu/(\nu-2)}}$$

has mean μ and variance σ^2 and its distribution is called the $t_\nu(\mu,\sigma^2)$ distribution. With this notation, the t_ν and $t_\nu\{0,\nu/(\nu-2)\}$ distributions are the same, except that $t_\nu\{0,\nu/(\nu-2)\}$ is undefined if $\nu = 1$ or 2.

The density of the t_ν-distribution is

$$\left[\frac{\Gamma\{(\nu+1)/2\}}{(\pi\nu)^{1/2}\Gamma(\nu/2)}\right]\frac{1}{\{1+(x^2/\nu)\}^{(\nu+1)/2}}. \tag{2.17}$$

Here Γ is the *gamma function* defined by

$$\Gamma(t) = \int_0^\infty x^{t-1}\exp(-x)dx.$$

The quantity in large square brackets in (2.17) is just a constant, though a somewhat complicated one.

The functions `tcdf`, `tpdf`, and `tinv` are available in MATLAB's Statistics Toolbox.

2.9.3 F-distributions

If U and W are independent and chi-squared distributed with n_1 and n_2 degrees of freedom, respectively, then the distribution of

$$\frac{U/n_1}{W/n_2}$$

is called the F-distribution with n_1 and n_2 degrees of freedom. The $(1-\alpha)$-quantile of this distribution is denoted by F_{α,n_1,n_2}. F_{α,n_1,n_2} is used as a critical value for F-tests in regression; see Chapter 6.

The degrees of freedom parameters of the chi-square, t-, and F-distributions are shape parameters.

The functions `fcdf`, `fpdf`, and `finv` are available in MATLAB's Statistics Toolbox.

---

[10] If $\nu = 1$, then the expected value of the t_ν-distribution does not exist and the variance is not defined. If $\nu = 2$ then the expected value is 0 and the variance is infinite.

2.10 Order Statistics and the Sample CDF

Suppose that $X_1, \ldots, X_n$ is a random sample from a probability distribution with CDF F. In this section we estimate F and its quantiles. The *sample* or *empirical CDF* $F_n(x)$ is defined to be the proportion of the sample that is less than or equal to x. For example, if 10 out of 40 ($= n$) elements of a sample are 3 or less, then $F_n(3) = 0.25$. More generally,

$$F_n(x) = \frac{\sum_{i=1}^n I\{X_i \leq x\}}{n},$$

where $I\{X_i \leq x\}$ is 1 if $X_i \leq x$ and is 0 otherwise. Figure 2.5 shows F_n for a sample of size 35 from an $N(0,1)$ distribution. The true CDF (Φ) is shown as well. The sample CDF differs from the true CDF only because of random variation.

The *order statistics* $X_{(1)}, X_{(2)}, \ldots, X_{(n)}$ are $X_1, \ldots, X_n$ ordered from smallest to largest. The subscripts of the order statistics are in parentheses to distinguish them from the unordered sample. For example, X_1 is simply the first observation in the original sample while $X_{(1)}$ is the smallest observation in that sample. The *sample quantiles* are defined in various ways by different authors, but roughly the q-sample quantile ($100q$th sample percentile) is $X_{(k)}$ where k is qn rounded to an integer. Some authors round up, some to the nearest value, and some round in both directions and use a weighted average of the two results.

Example 2.5. Suppose the sample is 6, 4, 8, 2, 3, 4. Then $n = 6$, the order statistics are 2, 3, 4, 4, 6, 8, and $F_n(x)$ equals 0 if $x < 2$, equals 1/6 if $2 \leq x < 3$, equals 2/6 if $3 \leq x < 4$, equals 4/6 if $4 \leq x < 6$, equals 5/6 if $6 \leq x < 8$, and equals 1 if $x \geq 8$. Suppose we want the 25th sample percentile. Note that $.25n = 1.5$ which could be rounded to either 1 or 2. Since 16.7% of the sample equals $X_{(1)}$ or less and 33.3% of the sample equals $X_{(2)}$ or less, either $X_{(1)}$ or $X_{(2)}$ or some number between them can be used as the 25th sample percentile.

2.10.1 Normal probability plots

Many statistical models assume that a random sample comes from a normal distribution. **Normal probability** plots are used to check this assumption. If the assumption is true, then the qth sample quantile will be approximately equal to $\mu + \sigma \Phi^{-1}(q)$, which is the population quantile. Therefore, except for sampling variation, a plot of the sample quantiles versus Φ^{-1} will be linear. The normal probability plot is a plot of $X_{(i)}$ versus $\Phi^{-1}\{i/(n+1)\}$ (these are the $i/(n+1)$ sample and population quantiles, respectively). Systematic deviation of the plot from a straight line is evidence of nonnormality. Figure

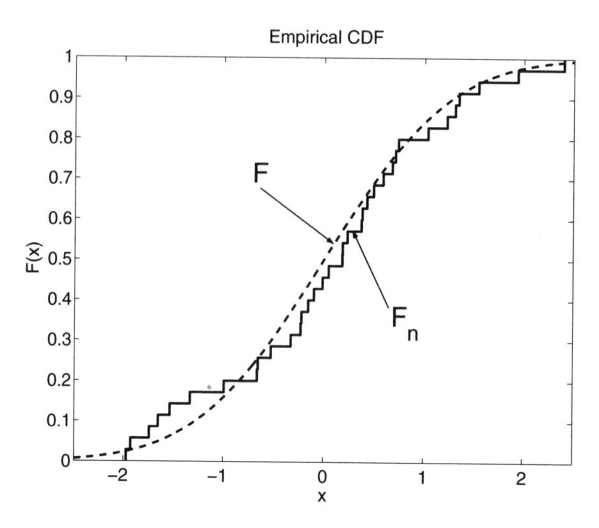

Fig. 2.5. *The sample CDF* (F_n) *and the true CDF* ($F = \Phi$) *from an* $N(0,1)$ *population. The sample size is* 35.

2.6 is a normal probability plot of the same data used in Figure 2.5. The plot is reasonably close to linear. Figure 2.7 is a normal probability plot of the a sample from a lognormal(0, 1) distribution. This distribution is very right skewed and the skewness is obvious in the normal plot.

It is often rather difficult to decide whether a normal plot is close enough to linear to conclude that the data are normally distributed. For example, even though the plot in Figure 2.6 is close to linear, there is some nonlinearity, particularly on the left side. Is this nonlinearity due to nonnormality or just due to random variation? If one did not know that the data were simulated from a normal distribution, then it would be difficult to tell unless one were very experienced with normal plots. In this case, a test of normality is very helpful. These tests are discussed in Section 2.20.5.

In MATLAB, `normplot(x)` produces a normal plot of the data in the vector `x`. If `x` is a matrix, then normal plots of the columns of `x` are overlaid on a single set of axes. In SAS, normal plots can be requested when using PROC UNIVARIATE; see Example 2.18.

2.11 Skewness and Kurtosis

Skewness and kurtosis measure the shape of a probability distribution. **Skewness** measures the degree of asymmetry, with symmetry implying zero skewness, positive skewness indicating a relatively long right tail compared to the

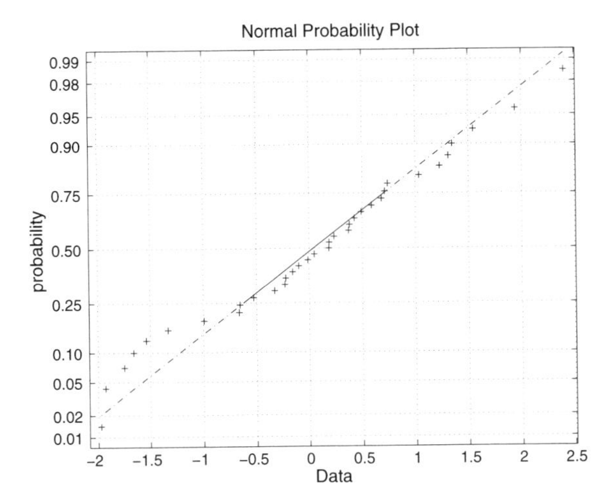

Fig. 2.6. *Normal probability plot of a random sample of size* 35 *from an* $N(0,1)$ *population.*

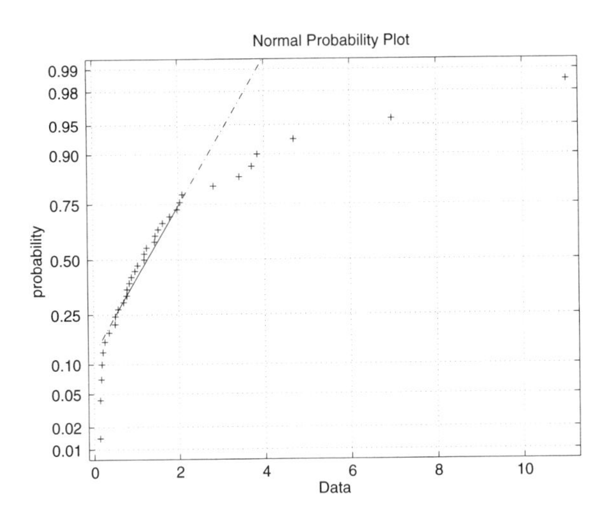

Fig. 2.7. *Normal probability plot of a random sample of size* 35 *from a lognormal population with* $\mu = 0$ *and* $\sigma = 1$.

left, and negative skewness indicating the opposite. By the *tails* of a distribution is meant the regions far from the center. Reasonable definitions of the "tails" would be that the left tail is the region from $-\infty$ to $\mu - 2\sigma$ and the right tail is the region from the $\mu + 2\sigma$ to $+\infty$, though the choices of $\mu - 2\sigma$ and $\mu + 2\sigma$ are somewhat arbitrary. *Kurtosis* indicates the extent to which probability is concentrated in the center and especially the tails of the distribution rather than in the "shoulders" which are the regions between the center and the tails. Reasonable definitions of *center* and *shoulder* would be that the center is the region from the $\mu - \sigma$ to the $\mu + \sigma$, the left shoulder is from the $\mu - 2\sigma$ to $\mu - \sigma$, and the right shoulder is from the $\mu + \sigma$ to $\mu + 2\sigma$. See the upper plot in Figure 2.9. Because skewness and kurtosis measure shape, they do not depend on the values of location and scale parameters.

The skewness of a random variable X is

$$Sk = E\left\{\frac{X - E(X)}{\sigma}\right\}^3 = \frac{E\{X - E(X)\}^3}{\sigma^3}.$$

To appreciate the meaning of the skewness, it is helpful to look at an example and the binomial distribution is convenient for that purpose. The parameters n and p of the binomial distribution are shape parameters and will determine skewness and kurtosis. The skewness of the Binomial(n, p) distribution is

$$Sk(n,p) = \frac{1 - 2p}{\sqrt{np(1-p)}}.$$

Figure 2.8 shows the binomial probability distribution and its skewness for $n = 10$ and four values of p. Notice that

- The skewness is positive if $p < 0.5$, negative if $p > 0.5$, and 0 if $p = 0.5$.
- The absolute skewness becomes larger as p moves closer to either 0 or 1.

Positive skewness is called right skewness and negative skewness is called left skewness. A distribution is symmetric about a point θ if $P(X > \theta + x) = P(X < \theta - x)$ for all $x > 0$. In this case, θ is a location parameter and equals $E(X)$. The skewness of any symmetric distribution is 0.

The kurtosis of a random variable X is

$$K = E\left(\frac{X - E(X)}{\sigma}\right)^4 = \frac{E\{X - E(X)\}^4}{\sigma^4}.$$

The kurtosis of a Binomial(n, p) distribution is

$$K(n,p) = 3 + \frac{1 - 6p(1-p)}{np(1-p)}.$$

Figure 2.8 also gives the kurtosis of the distributions in that figure. However, it is difficult to interpret the kurtosis of an asymmetric distribution because for such distributions kurtosis may measure both asymmetry and tail-weight, so the binomial is not a good example for understanding kurtosis. For that purpose we will look instead at t-distributions because they are symmetric. Figure 2.9 compares the $N(0, 1)$ density with the $t_5(0, 1)$-density. Both have a mean of 0 and a standard deviation of 1. The mean and standard deviation are location and scale parameters, respectively, and do not affect kurtosis. The parameter ν of the t-distribution is a shape parameter. The kurtosis of a t_ν-distribution is finite if $\nu \geq 5$ and then the kurtosis is

$$K(\nu) = 3 + \frac{6}{\nu - 4}. \tag{2.18}$$

Thus, the kurtosis is 9 for a t_5-distribution. Since the densities in Figure 2.9 have the same mean and standard deviation, they also have the same tails, center, and shoulders, at least according to our definitions of these regions, and these regions are indicated on the top plot. The bottom plot zooms in on the right tail. Notice that the t_5-density has more probability in the tails and center than the $N(0,1)$ density. This behavior of t_5 is typical of distributions with high kurtosis.

Every normal distribution has a skewness coefficient of 0 and a kurtosis of 3. The skewness and kurtosis must be constant because the normal distribution has only location and scale parameters, no shape parameters. The kurtosis of 3 agrees with formula (2.18) since a normal distribution is a t-distribution with $\nu = \infty$. The "excess kurtosis" of a distribution is $K - 3$ and measures the deviation of that distribution's kurtosis from the kurtosis of a normal distribution. From (2.18) we see that the excess kurtosis of a t_ν-distribution is $6/(\nu - 4)$.

An exponential distribution has a skewness equal to 2 and a kurtosis of 9. A double exponential distribution has skewness 0 and kurtosis 6. Since the exponential distribution has only a scale parameter and the double exponential has only a location and a scale parameter, their skewness and kurtosis must be constant.

The Lognormal(μ, σ^2) distribution has a skewness coefficient of

$$(e^{\sigma^2} + 2)\sqrt{e^{\sigma^2} - 1}. \tag{2.19}$$

The parameter μ is a scale parameter and has no effect on the skewness.

Estimation of the skewness and kurtosis of a distribution is relatively straightforward if we have a sample, $X_1, \ldots, X_n$, from that distribution. Let the sample mean and standard deviation be $\overline{X}$ and s. Then the sample skewness, denoted by $\widehat{Sk}$, is

$$\widehat{Sk} = \frac{1}{n}\sum_{i=1}^{n}\left(\frac{X_i - \overline{X}}{s}\right)^3,$$

and the sample kurtosis, denoted by $\widehat{K}$, is

$$\widehat{K} = \frac{1}{n}\sum_{i=1}^{n}\left(\frac{X_i - \overline{X}}{s}\right)^4.$$

Both the sample skewness and the excess kurtosis should be near 0 if a sample is from a normal distribution. Deviations of the sample skewness and kurtosis from these values are an indication of nonnormality.

A word of caution is in order. Skewness and kurtosis are highly sensitive to outliers. Sometimes outliers are due to *contaminants* which are data not from the population being sampled. An example would be a data recording error. A sample from a normal distribution with even a single contaminant that is sufficiently outlying will appear highly nonnormal according to the sample skewness and kurtosis. In such a case, a normal plot *will* look linear, except that the single contaminant will stick out. See Figure 2.10, which is a normal plot of a sample of 999 $N(0,1)$ data points plus a contaminant equal to 30. This figure shows clearly that the sample is nearly normal but with an outlier. The sample skewness and kurtosis, however, are 10.8473 and 243.0405, which might give the false impression that the sample is far from normal. Also, even if there were no contaminants, a distribution could be extremely close to a normal distribution and yet have a skewness or excess kurtosis that is very different from 0; see Problem 11.

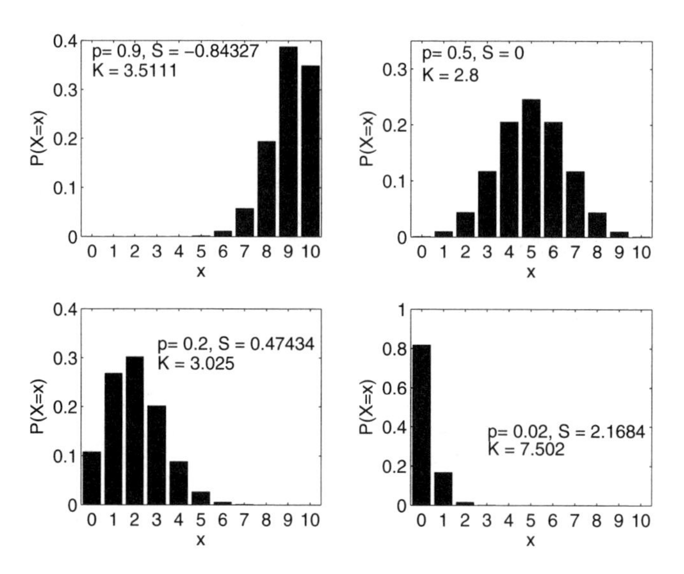

Fig. 2.8. *Several binomial probability distributions with* $n = 10$ *and their skewness determined by the shape parameter* p.

2.12 Heavy-Tailed Distributions

Distributions with high tail probabilities compared to a normal distribution with the same mean and standard deviation are called **heavy-tailed**. Because kurtosis is particularly sensitive to tail-weight, high kurtosis is nearly

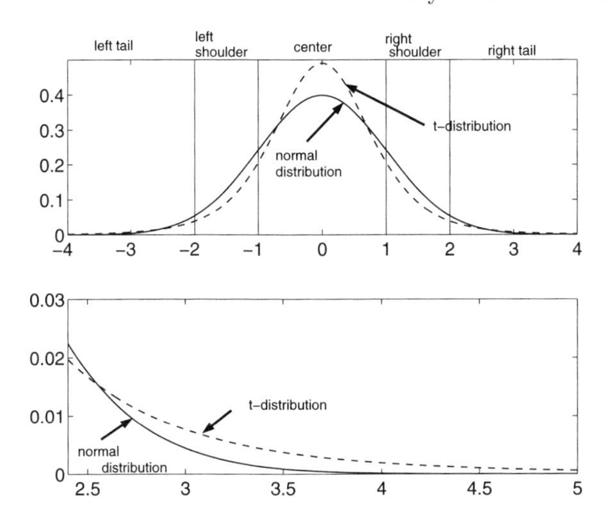

Fig. 2.9. *Comparison of a normal density and a t-density with* 5 *degrees of freedom. Both densities have mean* 0 *and standard deviation* 1.

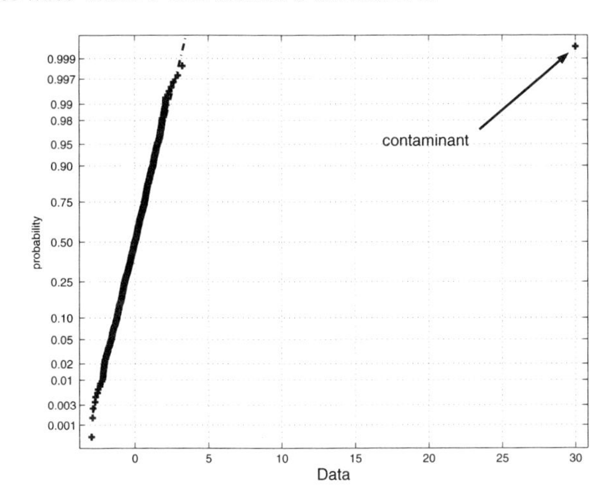

Fig. 2.10. *Normal plot of a sample of* 999 $N(0,1)$ *data plus a contaminant.*

synonymous with having heavy tails. Heavy-tailed distributions are important models in finance, because stock return distributions have often been observed to have heavy tails. A heavy-tailed distribution is more prone to extreme values, which are sometimes called outliers. In finance applications, one is especially concerned when the return distribution has heavy tails because of the possibility of an extremely large negative return which could, for example, entirely deplete the capital reserves of a firm.

2.12.1 Double exponential distributions

Double exponential distributions have slightly heavier tails than normal distributions. This fact can be appreciated by comparing their densities. The density of the double exponential with scale parameter θ is proportional to $\exp(-|x/\theta|)$ and the density of the $N(0, \sigma^2)$ distribution is proportional to $\exp\{-0.5(x/\sigma)^2\}$. The term $-x^2$ converges to $-\infty$ much faster than $-|x|$ as $|x| \to \infty$. Therefore, the normal density converges to 0 much faster than the double exponential density as $x \to \infty$. In fact, any density that is proportional to

$$\exp\left(-\left|x/\theta\right|^{\alpha}\right), \tag{2.20}$$

where $0 < \alpha < 2$ is a shape parameter and θ is a scale parameter, will have heavier tails than a normal distribution, with smaller values of α implying heavier tails. However, no density of form (2.20) will have truly heavy tails, and in particular $E(|X|^n) < \infty$ for all n no matter how large, and in addition there is a finite mean and variance. To achieve very heavy tails, the density must behave like $|x|^{-(a+1)}$ for some $a > 0$, which we call *polynomial tails*, rather than like (2.20), which we call *exponential tails*. A polynomial tail is also called a *Pareto tail* after the Pareto distribution, which we discuss shortly. The parameter a of a polynomial tail is called the *tail index*.

2.12.2 t-distributions have heavy tails

The t-distributions are a class of heavy-tailed distributions and can be used to model heavy-tail return distributions. For t-distributions both the kurtosis and the weight of the tails increase as ν gets smaller. When $\nu \le 4$ the tail weight is so high that the kurtosis is infinite.

By (2.17), the t-distribution's density is proportional to

$$\frac{1}{1 + (x^2/\nu)^{(\nu+1)/2}}$$

which for large values of $|x|$ is approximately

$$\frac{1}{(x^2/\nu)^{(\nu+1)/2}} \propto |x|^{-(\nu+1)}.$$

Therefore, the t-distribution has polynomial tails with $a = \nu$. Any distribution with polynomial tails has heavy tails, and the smaller the value of a the heavier the tails. Thus, the t-distribution's tails become heavier as ν decreases. From a modeling perspective, a problem with the t-distribution is that the tail index is integer valued, rather than a continuous parameter, which limits the flexibility of the t-distribution as a model for financial markets data.[11]

[11] However, there is a way to define the t-distributions so that noninteger values of the degrees of freedom parameter are possible.

2.12.3 Mixture models

Another class of heavy-tailed distributions is *mixture models*. Consider a distribution which is 90% $N(0,1)$ and 10% $N(0,25)$. This is an example of a *normal mixture distribution* since it is a mixture of two different normal distributions called the *components*. The variance of this distribution is $(0.9)(1) + (0.1)(25) = 3.4$ so its standard deviation is $\sqrt{3.4} = 1.84$. This distribution is much different than an $N(0,3.4)$ distribution, even though both distributions have the same mean and variance. To appreciate this, look at Figure 2.11.

You can see in the top left panel that the two densities look quite different. The normal density looks much more dispersed than the normal mixture, but we know that they actually have the same variances. What's happening? Look at the detail of the right tails in the top right panel. The normal mixture density is much higher than the normal density when x is greater than 6. This is the "outlier" region (along with $x < -6$). The normal mixture has far more outliers than the normal distribution and the outliers come from the 10% of the population with a variance of 25. Remember that ± 6 is only $6/5$ standard deviations from the mean, using the standard deviation 5 of the component from which they come. Thus, these observations are not outlying relative to their component's standard deviation of 5, only relative to the population standard deviation of $\sqrt{3.4} = 1.84$ since $6/1.84 = 3.25$ and 3 or more standard deviations from the mean is generally considered to be rather outlying.

Outliers have a powerful effect on the variance and this small fraction of outliers inflates the variance from 1.0 (the variance of 90% of the population) to 3.4.

Let's see how much more probability the normal mixture distribution has in the outlier range $|x| > 6$ compared to the normal distribution.[12] For a $N(0, \sigma^2)$ random variable X,

$$P\{|X| > x\} = 2\{1 - \Phi(x/\sigma)\}.$$

Therefore, for the normal distribution with variance 3.4,

$$P\{|X| > 6\} = 2\{1 - \Phi(6/\sqrt{3.4})\} = 0.0011.$$

For the normal mixture population which has variance 1 with probability 0.9 and variance 25 with probability 0.1 we have that

$$P\{|X| > 6\} = 2\Big(0.9\{1 - \Phi(6)\} + 0.1\{1 - \Phi(6/5)\}\Big)$$

[12] There is nothing special about "6" to define the boundary of the outlier range, but a specific number was needed to make numerical comparisons. Clearly, $|x| > 7$ or $|x| > 8$, say, would have been just as appropriate as outlier ranges.

$$= 2\{(0.9)(0) + (0.1)(0.115)\} = 0.023.$$

Since $0.023/0.0011 \approx 21$, the normal mixture distribution is 21 times more likely to be in this outlier range than the $N(0, 3.4)$ population, even though both have a variance of 3.4. In summary, the normal mixture is much more prone to outliers than a normal distribution with the same mean and standard deviation. So we should be much more concerned about large negative returns if the return distribution is more like the normal mixture distribution than like a normal distribution.

It is not difficult to compute the kurtosis of this normal mixture. Because a normal distribution has kurtosis equal to 3, if Z is $N(\mu, \sigma^2)$, then $E(Z-\mu)^4 = 3\sigma^4$. Therefore, if X has this normal mixture distribution, then

$$E(X^4) = 3(0.9 + (0.1)25^2) = 190.2$$

and the kurtosis of X is $190.2/3.4^2 = 16.45$.

Normal probability plots of samples of size 200 from the normal and the normal mixture distributions are shown in the bottom panels of Figure 2.11. Notice how the outliers in the normal mixture sample give the probability plot a nonlinear, almost S-shaped, pattern. The deviation of the plot of the normal sample from linearity is small and is due entirely to randomness.

In this example, the variance is conditional upon the specific component of the mixture from which an observation comes. The conditional variance is 1 with probability 0.9 and 25 with probability 0.1. Because there are only two components, the conditional variance is discrete, in fact, with only two possible values, and the example was easy to analyze. The marginal distributions of the GARCH processes studied in Chapter 12 are also normal mixtures, but with infinitely many components and a continuous distribution of the conditional variance. Although GARCH processes are more complex than the simple mixture model in this section, the same theme applies — a nonconstant conditional variance of a mixture distribution induces heavy-tailed marginal distributions even though the conditional distributions are normal distributions, which have relatively light tails.

Despite the heavy tails of a normal mixture, the tails are exponential, not polynomial.

2.12.4 Pareto distributions

The Swiss economics professor Vilfredo Pareto (1848–1923) formulated an eponymous law which states that the fraction of a population with income exceeding an amount x is equal to

$$Cx^{-a} \tag{2.21}$$

for all x, where C and a are positive constants independent of x but depending on the population. Pareto believed that his law was true for all populations regardless of economic and political conditions.[13] If $F(x)$ is the CDF of the income distribution, then (2.21) implies that

$$F(x) = 1 - \left(\frac{c}{x}\right)^a, \quad x > c, \tag{2.22}$$

where $c > 0$ is the minimum income. CDFs satisfying (2.22) are called *Pareto distributions of the first kind.* As the name implies, there are other types of Pareto distributions, but in this book, "Pareto distribution" always means of the first kind. We denote the distribution given in (2.22) by Pareto(a, c). Pareto distributions are believed to fit many economic variables and are sometimes used to model the distribution of losses from investing.

The PDF of the distribution in (2.22) is

$$f(x) = \frac{ac^a}{x^{a+1}}, \quad x > c, \tag{2.23}$$

so a Pareto distribution has polynomial tails. As mentioned before, the constant a is called the *tail index.* It is also called the *Pareto constant.* An advantage of the Pareto distribution over the t-distribution as a statistical model is that the tail index of the Pareto can be any positive value whereas for the t-distribution the tail index is a positive integer.

The survival function of a random variable X is

[13] Johnson, Kotz, and Balakrishnan (1994).

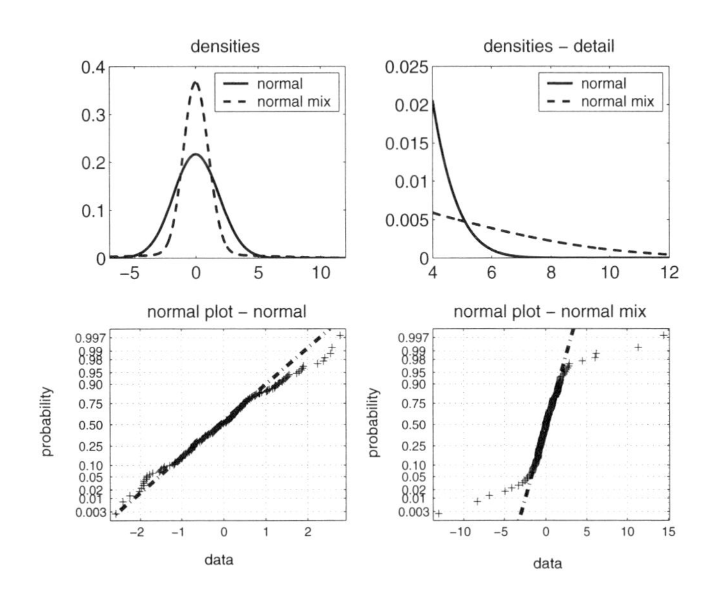

Fig. 2.11. *Comparison on normal and heavy-tailed distributions.*

$$P(X > x) = 1 - F(x), \tag{2.24}$$

where F is the CDF of X. If X is a loss, then the survival function $1 - F(x)$ is the probability of a loss greater than x and so is of obvious interest to an investor. The survival function of a Pareto distribution is

$$1 - F(x) = \left(\frac{c}{x}\right)^a \quad x > c. \tag{2.25}$$

As $x \to \infty$, the survival function of a Pareto converges to 0 at a slow polynomial rate rather than a fast exponential rate, which means that Pareto distributions have a heavy right tail, and the smaller the value of a the heavier the tail. Figure 2.12 shows the survival function of a Pareto distribution with $c = 0.25$ and $a = 1.1$. For comparison, the survival functions of normal and exponential distributions, conditional on being greater than 0.25, are also shown. The normal has mean 0 and its standard deviation $\sigma = 0.3113$, and the exponential has $\theta = c/a = 0.25/1.1$. These parameters were chosen so that the Pareto, normal, and exponential densities, conditional on being greater than 0.25, have the same height at 0.25, which implies that their survival functions have the same slope at 0.25 so the three survival functions *start* to decrease to 0 at the same rates; see Problem 12. Notice that despite their initial rates of decrease being equal, the normal survival functions converge to 0 much faster than the Pareto as $x \to \infty$. This means that extreme losses are much more likely if the loss distribution is Pareto rather than normal. The exponential survival function is intermediate between the normal and Pareto survival functions.

Pareto distributions can be used to model the probability of a large loss as follows. Let $X = -$return. When $X > 0$ there is a loss. We assume that for some $c > 0$, the distribution of X conditional on $X > c$ is Pareto with parameters c and a. The value of c can be selected by the analyst and might be the smallest loss which is of real interest, that is, such that losses smaller than c are too small to be of much concern. The parameter a can be estimated from a set of loss data; see Example 2.14.

2.12.5 Distributions with Pareto tails

A cumulative distribution function $F(x)$ is said to have a *Pareto right tail* if its survival function satisfies

$$1 - F(x) = L(x)x^{-a} \tag{2.26}$$

for some $a > 0$ where $L(x)$ is slowly varying at ∞. A Pareto left tail can be defined similarly; e.g., X has a Pareto left tail if $-X$ has a Pareto right tail. A function L is said to be *slowly varying at* ∞ if

$$\frac{L(\lambda x)}{L(x)} \to 1 \text{ as } x \to \infty, \tag{2.27}$$

for all $\lambda > 0$. To appreciate what (2.27) is saying, consider the case of $\lambda = 1/2$. Then for all large enough x, $L(x/2)$ and $L(x)$ are nearly equal so doubling the argument of L from $x/2$ to x hardly changes the value of L. Thus, for many practical purposes the function L can be treated as constant for large values of x. This means that survival function (2.26) will behave much like the Pareto survival function.

It can be useful to assume a Pareto tail when estimating value-at-risk (VaR), which is the maximum loss one expects to see with a given degree of confidence. More precisely, if the confidence is $1 - \alpha$ and the holding period is t then with probability $1 - \alpha$ the actual loss over a period t will be less than the VaR. VaR is discussed more fully in Chapter 11 and estimation of VaR when the loss distribution has Pareto tails is covered in Section 11.2.3.

A slightly stronger assumption than (2.26) is that F has a density f such that for some $A > 0$

$$f(x) \sim Ax^{-(a+1)}, \tag{2.28}$$

where $\sim$ means that the ratio of the two sides converges to 1 as $x \to \infty$, so as an approximation $f(x) = Ax^{-(a+1)}$ for large values of x.

It was mentioned that the closer a gets to 0 the heavier the tail of a Pareto. Let's see what the value of a implies about the existence of an expected value and finite variance. Under assumption (2.28)

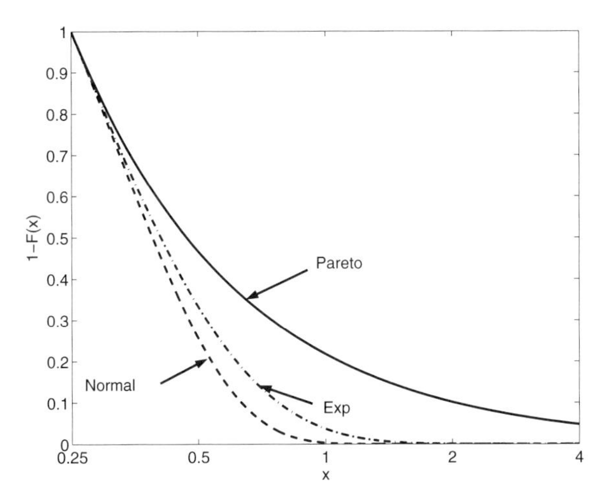

Fig. 2.12. *Survival function of a Pareto distribution with* $c = 0.25$ *and* $a = 1.1$ *and of normal and exponential distributions conditional on being greater than* 0.25. *The normal distribution has mean* 0 *and standard deviation* 0.3113. *The exponential distribution has* $\theta = c/a$.

$$\int_0^\infty x^b f(x)dx = \infty$$

if $b \geq a$. Thus, if X has a CDF F with Pareto left and right tails with index a in both cases, then $E(X)$ does not exist if $a = 1$ but $E(X)$ exists and is finite if $a > 1$. If $1 < a \leq 2$ then the variance of X is defined and infinite, but if $a > 2$ then the variance is finite.

If X is nonnegative and has a CDF with a Pareto right tail, then $E(X) = \infty$ if $a = 1$. If $a > 1$, then $E(X) < \infty$ and $\text{Var}(X)$ is infinite if $a \leq 2$ and finite if $a > 2$.

A symmetric Pareto tail distribution mentioned by Gourieroux and Jasiak (2001) has CDF

$$\begin{aligned} F(x) &= \frac{1}{2}\frac{1}{(1-x)^a}, \quad x < 0, \\ &= 1 - \frac{1}{2}\frac{1}{(1+x)^a}, \quad x \geq 0. \end{aligned} \tag{2.29}$$

The PDF is

$$f(x;a) = \frac{a}{2(1+|x|)^{a+1}}.$$

It is easy to add a location parameter μ and scale parameter σ to the shape parameter a to get a three-parameter family of distributions with densities

$$f(x;\mu,\sigma,a) = \frac{a}{2\sigma(1+|(x-\mu)/\sigma|)^{a+1}}. \tag{2.30}$$

The *stable distributions* are a class of distributions with Pareto tails. In the past, there was considerable interest in stable distributions as models for financial markets data, e.g., Mandelbrot (1963) and Fama (1965), but these distributions are less popular now, probably because they are difficult to work with since it is not possible to write the PDF or CDF of a stable distribution in closed form, except in some special cases such as the normal distribution and the t_1-distribution, which is also known as the Cauchy distribution.

2.13 Law of Large Numbers and Central Limit Theorem

Suppose that $\overline{X}_n$ is the mean of an i.i.d. sample $X_1, \ldots, X_n$. We assume that their common expected value $E(X_1)$ exists and call it μ. The **law of large numbers** states that

$$P(\overline{X}_n \to \mu \;\text{ as }\; n \to \infty) = 1.$$

Thus, the sample mean will be close to the population mean for large enough sample sizes. However, even more is true. The famous **Central Limit Theorem** (CLT) states that if the common variance σ^2 of $X_1, \ldots, X_n$ is finite, then the probability distribution of $\overline{X}_n$ gets closer to a normal distribution as n converges to ∞. More precisely, the CLT states that

$$P\{\sqrt{n}(\overline{X}_n - \mu) \leq x\} \to \Phi(x/\sigma) \text{ as } n \to \infty \text{ for all } x.$$

Students often misremember or misunderstand the CLT. A common misconception is that a large *population* is approximately normally distributed. The CLT says nothing about the distribution of a population; it is only a statement about the distribution of a sample mean. Also, the CLT does not assume that the population is large; it is the size of the sample that is converging to infinity. Assuming that the sampling is with replacement, the population could be quite small, in fact, with only two elements.

Although the CLT was first discovered for the sample mean, other estimators are now known to also have approximate normal distributions for large sample sizes. In particular, there are central limit theorems for the maximum likelihood estimators of Section 2.18.1 and the least squares estimators discussed in Chapter 6. This is very important, since most estimators we use will be maximum likelihood estimators or least squares estimators. So, if we have a reasonably large sample, we can assume that these estimators have an approximately normal distribution and the normal distribution can be used for testing and constructing confidence intervals.

When the variance of $X_1, \ldots, X_n$ is infinite, then the limit distribution of $\overline{X}_n$ may still exist but will be a nonnormal stable distribution.

2.14 Multivariate Distributions

Often we are not interested merely in a single random variable but rather in the joint behavior of several random variables, for example, the interrelationships among several asset prices and a market index. A multivariate distribution describes such joint behavior.

A pair of continuously distributed random variables, (X, Y), has a *bivariate probability density function* $f_{XY}(x, y)$ if

$$P\{(X, Y) \in A\} = \int \int_A f(x, y) dx\, dy$$

for all sets A. Similarly, a multivariate probability density (PDF) function for random variables $X_1, \ldots, X_n$ is a function $f_{X_1,\ldots,X_n}(x_1, \ldots, x_n)$ such that the probability that $X_1, \ldots, X_n$ is in any set A is found by integrating $f_{X_1,\ldots,X_n}(x_1, \ldots, x_n)$ over A. If $X_1, \ldots, X_n$ are independent, then

$$f_{X_1,\ldots,X_n}(x_1, \ldots, x_n) = f_{X_1}(x_1) \cdots f_{X_n}(x_n), \tag{2.31}$$

where f_{X_j} is the PDF of X_j.

If $X_1, \ldots, X_n$ are discrete then their joint probability distribution gives $P\{X_1 = x_1, \ldots, X_n = x_n\}$ for all numbers $x_1, \ldots, x_n$. If $X_1, \ldots, X_n$ are independent then

$$P\{X_1 = x_1, \ldots, X_n = x_n\} = P\{X_1 = x_1\} \cdots P\{X_n = x_n\}. \tag{2.32}$$

The joint CDF of $X_1, \ldots, X_n$, whether they are continuous or discrete, is

$$F_{X_1,\ldots,X_n}(x_1, \ldots, x_n) = P(X_1 \le x_1, \ldots, X_n \le x_n).$$

2.14.1 Correlation and covariance

Expectations and variances summarize the behavior of individual random variables. If we have two random variables, X and Y, then it is convenient to have something to summarize their joint behavior — correlation and covariance do this.

The **covariance** between two random variables X and Y is

$$\sigma_{XY} = E\Big[\{X - E(X)\}\{Y - E(Y)\}\Big].$$

If (X, Y) are continuously distributed, then

$$\sigma_{XY} = \int \{x - E(X)\}\{y - E(Y)\} f_{XY}(x, y)\, dx\, dy.$$

The following are useful formulas

$$\sigma_{XY} = E(XY) - E(X)E(Y). \tag{2.33}$$

$$\sigma_{XY} = E[\{X - E(X)\}Y]. \tag{2.34}$$

$$\sigma_{XY} = E[\{Y - E(Y)\}X]. \tag{2.35}$$

$$\sigma_{XY} = E(XY) \text{ if } E(X) = 0 \text{ or } E(Y) = 0. \tag{2.36}$$

The covariance between two variables measures the linear association between them, but it is also affected by their variability; all else equal, random variables with larger standard deviations have a larger covariance. Correlation is covariance with this size effect removed, so that correlation is pure measure of how closely two random variables are related, or more precisely, linearly related. The **correlation coefficient** between X and Y is

$$\rho_{XY} = \sigma_{XY} / \sigma_X \sigma_Y.$$

Given a bivariate sample $\{(X_i, Y_i)\}_{i=1}^n$, the sample covariance, denoted by s_{XY} or $\widehat{\sigma}_{XY}$, is

$$s_{XY} = \widehat{\sigma}_{XY} = (n-1)^{-1} \sum_{i=1}^{n} (X_i - \overline{X})(Y_i - \overline{Y}), \tag{2.37}$$

where $\overline{X}$ and $\overline{Y}$ are the sample means. Often the factor $(n-1)^{-1}$ is replaced by n^{-1}, but this change has little effect relative to the random variation in $\widehat{\sigma}_{XY}$. The **sample correlation** is

$$\widehat{\rho}_{XY} = r_{XY} = \frac{\widehat{\sigma}_{XY}}{s_X s_Y}, \tag{2.38}$$

where s_X and s_Y are the sample standard deviations.

To provide the reader with a sense of what a particular correlation coefficient implies about the relationship between two random variables, Figure 2.13 shows scatterplots and the sample correlation coefficients for eight random samples. A *scatterplot* is just a plot of a bivariate sample, $\{(X_i, Y_i)\}_{i=1}^n$. Notice that

- An absolute correlation of 0.25 or less is very weak — see the top row;
- An absolute correlation of 0.5 is only moderately strong — see the second row, left plot;
- An absolute correlation of 0.95 is rather strong — see the second row, right plot;
- An absolute correlation of 1 implies a linear relationship — see the fourth row, right plot;
- A strong nonlinear relationship may or may not imply a high correlation — see the plots in the third row;
- Positive correlations imply an increasing relationship (as X increases, Y increases on average);
- Negative correlations imply a decreasing relationship (as X increases, Y decreases on average).

If the correlation between two random variables is equal to 0, then we say that they are *uncorrelated.*

Example 2.6. Here is an example where sample correlations are computed in SAS. SAS programs contain a sequence of *data steps*, which input, manipulate, and output data, and *proc (or procedure) steps*, which do statistical computing and plotting and can create data sets containing the output of their computations.

In the data step, "`data ford`" names the data set being created, the "`infile`" statement gives the path to a text file containing the data, and the "`input`" statement names the variables being read. They are `tbill`, `sp500`, `ge`, `ford`, `day`, `month`, and `year` which are the daily closing three-month T-bill rate, the daily closing prices of the S&P 500 index, General Electric, and Ford Motor Company, and the day, month, and year of the observation. The statements following the "`input`" statement compute the returns on the S&P 500 index, GE, and Ford. Recall that in Chapter 1 the net return on an asset between times t_1 and t_2 was defined as the change in price divided by the initial price. In those statements, "`dif`" takes difference and "`lag(variable_name)`" is the previous value of a variable. Thus, for the ith observation, "`dif(sp500)`" is the ith minus the $i-1$st S&P 500 closing price and "`lag(sp500)`" is the $i-1$st S&P 500 closing price.

In the proc step, PROC CORR is used to compute the correlations between the variables specified in the "`var`" statement.

A SAS program can begin with options statements, such as "`options linesize = 68`" which specifies the length of the lines in the output.

```
options linesize = 68 ;
data ford ;
infile 'C:\book\SAS\prices.txt' ;
input tbill sp500 ge ford day month year ;
return_sp500 = dif(sp500)/lag(sp500) ;
return_ge = dif(ge)/lag(ge) ;
return_ford = dif(ford)/lag(ford) ;
proc corr ;
var return_sp500 return_ge return_ge ;
run ;
```

Here is the output.

```
                Pearson Correlation Coefficients, N = 2362
                        Prob > |r| under H0: Rho=0

                          return_                          return_
                            sp500         return_ge           ford

     return_sp500         1.00000           0.58708        0.39170
                                             <.0001         <.0001

     return_ge            0.58708           1.00000        0.74796
                           <.0001                           <.0001

     return_ford          0.39170           0.74796        1.00000
                           <.0001            <.0001
```

In the output we see that the sample correlation between the S&P 500 and GE returns is 0.58708, the sample correlation between the S&P 500 and Ford returns is 0.39170, and the sample correlation between the GE and Ford returns is 0.74796.

The output is in the form of a sample correlation matrix in which the i,jth entry is the sample correlation between the ith and jth variables. Similarly, the sample covariance matrix has covariances instead of correlations as its entries.

PROC CORR will output the sample covariance matrix, as well as the correlations, if the keyword "cov" is added to the proc statement as in

```
proc corr cov ;
```

2.14.2 Independence and covariance

If X and Y are independent then for all functions g and h,

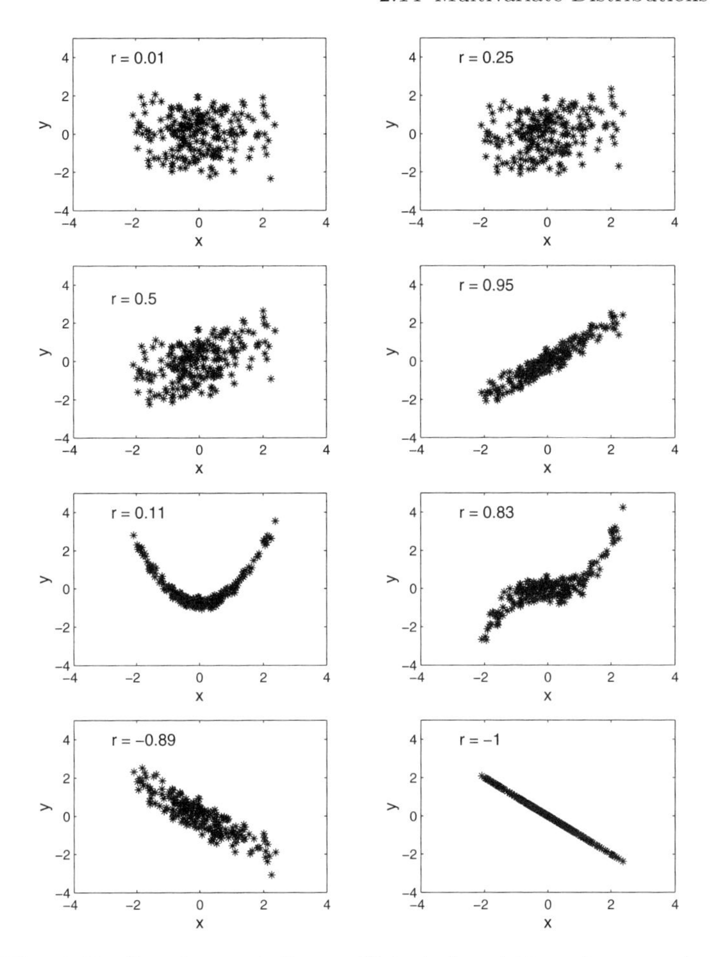

Fig. 2.13. *Sample correlation coefficients for eight random samples.*

$$E\{g(X)h(Y)\} = E\{g(X)\}E\{h(Y)\}. \tag{2.39}$$

This fact can be used to prove that if X and Y are independent, then $\sigma_{XY} = 0$; see Problem 7. The opposite is not true. For example, if X is uniformly distributed on $[-1, 1]$ and $Y = X^2$, then some calculation shows that $\sigma_{XY} = 0$ (see Problem 9), but the two random variables are not independent. The key point here is that Y is related to X, in fact, completely determined by X, but the relationship is highly nonlinear.

2.14.3 The multivariate normal distribution

The random variables $X_1, \ldots, X_n$ are said to have a *multivariate normal distribution* if for *every* set of constants $c_1, \ldots, c_n$, the weighted average (linear combination) $c_1 X_1 + \cdots + c_n X_n$ has a normal distribution. This implies, of course, that each of $X_1, \ldots, X_n$ is normally distributed; take $c_j = 1$ and the other c_i equal to 0 to show that X_j has a normal distribution.

The assumption of multivariate normality facilitates many useful probability calculations. For example, consider returns on assets, such as stocks, and on portfolios of assets. A *portfolio* is simply a weighted average of the assets with weights that sum to one. The weights specify what fractions of the total investment are allocated to the assets. If the returns on a set of assets have a multivariate normal distribution, then the return on any portfolio formed from these assets will be normally distributed. This is because the return on the portfolio is the weighted average of the returns on the assets. Therefore, the normal distribution could be used, for example, to find the probability of a loss of some size of interest, say, 50% or more, on the portfolio. Such calculations have important applications in finding a value-at-risk; see Chapter 11.

2.15 Prediction

Often we observe a random variable X and we want to predict another random variable Y that has not yet been observed but is related to X. For example, Y could be the future price of an asset and X might be the most recent change in that asset's price. If X and Y are dependent, then knowing X can help us predict Y. Prediction has many practical uses, but it is also useful in theoretical studies. For example, in Chapter 14 prediction theory is used to help decide between competing economic theories of the behavior of financial markets.

2.15.1 Best linear prediction

Linear prediction is theoretically simple and is closely related to the statistical technique of linear regression that is discussed in Chapter 6.

A linear predictor of Y based on X is a function $\beta_0 + \beta_1 X$ where β_0 and β_1 are parameters that we can choose. **Best linear prediction** means finding β_0 and β_1 so that expected squared prediction error, which is given by

$$E\{Y - (\beta_0 + \beta_1 X)\}^2, \tag{2.40}$$

is minimized to make the predictor as close as possible, on average, to Y. The expected squared prediction error can be rewritten as

$$E\{Y - (\beta_0 + \beta_1 X)\}^2 = E(Y^2) - 2\beta_0 E(Y) - 2\beta_1 E(XY) + E(\beta_0 + \beta_1 X)^2.$$

To find the minimizers, we set the partial derivatives of this expression to zero and get

$$0 = -E(Y) + \beta_0 + \beta_1 E(X) \text{ and} \tag{2.41}$$

$$0 = -E(XY) + \beta_0 E(X) + \beta_1 E(X^2). \tag{2.42}$$

After some algebra (see Problem 4) we find that

$$\beta_1 = \sigma_{XY}/\sigma_X^2 \tag{2.43}$$

and

$$\beta_0 = E(Y) - \beta_1 E(X) = E(Y) - \sigma_{XY}/\sigma_X^2 \, E(X). \tag{2.44}$$

One can check that the matrix of second derivatives of (2.40) is positive definite so that the solution (β_0, β_1) to (2.41) and (2.42) minimizes (2.40). Thus, the best linear predictor of Y is

$$\widehat{Y} = \beta_0 + \beta_1 X = E(Y) + \frac{\sigma_{XY}}{\sigma_X^2}\{X - E(X)\}. \tag{2.45}$$

In practice, (2.45) cannot be used directly unless $E(X)$, $E(Y)$, σ_{XY}, and σ_X^2 are known, which is often not the case. Linear regression analysis is essentially the use of (2.45) with these unknown parameters replaced by estimates.

2.15.2 Prediction error in linear prediction

The *prediction error* is $Y - \widehat{Y}$. It is easy to show (Problem 8) that $E\{Y - \widehat{Y}\} = 0$ so that the prediction is "unbiased." With a little algebra we can show that the expected squared prediction error is

$$E\{Y - \widehat{Y}\}^2 = \sigma_Y^2 - \frac{\sigma_{XY}^2}{\sigma_X^2} = \sigma_Y^2(1 - \rho_{XY}^2). \tag{2.46}$$

How much does X help us predict Y? To answer this question, notice first that if we do not observe X then we must predict Y using a constant, which we denote by c. Then the best predictor is to use c equal to $E(Y)$. To appreciate this fact, notice first that the expected squared prediction error is $E(Y - c)^2$. Some algebra shows that

$$E(Y - c)^2 = \text{Var}(Y) + \{c - E(Y)\}^2, \tag{2.47}$$

which, since $\text{Var}(Y)$ does not depend on c, shows that the expected squared prediction error is minimized by $c = E(Y)$. Thus, when X is unobserved the best predictor of Y is $E(Y)$ and the expected squared prediction error is σ_Y^2, but when X is observed then the expected squared prediction error is smaller, $\sigma_Y^2(1 - \rho_{XY}^2)$. Therefore, ρ_{XY}^2 is the fraction by which the prediction error is reduced when X is known. This is an important fact that we will see again.

Example 2.7. If $\rho_{XY} = .5$, then the prediction error is reduced by 25% by observing X. If $\sigma_Y^2 = 3$, then the expected squared prediction error is 3 if X is unobserved but only $2.25 = 3\{1 - (0.5)^2\}$ if X is observed.

We just saw that $E(Y)$ is the best predictor of Y if we observe no other random variables. The following related result is used repeatedly in Chapter 4.

Result 2.15.1 *Prediction when Y is independent of all available information*
If Y is independent of all presently available information, that is, Y is independent of all random variables that have been observed, then the best predictor of Y is $E(Y)$ and the expected value of the squared prediction error is Var(Y). *We say that Y "cannot be predicted" when there exists no predictor better than its expected value.*

2.15.3 Multivariate linear prediction

So far we have assumed that there is only a single random variable, X, available to predict Y. More commonly, Y is predicted using a set of observed random variables, $X_1, \ldots, X_n$. We do not discuss this topic here, but the sample analogue of multivariate linear prediction is multiple regression, which is discussed in detail in Chapter 6.

2.16 Conditional Distributions

Let $f_{XY}(x, y)$ be the joint density of a pair of random variables (X, Y). Then, the *marginal density* of X is obtained by "integrating out" Y:

$$f_X(x) = \int f_{XY}(x, y)dy,$$

and similarly $f_Y(y) = \int f_{XY}(x, y)dx$.

The **conditional density** of Y given X is

$$f_{Y|X}(y|x) = \frac{f_{XY}(x, y)}{f_X(x)}. \tag{2.48}$$

Equation (2.48) can be rearranged to give the joint density of X and Y as the product of a marginal density and a conditional density:

$$f_{XY}(x, y) = f_X(x)f_{Y|X}(y|x) = f_Y(y)f_{X|Y}(x|y). \tag{2.49}$$

The *conditional expectation* of Y given X is just the expectation calculated using $f_{Y|X}(y|x)$:

$$E(Y|X = x) = \int y f_{Y|X}(y|x)dy,$$

which is, of course, a function of x. The conditional variance of Y given X is

$$\text{Var}(Y|X = x) = \int \{y - E(Y|X = x)\}^2 f_{Y|X}(y|x)dy.$$

Example 2.8. Suppose $f_{XY}(x, y) = 2$ if $0 < x < 1$ and $x < y < 1$ and is 0 otherwise. Then the marginal density of X is

$$f_X(x) = \int_x^1 2\,dy = 2(1-x).$$

The conditional density of Y given X is

$$f_{Y|X}(y|x) = \frac{f_{XY}(x,y)}{f_X(x)} = \frac{2}{2(1-X)} = (1-x)^{-1}$$

for $x < y < 1$. Thus, we have shown that, given $X = x$, Y is uniformly distributed on the interval $[x, 1]$. Using the results about uniform distributions in Section 2.8.1, the conditional expectation of Y is

$$E(Y|X = x) = \frac{1+x}{2},$$

and the conditional variance of Y is

$$\mathrm{Var}(Y|X = x) = \frac{(1-x)^2}{12}.$$

2.16.1 Best prediction

Suppose that $\widehat{Y}$ can be any function of X, not necessarily a linear function as in Section 2.15.1. The *best predictor* is theoretically simple — it is the conditional expectation of Y given X. That is, $E(Y|X)$ is the best predictor of Y in the sense of minimizing $E\{Y - \widehat{Y}\}^2$ among *all* possible choices of $\widehat{Y}$ that are arbitrary functions of X.

However, in practice using $E(Y|X)$ for prediction is not trivial. The problem is that $E(Y|X)$ may be difficult to estimate whereas the best linear predictor can be estimated by linear regression as described in Chapter 6. However, the newer technique of **nonparametric regression** can be used to estimate $E(Y|X)$. Nonparametric regression by splines is discussed in Chapter 13. Artificial neural networks, another method of nonparametric regression, have been used for modeling financial data.

2.16.2 Normal distributions: Conditional expectations and variance

The calculation of conditional expectations and variances can be difficult for some probability distributions, but it is quite easy for a pair (X, Y) that has a bivariate normal distribution.[14]

For a bivariate normal pair, the conditional expectation of Y given X equals the best linear predictor of Y given X:

[14] If $X_1, \ldots, X_n$ are multivariate normal, then each pair of them is bivariate normal.

$$E(Y|X=x) = E(Y) + \frac{\sigma_{XY}}{\sigma_X^2}\{x - E(X)\}.$$

Therefore, for normal random variables, best linear prediction is the same as best prediction. Also, the conditional variance of Y given X is the expected squared prediction error:

$$\text{Var}(Y|X=x) = \sigma_Y^2(1-\rho_{XY}^2). \tag{2.50}$$

In general, $\text{Var}(Y|X=x)$ is a function of x but we see in (2.50) that for the special case of a bivariate normal distribution $\text{Var}(Y|X=x)$ is constant, that is, independent of x.

2.17 Linear Functions of Random Variables

Often we are interested in finding the expectation and variance of a linear combination (weighted average) of random variables. For example, because the return on a portfolio is a linear combination of the returns on the individual assets in the portfolio, the material in this section is used extensively in the portfolio theory of Chapters 5 and 7.

First, we look at a linear function of a single random variable. If Y is a random variable and a and b are constants, then

$$E(aY+b) = aE(Y) + b.$$

Also,

$$\text{Var}(aY+b) = a^2\text{Var}(Y).$$

Next, we consider linear combinations of two random variables. If X and Y are random variables and w_1 and w_2 are constants, then $w_1X + w_2Y$ is itself a random variable,

$$E(w_1X + w_2Y) = w_1E(X) + w_2E(Y),$$

and

$$\text{Var}(w_1X + w_2Y) = w_1^2\text{Var}(X) + 2w_1w_2\text{Cov}(X,Y) + w_2^2\text{Var}(Y). \tag{2.51}$$

Check that

$$\text{Var}(w_1X + w_2Y) = \begin{pmatrix} w_1 & w_2 \end{pmatrix} \begin{pmatrix} \text{Var}(X) & \text{Cov}(X,Y) \\ \text{Cov}(X,Y) & \text{Var}(Y) \end{pmatrix} \begin{pmatrix} w_1 \\ w_2 \end{pmatrix}. \tag{2.52}$$

Although (2.52) may seem unnecessarily complicated, we show that this equation generalizes in an elegant way to more than two random variables. The matrix in (2.52) is called the **covariance matrix** of the random vector $(X \quad Y)^{\mathsf{T}}$.

Now we generalize to a arbitrary number of random variables. Let $\boldsymbol{X} = (X_1, \ldots, X_N)^\mathsf{T}$ be a random vector, that is, a vector whose elements are random variables. We define the expectation vector of $\boldsymbol{X}$ to be

$$E(\boldsymbol{X}) = \begin{pmatrix} E(X_1) \\ \vdots \\ E(X_N) \end{pmatrix}.$$

The covariance matrix of $\boldsymbol{X}$ is

$$\mathrm{COV}(\boldsymbol{X}) = \begin{pmatrix} \mathrm{Var}(X_1) & \mathrm{Cov}(X_1, X_2) & \cdots & \mathrm{Cov}(X_1, X_N) \\ \mathrm{Cov}(X_2, X_1) & \mathrm{Var}(X_2) & \cdots & \mathrm{Cov}(X_2, X_N) \\ \vdots & \vdots & \ddots & \vdots \\ \mathrm{Cov}(X_N, x_1) & \mathrm{Cov}(X_N, X_2) & \cdots & \mathrm{Var}(X_N) \end{pmatrix}.$$

Let $\boldsymbol{w} = (w_1, \ldots, w_N)^\mathsf{T}$ be a vector of weights. Then

$$\boldsymbol{w}^\mathsf{T}\boldsymbol{X} = \sum_{i=1}^{N} w_i X_i$$

is a weighted average of the components of $\boldsymbol{X}$. Of course, $\boldsymbol{w}^\mathsf{T}\boldsymbol{X}$ is itself a random variable so it has an expected value and variance. One can show that

$$E(\boldsymbol{w}^\mathsf{T}\boldsymbol{X}) = \boldsymbol{w}^\mathsf{T}\{E(\boldsymbol{X})\}$$

and

$$\mathrm{Var}(\boldsymbol{w}^\mathsf{T}\boldsymbol{X}) = \sum_{i=1}^{N}\sum_{j=1}^{N} w_i\, w_j\, \mathrm{Cov}(X_i, X_j).$$

This last result can be expressed more simply using vector/matrix notation:

$$\mathrm{Var}(\boldsymbol{w}^\mathsf{T}\boldsymbol{X}) = \boldsymbol{w}^\mathsf{T}\mathrm{COV}(\boldsymbol{X})\boldsymbol{w}. \tag{2.53}$$

Example 2.9. Suppose that $\boldsymbol{X} = (X_1\ X_2\ X_3)^\mathsf{T}$, $\mathrm{Var}(X_1) = 2$, $\mathrm{Var}(X_2) = 3$, $\mathrm{Var}(X_3) = 5$, $\rho_{X_1,X_2} = 0.6$, and that X_1 and X_2 are independent of X_3. Find $\mathrm{Var}(X_1 + X_2 + 1/2\, X_3)$.

Answer: The covariance between X_1 and X_3 is 0 by independence, and the same is true of X_2 and X_3. The covariance between X_1 and X_2 is $(.6)\sqrt{(2)(3)} = 1.47$. Therefore,

$$\mathrm{COV}(\boldsymbol{X}) = \begin{pmatrix} 2 & 1.47 & 0 \\ 1.47 & 3 & 0 \\ 0 & 0 & 5 \end{pmatrix},$$

and by (2.53),

$$\begin{aligned}\operatorname{Var}(X_1+X_2+X_3/2) &= \begin{pmatrix}1 & 1 & \frac{1}{2}\end{pmatrix}\begin{pmatrix}2 & 1.47 & 0\\ 1.47 & 3 & 0\\ 0 & 0 & 5\end{pmatrix}\begin{pmatrix}1\\ 1\\ \frac{1}{2}\end{pmatrix}\\ &= \begin{pmatrix}1 & 1 & \frac{1}{2}\end{pmatrix}\begin{pmatrix}3.47\\ 4.47\\ 2.5\end{pmatrix}\\ &= 9.19.\end{aligned}$$

2.17.1 Two linear combinations of random variables

More generally, suppose that $\boldsymbol{w}_1^{\mathsf{T}}\boldsymbol{X}$ and $\boldsymbol{w}_2^{\mathsf{T}}\boldsymbol{X}$ are two weighted averages of the components of $\boldsymbol{X}$, e.g., returns on two different portfolios composed from the same set of assets. Then

$$\operatorname{Cov}(\boldsymbol{w}_1^{\mathsf{T}}\boldsymbol{X}, \boldsymbol{w}_2^{\mathsf{T}}\boldsymbol{X}) = \boldsymbol{w}_1^{\mathsf{T}}\mathrm{COV}(\boldsymbol{X})\boldsymbol{w}_2. \tag{2.54}$$

Example 2.10. (Example 2.9 *continued)* Suppose $\boldsymbol{X} = \begin{pmatrix}X_1 & X_2 & X_3\end{pmatrix}^{\mathsf{T}}$ have the same properties as in the previous example and are the returns on three assets. Find the covariance between a portfolio that allocates 1/3 to each of the three assets and a second portfolio that allocates 1/2 to each of the first two assets. That is, find the covariance between $(X_1+X_2+X_3)/3$ and $(X_1+X_2)/2$.

Answer: Let

$$\boldsymbol{w}_1 = \begin{pmatrix}\frac{1}{3} & \frac{1}{3} & \frac{1}{3}\end{pmatrix}^{\mathsf{T}}$$

and

$$\boldsymbol{w}_2 = \begin{pmatrix}\frac{1}{2} & \frac{1}{2} & 0\end{pmatrix}^{\mathsf{T}}.$$

Then

$$\begin{aligned}&\operatorname{Cov}\left\{\frac{X_1+X_2}{2}, \frac{X_1+X_2+X_3}{3}\right\}\\ &\quad= \boldsymbol{w}_1^{\mathsf{T}}\mathrm{COV}(\boldsymbol{X})\boldsymbol{w}_2\\ &\quad= \begin{pmatrix}1/3 & 1/3 & 1/3\end{pmatrix}\begin{pmatrix}2 & 1.47 & 0\\ 1.47 & 3 & 0\\ 0 & 0 & 5\end{pmatrix}\begin{pmatrix}1/2\\ 1/2\\ 0\end{pmatrix}\\ &\quad= \begin{pmatrix}1.157 & 1.490 & 1.667\end{pmatrix}\begin{pmatrix}1/2\\ 1/2\\ 0\end{pmatrix}\\ &\quad= 1.323.\end{aligned}$$

2.17.2 Independence and variances of sums

If $X_1, \ldots, X_n$ are independent, or at least uncorrelated (all correlation coefficients are 0), then

$$\mathrm{Var}(\boldsymbol{w}^{\mathsf{T}}\boldsymbol{X}) = \sum_{i=1}^{n} w_i^2 \mathrm{Var}(X_i). \tag{2.55}$$

When $\boldsymbol{w}^{\mathsf{T}} = (\,1/n \cdots 1/n\,)$ so that $\boldsymbol{w}^{\mathsf{T}}\boldsymbol{X} = \overline{X}$, then we obtain that

$$\mathrm{Var}(\overline{X}) = \frac{1}{n^2}\sum_{i=1}^{n} \mathrm{Var}(X_i). \tag{2.56}$$

In particular, if $\mathrm{Var}(X_i) = \sigma^2$ for all i, then we obtain the well-known result that

$$\mathrm{Var}(\overline{X}) = \frac{\sigma^2}{n}. \tag{2.57}$$

Another useful fact that follows from (2.55) is that if X_1 and X_2 are uncorrelated then

$$\mathrm{Var}(X_1 - X_2) = \mathrm{Var}(X_1) + \mathrm{Var}(X_2). \tag{2.58}$$

2.17.3 Application to normal distributions

Recall from Section 2.14.3 that if $\boldsymbol{X}$ has a multivariate normal distribution, then $\boldsymbol{w}^{\mathsf{T}}\boldsymbol{X}$ is a normally distributed random variable. We can use the previous results to find its expectation and variance and then probabilities can be found easily.

Example 2.11. Suppose that $E(X_1) = 1$, $E(X_2) = 1.5$, $\sigma_{X_1}^2 = 1$, $\sigma_{X_2}^2 = 2$, and $\mathrm{Cov}(X_1, X_2) = 0.5$. Find $E(0.3X_1 + 0.7X_2)$ and $\mathrm{Var}(0.3X_1 + 0.7X_2)$. If $(X_1\ X_2)^{\mathsf{T}}$ is bivariate normal, find $P\{0.3X_1 + 0.7X_2 < 2\}$.

Answer: First, $E(0.3X_1 + 0.7X_2) = (0.3)(1) + (0.7)(1.5) = 1.35$. Then, by (2.51) $\mathrm{Var}(0.3X_1 + 0.7X_2) = (0.3)^2(1) + 2(0.3)(0.7)(0.5) + (0.7)^2(2) = 1.28$. Finally, $P\{0.3X_1 + 0.7X_2 < 2\} = \Phi\{(2 - 1.35)/\sqrt{1.28}\} = \Phi(0.5745) = 0.7172$.

2.18 Estimation

One of the major areas of statistical inference is estimation of unknown parameters, such as a population mean, from data. An estimator is defined to be any function of the observed data. The key question is which of many possible estimators should be used. If θ is an unknown parameter and $\widehat{\theta}$ is an estimator, then $E(\widehat{\theta}) - \theta$ is called the *bias* and $E\{\widehat{\theta} - \theta\}^2$ is called the *mean squared error* (MSE). One seeks estimators that are efficient, that is, having the smallest possible value of the MSE (or of some other measure of inaccuracy). It can be shown from simple algebra that the MSE is the squared bias plus the variance, that is,

$$E\{\widehat{\theta} - \theta\}^2 = \{E(\widehat{\theta}) - \theta\}^2 + \mathrm{Var}(\widehat{\theta}), \tag{2.59}$$

so an efficient estimator will have both a small bias and a small variance. An estimator with a zero bias is called *unbiased.* However, it is not necessary to use an unbiased estimator — we only want the bias to be small, not necessarily exactly zero. One should be willing to accept a small bias if this leads to a significant reduction in variance.

2.18.1 Maximum likelihood estimation

Maximum likelihood is the most important and widespread method of estimation. Many well-known estimators such as the sample mean and the least squares estimator in regression are maximum likelihood estimators. Maximum likelihood estimation is very useful in practice and tends to give more efficient estimates than other methods of estimation. Parameters in the widely used ARIMA and GARCH time series models studied in Chapters 4 and 12 are usually estimated by maximum likelihood.

Let $\boldsymbol{Y} = (Y_1, \ldots, Y_n)^{\mathsf{T}}$ be a vector of data and let $\boldsymbol{\theta} = (\theta_1, \ldots, \theta_p)^{\mathsf{T}}$ be a vector of parameters. Let $f(\boldsymbol{Y}; \boldsymbol{\theta})$ be the density of $\boldsymbol{Y}$ that depends on the parameters.

Example 2.12. Suppose that $Y_1, \ldots, Y_n$ are i.i.d. $N(\mu, \sigma^2)$. Then $\boldsymbol{\theta} = (\mu, \sigma^2)$. Also, by (2.14) and (2.31),

$$f(\boldsymbol{Y}; \boldsymbol{\theta}) = \prod_{i=1}^{n} \frac{1}{\sigma} \phi\left(\frac{Y_i - \mu}{\sigma}\right) = \frac{1}{\sigma^n (2\pi)^{n/2}} \exp\left\{\frac{-1}{2\sigma^2} \sum_{i=1}^{n} (Y_i - \mu)^2\right\}.$$

The function $L(\boldsymbol{\theta}) = f(\boldsymbol{Y}; \boldsymbol{\theta})$ viewed as a function of $\boldsymbol{\theta}$ with $\boldsymbol{Y}$ fixed at the observed data is called the **likelihood function**. It tells us the likelihood of the sample that was actually observed. The **maximum likelihood estimator** (MLE) is the value of $\boldsymbol{\theta}$ that maximizes the likelihood function. In other words, the MLE is the value of $\boldsymbol{\theta}$ where the likelihood of the data is largest. We denote the MLE by $\widehat{\boldsymbol{\theta}}_{ML}$. Often it is mathematically easier to maximize $\log\{L(\boldsymbol{\theta})\}$. If the data are independent then the likelihood is the product of the marginal densities and products are cumbersome to differentiate. Taking the logarithm converts the product into an easily differentiated sum. Since the log function is increasing, maximizing $\log\{L(\boldsymbol{\theta})\}$ is equivalent to maximizing $L(\boldsymbol{\theta})$.

Example 2.13. In the example above, the log-likelihood is

$$\log\{L(\boldsymbol{\theta})\} = -\frac{n}{2}\{\log(\sigma^2) + \log(2\pi)\} - \frac{1}{2\sigma^2} \sum_{i=1}^{n} (Y_i - \mu)^2.$$

This shows that the MLE of μ minimizes $\sum_{i=1}^{n}(Y_i - \mu)^2$. It is an easy calculus exercise to verify that $\widehat{\mu}_{ML} = \overline{Y}$. Also, with μ fixed at its MLE, the MLE of σ^2 solves

$$0 = \frac{d}{d\sigma^2}\log\{L(\widehat{\mu}_{ML}, \sigma^2)\} = -\frac{n}{2\sigma^2} + \frac{1}{2\sigma^4}\sum_{i=1}^{n}(Y_i - \widehat{\mu}_{ML})^2.$$

The solution to this equation is

$$\widehat{\sigma}^2_{ML} = \frac{1}{n}\sum_{i=1}^{n}(Y_i - \overline{Y})^2.$$

It is possible to prove that $E(\widehat{\sigma}^2_{ML}) = (n-1)\,\sigma^2/n$, not σ^2, so that the MLE has a small bias. The "bias-corrected" MLE is the so-called "sample variance" defined before as

$$s_Y^2 = \frac{1}{n-1}\sum_{i=1}^{n}(Y_i - \overline{Y})^2. \tag{2.60}$$

Example 2.14. (MLE of the Pareto index and the Hill estimator) Assume that $X_1, \ldots, X_n$ are i.i.d. Pareto(a, c). The parameter c is often known and if not, one can use the minimum of $X_1, \ldots, X_n$ as an estimate of c. The tail index a will generally not be known, but a can be estimated easily by maximum likelihood as is now shown. For simplicity, in the following we assume that c is known but c would be replaced by an estimate if not. By (2.23), the likelihood is

$$L(a) = \left(\frac{ac^a}{X_1^{a+1}}\right)\left(\frac{ac^a}{X_2^{a+1}}\right)\cdots\left(\frac{ac^a}{X_n^{a+1}}\right),$$

so the log-likelihood is

$$\log\{L(a)\} = \sum_{i=1}^{n}\{\log(a) + a\log(c) - (a+1)\log(X_i)\}. \tag{2.61}$$

Differentiating (2.61) with respect to a and setting the derivative equal to 0 gives the equation

$$\frac{n}{a} = \sum_{i=1}^{n}\log(X_i/c).$$

Therefore the MLE of a is

$$\widehat{a} = \frac{n}{\sum_{i=1}^{n}\log(X_i/c)}. \tag{2.62}$$

If $X_1, \ldots, X_n$ are i.i.d. from a distribution with Pareto tails rather than being exactly Pareto, then one does not want to compute (2.62) using all of the data but rather only the data in the tail. Otherwise, there could be sizeable bias. Therefore, one should choose a constant c and use only the data greater than c. This estimator is called the *Hill estimator*. Typically, c is one of the X_i. The Hill estimator can be written as

$$\widehat{a}_{\text{Hill}}(c) = \frac{n(c)}{\sum_{X_i > c} \log(X_i/c)}, \tag{2.63}$$

where $n(c)$ is the number of X_i greater than c. The difficulty is how to choose c or, equivalently, $n(c)$. The *Hill plot* is a plot of $\widehat{a}_{\text{Hill}}(c)$ versus $n(c)$. We expect that $\widehat{a}_{\text{Hill}}(c)$ will be unstable when $n(c)$ is small due to random variability. If $n(c)$ gets too large, one is using nearly all of the data and $\widehat{a}_{\text{Hill}}(c)$ may suffer from bias. One hopes that the Hill plot will show some stability for $n(c)$ neither too small nor too large and one can then use a value of $n(c)$ in this region of stability. However, the Hill plot is not always helpful for choosing c because a region of stability is often not seen in the plot. The problem is that a Pareto tail does not mean that the tail is *exactly* like that of a Pareto distribution since L is only slowly varying, not constant; see Drees, de Haan, and Resnick (2000) and Resnick (2001). The application of the Hill estimator to VaR and an alternative estimator of the tail index are discussed in Chapter 11.

In "textbook examples" such as the ones above, it is possible to find an explicit formula for the MLE. With more complex models, typically there is no explicit formula for the MLE. Rather, one must write a program to compute $\log\{L(\boldsymbol{\theta})\}$ for any value of $\boldsymbol{\theta}$ and then use optimization software to maximize this function numerically. For many important models such as the ARIMA and GARCH time series models discussed in Chapter 4, there are software packages, e.g., SAS, that compute the MLE.

2.18.2 Standard errors

When a estimator is calculated from a random sample it is a random variable, but this fact is often not appreciated. Students tend not to think of estimators such as a sample mean as random. If we have only a single sample, then the sample mean does not *appear* random. However, if we realize that the observed sample is only one of many possible samples that could have been drawn, and that each sample has a different sample mean, then we see that the mean is in fact random.

Since an estimator is a random variable, it has an expectation and a standard deviation. We have already seen that the difference between its expectation and the parameter is called the bias. The standard deviation of an estimator is called its **standard error**. If there are unknown parameters in the formula for this standard deviation, then they are replaced by estimates. If $\widehat{\theta}$ is an estimator of θ, then $s_{\widehat{\theta}}$ will denote its standard error with any unknown parameters replaced by estimates.

Example 2.15. Suppose that $X_1, \ldots, X_n$ are i.i.d. with mean μ and variance σ^2. Then it follows from (2.57) that the standard deviation of $\overline{X}$ is $\sigma/\sqrt{n}$.

Thus, $\sigma/\sqrt{n}$, or when σ is unknown $s_X/\sqrt{n}$, is called the standard error of the sample mean. That is, $s_{\overline{X}}$ is $\sigma/\sqrt{n}$ or $s_X/\sqrt{n}$ depending on whether σ is known.

2.18.3 Fisher information

Standard errors are essential for gauging the accuracy of estimators. We have a formula for the standard error of $\overline{X}$, but what about standard errors for other estimators? Fortunately, there is a simple method for calculating the standard error of a maximum likelihood estimator. We assume for now that θ is one-dimensional. The *Fisher information* is defined to be minus the expected second derivative of the log-likelihood, so if $\mathcal{I}(\theta)$ denotes the Fisher information, then

$$\mathcal{I}(\theta) = -E\left[\frac{d^2}{d\,\theta^2}\log\{L(\theta)\}\right]. \tag{2.64}$$

The standard error of $\widehat{\theta}$ is simply the inverse square root of the Fisher information, with the unknown θ replaced by $\widehat{\theta}$:

$$s_{\widehat{\theta}} = \frac{1}{\sqrt{\mathcal{I}(\widehat{\theta})}}. \tag{2.65}$$

Example 2.16. Suppose that $X_1, \ldots, X_n$ are i.i.d. $N(\mu, \sigma^2)$ with σ^2 known. The log-likelihood for the unknown parameter μ is

$$\log\{L(\mu)\} = -\frac{n}{2}\{\log(\sigma^2) + \log(2\pi)\} - \frac{1}{2\sigma^2}\sum_{i=1}^{n}(Y_i - \mu)^2.$$

Therefore,

$$\frac{d}{d\mu}\log\{L(\mu)\} = \frac{1}{\sigma^2}\sum_{i=1}^{n}(Y_i - \mu),$$

and

$$\frac{d^2}{d\mu^2}\log\{L(\mu)\} = -\frac{\sum_{i=1}^{n} 1}{\sigma^2} = -\frac{n}{\sigma^2}.$$

Therefore $\mathcal{I}(\widehat{\mu}) = n/\sigma^2$ and $s_{\widehat{\mu}} = \sigma/\sqrt{n}$. Since the MLE is $\widehat{\mu} = \overline{X}$, this result is the familiar fact that when σ is known then $s_{\overline{X}} = \sigma/\sqrt{n}$.

So far in the discussion of Fisher information, θ has been assumed to be one-dimensional. If $\boldsymbol{\theta}$ is an m-dimensional parameter vector, then the Fisher information is an $m \times m$ square matrix and is equal to minus the matrix of expected second-order partial derivatives of $\log\{L(\boldsymbol{\theta})\}$.[15] In other words, the i, jth entry of the Fisher information matrix is

[15] The matrix of second partial derivatives is called the *Hessian matrix.*

$$\mathcal{I}_{ij}(\boldsymbol{\theta}) = -E\left[\frac{\partial^2}{\partial\theta_i\,\partial\theta_j}\log\{L(\boldsymbol{\theta})\}\right]. \tag{2.66}$$

The standard errors are the square roots of the diagonal entries of the inverse of the Fisher information matrix.

The calculation of standard errors of maximum likelihood estimators by computing and then inverting the Fisher information matrix is routinely programmed into statistical software. Therefore, we need not concern ourselves with the details of computing standard errors. The key point is that there is an explicit method of calculating standard errors for maximum likelihood estimators.

2.18.4 Bayes estimation*

Bayesian statistics has a different philosophy than other approaches to statistical inference. In Bayesian statistics, all unknowns are considered to be random variables and their probability distributions specify our beliefs about their likely values. Assume we are interested in a parameter θ. Then we start with a *prior distribution* with density $\pi(\theta)$ that gives our beliefs about θ prior to observing data. The likelihood function is interpreted as the conditional density of the data $\boldsymbol{Y}$ given θ and written as $f(\boldsymbol{y}|\theta)$. Then using equation (2.49) the joint density of θ and $\boldsymbol{Y}$ is the product of the prior and the likelihood; that is,

$$f(\boldsymbol{y},\theta) = \pi(\theta)f(\boldsymbol{y}|\theta). \tag{2.67}$$

The marginal density of $\boldsymbol{Y}$ is

$$f(y) = \int \pi(\theta)f(\boldsymbol{y}|\theta)d\theta \tag{2.68}$$

and the conditional density of θ given $\boldsymbol{Y}$ is

$$f(\theta|\boldsymbol{y}) = \frac{\pi(\theta)f(\boldsymbol{y}|\theta)}{\int \pi(\theta)f(\boldsymbol{y}|\theta)d\theta}. \tag{2.69}$$

Equation (2.69) is another form of Bayes' law that was introduced in Section 2.2.2. The density on the left-hand side of (2.69) is called the *posterior density* and gives the probability distribution of θ after observing the data $\boldsymbol{Y}$. Bayes estimators are based upon the posterior. The most common Bayes estimators are the mode and the mean of the posterior density. The mode is called the maximum a posteriori estimator or *MAP estimator*. The mean of the posterior is

$$E(\theta|\boldsymbol{y}) = \int \theta f(\theta|\boldsymbol{y})d\theta = \frac{\int \theta\,\pi(\theta)f(\boldsymbol{y}|\theta)d\theta}{\int \pi(\theta)f(\boldsymbol{y}|\theta)d\theta}. \tag{2.70}$$

Example 2.17. We continue Example 2.1 but change the simple, but unrealistic, prior that said that θ was either 0.4 or 0.6 to a more plausible prior where

θ could be any value in the interval [0, 1] but with values near 1/2 more likely. Specifically, we use the prior

$$\pi(\theta) = 6\theta(1-\theta), \quad 0 < \theta < 1.$$

Let Y be the number of times the stock price increases on three consecutive days. Then Y is Binomial(n, θ) and the likelihood is

$$f(y|\theta) = \binom{3}{y} \theta^y (1-\theta)^{3-y}, \ y = 0, 1, 2, 3.$$

Since we observed that $Y = 3$, the posterior density is

$$f(\theta|3) = \frac{6\,\theta^4(1-\theta)}{\int 6\,\theta^4(1-\theta)d\theta} = 30\,\theta^4(1-\theta).$$

The prior and posterior densities are shown in Figure 2.14. The posterior density is shifted towards the right compared to the prior because three consecutive days saw increased prices. The posterior expectation is

$$\int 30\,\theta^5(1-\theta)d\theta = \frac{30}{42} = 0.7143. \tag{2.71}$$

The MAP estimate is $4/5 = 0.8$; see Problem 10.

The posterior CDF is

$$F(\theta|Y=3) = 30\left(\frac{\theta^5}{5} - \frac{\theta^6}{6}\right), \quad 0 \le \theta \le 1.$$

The CDF is plotted in Figure 2.15 and the 0.05 and 0.95 posterior quantiles of θ are shown as well. These quantiles were found by extrapolation and are 0.4182 and 0.9372, respectively. Thus, there is 90% posterior probability that θ is between 0.4182 and 0.9372. The interval [0.4182, 0.9372] is called a 90% *credible interval* and provides us with a good idea about the likely values of θ.

Credible intervals are the Bayesian analogues of confidence intervals, which are discussed in Section 2.19.

Although the Bayesian calculations in this simple example are straightforward, this is generally not true for problems of practical interest. Frequently, the integral in the denominator of posterior density (2.69) is impossible to calculate analytically. The same is true of the integral in the expression for the posterior mean given by (2.70). Therefore, until recently Bayesian statistics was mostly of theoretical interest. Fortunately, simulation methods for approximating posterior densities and expectations have been developed. They have been a tremendous advance and not only have they made Bayesian methods practical, but they have led to the solution of applied problems that heretofore could not be tackled. The software package WinBUGS, which currently is available without charge over the Internet, implements these simulation methods and is now widely used.

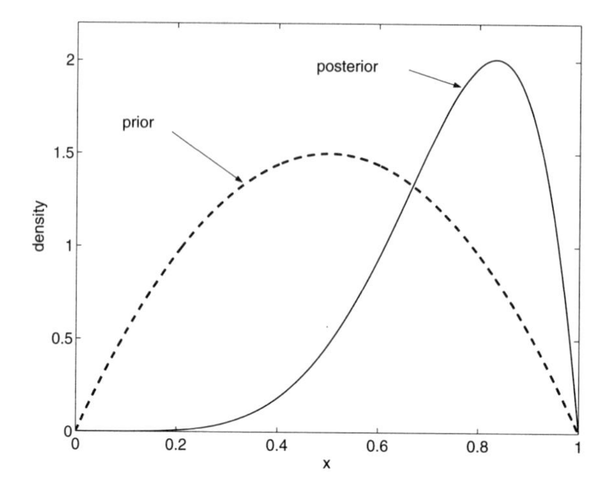

Fig. 2.14. *Prior and posterior densities in Example* 2.17.

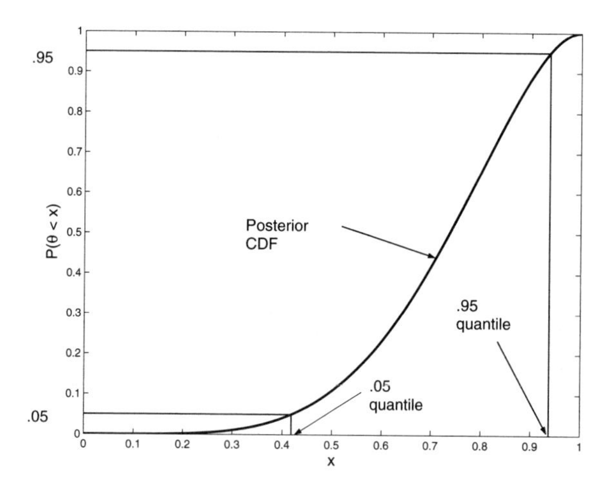

Fig. 2.15. *Posterior CDF in Example* 2.17.

2.18.5 Robust estimation*

Although maximum likelihood estimators have many attractive properties, they have one serious drawback of which anyone using them must be aware. Maximum likelihood estimators are very sensitive to the assumptions of the statistical model. For example, we saw in Example 2.13 that the MLE of the mean of a normal population is the sample mean and the MLE of σ^2 is the sample variance, except with the minor change of a divisor of n rather than $n-1$. The sample mean and variance are efficient estimators when the population is truly normally distributed, but these estimators are very sensitive to outliers. Because these estimators are averages of the data and the squared deviations from the mean, respectively, a single outlier in the sample can drive the sample mean and variance to wildly absurd values if it is far enough removed from the other data. Extreme outliers are nearly impossible with exactly normally distributed data, but if the data are only approximately normal with heavier tails than the normal distribution, then outliers are more probable and more likely to be extreme so the sample mean and variance can be very inefficient. Statisticians say that the MLE is not *robust* to mild deviations from the assumed model. This is bad news and has led researchers to find estimators that are robust.

An alternative to the sample mean that is robust is the *trimmed mean.* An α-trimmed mean is defined as follows. Order the sample from smallest to largest, remove the fraction α of the smallest and the same fraction of the largest observations and then take the mean of the rest. The idea behind trimming is simple and should be obvious. The sample is trimmed of extreme values before the mean is calculated. Here is a mathematical formulation of the α-trimmed mean. Let $k = n\alpha$ rounded[16] to an integer; k is the number of observations removed from both ends of the sample. Then the α-trimmed mean is

$$\overline{X}_\alpha = \frac{\sum_{i=k+1}^{n-k} X_{(i)}}{n-2k},$$

where $X_{(i)}$ is the ith order statistic. Typical values of α are 0.1, 0.15, 0.2, and 0.25. As α approaches 0.5, the α-trimmed mean approaches the sample median, which is the 0.5-sample quantile.

Dispersion means the variation in a distribution or sample. The sample standard deviation is the most common estimate of dispersion, but as stated it is nonrobust. A robust estimator of dispersion is the so-called *MAD* (*median absolute deviation*) estimator defined as

$$\text{median}\{|X_i - \text{median}(X_i)|\}.$$

This formula should be interpreted as follows. The expression "median(X_i)" is the sample median, $|X_i - \text{median}(X_i)|$ is the absolute deviation of X_i from the median, that is, the distance of the ith observation from the median. The measure of dispersion is the median of these absolute deviations of the data from the sample median, or in other words, the median distance of the data from the median. For normally distributed data, the MAD estimates not σ but rather $\Phi^{-1}(0.75)\sigma = 0.7734\sigma$, because for normally distributed data the MAD will converge to 0.7734σ as the sample size increases. Therefore, a robust estimator of σ is MAD/0.7734. SAS does not use MAD/0.7734 but

[16] Definitions vary and the rounding could be either upward or to the nearest integer.

a more sophisticated estimator that has a small-sample bias correction built in. Neither MAD/0.7734, SAS's version of the MAD, nor any other estimator based on the MAD will estimate σ for a nonnormal population. They do measure dispersion, but not dispersion as measured by the standard deviation. But that is just the point. For nonnormal populations the standard deviation is very sensitive to the tails of the distribution and does not tell us much about the dispersion in the central range of the distribution, just in the tails.

Example 2.18. This example uses Ford Motor Company daily closing prices from Nov 1, 1993 to Apr 3, 2003. Here is a SAS program to compute a number of statistics from the sample of returns. The first part of the program is the data step that reads the data from a file and then computes the returns. Here is the data step.

```
options linesize = 68 ;
data ford ;
infile 'C:\book\SAS\ford.txt' ;
input price ;
return = dif(price)/lag(price) ;
run ;
title 'Ford - Daily prices, Nov 1, 1993 to Apr 3, 2003' ;
```

The remainder of the program is the PROC step that calls PROC UNIVARIATE which computes a number of statistics from a sample. The keyword "normal" following "proc univariate" requests tests of a normal distribution, the "trimmed=.1" requests a 0.1-trimmed mean, and "robustscale" requests several robust estimates of dispersion, which is also called scale. The statement "var return" tells PROC UNIVARIATE the name of the variable to use; there could be several variable names in other applications. Finally "probplot" requests a normal probability plot.

```
proc univariate normal trimmed=.1 robustscale;
var return ;
probplot ;
run ;
```

Here is the output. The excess kurtosis is $\widehat{K} = 4.08$. The excess kurtosis is labeled simply "kurtosis" which can be misleading. The excess kurtosis is somewhat larger than the value 0 of a normal distribution. However, the p-values of the normality tests are large, so there is no reason to reject the null hypothesis of nonnormality. See Section 2.20 for a discussion of p-values and normality tests.

```
        Ford - Daily prices, Nov 1, 1999 to Apr 3, 2003

                    The UNIVARIATE Procedure
                       Variable:  return

                            Moments

N                        2362    Sum Weights               2362
Mean               0.00024902    Sum Observations    0.58819143
Std Deviation      0.02301418    Variance            0.00052965
Skewness           0.31946491    Kurtosis            4.08872113
Uncorrected SS     1.25065602    Corrected SS        1.25050955
Coeff Variation    9241.80314    Std Error Mean      0.00047354


                  Basic Statistical Measures

        Location                    Variability

     Mean     0.000249     Std Deviation            0.02301
     Median   0.000000     Variance               0.0005297
     Mode     0.000000     Range                    0.30268
                           Interquartile Range      0.02562


                  Tests for Location: Mu0=0

        Test           -Statistic-    -----p Value------

        Student's t    t  0.525876    Pr > |t|    0.5990
        Sign           M     -42.5    Pr >= |M|   0.0756
        Signed Rank    S   -7437.5    Pr >= |S|   0.8075


                      Tests for Normality

   Test                  --Statistic---     -----p Value------

   Kolmogorov-Smirnov    D     0.053255     Pr > D      <0.0100
   Cramer-von Mises      W-Sq  2.336439     Pr > W-Sq   <0.0050
   Anderson-Darling      A-Sq  13.96599     Pr > A-Sq   <0.0050


                         Trimmed Means

Percent    Number                    Std Error
Trimmed   Trimmed        Trimmed      Trimmed       95% Confidence
in Tail   in Tail           Mean         Mean           Limits

  10.03       237       -0.00007     0.000422    -0.00090  0.000761

                    The UNIVARIATE Procedure
                       Variable:  return

                         Trimmed Means

           Percent
           Trimmed                t for H0:
           in Tail         DF      Mu0=0.00   Pr > |t|

             10.03       1887      -0.15908     0.8736
```

Robust Measures of Scale

Measure	Value	Estimate of Sigma
Interquartile Range	0.025617	0.018990
Gini's Mean Difference	0.024556	0.021762
MAD	0.012813	0.018997
Sn	0.019791	0.019791
Qn	0.019935	0.019903

The 0.1-trimmed mean is -0.00007, close to $\overline{X} = 0.00024902$. The sample standard deviation is 0.02301 while the MAD estimate of σ is smaller, 0.01900. SAS computes four other robust scale measures, but the MAD is sufficient for our purposes and the others are not discussed here.

2.19 Confidence Intervals

Instead of estimating an unknown parameter by a single number, it is often better to provide a range of numbers that give a sense of the uncertainty of the estimate. Such ranges are called *interval estimates*. One type of interval estimate, the Bayesian credible interval, has already been mentioned. Another type of interval estimate is the **confidence interval**. A confidence interval is defined by the requirement that the probability that the interval will include the true parameter is a specified value called the *confidence coefficient*, so, for example, if a large number of independent 90% intervals are constructed then approximately 90% of them will contain the parameter.

Credible intervals have a different philosophical basis than confidence intervals. The theory of confidence intervals views the parameter as fixed and the interval as random because it is based on a random sample. Thus, when we say "the probability that the confidence interval will include the true parameter" it is the probability distribution of the interval, not the parameter, that is being considered. In the Bayesian theory of credible intervals, the opposite is true. In Bayesian statistics, the sample is considered fixed since we compute probabilities conditional on the data. Thus, the credible interval is considered a fixed quantity. But in Bayesian statistics, parameters are treated as random. Therefore, when we say "the probability that the credible interval will include the true parameter" the probability distribution being considered is the posterior distribution of the parameter.

2.19.1 Confidence interval for the mean

If $\overline{X}$ is the mean of a sample from a normal population, then

$$\overline{X} \pm t_{\alpha/2,n-1}\, s_{\overline{X}} \tag{2.72}$$

is a confidence interval with $(1-\alpha)$ confidence. This confidence interval is derived in Section 10.2. If $\alpha = 0.05$ (0.95 or 95% confidence) and if n is reasonably large, then $t_{\alpha/2,n-1}$ is approximately 2, so $\overline{X} \pm 2\, s_{\overline{X}}$ is often used as an approximate 95% confidence interval. Since $s_{\overline{X}} = s_X/\sqrt{n}$, the confidence can also be written as $\overline{X} \pm 2\, s_X/\sqrt{n}$. When n is reasonably large, say 20 or more, then $\overline{X}$ will be approximately normally distributed by the central limit theorem, and the assumption that the population itself is normal can be dropped.

Example 2.19. Suppose we have a sample of size 25 from a normal distribution, $s_X^2 = 2.7$, $\overline{X} = 16.1$, and we want a 99% confidence interval for μ. We need $t_{0.005,24}$. This quantile can be found, for example, using the MATLAB function `tinv`[17] that computes the inverse of the t-distribution's CDF so that `tinv(x,v)` is $t_{1-x,v}$. Using this MATLAB function we obtain:

```
>> tinv(0.995,24)
ans = 2.7969
```

Thus, $t_{0.005,24} = 2.797$. Then the 99% confidence interval for μ is

$$16.1 \pm \frac{(2.797)\sqrt{2.7}}{\sqrt{25}} = 16.1 \pm 0.919 = [15.18,\ 17.02].$$

Since $n = 25$ is reasonably large, this interval has approximately 99% confidence even if the population is not normally distributed.

Just how large a sample is needed for $\overline{X}$ to be nearly normally distributed depends on the population. If the population is symmetric, then approximate normality is often achieved with n around 10. For very skewed populations 50 or 100 observations may be needed, and even more in extreme cases.

2.19.2 Confidence intervals for the variance and standard deviation

A $(1-\alpha)$ confidence interval for the variance of a normal distribution is given by

$$\left[\frac{(n-1)s_X^2}{\chi^2_{\alpha/2,n-1}},\ \frac{(n-1)s_X^2}{\chi^2_{1-\alpha/2,n-1}}\right],$$

where n is the sample size, s_X^2 is the sample variance given by equation (2.13), and, as defined in Section 2.9.1, $\chi^2_{\gamma,n-1}$ is the $(1-\gamma)$-quantile of the chi-square distribution with $n-1$ degrees of freedom.

Example 2.20. Suppose we have a sample of size 25 from a normal distribution, $s_X^2 = 2.7$, and we want a 90% confidence interval for σ^2. The quantiles we need for constructing the interval are $\chi^2_{0.95,24} = 13.848$ and $\chi^2_{0.05,24} = 36.415$. The values can be found in statistical tables or computed using software such

[17] The function `tinv` is part of the MATLAB statistics toolbox and so can only be used if that toolbox is installed.

as SAS or MATLAB. In MATLAB, `chi2inv(x,v)`[18] will return the value of $\chi^2_{1-x,v}$ where `x` can be a vector, e.g.,

```
>> chi2inv([0.05 0.95],24)
ans = 13.8484   36.4150
```

Here the first line was entered from the keyboard into MATLAB and the second line was returned by MATLAB. The 90% confidence interval for σ^2 is

$$\left[\frac{(2.7)(24)}{36.4150}, \frac{(2.7)(24)}{13.8484}\right] = [1.78,\ 4.68].$$

Taking square roots of both endpoints, we get $1.33 < \sigma < 2.16$ as a 90% confidence interval for the standard deviation.

Unfortunately, the assumption that the population is normally distributed cannot be dispensed with, even if the sample size is large. If a normal probability plot or test of normality (see Section 2.20.5) suggests that the population might be nonnormally distributed, then one might instead construct a confidence interval for σ using the bootstrap; see Chapter 10, Problem 2.

2.19.3 Confidence intervals based on standard errors

Many estimators are approximately unbiased and approximately normally distributed. Then an approximate 95% confidence interval is the estimator plus or minus twice its standard error, that is,

$$\widehat{\theta} \pm 2\, s_{\widehat{\theta}}$$

is an approximate 95% confidence interval for θ.

2.20 Hypothesis Testing

2.20.1 Hypotheses, types of errors, and rejection regions

Statistical hypothesis testing uses data to decide whether a certain statement called the **null hypothesis** is true. The negation of the null hypothesis is called the **alternative hypothesis**. For example, suppose that $X_1, \ldots, X_n$ are i.i.d. $N(\mu, 1)$ and μ is unknown. The null hypothesis could be that μ is 1. Then we write H_0: $\mu = 1$ and H_1: $\mu \neq 1$ to denote the null and alternative hypotheses. Later we show null hypotheses that are more relevant to finance, for example, the random walk hypothesis which states that future returns on an asset are independent of present and past returns.

[18] The function `chi2inv` is part of MATLAB's statistics toolbox.

There are two types of errors that we hope to avoid. If the null hypothesis is true but we reject it, then we are making a *type I error*. Conversely, if the null hypothesis is false and we accept it then we are making a *type II error*.

The *rejection region* is the set of possible samples that lead us to reject H_0. For example, suppose that μ_0 is a hypothesized value of μ and the null hypothesis is H_0: $\mu = \mu_0$ and the alternative is H_1: $\mu \neq \mu_0$. One rejects H_0 if $|\overline{X} - \mu_0|$ exceeds an appropriately chosen cutoff value c called a *critical value*. The rejection region is chosen to keep the probability of a type I error below a prespecified small value called the *level* and often denoted by α. Typical values of α used in practice are 0.01, 0.05, or 0.1. As α is made smaller, the rejection region must be made smaller. In the example, since we reject the null hypothesis when $|\overline{X} - \mu_0|$ exceeds c, the critical value c gets larger as the α gets smaller. The value of c is easy to determine. Assuming that σ is known, c is $z_{\alpha/2}\,\sigma/\sqrt{n}$ where, as defined in Section 2.8.2, $z_{\alpha/2}$ is the $1-\alpha/2$-quantile of the standard normal distribution. If σ is unknown, then σ is replaced by s_X and $z_{\alpha/2}$ is replaced by $t_{\alpha/2,n-1}$ where, as defined in Section 2.9.2, $t_{\alpha/2,n-1}$ is the $1 - \alpha/2$-quantile of the t-distribution with $n-1$ degrees of freedom. The test using the t-quantile is called the *one sample t-test.*

2.20.2 *P*-values

Rather than specifying α and deciding whether to accept or reject the null hypothesis at that α, we might ask "for what values of α do we reject the null hypothesis?" The p-**value** for a sample is defined as the smallest value of α for which the null hypothesis is rejected. Stated differently, to perform the test using a given sample we first find the p-value of that sample and then H_0 is rejected if we decide to use α larger than the p-value and H_0 is accepted if we use α smaller than the p-value. Thus,

- a small p-value is evidence *against* the null hypothesis

while

- a large p-value shows that the *data are consistent* with the null hypothesis.

Example 2.21. If the p-value of a sample is 0.0331, then we reject H_0 if we use α equal to 0.05 or 0.1 but we accept H_0 if we use $\alpha = 0.01$.

The p-value not only tells us whether the null hypothesis should be accepted or rejected, but it also tells us whether the decision to accept or reject H_0 is a close call. For example, if we are using $\alpha = 0.05$ and the p-value were 0.047, then we would reject H_0 but we would know the decision was close. If instead the p-value were 0.001, then we would know the decision was not so close.

When performing hypothesis tests, statistical software routinely calculates p-values. Doing this is much more convenient than asking the user to specify α and then reporting whether the null hypothesis is accepted or rejected for that α.

2.20.3 Two-sample t-tests

Two-sample t-tests are used to test hypotheses about the difference between two population means. The independent-samples t-test is used when we sample independently from the two populations. Let μ_i, $\overline{X}_i$, s_i, and n_i be the population mean, sample mean, sample standard deviation, and sample size for the ith sample, $i = 1, 2$. Let Δ_0 be a hypothesized value of $\mu_1 - \mu_2$. We assume that the two populations have the same standard deviation and estimate this parameter by the so-called *pooled standard deviation* which is

$$s_{\text{pool}} = \left\{ \frac{(n_1 - 1)s_1^2 + (n_2 - 1)s_2^2}{n_1 + n_2 - 2} \right\}^{1/2}. \tag{2.73}$$

The independent-samples t-statistic is

$$t = \frac{\overline{X}_1 - \overline{X}_2 - \Delta_0}{s_{\text{pool}} \sqrt{\frac{1}{n_1} + \frac{1}{n_2}}}.$$

If the hypotheses are H_0: $\mu_1 - \mu_2 = \Delta_0$ and H_1: $\mu_1 - \mu_2 \neq \Delta_0$, then H_0 is rejected if $|t| > t_{\alpha/2; n_1+n_2-2}$. If the hypotheses are H_0: $\mu_1 - \mu_2 \leq \Delta_0$ and H_1: $\mu_1 - \mu_2 > \Delta_0$, then H_0 is rejected if $t > t_{\alpha; n_1+n_2-2}$ and if they are H_0: $\mu_1 - \mu_2 \geq \Delta_0$ and H_1: $\mu_1 - \mu_2 < \Delta_0$, then H_0 is rejected if $t < -t_{\alpha; n_1+n_2-2}$.

Sometimes the samples are paired rather than independent. For example, suppose we wish to compare returns on small-cap versus large-cap[19] stocks and for each of n years we have the returns on a portfolio of small-cap stocks and on a portfolio of large-cap stocks. Let $d_i = X_{i,1} - X_{i,2}$ be the difference between the observations from populations 1 and 2 for the ith pair, and let $\overline{d}$ and s_d be the sample mean and standard deviation of $d_1, \ldots, d_n$. The paired-sample t-statistics is

$$t = \frac{\overline{d} - \Delta_0}{\frac{s_d}{\sqrt{n}}}. \tag{2.74}$$

The rejection regions are the same as for the independent-samples t-tests except that the degrees of freedom parameter for the t-quantiles is $n - 1$ rather than $n_1 + n_2 - 2$.

The power of a test is the probability of correctly rejecting H_0 when H_1 is true. Paired samples are often used to obtain more power. In the example of comparing small- and large-cap stocks, the returns on both portfolios will have high year-to-year variation, but the d_i will be free of this variation so that s_d should be relatively small compared to s_1 and s_2. Small variation in the data means that $\mu_1 - \mu_2$ can be more accurately estimated and deviations of this parameter from Δ_0 are more likely to be detected.

[19] The market capitalization of a stock is the product of the share price and the number of shares outstanding. If stocks are ranked based on market capitalization, then all stocks below some specified quantile would be small-cap stocks and all above another specified quantile would be large-cap.

Since $\overline{d} = \overline{X}_1 - \overline{X}_2$, the numerators in (2.73) and (2.74) are equal. What differs are the denominators. The denominator in (2.74) will be smaller than in (2.73) when the correlation between observations $(X_{i,2}, X_{i,2})$ in a pair is positive; see Problem 16. It is the smallness of the denominator in (2.74) that gives the paired t-test increased power.

Suppose someone had a paired sample but incorrectly used the independent-samples t-test. If the correlation between $X_{i,1}$ and $X_{i,2}$ is zero, then the paired samples behave the same as independent samples and the effect of using the incorrect test would be small. Suppose that this correlation is positive. The result of using the incorrect test would be that if H_0 is false then the true p-value would be overestimated and one would be less likely to reject H_0 than if the paired-sample test had been used. However, if the p-value is small then one can be confident in rejecting H_0 because the p-value for the paired-sample test would be even smaller.[20] Unfortunately, statistical methods are often used by researchers without a solid understanding of the underlying theory and this can lead to misapplications. The hypothetical use just described of an incorrect test is often a reality, and it is sometimes necessary to evaluate whether the results that are reported can be trusted.

2.20.4 Statistical versus practical significance

When we reject a null hypothesis, we often say there is a *statistically significant effect*. In this context, the word "significant" is easily misconstrued. It does *not* mean that there is an effect of practical importance. For example, suppose we were testing the null hypothesis that the means of two populations are equal versus the alternative that they are unequal. Statistical significance simply means that the two sample means are sufficiently different that this difference cannot reasonably be attributed to mere chance. Statistical significance does *not* mean that the population means are so dissimilar that their difference is of any practical importance. When large samples are used, small and unimportant effects are likely to be statistically significant.

When determining practical significance, confidence intervals typically are more useful than tests. In the case of the comparison between two population means, it is important to construct a confidence interval and to conclude that there is an effect of practical significance only if *all* differences in that interval are large enough to be of practical importance. How large is "large enough" is *not* a statistical question but rather must be answered by a subject matter expert. For an example, suppose a difference between the two population means that exceeds 0.2 is considered important. If a 95% confidence interval were $[0.23, 0.26]$, then with 95% confidence we could conclude that there is an important difference. If instead the interval were $[0.13, 0.16]$ then we could conclude with 95% confidence that there is no important difference.

[20] An exception would be the rare situation where $X_{i,1}$ and $X_{i,2}$ are *negatively* correlated.

If the confidence interval were $[0.17, 0.23]$ then we could not state with 95% confidence whether the difference is important.

2.20.5 Tests of normality

When viewing a normal probability plot, it is often difficult to judge whether any deviation from linearity is systematic or merely due to sampling variation, so a test of normality is useful. The null hypothesis is that the sample comes from a normal distribution and the alternative is that the sample is from a nonnormal distribution. The Shapiro-Wilk test uses the normal probability plot to test these hypotheses. Specifically, the Shapiro-Wilk test is based on the correlation between $X_{(i)}$ and $\Phi^{-1}\{i/(n+1)\}$, which are the i/n quantiles of the sample and of the standard normal distribution, respectively. Under normality, the correlation should be close to 1 and the null hypothesis of normality is rejected for small values of the correlation coefficient. Some software packages, e.g., MINITAB which is often used in introductory statistics courses, use the Ryan-Joiner test which is very similar to the Shapiro-Wilk test.

Other tests of normality in common use are the Anderson-Darling, Cramér-von Mises, and Kolmogorov-Smirnov tests. These tests compare the sample CDF to the normal CDF with mean equal to $\overline{X}$ and variance equal to s_X^2. The Kolmogorov-Smirnov test statistic is the maximum absolute difference between these two functions, while the Anderson-Darling and Cramér-von Mises tests are based on a weighted integral of the squared difference. The p-values of the Shapiro-Wilk, Anderson-Darling, Cramér-von Mises, and Kolmogorov-Smirnov tests are routinely part of the output of statistical software. A small p-value is interpreted as evidence that the sample is not from a normal distribution.

Recall that it was somewhat unclear whether the plot in Figure 2.6 was sufficiently linear to conclude that the data were normal. A normality test can often resolve this ambiguity. The p-value of the Kolmogorov-Smirnov test for the data in that figure is 0.9, so we accept the null hypothesis of normality. For the highly skewed data in Figure 2.7, the p-value of the Kolmogorov-Smirnov test is very small, less than 10^{-9}, so the nonnormality is clear. Generally, the different tests of normality reach the same conclusion. In Example 2.18, all four tests in PROC UNIVARIATE output lead to the same conclusion that normality is accepted at any of the usual values of α since all the p-values are 0.15 or greater.

2.20.6 Likelihood ratio tests

Likelihood ratio tests, like maximum likelihood estimation, are based upon the likelihood function. Both are convenient, all-purpose tools that are widely used in practice. Suppose that $\boldsymbol{\theta}$ is a parameter vector and that the null hypothesis puts m restrictions on $\boldsymbol{\theta}$. For example, we might want to test that a population mean is zero; then $\boldsymbol{\theta} = (\mu, \sigma)^{\mathsf{T}}$ and $m = 1$ since the null hypothesis puts one restriction on $\boldsymbol{\theta}$, specifically that the first component is zero.

Let $\widehat{\boldsymbol{\theta}}_{ML}$ be the maximum likelihood estimator without restrictions and let $\widehat{\boldsymbol{\theta}}_{0,ML}$ be the value of $\boldsymbol{\theta}$ that maximizes $L(\boldsymbol{\theta})$ subject to the restrictions of the null hypothesis. If H_0 is true, then $\widehat{\boldsymbol{\theta}}_{0,ML}$ and $\widehat{\boldsymbol{\theta}}_{ML}$ should both be close to $\boldsymbol{\theta}$ and therefore $L(\widehat{\boldsymbol{\theta}}_{0,ML})$ should be similar to $L(\widehat{\boldsymbol{\theta}})$. If H_0 is false, then the constraints will keep $\widehat{\boldsymbol{\theta}}_{0,ML}$ far from $\widehat{\boldsymbol{\theta}}_{ML}$ so that $L(\widehat{\boldsymbol{\theta}}_{0,ML})$ should be noticeably *smaller* that $L(\widehat{\boldsymbol{\theta}})$.

The likelihood ratio test rejects H_0 if

$$2\Big[\log\{L(\widehat{\boldsymbol{\theta}}_{ML})\} - \log\{L(\widehat{\boldsymbol{\theta}}_{0,ML})\}\Big] \geq c, \tag{2.75}$$

where c is a critical value. Often, an *exact critical value* can be found; see Example 2.22. A critical value is exact if it gives a level that is exactly equal to α. When an exact critical value is unknown, then the usual choice of the critical value is

$$c = \chi^2_{\alpha,m}, \tag{2.76}$$

where, as defined in Section 2.9.1, $\chi^2_{\alpha,m}$ is the α upper-probability value of the chi-squared distribution with m degrees of freedom. The critical value (2.76) is only approximate and uses the fact that under the null hypothesis, as the sample size increases the distribution of twice the log-likelihood ratio converges to the chi-squared distribution with m degrees of freedom under certain assumptions. One of these assumptions is that the null hypothesis is *not* on the boundary of the parameter space. For example, if the null hypothesis is that a variance parameter is zero, then the null hypothesis is on the boundary of the parameter space since a variance must be zero or greater. In this case (2.75) should not be used; see Self and Liang (1987). Also, if the sample size is small, then approximation (2.75) is suspect and should be used with caution. An alternative is to use the bootstrap to determine the rejection region. The bootstrap is discussed in Chapter 10.

Example 2.22. Suppose that $Y_1, \ldots, Y_n$ are i.i.d. $N(\mu, \sigma^2)$, $\boldsymbol{\theta} = (\mu, \sigma^2)$, and we want to test that μ is zero. Note that

$$\log\{L(\mu, \sigma^2)\} = -\frac{n}{2}\log(2\pi) - \frac{n}{2}\log(\sigma^2) - \frac{1}{2\sigma^2}\sum_{i=1}^{n}(Y_i - \mu)^2.$$

Recall that the MLE of μ is $\overline{Y}$ and of σ^2 is

$$\widehat{\sigma}^2_{ML} = n^{-1}\sum_{i=1}^{n}(Y_i - \overline{Y})^2.$$

If we evaluate $\log\{L(\mu, \sigma^2)\}$ at the MLE, we get

$$\log\{L(\overline{Y}, \widehat{\sigma}^2_{ML})\} = -\frac{n}{2}\{1 + \log(2\pi) + \log(\widehat{\sigma}^2_{ML})\}.$$

In Problem 6 it is shown that the value of σ^2 that maximizes L when $\mu = 0$ is

$$\widehat{\sigma}^2_{0,ML} = \frac{1}{n}\sum_{i=1}^{n} Y_i^2.$$

Therefore,

$$2\Big[\log\{L(\overline{Y}, \widehat{\sigma}^2_{ML})\} - \log\{L(0, \widehat{\sigma}^2_{0,ML})\}\Big] = n\log\left\{\frac{\widehat{\sigma}^2_{0,ML}}{\widehat{\sigma}^2_{ML}}\right\}$$

$$= n\log\left\{\frac{\sum_{i=1}^n Y_i^2}{\sum_{i=1}^n (Y_i - \overline{Y})^2}\right\}.$$

The likelihood ratio test rejects H_0 if

$$n\log\left(\frac{\sum_{i=1}^n Y_i^2}{\sum_{i=1}^n (Y_i - \overline{Y})^2}\right) > \chi^2_{\alpha,1}. \tag{2.77}$$

To appreciate why (2.77) is a sensible test, first note that by simple algebra

$$\sum_{i=1}^{n} Y_i^2 = \sum_{i=1}^{n} (Y_i - \overline{Y})^2 + n(\overline{Y})^2.$$

Therefore, the left-hand side of (2.77) is

$$n\log\left(1 + \frac{n}{n-1}\frac{(\overline{Y})^2}{s_Y^2}\right),$$

so the likelihood ratio test rejects H_0 that $\mu = 0$ when $(\overline{Y})^2$ is large relative to s_Y^2, that is, when $|\overline{Y}/s_Y|$ exceeds a critical value. Since $s_{\overline{Y}} = s_Y/\sqrt{n}$, the likelihood ratio test rejects the null hypothesis when the absolute t-statistic $|\overline{Y}/s_{\overline{Y}}|$ exceeds a critical value. We know that using critical value $t_{\alpha/2,n-1}$ is exact and if this critical value is used then the likelihood ratio test is equivalent to the one-sample t-test described in Section 2.20.1.

2.21 Summary

A random variable can be described by its probability distribution, its probability density function (PDF), or its cumulative distribution function (CDF). Probability distributions can be summarized by expected values, variances (σ^2) or standard deviations (σ), and covariances and correlations.

For q between 0 and 1, $F^{-1}(q)$ is called the q-quantile or 100qth percentile. The probability that a continuous X is below its qth quantile is q; that is,

$$P\{X \le F^{-1}(q)\} = q.$$

The median is the 50% percentile or 0.5-quantile.

The covariance between X and Y is

$$\sigma_{XY} = E\big[\{X - E(X)\}\{Y - E(Y)\}\big].$$

The correlation coefficient between X and Y is

$$\rho_{XY} = \sigma_{XY}/\sigma_X\sigma_Y.$$

The best linear predictor of Y is

$$\widehat{Y} = E(Y) + \frac{\sigma_{XY}}{\sigma_X^2}\{X - E(X)\}.$$

The expected squared prediction error is

$$E\{Y - \widehat{Y}\}^2 = \sigma_Y^2 - \frac{\sigma_{XY}^2}{\sigma_X^2} = \sigma_Y^2(1 - \rho_{XY}^2).$$

If X is independent of Y, then $\widehat{Y} = E(Y)$ and the expected squared prediction error is σ_Y^2.

If $X \sim N(\mu, \sigma^2)$ then X has density

$$\frac{1}{\sigma}\phi\left(\frac{x-\mu}{\sigma}\right)$$

and

$$P\{X \leq x\} = \Phi\{(x - \mu)/\sigma\}.$$

The q-quantile of X is $\mu + \sigma\Phi^{-1}(q)$.

If Y is a random variable and a and b are constants then

$$E(aY + b) = aE(Y) + b \text{ and}$$

$$\text{Var}(aY + b) = a^2\text{Var}(Y).$$

If w_1 and w_2 are constants and X and Y are random variables, then

$$E(w_1X + w_2Y) = w_1E(X) + w_2E(Y), \text{ and}$$

$$\text{Var}(w_1X + w_2Y) = w_1^2\text{Var}(X) + 2w_1w_2\text{Cov}(X, Y) + w_2^2\text{Var}(Y).$$

Suppose

$$\boldsymbol{X} = \begin{pmatrix} X_1 \\ \vdots \\ X_N \end{pmatrix}.$$

Then the expectation of $\boldsymbol{X}$ is

$$E(\boldsymbol{X}) = \begin{pmatrix} E(X_1) \\ \vdots \\ E(X_N) \end{pmatrix}$$

and the covariance matrix of $\boldsymbol{X}$ is

$$\text{COV}(\boldsymbol{X}) = \begin{pmatrix} \text{Var}(X_1) & \text{Cov}(X_1, X_2) & \cdots & \text{Cov}(X_1, X_N) \\ \text{Cov}(X_2, X_1) & \text{Var}(X_2) & \cdots & \text{Cov}(X_2, X_N) \\ \vdots & \vdots & \ddots & \vdots \\ \text{Cov}(X_N, X_1) & \text{Cov}(X_N, X_2) & \cdots & \text{Var}(X_N) \end{pmatrix}.$$

An estimator is a random variable and its standard deviation is called the standard error. Also, $s_{\widehat{\theta}}$ denotes the standard error of $\widehat{\theta}$. An approximate 95% confidence interval is $\widehat{\theta} \pm 2s_{\widehat{\theta}}$.

$L(\boldsymbol{\theta}) = f(\boldsymbol{Y};\boldsymbol{\theta})$ is the "likelihood function." The maximum likelihood estimator is the value of $\boldsymbol{\theta}$ that maximizes $L(\boldsymbol{\theta})$. For some models such as the ARIMA time series models, there are software packages, e.g., SAS, that compute the MLE. The Fisher information can be used to compute the standard error of the MLE.

The *p-value* for a sample is defined as the smallest value of α for which the null hypothesis is rejected for that sample. The likelihood ratio test rejects H_0 if

$$2\log\left\{\frac{L(\widehat{\boldsymbol{\theta}}_{ML})}{L(\widehat{\boldsymbol{\theta}}_{0,ML})}\right\} = 2\Big[\log\{L(\widehat{\boldsymbol{\theta}}_{ML})\} - \log\{L(\boldsymbol{\theta}_{0,ML})\}\Big] \geq \chi^2_{\alpha,m}, \quad (2.78)$$

where m is the number of constraints on $\boldsymbol{\theta}$ under the null hypothesis.

2.22 Bibliographic Notes

Casella and Berger (2002) covers in greater detail most of the material in this chapter. Alexander (2001) is a recent introduction to financial econometrics and has a chapter on covariance matrices; her technical appendices cover maximum likelihood estimation, confidence intervals, and hypothesis testing, including likelihood ratio tests. Evans, Hastings, and Peacock (1993) provides a concise reference for the basic facts about commonly used distributions in statistics. Johnson, Kotz, and Balakrishnan (1993) discusses most of the common discrete distributions including the binomial. Johnson, Kotz, and Balakrishnan (1994, 1995) contains a wealth of information and extensive references about the normal, lognormal, chi-square, exponential, uniform, t, F, Pareto, and many other continuous distributions. Together, these works by Johnson, Kotz, Kemp, and Balakrishnan are almost an encyclopedia of statistical distributions.

Ideally, the reader already has some knowledge of MATLAB and SAS, or is taking a course that includes some instruction in these languages. A independent reader with no experience in MATLAB or SAS may need additional help. Both SAS and MATLAB have online help, but introductory books can be very useful when learning to use these packages. For the reader wishing to learn MATLAB, I recommend Hanselman and Littlefield (2000), and Delwiche and Slaughter (1998) is a good introduction to SAS.

There is not space in this book to cover Bayesian techniques, but there are many good books covering that topic. Gelman, Carlin, Stern, and Rubin (1995) is an introduction to Bayesian statistics written at about the same mathematical level as this book. Box and Tiao (1973) is a classical work on Bayesian statistics with a wealth of examples and still worth reading despite its age. Bernardo and Smith (1994) is a more recent book on Bayesian statistics. Congdon (2001, 2003) covers the more recent developments in Bayesian computing with an emphasis on WinBUGS software.

2.23 References

Alexander, C. (2001) *Market Models: A Guide to Financial Data Analysis*, Wiley, Chichester.

Bernardo, J. M. and Smith, A. F. M. (1994) *Bayesian Theory*, Wiley, Chichester.

Box, G. E. P. and Tiao, G. C. (1973) *Bayesian Inference in Statistical Analysis*, Addison-Wesley, Reading, MA.

Casella, G. and Berger, R. L. (2002) *Statistical Inference, 2nd Ed.*, Thomson Learning, Pacific Grove, CA.

Congdon, P. (2001) *Bayesian Statistical Modelling*, Wiley, Chichester.

Congdon, P. (2003) *Applied Bayesian Modelling*, Wiley, Chichester.

Delwiche, L. D. and Slaughter, S. J. (1998) *The Little SAS Book: A Primer, 2nd Ed.*, SAS Publishing, Cary, N. C.

Drees, H., de Haan, L., and Resnick, S. (2000) How to make a Hill plot, *The Annals of Statistics*, **28**, 254–274.

Edwards, W. (1982) Conservatism in human information processing. In *Judgement Under Uncertainty: Heuristics and Biases*, edited by Kahneman, D., Slovic, P., and Tversky, A., Cambridge University Press, New York.

Evans, M., Hastings, N., and Peacock, B. (1993) *Statistical Distributions, 2nd Ed.*, Wiley, New York.

Fama, E. (1965) The behavior of stock market prices, *Journal of Business*, **38**, 34–105.

Gelman, A., Carlin, J. B., Stern, H. S., and Rubin, D. B. (1995) *Bayesian Data Analysis*, Chapman & Hall, London.

Gourieroux, C. and Jasiak, J. (2001) *Financial Econometrics*, Princeton University Press, Princeton, NJ.

Hanselman, D. and Littlefield, B. R. (2000) *Mastering MATLAB 6*, Prentice-Hall, Upper Saddle River, NJ.

Johnson, N. L., Kotz, S., and Balakrishnan, N., (1994) *Continuous Univariate Distributions, Vol. 1, 2nd Ed.*, Wiley, New York.

Johnson, N. L., Kotz, S., and Balakrishnan, N. (1995) *Continuous Univariate Distributions, Vol. 2, 2nd Ed.*, Wiley, New York.

Johnson, N. L., Kotz, S., and Kemp, A. W. (1993) *Discrete Univariate Distributions, 2nd Ed.*, Wiley, New York.

Mandelbrot, B. (1963) The variation of certain speculative prices, *Journal of Business*, **36**, 394–419.

Resnick, S. I. (2001) *Modeling Data Networks*, School of Operations Research and Industrial Engineering, Cornell University, Technical Report #1345.

Self, S. G. and Liang, K. Y. (1987) Asymptotic properties of maximum likelihood estimators and likelihood ratio tests under non-standard conditions. *Journal of the American Statistical Association*, **82**, 605–610.

2.24 Problems

1. Suppose that $E(X) = 1$, $E(Y) = 1$, $\text{Var}(X) = 2$, $\text{Var}(Y) = 3$, and $\text{Cov}(X, Y) = 1$.
   (a) What are $E(0.1X + 0.9Y)$ and $\text{Var}(0.1X + 0.9Y)$?
   (b) For what value of w is $\text{Var}\{wX + (1-w)Y\}$ minimized? Suppose that X is the return on one asset and Y is the return on a second asset. Why would it be useful to minimize $\text{Var}\{wX + (1-w)Y\}$?
2. Let X_1, X_2, Y_1, and Y_2 be random variables.
   (a) Show that $\text{Cov}(X_1 + X_2, Y_1 + Y_2) = \text{Cov}(X_1, Y_1) + \text{Cov}(X_1, Y_2) + \text{Cov}(X_2, Y_1) + \text{Cov}(X_2, Y_2)$.
   (b) Show that if X, Y, and Z are independent, then $\text{Cov}(\alpha_1 X + Y, \alpha_2 X + Z) = \alpha_1 \alpha_2 \sigma_X^2$.
   (c) Generalize part (a) to an arbitrary number of X_is and Y_is.
3. Suppose that $Y_1, \ldots, Y_n$ are i.i.d. $N(\mu, \sigma^2)$ where μ is *known*. Show that the MLE of σ^2 is

$$n^{-1} \sum_{i=1}^{n} (Y_i - \mu)^2.$$

4. When we were finding the best linear predictor of Y given X we derived the equations:

$$0 = -E(Y) + \beta_0 + \beta_1 E(X)$$
$$0 = -E(XY) + \beta_0 E(X) + \beta_1 E(X^2).$$

   Show that their solution is:

$$\beta_1 = \frac{\sigma_{XY}}{\sigma_X^2}$$

and

$$\beta_0 = E(Y) - \beta_1 E(X) = E(Y) - \frac{\sigma_{XY}}{\sigma_X^2} E(X).$$

5. Verify the following results that were stated in Section 2.17

$$E(\boldsymbol{w}^\mathsf{T}\boldsymbol{X}) = \boldsymbol{w}^\mathsf{T}\{E(\boldsymbol{X})\}$$

and

$$\begin{aligned}\operatorname{Var}(\boldsymbol{w}^\mathsf{T}\boldsymbol{X}) &= \sum_{i=1}^{N}\sum_{j=1}^{N} w_i\, w_j \operatorname{Cov}(X_i, X_j) \\ &= \operatorname{Var}(\boldsymbol{w}^\mathsf{T}\boldsymbol{X}) = \boldsymbol{w}^\mathsf{T}\mathrm{COV}(\boldsymbol{X})\boldsymbol{w}.\end{aligned}$$

6. In Example 2.22 it was stated that if we evaluate $\log\{L(\mu, \sigma^2)\}$ at the MLE then we get

$$\log\{L(\overline{Y}, \widehat{\sigma}^2_{ML})\} = -\frac{n}{2}\{1 + \log(2\pi) + \log(\widehat{\sigma}^2_{ML})\}$$

and that the value of σ^2 that maximizes $L(\mu, \sigma^2)$ when $\mu = 0$ is

$$\widehat{\sigma}^2_{0,ML} = \frac{1}{n}\sum_{i=1}^{n} Y_i^2.$$

Verify these two statements.

7. (a) Show that

$$E\{X - E(X)\} = 0$$

for any random variable X.

(b) Use the result in part (a) and equation (2.39) to show that if two random variables are independent then they are uncorrelated.

8. (a) Show that in best linear prediction $E(Y - \widehat{Y}) = 0$.
(b) Verify equation (2.46).

9. Show that if X is uniformly distributed on $[-a, a]$ for any $a > 0$ and if $Y = X^2$, then X and Y are uncorrelated but they are not independent.

10. Show in Example 2.17 that the MAP estimator is 4/5.

11. (a) What is the kurtosis of a normal mixture distribution that is 95% $N(0, 1)$ and 5% $N(0, 10)$?
(b) Find a formula for the kurtosis of a normal mixture distribution that is $100p\%$ $N(0, 1)$ and $100(1-p)\%$ $N(0, \sigma^2)$ where p and σ are parameters. Your formula should give the kurtosis as a function of p and σ.
(c) Show that the kurtosis of the normal mixtures in part (b) can be made arbitrarily large by choosing p and σ appropriately. Find values of p and σ so that the kurtosis is 10,000 or larger.
(d) Let $M > 0$ be arbitrarily large. Show that for any $p_0 < 1$, no matter how close to 1, there is a $p > p_0$ and a σ, such that the normal mixture with these values of p and σ has a kurtosis at least M. This shows that there is a normal mixture arbitrarily close to a normal distribution with a kurtosis above any M.

12. Let X be $N(0, \sigma^2)$. Show that the CDF of the conditional distribution of X given that $X > c$ is

$$\frac{\Phi(x/\sigma) - \Phi(c/\sigma)}{1 - \Phi(c/\sigma)}, \qquad x > c$$

and that the PDF of this distribution is

$$\frac{\phi(x/\sigma)}{\sigma\{1 - \Phi(c/\sigma)\}}, \qquad x > c.$$

Show that if $c = 0.25$ and $\sigma = 0.3113$, then at 0.25 this PDF equals the PDF of a Pareto distribution with parameters $a = 1.1$ and $c = 0.25$.[21]

13. Suppose that $X_1, \ldots, X_n$ are i.i.d. exponential(θ). Show that the MLE of θ is $\overline{X}$.
14. Assume that X is Uniform$(0, 1)$.
    (a) Suppose that $Y = 3X - 2$. Find the CDF and PDF of Y. Find the median of Y.
    (b) Suppose instead that $Y = X^2$. Find the CDF and PDF of Y. Find the first and third quartiles of Y.
15. Suppose that X is Uniform$(1, 5)$.
    (a) What is the 0.1-quantile of X?
    (b) What is the 0.1-quantile of $1/X$?
16. This problem uses the notation of Section 2.20.3. Suppose one has a paired sample and computes both the independent-samples and paired-samples t-statistics. Define $s_{1,2}$ to be the sample covariance between the $X_{i,1}$ and the $X_{i,2}$.
    (a) Show that
$$s_d^2 = s_1^2 + s_2^2 - 2s_{1,2}.$$
    (b) Show that the independent-samples and paired-samples t-statistics are identical if $s_{1,2} = 0$. Show that the paired-samples t-statistic is larger than the independent-samples t-statistic if $\overline{d} - \Delta_0 > 0$ and $s_{1,2} > 0$.

[21] The value 0.3113 was originally found by interpolation.

3

Returns

3.1 Introduction

The goal of investing is, of course, to make a profit. The revenue from investing, or the loss in the case of a negative revenue, depends upon both the change in prices and the amounts of the assets being held. Investors are interested in revenues that are high relative to the size of the initial investments. **Returns** measure this, because returns on assets are changes in price expressed as a fraction of the initial price.

3.1.1 Net returns

Let P_t be the price of an asset at time t. Assuming no dividends the **net return** over the holding period from time $t-1$ to time t is

$$R_t = \frac{P_t}{P_{t-1}} - 1 = \frac{P_t - P_{t-1}}{P_{t-1}}.$$

The numerator, $P_t - P_{t-1}$, is the revenue or profit during the holding period, with a negative profit meaning a loss. The denominator, P_{t-1}, was the initial investment at the start of the holding period. Therefore, the net return can be viewed as the relative revenue or profit rate.

The revenue from holding an asset is

$$\text{revenue} = \text{initial investment} \times \text{net return}.$$

For example, an initial investment of \$1000 and a net return of 8% earns a revenue of \$80.

3.1.2 Gross returns

The simple **gross return** is

$$\frac{P_t}{P_{t-1}} = 1 + R_t.$$

Example 3.1. If $P_t = 2$ and $P_{t+1} = 2.1$, then $1 + R_{t+1} = 1.05$ or 105% and $R_{t+1} = 0.05$ or 5%.

The *gross return over the most recent k periods* is the product of the k single period gross returns (from time $t - k$ to time t):

$$\begin{aligned} 1 + R_t(k) = \frac{P_t}{P_{t-k}} &= \left(\frac{P_t}{P_{t-1}}\right)\left(\frac{P_{t-1}}{P_{t-2}}\right)\cdots\left(\frac{P_{t-k+1}}{P_{t-k}}\right) \\ &= (1 + R_t)\cdots(1 + R_{t-k+1}). \end{aligned}$$

Example 3.2. In this table, the prices at times $t - 2$ to $t + 1$ are given in the first row. The simple gross returns are in the second row, while gross returns over two or three periods are in the remaining rows.

Time	$t-2$	$t-1$	t	$t+1$
P	200	210	206	212
$1+R$		1.05	0.981	1.03
$1+R(2)$			1.03	1.01
$1+R(3)$				1.06

Returns are scale-free, meaning that they do not depend on units (dollars, cents, etc.). Returns are *not* unit-less. Their unit is time; they depend on the units of t (hour, day, etc.). In the example, if t is measured in years, then the net return is 5% per year.

3.1.3 Log returns

Continuously compounded returns, also known as **log returns**, are denoted by r_t and defined as

$$r_t = \log(1 + R_t) = \log\left(\frac{P_t}{P_{t-1}}\right) = p_t - p_{t-1},$$

where p_t is defined to be $\log(P_t)$ and is called the "log price."

Notation In this book, $\log(x)$ means the natural logarithm of x and $\log_{10}(x)$ is used to denote the logarithm to base ten.

Log returns are approximately equal to returns because if x is small, then $\log(1 + x) \approx x$ as can been seen in Figure 3.1 where $\log(1 + x)$ is plotted. Notice in that figure that $\log(1 + x)$ is very close to x is $|x| < 0.1$, e.g., for returns that are less than 10%.

Example 3.3. A 5% return equals a 4.88% log return since $\log(1 + .05) = 0.0488$. Also, a -5% return equals a -5.13% log return since $\log(1 - 0.05) = -0.0513$. In both cases, $r_t = \log(1 + R_t) \approx R_t$. Also, $\log(1 + 0.01) = 0.00995$ and $\log(1 - 0.01) = -0.01005$, so log returns of ± 1% are very close to the corresponding net returns.

One advantage of using log returns is simplicity of multiperiod returns. A k period log return is simply the sum of the single period log returns, rather than the product as for returns. To see this, note that

$$\begin{aligned}
r_t(k) &= \log\{1 + R_t(k)\} \\
&= \log\{(1 + R_t)\cdots(1 + R_{t-k+1})\} \\
&= \log(1 + R_t) + \cdots + \log(1 + R_{t-k+1}) \\
&= r_t + r_{t-1} + \cdots + r_{t-k+1}.
\end{aligned}$$

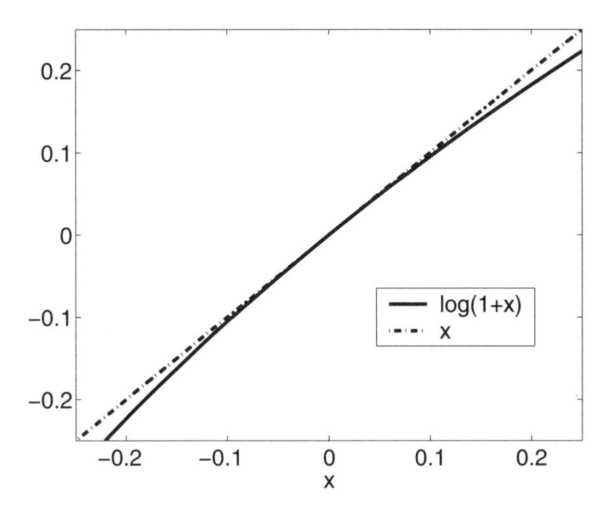

Fig. 3.1. *Comparison of functions* $\log(1 + x)$ *and* x.

3.1.4 Adjustment for dividends

Many stocks, especially those of mature companies, pay dividends that must be accounted for when computing returns. If a dividend D_t is paid prior to time t, then the gross return at time t is defined as

$$1 + R_t = \frac{P_t + D_t}{P_{t-1}}, \tag{3.1}$$

and so the net return is $R_t = (P_t + D_t)/P_{t-1} - 1$ and the log return is $r_t = \log(1+R_t) = \log(P_t+D_t) - \log(P_{t-1})$. Multiple period gross returns are products of single-period gross returns so that

$$
\begin{aligned}
1 + R_t(k) = &= \left(\frac{P_t + D_t}{P_{t-1}}\right)\left(\frac{P_{t-1} + D_{t-1}}{P_{t-2}}\right)\cdots\left(\frac{P_{t-k+1} + D_{t-k+1}}{P_{t-k}}\right) \\
&= (1 + R_t)(1 + R_{t-1})\cdots(1 + R_{t-k+1}),
\end{aligned} \tag{3.2}
$$

where, for any time s, $D_s = 0$ if there is no dividend between $s-1$ and s. Similarly, a k-period log return is

$$
\begin{aligned}
r_t(k) &= \log\{1 + R_t(k)\} = \log(1 + R_t) + \cdots + \log(1 + R_{t-k+1}) \\
&= \log\left(\frac{P_t + D_t}{P_{t-1}}\right) + \cdots + \log\left(\frac{P_{t-k+1} + D_{t-k+1}}{P_{t-k}}\right).
\end{aligned}
$$

3.2 Behavior Of Returns

What can we say about returns? First, they cannot be perfectly predicted, but rather they are random. This randomness implies that a return might be smaller than its expected value and even negative, which means that investing involves **risk**.[1] It took quite some time before it was realized that risk could be described by probability theory. In fact randomness is a concept that developed very slowly. For example, despite their tremendous intellectual developments in other areas, the ancient Greeks had little appreciation for probability. The Greeks would have thought of the returns as determined by the gods or Fates (three goddesses of destiny). The Greeks did not seem to realize that random phenomena do exhibit some regularities such as the law of large numbers and the central limit theorem. Probability theory eventually arose out of gambling during the Renaissance, well after other areas of mathematics were established. Peter Bernstein has written an interesting popular book *Against the Gods: The Remarkable Story of Risk* that chronicles the developments of probability theory and our understanding of risk.

Even after the applications of probability to games of chance were well appreciated, the application of probability to other human activities, in particular, to financial markets was undeveloped. In gambling, probabilities can be found by simple reasoning and an assumption of symmetry. If a die is fair, then each face has probability 1/6. Similarly the probability is 1/2 that a head will appear on the toss of a fair coin. There is no way to apply this type of reasoning to returns. University of Chicago economist Frank Knight made an important distinction between *measurable uncertainty* or *risk proper* (e.g., games of chance) where the probabilities are known and *unmeasurable uncertainty* where the probabilities are unknown. Unmeasurable uncertainty is much like the Greek concept of fate; there isn't much one can say about the future except that it is uncertain.

[1] Of course the return could be larger than expected, but to the investor is not risk.

As a simple example of measurable uncertainty, suppose there is a well-mixed bag of 30 blue and 70 red marbles and one is chosen at random. Consider a game that costs \$60 to play and where you win \$100 if the marble is red. What is the expected earnings? We know that the probability of a red marble is 0.7 and can calculate the expected earnings: $(0.7)(\$100) - \$60 = \$10$. So this seems like a good game to play, though that depends somewhat upon one's attitude towards risk. Now suppose that we only know that the bag contains 100 marbles, some blue and some red. Then there isn't much we can say about the probability of winning or the expected earnings, so the uncertainty is unmeasurable.

Measurable uncertainty is rather rare outside gambling and random sampling. Fortunately, there is a compromise between measurable uncertainty, where probabilities can be deduced by mathematical reasoning, and unmeasurable uncertainty, where probabilities are unknowable. Statistical inference is the science of *estimating* probabilities using data and assuming that the data are representative of what is to be estimated. In the example of the bag of marbles, we could apply statistical inference if we had a sample of marbles from the bag and were willing to assume that the sample was representative because the marbles had been chosen at random. We say a sample of size k was chosen at random if all subsets of size k of the population had the same probability of being selected.

Without any assumptions, finance would be in the realm of unmeasurable uncertainty with no way out. At time $t-1$, P_t and R_t would not only be unknown, but we would not know their probability distributions. Mathematical finance would not be possible in this case. However, we can estimate these probability distributions *if* we are willing to make the assumption that future returns will be similar to past returns, a condition called **stationarity**. With this assumption, the machinery of statistical inference can be applied and the probability distribution of P_t can be estimated from past data. Thus, finance like most of the sciences[2] is in a grey area between measurable and unmeasurable uncertainty. The uncertainty is unmeasurable if we are unwilling to make assumptions, but if we are willing to assume that returns are stationary, then we can act as if the uncertainty were measurable.

In this chapter we focus attention on a sample $\{R_1, R_2, \ldots, R_t\}$ of returns on a single asset at times $1, 2, \ldots, t$. Later we study asset pricing models (e.g., CAPM in Chapter 7) that use the joint distribution of a cross-section $\{R_{1t}, \ldots, R_{Nt}\}$ of returns on N assets at a single time t. Here R_{it} is the return on the ith asset at time t.

One of the major issues in finance is how we should model the probability distributions of returns. Specification of these distributions is essential for many purposes, for example, pricing of options. A contentious question is whether returns are predictable using past returns or other data.[3] The "stan-

[2] I believe that finance is a branch of economics that is a social science.

[3] By "predictable" I mean that there is a better predictor than the expected return.

dard" model in many areas of finance, e.g., derivatives pricing, is the geometric random walk in discrete time or its analogue, geometric Brownian motion, in continuous time. This model says that returns are not predictable. However, empirical evidence cumulated over the past few decades shows that returns are at least to some extent predictable.

3.3 The Random Walk Model

3.3.1 I.i.d. normal returns

Suppose that $R_1, R_2, \ldots$ are the returns from a single asset. A common model is that they are

1. mutually independent;
2. identically distributed, i.e., they have the same probability distributions and in particular the same mean and variance; and
3. normally distributed;

More succinctly, $R_1, R_2, \ldots$ are i.i.d. $N(\mu, \sigma^2)$ for some (unknown) μ and σ^2.

There are two problems with this model. First, because a normally distributed random variable can take any value between $-\infty$ and $+\infty$, the model implies the possibility of unlimited losses, but liability is usually limited; $R_t \geq -1$ since you can lose no more than your investment. Second, multiperiod returns are not normally distributed because $1 + R_t(k) = (1 + R_t)(1 + R_{t-1}) \cdots (1 + R_{t-k+1})$ is not normal — sums of normals are normal but not so with products.

3.3.2 The lognormal model

A second model assumes that the log returns are i.i.d. Recall that the log return is $r_t = \log(1 + R_t)$, where $1 + R_t$ is the simple gross return.

Thus, we assume that $\log(1+R_t)$ is $N(\mu, \sigma^2)$ so that $1+R_t$ is an exponential and therefore positive and thus $R_t \geq -1$. This solves the first problem. Also,

$$\begin{aligned} 1 + R_t(k) &= (1 + R_t) \cdots (1 + R_{t-k+1}) \\ &= \exp(r_t) \cdots \exp(r_{t-k+1}) \\ &= \exp(r_t + \cdots + r_{t-k+1}). \end{aligned}$$

Therefore, $\log\{1 + R_t(k)\} = r_t + \cdots r_{t-k+1}$. Since sums of normal random variables are themselves normal, the second problem is solved — normality of single-period log returns implies normality of multiple-period log returns.

The lognormal distribution goes back to Louis Bachelier's (1900) dissertation at the Sorbonne called *The Theory of Speculation.* Bachelier's work anticipated Einstein's (1905) theory of Brownian motion. In 1827, Robert Brown, a Scottish botanist, observed the erratic unpredictable motion of pollen grains under a microscope. In 1905 Einstein understood that the movement was due to bombardment by water molecules and he developed a mathematical theory. Later, Norbert Wiener, an M.I.T. mathematician, developed a more precise mathematical model of Brownian motion, now called the Wiener process.

Although Bachelier's work anticipated much of modern mathematical finance, his work was ignored at the time and he never found a decent academic job. Bachelier's thesis came to light accidently more than 50 years after he wrote it. The statistician and economist Jimmie Savage found a book by Bachelier in the University of Chicago library and asked other economists about it. Paul Samuelson found Bachelier's thesis in the MIT library.[4] The English translation was published in 1964 in *The Random Character of Stock Market Prices,* an edited volume containing a number of seminal papers.

Example 3.4. A simple gross return, $(1 + R)$, is Lognormal$(0,(0.1)^2)$, which means that $\log(1 + R)$ is $N\{0, (0.1)^2\}$. What is $P(1 + R < 0.9)$?

Answer: Since $\log(0.9) = -0.105$, $P(1 + R < 0.9) = P\{\log(1 + R) < \log(0.9)\} = \Phi\{(-0.105 - 0)/0.1\} = \Phi(-1.05) = 0.1469$. In MATLAB 6, `cdfn(-1.05) = 0.1469`.

Example 3.5. Assume again that $1 + R$ is Lognormal$\{0,(0.1)^2\}$. Also, assume that the returns are an independent sequence. Find the probability that a simple gross two-period return is less than 0.9.

Answer: The two-period gross return is Lognormal$\{0, 2(0.1)^2\}$ so this probability is $\Phi\left[\log(0.9)/\{\sqrt{2}\,(0.1)\}\right] = \Phi(-0.745) = 0.2281$.

Let's find a general formula for the kth period returns. Recall that $1 + R_t(k) = (1 + R_t) \cdots (1 + R_{t-k+1})$. Assume that $\log(1 + R_i)$ is $N(\mu, \sigma^2)$ for all i and that the $\{R_i\}$ are mutually independent. Then $\log\{1 + R_t(k)\}$ is the sum of k independent $N(\mu, \sigma^2)$ random variables, so that $\log\{1 + R_t(k)\}$ is $N(k\mu, k\sigma^2)$. Therefore

$$P\{1 + R_t(k) < x\} = \Phi\left\{\frac{\log(x) - k\mu}{\sqrt{k}\sigma}\right\}.$$

By (2.19), the skewness of $R_t(k)$ (or of $1 + R_t(k)$) equals

$$(e^{k\sigma^2} + 2)\sqrt{e^{k\sigma^2} - 1}$$

and grows very rapidly with k. Thus we see that if one-period log returns are normally distributed, then the return over a long holding period is very highly skewed.

[4] See Bernstein (1992).

3.3.3 Random walks

Let $Z_1, Z_2, \ldots$ be i.i.d. with mean μ and standard deviation σ. Let S_0 be an arbitrary starting point and

$$S_t = S_0 + Z_1 + \cdots + Z_t, \quad t \geq 1.$$

The process $S_0, S_1, \ldots$ is called a **random walk** and $Z_1, \ldots$ are its steps. The expectation and variance of S_t, conditional given S_0, are $E(S_t|S_0) = S_0 + \mu t$ and $\text{Var}(S_t|S_0) = \sigma^2 t$. The parameter μ is called the **drift** and determines the general direction of the random walk. The parameter σ is the **volatility** and determines how much the random walk fluctuates about the mean $S_0 + \mu t$. Since the standard deviation of S_t given S_0 is $\sigma\sqrt{t}$, $(S_0 + \mu t) \pm \sigma\sqrt{t}$ gives the mean plus and minus one standard deviation which gives a range containing 68% probability under the assumption of a normal distribution. The width of this range grows proportionally to $\sqrt{t}$, as is illustrated in Figure 3.2, showing that at time $t = 0$ we know far less about where the random walk will be in the distant future compared to where it will be in the immediate future.

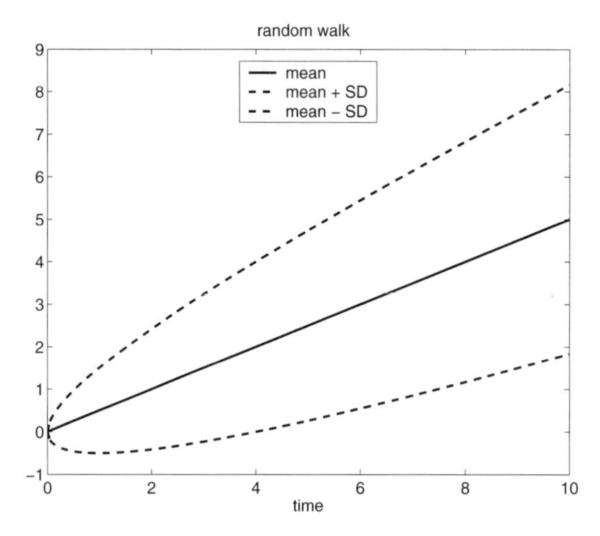

Fig. 3.2. *Mean and bounds (mean plus and minus one standard deviation) on a random walk with $S_0 = 0$, $\mu = 0.5$, and $\sigma = 1$. At any given time, the probability of being between the bounds (dashed curves) is 68% if the distribution of the steps is normal.*

3.3.4 Geometric random walks

Recall that $\log\{1 + R_t(k)\} = r_t + \cdots + r_{t-k+1}$. Therefore

$$\frac{P_t}{P_{t-k}} = 1 + R_t(k) = \exp(r_t + \cdots + r_{t-k+1}),$$

so taking $k = t$ we have

$$P_t = P_0 \exp(r_t + r_{t-1} + \cdots + r_1).$$

Therefore, if the log returns are assumed to be i.i.d. normals, then the process $\{P_t : t = 1, 2, \ldots\}$ is the exponential of a random walk. We call such a process a **geometric random walk** or an **exponential random walk.** If $r_1, r_2, \ldots$ are i.i.d. $N(\mu, \sigma^2)$ then the process is called a lognormal geometric random walk with parameters (μ, σ^2).

As the time between steps becomes shorter and the step sizes shrink in the appropriate way, a random walk converges to Brownian motion and a geometric random walk converges to geometric Brownian motion; see Chapter 8. Geometric Brownian motion is the "standard model" for stock prices used in option pricing.

3.3.5 The effect of the drift μ

The geometric random walk model implies that future price changes are independent of the past and therefore not possible to predict but it does *not* imply that one cannot make money in the stock market. Quite to the contrary, since μ is generally positive, there is an upward trend to the random walk. It is only the future *deviations* from the trend that cannot be predicted. The trend itself can be predicted once μ is estimated.

If the log return is $N(\mu, \sigma^2)$, then the return has a lognormal distribution. As explained in Section 2.8.3, the median of the lognormal distribution is $\exp(\mu)$ and its expected value is $\exp(\mu + \sigma^2/2)$ which is larger than the median because of right skewness. Using these facts, general formulas for the median price and expected price after k years are $P_0 \exp\{k\mu\}$ and $P_0 \exp\{k\mu + k\sigma^2/2\}$, respectively, where P_0 is the price at time 0.

Example 3.6. The log returns on the U.S. stock market as a whole have a mean of about 10% and a standard deviation of about 20%. To see what these values imply, we look at 20 years of a geometric random walk with $\mu = 0.1$ and $\sigma = 0.2$. The expected log return on each step is 0.1. The expected log return for 20 years would be $(20)(0.1) = 2.0$. This is also the median log return, since the log returns are normally distributed and therefore the mean equals the median.

The median gross 20-year return in our example is $\exp(2) = 7.38$. If the stock price starts at \$100, then the median price after 20 years is \$738. The expected price after 20 years is $\$100 \exp\{(20)(0.1) + (20)(0.2)^2/2\} = \$1102$.

Quantiles of the 20-year return distribution can be found using the normality of the log return. See Problem 3.

Figure 3.3 shows two independent simulations of 20 years of a geometric random walk with values of μ and σ as in Example 3.6. The prices start at 100 in both simulations. Notice that the two series of log returns look quite different. However, the log prices and prices look much more similar. The reason is that they both have $\mu = 0.1$ which over the long run gives an upward trend which is the same in both simulations. In the plots, the medians of the log returns, log prices, and prices are also shown.

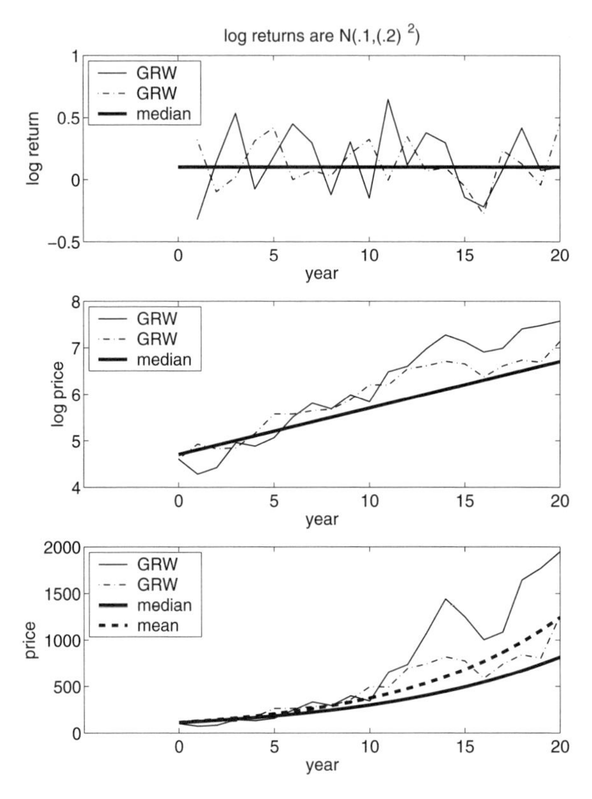

Fig. 3.3. *Two independent simulations of a geometric random walk (GRW) with* $\mu = 0.1$ *and* $\sigma = 0.2$ *(thinner solid and dashed lines). Medians are shown as thick solid curves. Means are shown as a thick dashed curve in the case of the lognormally distributed prices where the mean does not equal the mediam.*

3.3.6 Are log returns normally distributed?

There are several ways to check whether log returns are really normally distributed. One way is to look at a normal probability plot of the log returns.

Recall that a normal probability plot is a plot of the sample quantiles versus the quantiles of the $N(0,1)$ distribution. If the plot is roughly a straight line, then the sample appears normally distributed.

Another method of checking normality is to look at the sample skewness and kurtosis of the log returns to see if their values are near those of the normal distribution. Table 1.1 of Campbell et al. (1997, p. 21) gives $\widehat{Sk}$ and $\widehat{K} - 3$ for several market indices and common stocks. In that table, $\widehat{Sk}$ is generally close to zero, which indicates that log returns are not very skewed. However, the excess kurtosis is typically rather large for daily returns, e.g., 34.9 for a value-weighted index, and positive though not as large for monthly returns, e.g., 2.42 for the value-weighted index. By the central limit theorem, the distribution of log returns over longer periods should approach the normal distribution.[5] Therefore, the smaller excess kurtosis for monthly log returns, in contrast to daily log returns, is expected. The large kurtosis of daily returns indicates that they are heavy-tailed.

A normal probability plot can be supplemented by a test of normality, e.g., the Shapiro-Wilk, Anderson-Darling, Cramér-von Mises or Kolmogorov-Smirnov test. All four are available in PROC UNIVARIATE of SAS.

Example 3.7. (GE daily returns) To illustrate checking for normality we use daily returns for GE common stock from December 1999 to December 2000. Figure 3.4 shows plots of the prices, returns, log returns, and a normal probability plot of the log returns. There is also a plot of the volatility that is discussed later. As can be seen in this figure, the net returns R and the log returns r are *very* similar. The normal plot shown in the second row, second column is roughly linear, indicating at least approximate normality.

Here is a SAS program to read in the GE prices, create the log return variable, and to use PROC UNIVARIATE to compute basic statistics for that variable. Tests of normality are requested by the keyword "`normal`" after "`proc univariate`" and the statement "`probplot`" requests a normal plot. In SAS, "`log`" is the natural logarithm and "`dif`" is the differencing function.

```
options linesize = 72 ;
data ge ;
infile 'C:book\SAS\ge.dat' ;
input price ;
logreturn = dif(log(price)) ;
run ;
title 'GE - Daily prices, Dec 17, 1999 to Dec 15, 2000' ;
proc univariate normal;
var logreturn ;
probplot ;
run ;
```

Here are selected parts of the SAS output.

[5] Assuming that their variance is finite.

```
The UNIVARIATE Procedure
                         Variable:  logreturn

                               Moments

  N                          252    Sum Weights                252
  Mean                -0.0000115    Sum Observations    -0.0028973
  Std Deviation       0.01758546    Variance            0.00030925
  Skewness            -0.1500695    Kurtosis            0.14401017
  Uncorrected SS      0.07762136    Corrected SS        0.07762133
  Coeff Variation     -152954.42    Std Error Mean      0.00110778


                         Tests for Normality

     Test                  --Statistic---     -----p Value------

     Shapiro-Wilk          W      0.994705    Pr < W      0.5311
     Kolmogorov-Smirnov    D      0.041874    Pr > D     >0.1500
     Cramer-von Mises      W-Sq   0.068111    Pr > W-Sq  >0.2500
     Anderson-Darling      A-Sq   0.399967    Pr > A-Sq  >0.2500
```

The log returns have a sample mean, standard deviation, skewness, and excess kurtosis of -0.0000115, 0.0176, -0.15, and 0.144, respectively. The values of the sample skewness and excess kurtosis are close to zero and suggest that the log returns are approximately normally distributed in agreement with the normal plot.

The Shapiro-Wilk, Kolmogorov-Smirnov, Cramér-von Mises, and Anderson-Darling tests of normality have p-values of 0.15 or larger. Since each p-value exceeds 0.1, each test would accept the null hypothesis of normality at $\alpha = 0.1$ and of course at smaller values such as $\alpha = 0.05$.

3.3.7 Do the GE daily returns look like a geometric random walk?

Figure 3.5 shows five independent simulated geometric random walks with the same parameters as the GE daily log returns. Note that the geometric random walks *appear* to have "patterns" and "momentum" even though they do not. The GE log returns look similar to the geometric random walks.

With only one year of data, it is difficult to distinguish between a random walk and a geometric random walk. Figure 3.6 shows three independent simulated time series. For each pair, the log price series (a random walk) is plotted on the left while the price series (a geometric random walk) is plotted on the right. Note that the price and log price series in each pair look similar. However, there are subtle differences between the prices and the log prices, especially that the prices become less volatile as they get closer to 0 while the log prices have constant volatility.

The geometric random walk model is preferred to the random walk model as being more realistic, since the geometric random walk implies nonnegative prices and net returns that are at least -1.

It is interesting to note in Figure 3.5 that geometric random walks can give the appearance of having momentum and trends. Momentum means that once

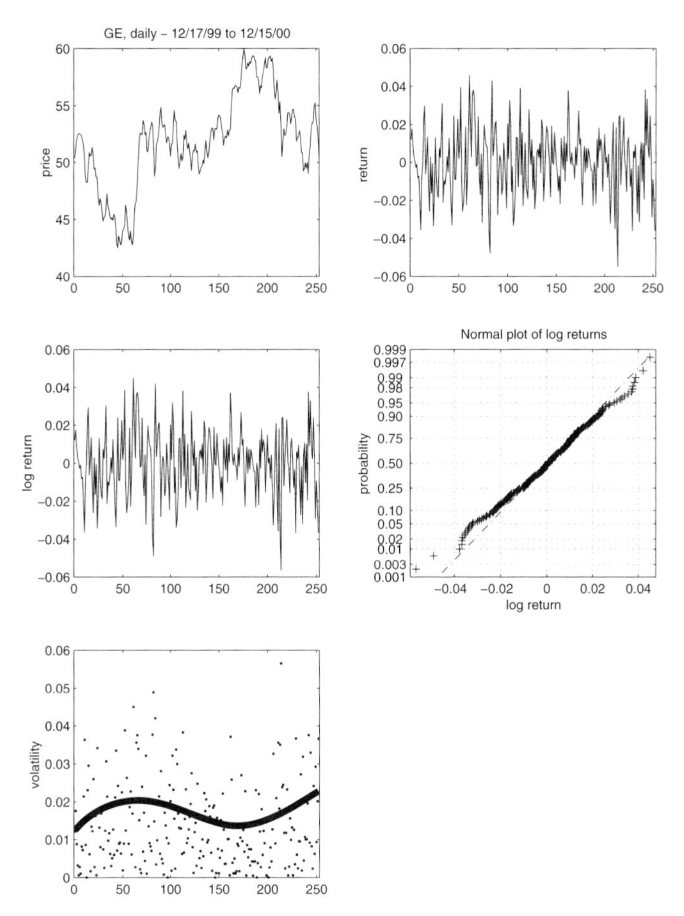

Fig. 3.4. *GE daily returns. The first plot is the prices. The second and third are the net returns and the log returns. The fourth plot is a normal probability plot of the log returns. The final plot is of the absolute log returns; there is a scatterplot smooth to determine whether the volatility is constant.*

prices start to either rise or fall they tend to continue to do so. This gives rise to an upward or downward trend. The top right plot seems to have downward momentum and the middle right plot appears to have upward momentum. However, we know that, in fact, there is no momentum in any of these simulated series since they were simulated to have independent steps. Whether momentum exists in real stock prices is still the subject of considerable debate, much of it summarized by Shefrin (2000).

This simple graphical comparison of GE prices to geometric random walks is not, by itself, much evidence in favor of the geometric random walk hypothesis. A more thorough comparison is needed to determine whether the GE prices are consistent with the geometric random walk hypothesis. This

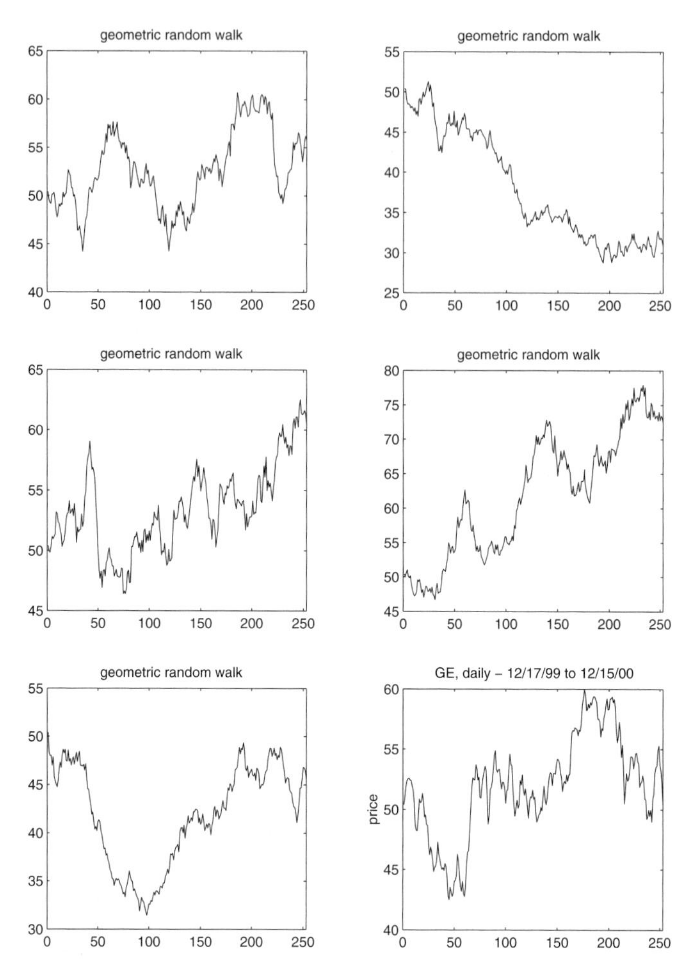

Fig. 3.5. *Five independent geometric random walks and GE daily prices. The geometric random walks have the same expected log return, volatility, and starting point as the GE prices.*

hypothesis implies that the log returns are *mutually independent* and, therefore, uncorrelated. Therefore we should check for evidence that the log returns are correlated. If we find no such evidence, then we have more reason to believe the geometric random walk hypothesis. We return to this issue in Chapter 4 after we have studied time series methods of detecting correlation. The result is that time series analysis shows at least some deviation from random walk behavior. In Chapter 14 on behavioral finance we find more evidence against the random walk hypothesis. Despite this evidence, the geometric random walk model is still widely used, perhaps due to lack of an alternative.

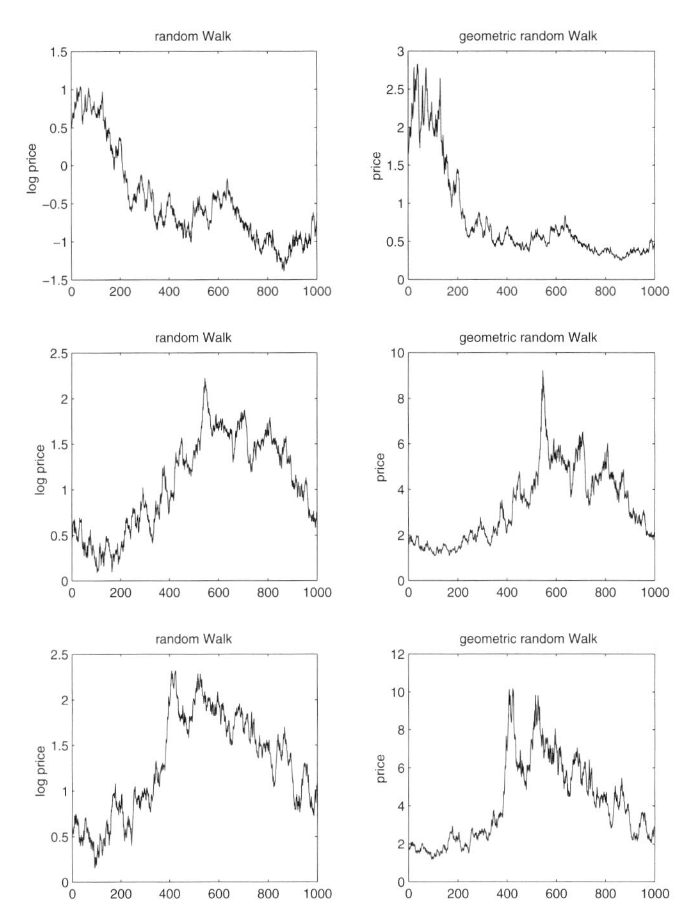

Fig. 3.6. *Three independent simulated price series.* ***On left:*** *log prices.* ***On right:*** *prices.*

3.4 Origins of the Random Walk Hypothesis

Many studies, starting with Bachelier's (1900) have led to the random walk model and only a very few are mentioned here. Much of this work addresses the question of whether returns are predictable. There are two classes of methods for attempting to forecast returns, fundamental analysis and technical analysis.

3.4.1 Fundamental analysis

Fundamental analysis is practiced by **securities analysts** who examine accounting data, interview management, and look at economic forecasts, interest rates, and political trends. Their ultimate goal is to predict the future earnings of stocks, which determine a stock's **fundamental value**.

In 1933 Alfred Cowles published "Can stock market forecasters forecast?" The article appeared in the brand-new journal *Econometrica*. *Econometrica* is now the leading journal in econometrics, which is the field of empirical modeling and analysis of economic data.

Cowles analyzed the track records of

- 16 leading financial services that furnished their subscribers with selected lists of common stocks;
- Purchases and sales of stock by 20 leading fire insurance companies;
- 24 publications including statements of financial services, financial weeklies, and bank letters; and
- Editorials in *The Wall Street Journal* by William Peter Hamilton, an expounder of the "Dow Theory" due to Charles Dow (the Dow of Dow-Jones). The Dow theory compared stock prices to tides and ocean waves. The tides were a metaphor to explain "price momentum."

Cowles found that only 6 of 16 financial services had achieved any measure of success. Even the best record could not be definitely attributed to skill rather than luck. One needs statistical analysis to reach such a conclusion. The null hypothesis to test is that the record of an investor is no better than would be expected by chance. In 1944, Cowles published a new study with basically the same conclusions.[6]

Fama (1965) corroborated Cowles's conclusions when he found that selecting stocks by fundamental analysis seems to do no better than choosing them at random, perhaps by using a dartboard with *The Wall Street Journal* tacked on. Of course, good management, favorable economic trends, and so forth do influence the prices of assets, but Fama claimed that this information is already fully reflected in stock prices because markets react instantaneously to information.

Security analysis may be essential in order for stocks to be priced correctly, but ironically the result of security analysis is that there are few discrepancies between actual prices and the values of stocks and therefore little opportunity to profit from fundamental analysis. William Sharpe discussed the antagonism of professional investors to the random walk theories of Fama and other academics. He stated that "Interestingly, professional economists seem to think more highly of professional investors than do other professional investors."[7] What Sharpe is saying here is that professional economists believe that professional investors are so good that stocks are properly priced, while professional investors think that other professional investors are sufficiently incompetent that there are bargains out there waiting to be purchased by a few smart investors like themselves.

There is no question as to whether one can make money in the stock market. Over the long haul, stocks have outperformed bonds which have outperformed savings accounts. The question is rather whether anyone can "beat

[6] This material on Cowles was adopted from Bernstein (1992).
[7] Bernstein (1992).

the market." "Beating the market" means selecting a portfolio that outperforms a market index such as the S&P 500. The outperformance should not be explainable as reasonably due to chance. There also should be some correction for the degree of risk the individual has assumed since riskier stocks, meaning stocks with a higher standard deviation of the return distribution, generally have higher expected returns in order to compensate investors for assuming risk.[8] The higher expected returns of risky stocks are called **risk premiums**. If an investor chooses risky stocks, then his expected return is higher than if he selects a less risky portfolio — see Chapter 7. The question is whether his expected return is higher than that of a market index with the same level of risk.[9]

3.4.2 Technical analysis

Technical analysis is practiced by so-called technical analysts or chartists. Technical analysts believe that future prices can be predicted from past patterns. Technical analysis uses only past data on stock prices and trading volume to predict future returns.

A very skeptical, though in my opinion accurate, view of technical analysis is found in *A Random Walk Down Wall Street* by Burton Malkiel, a professor of economics at Princeton. This book is a perennial best seller and has been revised several times. It contains much sensible advice for the small investor.

Malkiel describes many of the technical theories, including the Dow Theory, the Filter System, and the Relative-Strength system. The latter advises buying stocks that have done well recently. There is also the hemline theory which predicts price changes by the lengths of women's dresses and the Super Bowl indicator which forecasts the market based upon whether the Super Bowl winner was a former AFL team.

Then there is the odd-lot theory.[10] It is based on the impeccable logic that a person who is *always* wrong is a reliable source of information — just negate whatever that person says. The assumption of the odd-lot theory is that the odd-lot trader is precisely that sort of unfortunate individual. It turns out that the odd-lotter isn't such a dolt after all. Research suggests that odd-lotters do not do worse than average. Yes, you can make a ton of money if you can find an investor who is always wrong, but finding such an investor is just as difficult as finding one who is always correct. If stocks are correctly priced as empirical research suggests, then not only is it impossible to beat the market

[8] This is somewhat an oversimplication and is corrected later. As explained in Chapter 7, it is only the risk that cannot be removed by holding a diversified portfolio that should have a risk premium.

[9] Finding a market index with a given level of risk is unlikely to be possible, but we show in Chapter 7 that one can achieve any desired level of risk with a portfolio consisting of a market index and a risk-free asset.

[10] An odd-lot is a block of less than 100 shares, usually traded only by a small investor.

except by good luck, but it is impossible to do worse than the market except by bad luck.[11]

Technical theories seem to fall into disfavor quickly but new ones appear just as quickly. The ones mentioned above are old and probably no longer have many adherents.

Why are technical theories so appealing? Cognitive psychology suggests an answer to this question. Human nature finds it difficult to accept that many interesting patterns will occur simply by chance. Gilovich (1993) believes that we have an innate tendency to find patterns. Normally, this desire to organize facts into theory serves us well. However, this need to find structure is so strong that we find patterns even when they do not exist. Our natural tendency is to overfit and to generalize from merely random patterns. For example, sports fans have many theories of streaks in athletics, e.g., the "hot hand" theory of basketball. Extensive testing of basketball players' performances have shown no evidence of streaks beyond what would be expected by pure chance.[12] The point is that streaks in the financial markets will occur by chance, but you cannot make money on the basis of random streaks since they are unlikely to continue.

Eugene Fama is one of the leading finance economists today, though many disagree with his strongly held opinions. Peter Bernstein, in his book *Capital Ideas* sketches a biography of Fama. In college, Fama earned extra money working for Harry Ernst who published a stock market newsletter. Fama's job was to find useful buy and sell signals. Ernst believed that trends, once in place, would continue because of "price momentum." By studying stock price data, Fama was able to find trading rules that appeared profitable, but when tested on new data they could not outperform the overall market.

Although he did not realize it at the time, Fama was running into one of the most important problems in statistics and data mining. When we look at data, we often find patterns. The question is whether the patterns are due merely to chance or instead are a sign of something that is more systematic and therefore will continue to be present in new data. If we develop a trading rule or fit a statistical model to the data, we must decide whether the rule or model should be flexible enough to "fit" to the data's patterns. If the patterns are merely random, then fitting them is called **overfitting** and is a bad thing to do. The problem is that an overfit model does not *generalize* to new data since

[11] The random walk hypothesis implies that stupidity cannot lead you to pick "bad stocks" any more than genius can lead you to pick "good stocks." According to that hypothesis, neither good nor bad stocks exist. However, stupidity can get you a larger risk than necessary if you fail to diversify as explained in Chapters 5 and 7. There are no "bad stocks" but there are poor portfolios.

[12] See Gilovich, Vallone, and Tversky (1985). I don't mean to imply that there are no streaks in sports. For example, I do not know of any investigation of batting slumps in baseball and these may very well be nonrandom. In the case of basketball, however, there is a strong belief in the hot hand despite a lack of evidence. Sports fans will find patterns even if they are not there.

new data will have different random patterns. However, systematic patterns will reoccur in new data, so we want our models to include them. A model that misses patterns due to systematic effects is said to be **underfit**. Statisticians and data miners now understand overfitting and underfitting very well and have developed methods for achieving a correct fit, which filters out random noise but retains systematic patterns as best as possible. Unfortunately, an appreciation of the problem of overfitting appears not to have found its way into popular culture despite its importance to everyday phenomena such as the stock market.

Fama's frustrations inspired him to go to graduate school, and in 1964 he earned a doctorate at the University of Chicago. Fama stayed at Chicago where he taught finance and in 1965 published "The Behavior of Stock Market Prices" (his thesis) in the *Journal of Business.* A less technical version was published later as "Random Walks in Stock Market Prices" in the *Financial Analysts Journal.* Fama (1965) stated of technical analysis that "The chartist must admit that the evidence in favor of the random walk model is both consistent and voluminous, whereas there is precious little published in discussion of rigorous empirical tests of various technical theories."

3.5 Efficient Markets Hypothesis (EMH)

As evidence accumulated that stock prices fluctuated like random walks, economists sought a theory as to why that would be so. In 1965 Paul Samuelson published a paper "Proof that properly anticipated prices fluctuate randomly." Samuelson's idea is that random walk behavior is due to the very efficiency of the market.

A market is said to be information efficient if prices "fully reflect" available information. A market is "efficient with respect to an information set" if prices would be unchanged by revealing that information to *all* participants. This implies that it is impossible to make economic profits by trading on the basis of this information set. This last idea is the key to testing (empirically) the EMH.

Samuelson proposed that in an efficient market prices will change only when there is new and unanticipated information. Since the information, and therefore the price changes, are unanticipated, price changes will be random. The market does not react to events that are expected to happen, because prices already reflect what is expected. For example, in 2001 stock prices dropped instantaneously when markets reopened after the September 11th surprise attacks. A few weeks later when the United States started to bomb the Taliban there was little or no market reaction, perhaps because investors already believed that the bombing was coming.

3.5.1 Three types of efficiency

There are three forms of market efficiency:

- weak-form efficiency — the information set includes only the history of prices or returns;
- semi-strong efficiency — the information set includes all information that is publically available; and
- strong-form efficiency — the information set includes all information known to any market participant.

Weak-form efficiency implies that technical analysis will not make money. Semi-strong form efficiency implies that fundamental analysis will not help the investor.

3.5.2 Testing market efficiency

The research of Fama, Cowles, and others tests the various forms of the EMH and their work supports the semi-strong and perhaps the strong form of the EMH.

In their book *Investments*, Bodie, Kane, and Marcus (1999) discuss some of the issues involved when testing the EMH. One is the magnitude issue. No one believes that markets are perfectly efficient. The small inefficiencies might be important to the manager of a large portfolio. If one is managing a $5 billion portfolio, beating the market by 0.1% results in a $5 million increase in profit. This is clearly worth achieving. Yet, no statistical test is likely to undercover a 0.1% inefficiency amidst typical market fluctuations, since the latter are large. For example, the S&P 500 index has a 20% standard deviation in annual returns.

Another issue is selection bias. If there are investors who can consistently beat the market, they probably are keeping that a secret. We can only test market efficiency by testing methods of technical or fundamental analysis that are publicized. These may be the ones that don't reveal market inefficiencies.

Another problem is that for any time period, *by chance* there will be some investment managers who consistently beat the market. As an example, suppose that a portfolio of stocks chosen at random has a 50% chance of outperforming the S&P 500 in any given year. Suppose one chooses a portfolio at random in each of 10 years and assume the outcomes are independent from year to year. Then the probability is only $2^{-10} = 0.000976$, approximately one in a thousand, that you will outperform the market in 10 consecutive years. However, if 2000 people each choose portfolios for 10 consecutive years, it is likely that at least one will outperform the market in all 10 years since the expected number of people outperforming the market in all 10 years is $2000 * 2^{-10} = 1.95$. The probability that no one outperforms the market in each of 10 years is $(1 - 2^{-10})^{2000} = 0.1417$.[13] If one person among the 2000

[13] This is the binomial probability $\binom{n}{x} p^x(1-p)^{n-x}$ with $n = 2000$, $x = 0$, and $p = 2^{-10} = 1/1024$.

does outperform the market in each of 10 years, it would be a mistake to say that this person has skill in tossing heads.

Now apply this reasoning to actual financial markets. Peter Lynch's Magellan Fund outperformed the S&P 500 in 11 of 13 years ending in 1989. Was Lynch a skilled investment manager or just lucky? If he really was skilled, then this is evidence against the semi-strong form of the EMH. However, there are so many mutual funds, that it seems likely that at least one would do as well as Lynch's. Maybe he was just that lucky one.

The extent to which stock returns are predictable remains a area of intense study and there certainly is no consensus. At one extreme, some academics studying the financial markets data have come to the conclusion that security analysts can do no better than blindfolded monkeys who throw darts at the *Wall Street Journal.* The general public does seem to have a strong faith that listening to security analysts is worthwhile, at least when the analysts do not have conflicts of interest that dissuade them from stating their true opinions.

Behavioral finance is a relatively new subject, having started in the 1980s, that combines cognitive psychology and economics. The findings of behavioral finance often contradict the efficient market hypothesis, though proponents of the efficient market hypothesis such as Fama (1998) attribute these results to flaws in the studies or chance. We study behavioral finance later in Chapter 14.

3.6 Discrete and Continuous Compounding

One dollar invested for one year at a 5% rate of simple interest is worth \$1.05 at the end of one year. If instead the 5% interest is compounded semi-annually then the worth after one year is

$$\left(1 + \frac{0.05}{2}\right)^2 = 1.050625.$$

If the compounding is daily, then the worth after one year is

$$\left(1 + \frac{0.05}{365}\right)^{365} = 1.0512675.$$

If one compounded the interest every hour, then the worth after one year would be

$$\left(1 + \frac{0.05}{(24)(365)}\right)^{(24)(365)} = 1.0512709.$$

Discrete compounding means compounding at any fixed time interval. As discrete compounding becomes more and more frequent, the worth after one year has a limit:

$$\lim_{N \to \infty} \left(1 + \frac{0.05}{N}\right)^N = \exp(0.05) = 1.0512711.$$

This limit is called **continuous compounding**.

Table 3.1 shows the convergence to the limit. In the table, N is the number of times that interest is compounded in a year and D_1 is the value at the end of the year of an initial deposit of one thousand dollars. The second column contains the values of $(D_1 - 1050)$, which is the extra value (in dollars) of compounded over simple interest. To eight significant digits, one hundredth of a penny, there is no difference in gross returns between discrete compounding every minute and continuous compounding — both have a gross return of 1.0512711 to eight significant digits. Hourly and continuous compounding agree in the seventh digit, one tenth of a penny.

Table 3.1. D_1 is the value after one year of a deposit of \$1000 that has been compounded N times during the year at 5%.

N	$(D_1 - 1{,}050)$
1	0
2	0.6250
4	0.9433
12	1.1619
52	1.2458
365	1.2675
24*365	1.2709
60*24*365	1.2711
∞	1.2711

In general, if D dollars are invested at a continuously compounded rate r for one year, then the value after one year is $\exp(r)D$. The return is $\exp(r)$ so that the log return is r. This is the reason that the log return on an asset is called the continuously compounded rate.

Here is another way of looking at continuous compounding. Continuous increase can be described according to a differential equation for the rate of growth. Let t be time in years and let D_t be the value of a deposit that is growing at a constant rate r according to the differential equation

$$\frac{d\,D_t}{d\,t} = r D_t. \tag{3.3}$$

Then the solution to this differential equation is

$$D_t = D_0 \exp(rt). \tag{3.4}$$

3.7 Summary

Let P_t be the price of an asset at time t. Then P_t/P_{t-1} is the simple gross return and $R_t = P_t/P_{t-1} - 1$ is the simple net return. (Simple means one period.) The gross return over the last k periods is $1 + R_t(k) = P_t/P_{t-k}$. Let $p_t = \log(P_t)$. The (one-period) log return is $r_t = p_t - p_{t-1}$ and R_t is approximately equal to r_t.

Log returns are often modeled as geometric random walks. The geometric random walk model implies that log returns are mutually independent; one cannot predict future returns from past returns. The model also implies that R_t is lognormally distributed.

The geometric random walk model suggested the efficient market hypothesis (EMH) that states that all valuable information is reflected in the market prices; price changes occur because of unanticipated new information. There are three forms of the EMH, the weak form, the semi-strong form, and the strong form.

Early empirical research by Fama, Cowles, and other researchers supports the geometric random walk model and the efficient market hypothesis. Although more recent work has suggested problems with these models, the geometric random walk and geometric Brownian motion are still very widely used models and market efficiency still has supporters (Fama, 1998).

3.8 Bibliographic Notes

I learned much from reading Bernstein's two books and Chapter 12 of Bodie, Kane, and Marcus (1999). Bodie, Kane, and Marcus (1999) and Sharpe, Alexander, and Bailey (1999) are good introductions to market efficiency and the random walk hypothesis. A more advanced discussion of the random walk hypothesis is found in Chapter 2 of Campbell, Lo, and MacKinlay (1997). The returns notation used in this chapter follows Campbell, Lo, and MacKinlay (1997). Much empirical evidence for market efficiency is reviewed by Fama (1970, 1991). Fama (1998) addresses recent challenges to the EMH. Gilovich (1993) is an interesting discussion of the errors of human perception and interpretation and the hot hand theory is analyzed in his Chapter 2 on the misperception of random data. Recent findings that contradict market efficiency are discussed in Shefrin (2000) and Shleifer (2000).

3.9 References

Bachelier, L. (1900) Theory of speculation, Gaulthier-Vilar, Paris. (English translation by A. J. Boness reprinted in Cootner (1964).)

Bernstein, P. (1996) *Against the Gods: The Remarkable Story of Risk,* Wiley, New York.

Bernstein, P. (1992) *Capital Ideas: The Improbable Origins of Modern Wall Street,* Free Press, New York.

Bodie, Z., Kane, A., and Marcus, A. (1999) *Investments, 4th Ed.*, Irwin/ McGraw-Hill, Boston.

Campbell, J., Lo, A., and MacKinlay, A. (1997). *The Econometrics of Financial Markets*, Princeton University Press, Princeton, NJ.

Cootner, P. (ed.) (1964) *The Random Character of Stock Market Prices*, MIT Press, Cambridge, MA.

Cowles, A. (1933) Can stock market forecasters forecast?, *Econometrica*, **1**, 309–324.

Cowles, A. (1944) Stock market forecasting, *Econometrica*, **12**, 206–214.

Cowles, A. (1960) A revision of previous conclusions regarding stock price behavior, *Econometrica*, **28**, 909–915.

Fama, E. (1965a) The behavior of stock market prices, *Journal of Business*, **38**, 34–105.

Fama, E. (1965b) Random walks in stock market prices, *Financial Analysts Journal*, **21**, 55–59.

Fama, E. (1970) Efficient capital markets: a review of theory and empirical work, *Journal of Finance*, **25**, 383–417.

Fama, E. (1991) Efficient Capital Markets: II, *Journal of Finance*, **46**, 1575–1618.

Fama, E. (1998) Market efficiency, long-term returns, and behavioral finance, *Journal of Financial Economics*, **49**, 283–306.

Gilovich, T. (1993) *How We Know What Isn't So (Paperback Ed.)*, Free Press, New York.

Gilovich, T., Vallone, R., and Tversky, A. (1985) The hot hand in basketball: On the misperceptions of random sequences, *Cognitive Psychology*, **17**, 295–314.

Malkiel, B. G. (1999) *A Random Walk Down Wall Street (Updated and Revised Ed.)*, W. W. Norton, New York.

Samuelson, P. (1965) Proof that properly anticipated prices fluctuate randomly, *Industrial Management Review*, **6**, 41–50.

Sharpe, W. F., Alexander, G. J., and Bailey, J. V. (1999) *Investments, 6th Ed.*, Prentice-Hall, Upper Saddle River, NJ.

Shefrin, H. (2000) *Beyond Greed and Fear: Understanding Behavioral Finance and the Psychology of Investing*, Harvard Business School Press, Boston.

Shleifer, A. (2000) *Inefficient Markets: An Introduction to Behavioral Finance*, Oxford University Press, Oxford, UK.

3.10 Problems

1. The prices and dividends of a stock are given in the table below.
   (a) What is R_2?
   (b) What is $R_4(3)$?
   (c) What is r_3?

t	P_t	D_t
1	51	0.2
2	56	0.2
3	53	0.25
4	58	0.25

2. Let r_t be a log return. Suppose that $\ldots, r_{-1}, r_0, r_1, r_2, \ldots$ are i.i.d. $N(0.1, 0.6)$.
   (a) What is the distribution of $r_t(3) = r_t + r_{t-1} + r_{t-2}$?
   (b) What is the $P\{r_1(3) < 2\}$?
   (c) What is the covariance between $r_1(2)$ and $r_2(2)$?
   (d) What is the conditional distribution of $r_t(3)$ given $r_{t-2} = 0.8$?
3. In Example 3.6, what is the 0.05-quantile of the 20-year return? What is the 0.95-quantile? (Note that there is a 90% chance that the 20-year return will be between these two values, so that they give us a 90% prediction interval.)
4. (a) Give two examples of measurable uncertainty besides coin and dice tossing.
   (b) Give an example of unmeasurable uncertainty. Is there any way in your example that the probability distribution of the outcomes could be estimated from past data? Why or why not?
5. Suppose that $X_1, X_2, \ldots$ is a lognormal geometric random walk with parameters (μ, σ^2). More specifically, suppose that $X_k = X_0 \exp(r_1 + \cdots + r_k)$ where X_0 is a fixed constant and $r_1, r_2, \ldots$ are iid $N(\mu, \sigma_2)$.
   (a) What is the expected value of X_k^2 for any k? (Find a formula giving the expected value as a function of k.) Use (2.6) to find the variance of X_k for any k.
   (b) Use (2.10) to find the density of X_1.
   (c) What is the third quartile of X_k for any k?
6. Malkiel (1999) discusses the Super Bowl theory that a victory in January by a former NFL team predicts a bull market for the year and by a former AFL[14] team a bear market.[15] Malkiel mentions that the indicator failed only once or perhaps twice since the Super Bowl started. It failed in 1970 and perhaps failed in 1987 when the market dropped after a victory by the NY Giants, a former NFL team, but the market then recovered and ended higher for the year.

   Discuss how you might go about testing the Super Bowl theory. Assume that when Malkiel wrote he had seen data through 1998 but none since then. Would data before 1998 be useful for testing given that this data set was used to devise the theory? How would you define a bull and a bear market? Assuming that the market is equally likely to be a bull or bear in any year and that years are independent, how could you use the binomial distribution to perform a test of the theory?

[14] The leagues have merged and now all are in the NFL.
[15] A bull market means rising prices, a bear market falling prices.

4

Time Series Models

4.1 Time Series Data

A time series is a sequence of observations taken over time, for example, a sequence of daily log returns on a stock. In this chapter, we study statistical models for times series. These models are widely used in econometrics as well as in other areas of business and operations research. For example, time series models are routinely used in operations research to model the output of simulations and are used in supply chain management for forecasting demand.

We focus on two questions of interest in finance. The first is whether asset prices behave like the geometric random walk models discussed in Chapter 3. The second issue is whether, when the geometric random walk model does not fit asset prices well, there are other models that fit better.

A **stochastic process** is a sequence of random variables and can be viewed as the "theoretical" or "population" analogue of a time series — conversely, a time series can be considered a sample from the stochastic process. By a time series model, we mean a statistical model for a stochastic process.

A statistical model should have as few parameters as possible. One reason why having few parameters is good is that each unknown parameter is another quantity that must be estimated and each estimate is a source of estimation error. Estimation error, among other things, increases the forecast uncertainty when we use a time series model to forecast future values of the time series. On the other hand, a statistical time series model must have enough parameters to adequately describe the behavior of the time series data. A model with too few parameters can create biases because the model does not fit the data well. A statistical model without excess parameters is sometimes called "parsimonious." One of the most useful methods for obtaining parsimony in a time series model is to assume **stationarity**, a property that we discuss next.

4.2 Stationary Processes

Often we observe a time series whose fluctuations appear random but with the same type of random behavior from one time period to the next. For example, returns on stocks are random and the returns one year can be very different from the previous year, but the mean and standard deviation are often similar from one year to the next.[1] This does not mean that the returns themselves will be the same from one year to the next but rather that the statistical properties of the returns will be the same. Similarly, the demand for many consumer products, such as sunscreen, heavy coats, and electricity, has random as well as seasonal variation but each summer is similar to the past summers and each winter to past winters, at least over shorter time periods, though there are long-term trends if one goes back far enough because of economic growth and technological change.[2] Also, interest rates in the past are similar to those of the present and we expect future interest rates also to be similar to those of the past. *Stationary stochastic processes* are probability models for such time series with time-invariant behavior.

A process is said to be stationary if all aspects of its behavior are unchanged by shifts in time. Mathematically, stationarity is defined as the requirement that for every m and n, the distribution of $Y_1, \ldots, Y_n$ and $Y_{1+m}, \ldots, Y_{n+m}$ are the same; that is, the probability distribution of a sequence of n observations does not depend on their time origin. Stationarity is a very strong assumption because it requires that "all aspects" of behavior be constant in time. Generally, we can get by with assuming less, namely, weak stationarity. A process is *weakly stationary* if its mean, variance, and covariance are unchanged by time shifts. More precisely, $Y_1, Y_2, \ldots$ is a weakly stationary process if

- $E(Y_i) = \mu$ (a constant) for all i;
- $\text{Var}(Y_i) = \sigma^2$ (a constant) for all i; and
- $\text{Corr}(Y_i, Y_j) = \rho(|i - j|)$ for all i and j for some function $\rho(h)$.

Thus, the mean and variance do not change with time and the correlation between two observations depends only on the time distance between them. For example, if the process is stationary then the correlation between

[1] It is the returns, not the stock prices, that have this time-invariant behavior. Stock prices themselves tend to increase over time so this year's stock prices tend to be much higher than those a decade or two ago.

[2] Although many economic time series have seasonal variation, seasonal effects are rarer in financial markets. This is not surprising since seasonal effects are predictable and, if they existed, would lead to trading strategies that would cause their own elimination. This appears to be what happened with the so-called January effect mentioned in Section 14.5.2. Time series models that incorporate seasonal effects are very important in many business applications such as the forecasting of demand for products, but seasonal time series are not needed for most applications in finance and so are not discussed in this book.

Y_2 and Y_5 is the same as the correlation between Y_7 and Y_{10}, since each pair is separated from the other by three units of time. The adjective "weakly" in "weakly stationary" refers to the fact that we are only assuming that means, variance, and covariances, not other distributional characteristics such as quantile, skewness, and kurtosis, are stationary.

The function ρ is called the **autocorrelation function** of the process. Note that $\rho(h) = \rho(-h)$. Why?

The covariance between Y_t and Y_{t+h} is denoted by $\gamma(h)$ and $\gamma(\cdot)$ is called the **autocovariance function**. Note that $\gamma(h) = \sigma^2\rho(h)$ and that $\gamma(0) = \sigma^2$ since $\rho(0) = 1$. Also, $\rho(h) = \gamma(h)/\sigma^2 = \gamma(h)/\gamma(0)$.

Many financial time series, for example, stock prices, do not appear stationary, but the *changes* in these time series do appear stationary and can be modeled as stationary processes. For this reason, stationary time series models are far more applicable than they might appear. From the viewpoint of statistical modeling, it is not important whether it is the time series itself or changes in the time series that are stationary, because either way we get a parsimonious model.

The beauty of a stationary process, at least from the statistical viewpoint, is that it can be modeled with relatively few parameters. For example, we do not need a different expectation for each Y_t; rather they all have a common expectation, μ. This means that μ can be estimated accurately by $\overline{X}$. If instead, for example, we did not assume stationarity and each Y_t had its own unique expectation, μ_t, then it would not be possible to estimate μ_t accurately — μ_t could only be estimated by the single observation Y_t itself.

When a time series is observed, a natural question is whether it appears to be stationary. This is not an easy question to address, and we can never be absolutely certain of the answer. However, methods for checking for stationarity are available and are discussed in Section 4.8.

4.2.1 Weak white noise

White noise is the simplest example of a stationary process. The sequence $Y_1, Y_2, \ldots$ is a *weak white noise process* with mean μ, denoted WhiteNoise(μ, σ^2), if

- $E(Y_i) = \mu$ for all i;
- $\text{Var}(Y_i) = \sigma^2$ (a constant) for all i; and
- $\text{Corr}(Y_i, Y_j) = 0$ for all $i \neq j$.

If in addition $Y_1, Y_2 \ldots$ are independent normal random variables, then the process is called a *Gaussian white noise process*. (The normal distribution is sometimes called the Gaussian distribution.)

A weak white noise process is weakly stationary with

$$\rho(0) = 1$$
$$\rho(h) = 0 \ \text{ if } \ h \neq 0$$

so that

$$\gamma(0) = \sigma^2$$
$$\gamma(h) = 0 \text{ if } h \neq 0.$$

In this book, "white noise" means weak white noise, which includes Gaussian white noise as a special case. White noise (either weak or Gaussian) is not very interesting in itself but is the building block of important time series models used for economic data.

4.2.2 Predicting white noise

Because of the lack of correlation, the future of a white noise cannot be predicted from past values. More precisely, suppose that $\ldots, Y_1, Y_2, \ldots$ is a Gaussian WhiteNoise(μ, σ^2) process. Then

$$E(Y_{i+t}|Y_1, \ldots, Y_i) = \mu \text{ for all } t \geq 1. \tag{4.1}$$

What this equation is saying is that you cannot predict the future deviations of a white noise process from its mean, because its future is independent of its past and present, and therefore the best predictor of any future value of the process is simply the mean μ, what you would use even if $Y_1, \ldots, Y_i$ had not been observed. For white noise that is not necessarily Gaussian, (4.1) need not be true, but it is still true that the best linear predictor of Y_{i+t} given $Y_1, \ldots, Y_i$ is μ.

4.2.3 Estimating parameters of a stationary process

Suppose we observe $Y_1, \ldots, Y_n$ from a stationary process. To estimate the mean μ and variance σ^2 of the process we use the sample mean $\overline{Y}$ and sample variance s^2 defined in equation (2.60).

To estimate the autocovariance function we use the *sample autocovariance function*

$$\widehat{\gamma}(h) = n^{-1} \sum_{j=1}^{n-h} (Y_{j+h} - \overline{Y})(Y_j - \overline{Y}). \tag{4.2}$$

Equation (4.2) is an example of the usefulness of parsimony induced by the stationarity assumption. Because the correlation between Y_t and Y_{t+h} is independent of t, *all* $n - h$ pairs of data points that are separated by a lag of h time units can be used to estimate $\gamma(h)$. Some authors define $\widehat{\gamma}(h)$ with the factor n^{-1} in (4.2) replaced by $(n - h)^{-1}$, but this change has little effect if n is reasonably large and h is small relative to n, as is typically the case.

To estimate $\rho(\cdot)$ we use the *sample autocorrelation function* (*SACF*) defined as

$$\widehat{\rho}(h) = \frac{\widehat{\gamma}(h)}{\widehat{\gamma}(0)}.$$

Although a stationary process is somewhat parsimonious with parameters, at least relative to a general nonstationary process, a stationary process is still not sufficiently parsimonious for our purposes. The problem is that there are still an infinite number of parameters, $\rho(1), \rho(2), \ldots$. What we need is a class of stationary time series models with only a finite, preferably small, number of parameters. The ARIMA models of this chapter are precisely such a class. The simplest ARIMA models are autoregressive (AR) models and we turn to these first.

4.3 AR(1) Processes

Although white noise is not an appropriate model for time series with correlation, time series models with correlation can be built out of white noise. The simplest stationary processes with correlation are **autoregressive processes** where Y_t is modeled as a weighted average of past observations plus a white noise "error." In other words, autoregressive processes follow a regression model where Y_t is the "response" or "outcome" and where past values of the process are the "independent" or "predictor" variables. We start with AR(1) processes, the simplest autoregressive processes.

Let $\epsilon_1, \epsilon_2, \ldots$ be white noise with mean 0, that is, WhiteNoise($0,\sigma_\epsilon^2$). We say that $Y_1, Y_2, \ldots$ is an **AR(1) process** if for some constant parameters μ and ϕ,

$$Y_t - \mu = \phi(Y_{t-1} - \mu) + \epsilon_t \tag{4.3}$$

for all t. The parameter μ is the mean of the process. Think of the term $\phi(Y_{t-1} - \mu)$ as representing "memory" or "feedback" of the past into the present value of the process. The process $\{Y_t\}_{t=-\infty}^{+\infty}$ is correlated because the deviation of Y_{t-1} from its mean is fed back into Y_t which induces a correlation between Y_t and the past. The parameter ϕ determines the amount of feedback, with a larger absolute value of ϕ resulting in more feedback and $\phi = 0$ implying that $Y_t = \mu + \epsilon_t$ so that Y_t is WhiteNoise(μ, σ_ϵ^2). Also, you can think of ϵ_t as representing the effect of "new information." For example, if Y_t is the log return on an asset at time t, then ϵ_t represents the effect on the asset's price of new information that is revealed at time t. Information that is truly new cannot be anticipated so that the effects of today's new information should be independent of the effects of yesterday's news. This is why we model the effects of new information as white noise with mean zero.

If $|\phi| < 1$, then $Y_1, \ldots$ is a weakly stationary process. Its mean is μ. Simple algebra shows that (4.3) can be rewritten as

$$Y_t = (1 - \phi)\mu + \phi Y_{t-1} + \epsilon_t. \tag{4.4}$$

Recall the linear regression model $Y_t = \beta_0 + \beta_1 X_t + \epsilon_t$ from your statistics courses or peek ahead to Chapter 6 for an introduction to regression analysis.

Equation (4.4) is just a linear regression model with $\beta_0 = (1-\phi)\mu$ and $\beta_1 = \phi$. If it is assumed that the process has a zero mean, that is, that $\mu = 0$, then $\beta_0 = 0$ as well. Linear regression with $\beta_0 = 0$ is the "linear regression through the origin" model. The term *autoregression* refers to the regression of the process on its own past values.

When $|\phi| < 1$ then repeated use of equation (4.3) shows that

$$Y_t = \mu + \epsilon_t + \phi\epsilon_{t-1} + \phi^2\epsilon_{t-2} + \cdots = \mu + \sum_{h=0}^{\infty} \phi^h \epsilon_{t-h}, \tag{4.5}$$

and assumes that time parameter t of Y_t and ϵ_t can be extended to negative values so that the white noise process is $\ldots, \epsilon_{-2}, \epsilon_{-1}, \epsilon_0, \epsilon_1, \ldots$ and (4.3) is true for all integers t. Equation (4.5) is called the infinite moving average (MA(∞)) represention of the process. This equation shows that Y_t is a weighted average of *all* past values of the white noise process. This representation should be compared to the AR(1) representation that shows Y_t as depending on Y_{t-1} and ϵ_t. Since $|\phi| < 1$, $\phi^h \to 0$ as the lag $h \to \infty$. Thus, the weights given to the distant past are small. In fact, they are quite small. For example, if $\phi = 0.5$ then $\phi^{10} = 0.00098$ of ϵ_{t-10} has virtually no effect on Y_t.

4.3.1 Properties of a stationary AR(1) process

When the AR(1) process is stationary, which implies that $|\phi| < 1$, then

$$E(Y_t) = \mu \quad \forall t, \tag{4.6}$$

$$\gamma(0) = \text{Var}(Y_t) = \frac{\sigma_\epsilon^2}{1-\phi^2} \quad \forall t, \tag{4.7}$$

$$\gamma(h) = \text{Cov}(Y_t, Y_{t+h}) = \frac{\sigma_\epsilon^2 \phi^{|h|}}{1-\phi^2} \quad \forall t \text{ and } \forall h, \tag{4.8}$$

and

$$\rho(h) = \text{Corr}(Y_t, Y_{t+h}) = \phi^{|h|} \quad \forall t \text{ and } \forall h. \tag{4.9}$$

It is important to remember that formulas (4.6) to (4.9) hold only if $|\phi| < 1$ and only for AR(1) processes. Moreover, for Y_t to be stationary, Y_0 must start in the stationary distribution so that $E(Y_0) = \mu$ and $\text{Var}(Y_0) = \sigma_\epsilon^2/(1-\phi^2)$ or the process must have started in the infinite past so that (4.3) is true for all integers t.

These formulas can be proved using (4.5). For example using (2.55),

$$\text{Var}(Y_t) = \text{Var}\left(\sum_{h=0}^{\infty} \phi^h \epsilon_{t-h}\right) = \sigma_\epsilon^2 \sum_{h=0}^{\infty} \phi^{2h} = \frac{\sigma_\epsilon^2}{1-\phi^2}, \tag{4.10}$$

which proves (4.7). In (4.10) the formula for summation of a geometric series was used. This formula is

$$\sum_{i=0}^{\infty} r^i = \frac{1}{1-r} \quad \text{if } |r| < 1. \tag{4.11}$$

Also, for $h > 0$

$$\text{Cov}\left(\sum_{i=0}^{\infty} \epsilon_{t-i}\phi^i, \sum_{j=0}^{\infty} \epsilon_{t+h-j}\phi^j \right) = \frac{\sigma_\epsilon^2 \phi^{|h|}}{1-\phi^2}, \tag{4.12}$$

thus verifying (4.8); see Problem 7. Then (4.9) follows by dividing (4.8) by (4.7).

Be sure to distinguish between σ_ϵ^2 which is the variance of the stationary white noise process $\epsilon_1, \epsilon_2, \ldots$ and $\gamma(0)$ which is the variance of the AR(1) process $Y_1, Y_2, \ldots$. We can see from the result above that $\gamma(0)$ is larger than σ_ϵ^2 unless $\phi = 0$ in which case $Y_t = \mu + \epsilon_t$ so that Y_t and ϵ_t have the same variance.

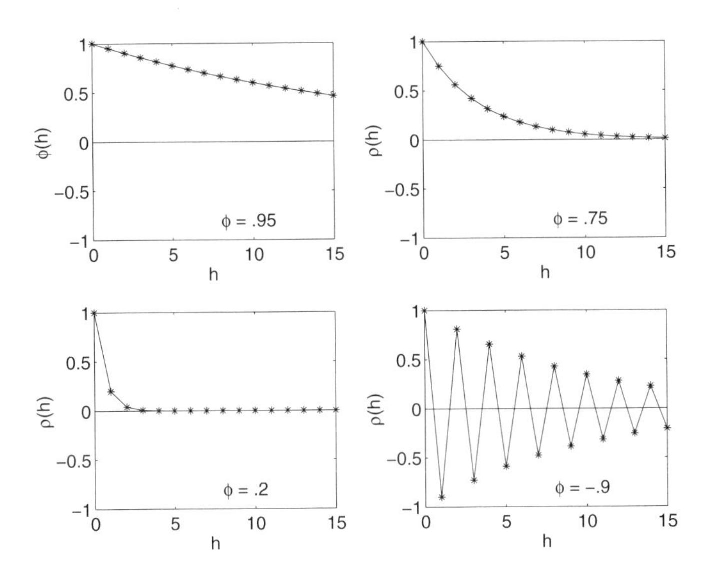

Fig. 4.1. *Autocorrelation functions of AR(1) processes with ϕ equal to* 0.95, 0.75, 0.2, *and* −0.9.

The ACF (autocorrelation function) of an AR(1) process depends upon only one parameter, ϕ. This is a remarkable amount of parsimony, but it comes at a price. The ACF of an AR(1) process has only a limited range of shapes as can be seen in Figure 4.1. The magnitude of its ACF decays geometrically to zero, either slowly as when $\phi = 0.95$, moderately slowly as when $\phi = 0.75$, or rapidly as when $\phi = 0.2$. If $\phi < 0$, then the sign of the ACF oscillates as its magnitude decays geometrically. If the SACF of the data does not behave in one of these ways, then an AR(1) model is unsuitable. The remedy is to use more AR parameters, to switch to another class of models such as the moving average (MA) models, or to combine the AR and MA models into a so-called ARMA model. We investigate each of these possibilities later in this chapter.

4.3.2 Convergence to the stationary distribution

Suppose that Y_0 is an arbitrary value and that (4.3) holds for $t = 1, \ldots$. Then the process is not stationary, but converges to the stationary distribution satisfying (4.6) to (4.9) as $t \to \infty$.[3] For example, since $Y_t - \mu = \phi(Y_{t-1} - \mu) + \epsilon_t$, $E(Y_1) - \mu = \phi\{E(Y_0) - \mu\}$, $E(Y_2) - \mu = \phi^2\{E(Y_0) - \mu\}$, and so forth so that

$$E(Y_t) = \mu + \phi^t\{E(Y_0) - \mu\} \text{ for all } t > 0. \tag{4.13}$$

Since $|\phi| < 1$, $\phi^t \to 0$ and $E(Y_t) \to \mu$ as $t \to \infty$. The convergence of $\mathrm{Var}(Y_t)$ to $\sigma_\epsilon^2/(1-\phi^2)$ can be proved in a somewhat similar manner. The convergence to the stationary distribution can be very rapid when $|\phi|$ is not too close to 1. For example, if $\phi = 0.5$, then $\phi^{10} = 0.00097$, so by (4.13) $E(Y_{10})$ is very close to μ unless $E(Y_0)$ was quite far from μ.

4.3.3 Nonstationary AR(1) processes

If $|\phi| \geq 1$, then the AR(1) process is nonstationary, and the mean, variance, and correlation are not constant.

Random Walk ($\phi = 1$)

If $\phi = 1$ then

$$Y_t = Y_{t-1} + \epsilon_t$$

and the process is *not* stationary. This is the random walk process we saw in Chapter 3.

Suppose we start the process at an arbitrary point Y_0. It is easy to see that

$$Y_t = Y_0 + \epsilon_1 + \cdots + \epsilon_t.$$

Then $E(Y_t|Y_0) = Y_0$ for all t, which is constant but depends entirely on the arbitrary starting point. Moreover, $\mathrm{Var}(Y_t|Y_0) = t\sigma_\epsilon^2$ which is not stationary but rather increases linearly with time. The increasing variance makes the random walk "wander" in that Y_t takes increasingly longer excursions away from its mean of Y_0.

[3] However, it must be assumed that Y_0 has a finite mean and variance, since otherwise Y_t will not have a finite mean and variance for any $t > 0$.

AR(1) processes when $|\phi| > 1$

When $|\phi| > 1$, an AR(1) process has explosive behavior. This can be seen in Figure 4.2. This figure shows simulations of 200 observations from AR(1) processes with various values of ϕ. The explosive case where $\phi = 1.02$ clearly is different from the other cases where $|\phi| \leq 1$. However, the case where $\phi = 1$ is not that much different from $\phi = 0.9$ even though the former is nonstationary while the latter is stationary.

The ability to distinguish the three types of AR(1) processes (stationary, random walk, and explosive) depends on the length of the observed series. With only a short sequence of data from an AR(1) process, it is very difficult to tell if the process is stationary, random walk, or explosive. For example, in Figure 4.3, we see 30 observations from processes with ϕ equal to 1 and 1.2. Notice that unlike in Figure 4.2, in Figure 4.3 the random walk process with $\phi = 1$ appears very similar to the explosive process with $\phi = 1.02$.

If we observe the AR processes for longer than 200 observations, then the behavior of $\phi = 0.9$ and $\phi = 1$ processes would not look as similar as in Figure 4.2. For example, in Figure 4.4 there are 1000 observations from the processes with ϕ equal to 0.9 and 1. Now these processes look dissimilar, unlike in Figure 4.2 with only 200 observations. The stationary process with $\phi = 0.9$ continues to return to its mean of zero. The random walk ($\phi = 1$) wanders without tending to return to any particular value.

Suppose an explosive AR(1) process starts at $Y_0 = 0$ and has $\mu = 0$. Then

$$\begin{aligned} Y_t &= \phi Y_{t-1} + \epsilon_t = \phi(\phi Y_{t-2} + \epsilon_{t-1}) + \epsilon_t = \phi^2 Y_{t-2} + \phi\epsilon_{t-1} + \epsilon_t = \cdots \\ &= \epsilon_t + \phi\epsilon_{t-1} + \phi^2\epsilon_{t-2} + \cdots + \phi^{t-1}\epsilon_1 + \phi^t Y_0. \end{aligned}$$

Therefore, $E(Y_t) = \phi^t Y_o$ and

$$\text{Var}(Y_t) = \sigma^2(1 + \phi^2 + \phi^4 + \cdots + \phi^{2(t-1)}) = \sigma^2 \frac{\phi^{2t} - 1}{\phi^2 - 1}$$

by the summation formula for a finite geometric series

$$\sum_{i=1}^{n} r^i = \frac{1 - r^{n+1}}{1 - r} \text{ if } r \neq 0. \tag{4.14}$$

Since $|\phi| > 1$, this variance increases geometrically fast as $t \to \infty$. For example, if $\sigma = 1$ and $\phi = 1.02$, so that ϕ is only very slightly larger than 1, then $\text{Var}(Y_t)$ equals 2.02, 24.3, 60.4, and 312 for t equal to 1, 10, 20, and 50, respectively. The growth of $\text{Var}(Y_t)$ is even more rapid if ϕ is larger.

Explosive AR processes are not widely used in econometrics since economic growth is usually not explosive.

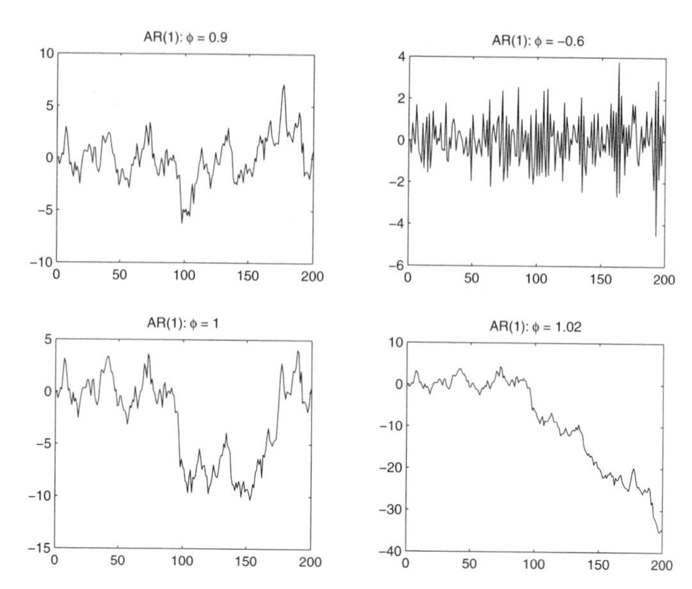

Fig. 4.2. *Simulations of* 200 *observations from* $AR(1)$ *processes with various values of* ϕ *and* $\mu = 0$. *The white noise process* $\epsilon_1, \epsilon_2, \ldots$ *is the same for all four* $AR(1)$ *processes.*

4.4 Estimation of AR(1) Processes

Depending upon the application, one will want to fit an AR(1) model to either one of the variables in the raw data or a variable that has been constructed from the raw data. In finance applications, one often has asset prices as the raw data but wants to fit an AR(1) to the log returns. To create the log returns, one first log-transforms the prices and then differences the log prices. SAS has functions `log` and `dif` to take logarithms and to do differencing.

Let's assume we have a time series $Y_1, \ldots, Y_n$ and we want to fit an AR(1) model to this series. Since an AR(1) model is a linear regression model, it can be analyzed using linear regression software. One creates a lagged variable in Y_t and uses this as the "x-variable" in the regression. SAS supports lagging with the `lag` function. However, SAS has special procedures, AUTOREG and ARIMA, for fitting AR models that are more convenient than regression software.

The least squares estimates of ϕ and μ minimize

$$\sum_{t=2}^{n} \Big[\{Y_t - \mu\} - \{\phi(Y_{t-1} - \mu)\} \Big]^2 .$$

If the errors $\{\epsilon_1, \ldots, \epsilon_n\}$ are *Gaussian* white noise then the least squares estimates are also the MLEs.

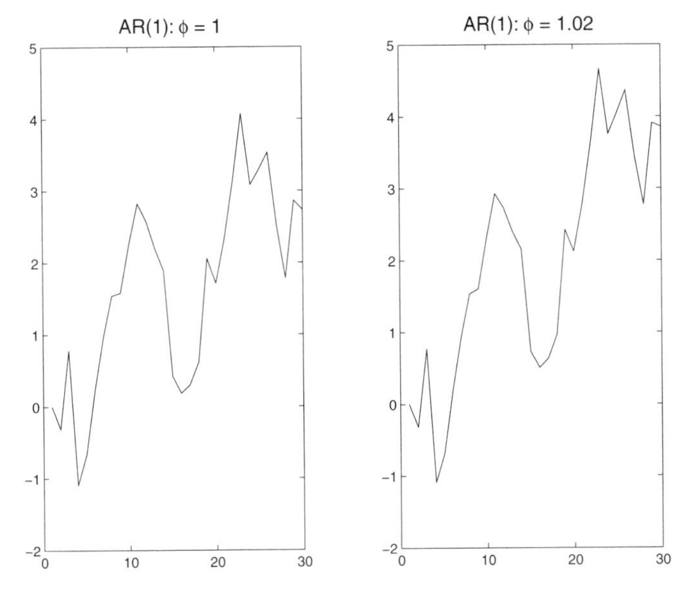

Fig. 4.3. *Simulations of* 30 *observations from* $AR(1)$ *processes with* ϕ *equal to* 1 *and* 1.02 *and* $\mu = 0$. *The white noise process* $\epsilon_1, \epsilon_2, \ldots$ *is the same for both* $AR(1)$ *processes.*

4.4.1 Residuals and model checking

Once μ and ϕ have been estimated, one can estimate the white noise process $\epsilon_1, \ldots, \epsilon_n$. Rearranging equation (4.3) we have

$$\epsilon_t = (Y_t - \mu) - \phi(Y_{t-1} - \mu). \tag{4.15}$$

In analogy with (4.15), the residuals, $\widehat{\epsilon}_1, \widehat{\epsilon}_2, \ldots, \widehat{\epsilon}_n$, are defined as

$$\widehat{\epsilon}_t = (Y_t - \widehat{\mu}) - \widehat{\phi}(Y_{t-1} - \widehat{\mu}) \tag{4.16}$$

and estimate $\epsilon_1, \epsilon_2, \ldots, \epsilon_n$. The residuals can be used to check the assumption that $Y_1, Y_2, \ldots, Y_n$ is an AR(1) process; any autocorrelation in the residuals is evidence against the assumption of an AR(1) process.

To appreciate why residual autocorrelation indicates a possible problem with the model, suppose that we are fitting an AR(1) model but the true model is a AR(2) process[4] given by

$$(Y_t - \mu) = \phi_1(Y_{t-1} - \mu) + \phi_2(Y_{t-2} - \mu) + \epsilon_t.$$

Since we are fitting the wrong model, there is no hope of estimating ϕ_2. Moreover, $\widehat{\phi}$ does not necessarily estimate ϕ_1 because of bias caused by model

[4] We discuss higher-order AR models in more detail soon.

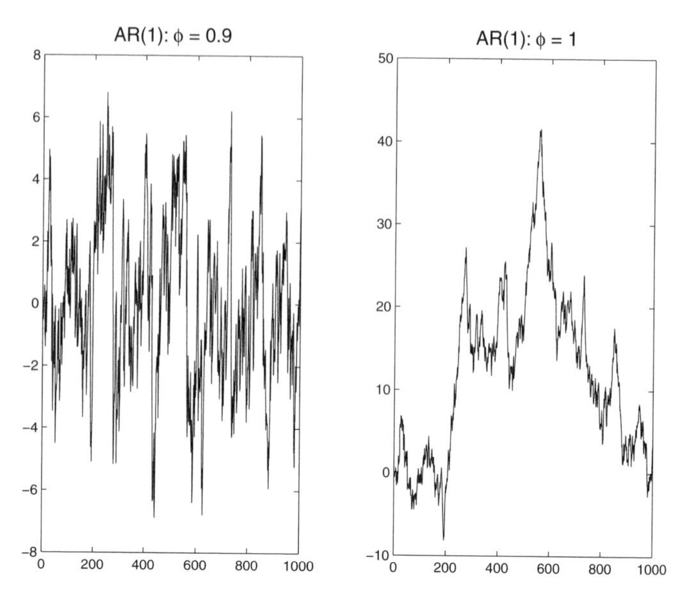

Fig. 4.4. *Simulations of* 1000 *observations from* $AR(1)$ *processes with* ϕ *equal to* 0.9 *and* 1 *and* $\mu = 0$. *The white noise process* $\epsilon_1, \epsilon_2, \ldots$ *is the same for both* $AR(1)$ *processes.*

misspecification. Let ϕ^* be the expected value of $\widehat{\phi}$. For the purpose of illustration, assume that $\widehat{\mu} \approx \mu$ and $\widehat{\phi} \approx \phi^*$. This is a sensible approximation if the sample size n is large enough. Then

$$\begin{aligned}\widehat{\epsilon}_t &\approx (Y_t - \mu) - \phi^*(Y_{t-1} - \mu) \\ &= \phi_1(Y_{t-1} - \mu) + \phi_2(Y_{t-2} - \mu) + \epsilon_t - \phi^*(Y_{t-1} - \mu) \\ &= (\phi_1 - \phi^*)(Y_{t-1} - \mu) + \phi_2(Y_{t-2} - \mu) + \epsilon_t.\end{aligned}$$

Thus, the residuals do not estimate the white noise process as they would if the correct AR(2) model were used. Even if there is no bias in the estimation of ϕ so that $\phi_1 = \phi^*$ and the term $(\phi_1 - \phi^*)(Y_{t-1} - \mu)$ drops out, the presence of $\phi_2(Y_{t-2} - \mu)$ causes the residuals to be autocorrelated.

To test for residual autocorrelation one can use the *test bounds* provided by SAS's autocorrelation plots. Any residual SACF value outside the test bounds is significantly different from 0 at the 0.05 level. A danger here is that some SACF values will be significant merely by chance. For example, if we look at the first 20 SACF values, then even if the residuals really are white noise we can expect 1 (.05 of 20) of these to be significant by chance. To guard against this danger, one can use the Ljung-Box test that *simultaneously* tests that all autocorrelations up to a specified lag are zero. For example, if the residuals really are white noise and the specified lag is 12, then there is only a 1 in 20 chance of declaring *any* of the first 12 SACF values significant.

4.4.2 AR(1) model for GE daily log returns

An AR(1) model can be fit with SAS using either the AUTOREG procedure or the ARIMA procedure. Here is a SAS program that fits an AR(1) model using PROC AUTOREG. The raw data are daily GE closing prices. The data step inputs data from the ASCII file "ge.dat" and then creates the variable `logR` which is the log return. The two title statements annotate the output. Finally, PROC AUTOREG with `nlag` equal to 1 fits an AR(1) model.

```
options linesize=70 ;
data ge ;
infile 'c:\book\sas\ge.dat' ;
input close ;
logP = log(close) ;
logR = dif(logP) ;
run ;
title 'GE - Daily prices, Dec 17, 1999 to Dec 15, 2000' ;
title2 'AR(1)' ;
proc autoreg ;
model logR =/nlag = 1 ;
run ;
```

Here is the SAS output.

```
GE - Daily prices, Dec 17, 1999 to Dec 15, 2000            1
                   AR(1)
                     The AUTOREG Procedure
                   Dependent Variable    logR

                Ordinary Least Squares Estimates

SSE                    0.07762133   DFE                      251
MSE                     0.0003092   Root MSE             0.01759
SBC                    -1316.8318   AIC               -1320.3612
Regress R-Square           0.0000   Total R-Square        0.0000
Durbin-Watson              1.5299

                                     Standard                Approx
Variable        DF      Estimate        Error    t Value    Pr > |t|

Intercept        1     -0.000011     0.001108      -0.01      0.9917

                 Estimates of Autocorrelations
                Lag     Covariance      Correlation

                  0       0.000308         1.000000
                  1       0.000069         0.225457

                 Estimates of Autocorrelations

         Lag    -1 9 8 7 6 5 4 3 2 1 0 1 2 3 4 5 6 7 8 9 1

           0     |                    |********************|
           1     |                    |*****               |

                 Preliminary MSE     0.000292

              Estimates of Autoregressive Parameters

                                      Standard
            Lag      Coefficient         Error     t Value

              1        -0.225457      0.061617       -3.66
```

```
                    The AUTOREG Procedure

                    Yule-Walker Estimates

  SSE                    0.07359998   DFE                    250
  MSE                     0.0002944   Root MSE           0.01716
  SBC                    -1324.6559   AIC             -1331.7148
  Regress R-Square           0.0000   Total R-Square      0.0518
  Durbin-Watson              1.9326
                                   Standard                  Approx
Variable      DF      Estimate        Error    t Value    Pr > |t|

Intercept      1     -0.000040     0.001394      -0.03      0.9773
```

The AUTOREG estimate of ϕ is -0.2254. AUTOREG uses the model

$$Y_t = \beta_0 - \phi Y_{t-1} + \epsilon_t \tag{4.17}$$

so AUTOREG's ϕ is the negative of ϕ as we define it.[5] Thus, $\widehat{\phi} = 0.2254$ using the notation of this book, and of many other books and software. The standard error of $\widehat{\phi}$ is 0.0616 so an approximate 95% confidence interval for ϕ is $0.2254 \pm (2)(0.0616) = (0.1022, 0.3486)$. The confidence interval includes only positive values, so we can conclude with 95% confidence that ϕ is positive which implies positive autocorrelation. Thus, we have evidence against the geometric random walk hypothesis. However, $\phi = 0.2254$ is not large. Since $\rho(h) = \phi^h$, the correlation between successive log returns is 0.2254 and the squared correlation is only 0.0508. As discussed in Section 2.15.2 the squared correlation is the fraction of variation that is predictable so we see that only about 5% of the variation in a log return can be predicted by the previous day's return. The t-value for testing H_0: $\phi = 0$ versus H_1: $\phi \neq 0$ is -3.66 and the p-value is 0.000 (zero to three decimals).

The parameter β_0 in equation (4.17) is called the "constant" or "intercept" of the model and β_0 equals $(1 - \phi)\mu$. Since $\phi < 1$ for a stationary process, $\mu = 0$ if and only if β_0 is zero. Notice that the intercept has a p-value equal to 0.9917. This is the p-value for testing the null hypothesis that the intercept is zero versus the alternative that it is not zero. Since the p-value is large, we accept the null hypothesis.

The AR(1) can also be fit using PROC ARIMA with the following code.

```
proc arima ;
identify var=logR ;
estimate p=1 ;
run ;
```

Here is the output.

```
                 Conditional Least Squares Estimation

                             Standard                 Approx
Parameter       Estimate        Error    t Value    Pr > |t|     Lag
MU            -0.0000361    0.0014009      -0.03      0.9795       0
AR1,1            0.22943      0.06213       3.69      0.0003       1
```

[5] Another SAS procedure, ARIMA, agrees with our notation.

```
           Constant Estimate       -0.00003
           Variance Estimate       0.000294
           Std Error Estimate      0.017159
           AIC                     -1331.75
           SBC                     -1324.69
           Number of Residuals          252
       * AIC and SBC do not include log determinant.


              Correlations of Parameter
                      Estimates

            Parameter         MU      AR1,1

            MU             1.000     -0.010
            AR1,1         -0.010      1.000
       GE - Daily prices, Dec 17, 1999 to Dec 15, 2000            4
                                       21:36 Sunday, July 27, 2003

                 The ARIMA Procedure
              Autocorrelation Check of Residuals

 To      Chi-        Pr >
Lag    Square   DF  ChiSq -------------Autocorrelations------------

  6     15.71    5 0.0077  0.022 -0.102  0.030 -0.025  0.028 -0.218
 12     22.95   11 0.0179 -0.005  0.023 -0.026 -0.021 -0.111  0.116
 18     27.18   17 0.0555  0.065 -0.081  0.018  0.060  0.018  0.023
 24     33.59   23 0.0714 -0.010 -0.024 -0.091  0.064  0.100 -0.013
 30     43.13   29 0.0443 -0.002  0.106  0.052 -0.077 -0.117  0.001
 36     47.10   35 0.0831  0.022 -0.031 -0.083 -0.002  0.071  0.010
 42     71.93   41 0.0020 -0.057  0.100  0.057 -0.038 -0.221 -0.123
 48     78.64   47 0.0026  0.040 -0.070  0.009  0.032  0.087  0.081
```

One difference between ARIMA and AUTOREG is that AUTOREG outputs the estimated "constant" in the model (called $\beta_0 = (1 - \phi)\mu$ earlier) whereas ARIMA outputs an estimate of the mean μ. Another difference is that ARIMA agrees with our notation about the signs of the coefficients.

The SACF of the residuals from the GE daily log returns seen in Figure 4.5 shows high negative autocorrelation at lag 6; $\widehat{\rho}(6)$ is outside the test limits and so is "significant" at $\alpha = 0.05$. This is somewhat worrisome. Moreover, the more conservative Ljung-Box "simultaneous" tests that $\rho(1) = \cdots = \rho(12) = 0$ has $p = 0.0179$. The results of the Ljung-Box test are in the ARIMA output and labeled as `Autocorrelation Check of Residuals`. Since the AR(1) model does not fit well, one might consider more complex models. These are discussed in the following sections.

4.5 AR(p) Models

We have seen that the ACF of an AR(1) process decays geometrically to zero and also oscillates in sign if $\phi < 0$. This is a limited range of behavior and many time series do not behave in this way. To get a more flexible class of models, but one that still is parsimonious with parameters, we can use a model

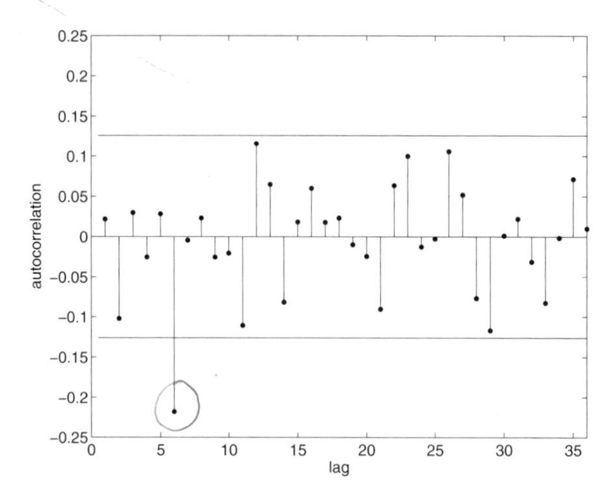

Fig. 4.5. *SACF of residuals from an AR(1) fit to the GE daily log returns. Notice the large negative residual autocorrelation at lag 6. This is a sign that the AR(1) model may not fit well.*

that regresses the current value of the process on several of the recent past values, not just the most recent. In other words, we let the last p values of the process, $Y_{t-1}, \ldots, Y_{t-p}$, feed back into the current value Y_t.

Here's a formal definition. The stochastic process Y_t is an **AR(p) process** if

$$Y_t - \mu = \phi_1(Y_{t-1} - \mu) + \phi_2(Y_{t-2} - \mu) + \cdots + \phi_p(Y_{t-p} - \mu) + \epsilon_t,$$

where $\epsilon_1, \ldots, \epsilon_n$ is WhiteNoise$(0, \sigma_\epsilon^2)$.

This is a multiple linear regression model with lagged values of the time series as the "x-variables." The model can be reexpressed as

$$Y_t = \beta_0 + \phi_1 Y_{t-1} + \cdots + \phi_p Y_{t-p} + \epsilon_t,$$

where $\beta_0 = \{1-(\phi_1 + \cdots + \phi_p)\}\mu$. The parameter β_0 is called the "constant" or "intercept" as in an AR(1) model. It can be proved that $\{1-(\phi_1+\cdots+\phi_p)\} > 0$ for a stationary process, so $\mu = 0$ if and only if β_0 is zero. Therefore, analogously as for an AR(1) model, the p-value in SAS output for testing that the "constant" is zero is also the p-value for testing that the mean is zero.

The ACF of an AR(p) process with $p > 1$ is more complicated than for an AR(1) process. This is a topic discussed in any time series textbook, and is not be treated here. See the references at the end of the chapter for further information.

The least squares estimator minimizes

$$\sum_{t=p+1}^{n} \{Y_t - (\beta_0 + \phi_1 Y_{t-1} + \cdots + \phi_p Y_{t-p})\}^2.$$

The least squares estimator can be calculated using SAS's AUTOREG or ARIMA procedures.

Most of the concepts we have discussed for AR(1) models generalize easily to AR(p) models. Each of the coefficients $\phi_1, \ldots, \phi_p$ will have an estimate, a standard error, and a p-value for testing the null hypothesis that it is zero. The residuals are defined by

$$\widehat{\epsilon}_t = Y_t - \{\widehat{\beta}_0 + \widehat{\phi}_1 Y_{t-1} + \cdots + \widehat{\phi}_{t-p} Y_{t-p}\}.$$

If the AR(p) model fits the time series well, then the residuals should look like white noise. Residual autocorrelation can be detected by examining the SACF of the residuals and using the Ljung-Box test. Any significant residual autocorrelation is a sign that the AR(p) model does not fit well.

4.5.1 AR(6) model for GE daily log returns

To fit an AR(6) model to the GE daily log returns, the SAS program shown above was rerun with

```
model logR =/nlag = 1
```

replaced by

```
model logR =/nlag = 6 .
```

Here are selected portions of the output.

```
                 The AUTOREG Procedure
               Dependent Variable    logR
          Estimates of Autoregressive Parameters

                                    Standard
         Lag     Coefficient           Error    t Value

           1       -0.253149        0.062283      -4.06
           2        0.125685        0.064263       1.96
           3       -0.071448        0.064586      -1.11
           4        0.074783        0.064586       1.16
           5       -0.051952        0.064263      -0.81
           6        0.222743        0.062283       3.58

                 Yule-Walker Estimates

   SSE                   0.06877662   DFE                      245
   MSE                    0.0002807   Root MSE             0.01675
   SBC                   -1313.7482   AIC               -1338.4542
   Regress R-Square          0.0000   Total R-Square        0.1139
   Durbin-Watson             1.9479

                                     Standard                Approx
   Variable        DF    Estimate       Error    t Value    Pr > |t|

   Intercept        1   -2.106E-6    0.001013      -0.00      0.9983
```

The autoregression coefficients (the ϕ_i) are "significant" at lags 1 and 6 but not at lags 2 through 5. Here "significant" means at $\alpha = 0.05$ which corresponds to an absolute t-value bigger than 2. There is no known economic reason why there should be autocorrelation at a lag of 6, and the large estimated autocorrelation at that lag might be due simply to chance. Even though the correlation is "significant," we have tested a number of hypotheses so some of the p-values could be expected to be small by chance.

4.6 Moving Average (MA) Processes

4.6.1 MA(1) processes

As we have seen, the idea behind AR processes is to feed past data back into the current value of the process. This induces correlation between the past and present. The effect is to have at least some correlation at *all* lags. Sometimes data show correlation at only short lags, for example, only at lag 1 or only at lags 1 and 2. AR processes do not behave this way and will not fit such data well. In such situations, a useful alternative to an AR model is a moving average (MA) model. A process Y_t is a moving average process if Y_t can be expressed as a weighted average (moving average) of the past values of the white noise process ϵ_t, rather than of past values Y_t itself as happens in an AR process.

The **MA(1)** (moving average of order 1) process is

$$Y_t - \mu = \epsilon_t - \theta\epsilon_{t-1},$$

where as before the ϵ_ts are WhiteNoise$(0, \sigma_\epsilon^2)$.

One can show that

$$\begin{aligned}
E(Y_t) &= \mu, \\
\mathrm{Var}(Y_t) &= \sigma_\epsilon^2(1+\theta^2), \\
\gamma(1) &= -\theta\sigma_\epsilon^2, \\
\gamma(h) &= 0 \text{ if } |h| > 1, \\
\rho(1) &= -\frac{\theta}{1+\theta^2}, \text{ and} \qquad (4.18) \\
\rho(h) &= 0 \text{ if } |h| > 1. \qquad (4.19)
\end{aligned}$$

Notice the implication of (4.18) and (4.19) — an MA(1) has zero correlation at all lags except lag 1 (and of course lag 0). It is relatively easy to derive these formulas and I recommend that you try to do this yourself as an exercise and a check of your understanding.

4.6.2 General MA processes

The **MA(q)** process is

$$Y_t = \mu + \epsilon_t - \theta_1 \epsilon_{t-1} - \cdots - \theta_q \epsilon_{t-q}. \tag{4.20}$$

One can show that $\gamma(h) = 0$ and $\rho(h) = 0$ if $|h| > q$. Formulas for $\gamma(h)$ and $\rho(h)$ when $|h| \leq q$ are given in time series textbooks but are not needed by us.

Unlike AR(p) models where the "constant" in the model is not the same as the mean in an MA(q) model μ, the mean of the process, is the same as β_0, the "constant" in the model. This fact can be appreciated by examining the right-hand side of equation (4.20) where μ is the "intercept" or "constant" in the model and is also the mean of Y_t because $\epsilon_t, \ldots, \epsilon_{t-q}$ have mean zero.

Fitting MA models is complicated. Although (4.20) looks like a regression model with $\epsilon_{t-1}, \ldots, \epsilon_{t-p}$ as the predictor variables, it is not. The problem is that $\epsilon_{t-1}, \ldots, \epsilon_{t-p}$ are *unobserved* and therefore must be estimated at the same time the MA coefficients are estimated. As for an AR process the estimate of ϵ_t is called the **residual** $\widehat{\epsilon}_t$. We do not discuss the theory of estimation of MA models here but instead show how the estimation can be done with software packages such as SAS.

4.6.3 MA(2) model for GE daily log returns

An MA(2) model was fit to the GE daily log returns using PROC ARIMA in SAS. Here is part of the SAS program.

```
proc arima ;
identify var = logR ;
estimate q=2 ;
run ;
```

The `estimate q=2` statement specifies an MA(2) model. Here are portions of the output. We see that $\widehat{\theta}_1 = -0.26477$ and this coefficient is highly significant ($p < 0.0001$). The second MA coefficient is smaller and not significant. The residuals do not pass the white noise test so we might consider using more MA terms or adding AR terms. This lack of fit to an MA(2) model is due to the same problem discussed earlier when we fit AR models to this time series. There is significant autocorrelation at lag 6 in the data and the residuals will pass a white noise test only if the model can accommodate this autocorrelation. Therefore, an MA(6) model would be needed. Since there is no economic reason why the autocorrelation at lag 6 is "real," I would be reluctant to use such a nonparsimonious model.

```
          GE - Daily prices, Dec 17, 1999 to Dec 15, 2000
                Conditional Least Squares Estimation
                              Standard                 Approx
Parameter      Estimate          Error    t Value    Pr > |t|     Lag

MU           -0.0000247      0.0012775      -0.02      0.9846       0
MA1,1          -0.26477        0.06362      -4.16      <.0001       1
MA1,2           0.07617        0.06385       1.19      0.2340       2

                    Constant Estimate       -0.00002
                    Variance Estimate       0.000291
```

```
                   Std Error Estimate       0.017069
                   AIC                      -1333.41
                   SBC                      -1322.82
                   Number of Residuals           252
              * AIC and SBC do not include log determinant.

                  Autocorrelation Check of Residuals
 To      Chi-          Pr >
Lag     Square  DF  ChiSq  ------------Autocorrelations-----------

  6      14.47   4 0.0059 -0.004  0.022  0.021 -0.055  0.037 -0.224
 12      21.25  10 0.0194  0.004 -0.005 -0.036 -0.002 -0.109  0.110
 18      24.12  16 0.0868  0.050 -0.059  0.024  0.055  0.017  0.028
```

4.7 ARIMA Processes

Stationary time series with complex autocorrelation behavior often are more parsimoniously modeled by mixed autoregressive and moving average (ARMA) processes than by either a pure AR or pure MA process. For example, it is sometimes the case that a model with one AR parameter and one MA parameter, called an *ARMA(1,1)* model, will fit the data rather well and a pure AR or pure MA model will need more than two parameters to provide as good a fit as the ARMA(1,1) model. This section introduces ARMA processes. Also discussed are ARIMA (autoregressive, integrated, moving average) processes which are based on ARMA processes and are models for nonstationary time series.

4.7.1 The backwards operator

The *backwards operator* B is a bit of notation that is useful for describing ARMA and ARIMA models. The backwards operator is defined by

$$B\,Y_t = Y_{t-1}$$

and, more generally,

$$B^k\,Y_t = Y_{t-k}.$$

Thus, B backs up time one unit while B^k does this repeatedly so that time is backed up k time units. Note that $B\,c = c$ for any constant c since a constant does not change with time.

4.7.2 ARMA processes

An **ARMA**(p, q) process combines both AR and MA terms and is defined succinctly using the backward operator by the equation

$$(1 - \phi_1 B - \cdots - \phi_p B^p)(Y_t - \mu) = (1 - \theta_1 B - \ldots - \theta_q B^q)\epsilon_t. \tag{4.21}$$

Equation (4.21) can be rewritten without the backwards operator as

$$(Y_t - \mu) = \phi_1(Y_{t-1} - \mu) + \cdots + \phi_p(Y_{t-p} - \mu) + \epsilon_t - \theta_1\epsilon_{t-1} - \cdots - \theta_q\epsilon_{t-q},$$

showing how Y_t depends on lagged values of itself and lagged values of the white noise. A white noise process is ARMA(0,0) since if $p = q = 0$, then (4.21) reduces to

$$(Y_t - \mu) = \epsilon_t.$$

4.7.3 Fitting ARMA processes: GE daily log returns

PROC ARIMA in SAS can fit an ARMA model to data. The estimate statement has two parameters, p which is the AR order and q which is the MA order. Here is part of a SAS program to fit an ARMA(2,1) model to the GE daily log returns. The ARMA(2,1) is used simply for illustration, not because it is the best model to use for this data set. The first part of the program reads in data and is identical to the data step of the program on page 113.

```
(data step omitted)
proc arima ;
identify var = logR ;
estimate p=2 q=1 ;
```

Here is the output. Note that the first AR coefficient is $\widehat{\phi}_1 = -0.533$ and is significant with a p-value of 0.0012, but the second AR coefficient is smaller, $\widehat{\phi}_2 = 0.078$, with large p-value (0.3841). Also, the MA coefficient is $\widehat{\theta}_1 = -0.806$ and significant with a p-value less than 0.0001. The residuals do not pass the white noise test and other models might be considered. The problem is the relatively large autocorrelation at lag 6 that we have seen before in this data set.

```
                 Conditional Least Squares Estimation
                                Standard                 Approx
Parameter       Estimate           Error    t Value    Pr > |t|     Lag

MU            -0.0000217       0.0013272      -0.02      0.9870       0
MA1,1           -0.80566         0.14654      -5.50      <.0001       1
AR1,1           -0.53313         0.16319      -3.27      0.0012       1
AR1,2            0.07806         0.08953       0.87      0.3841       2

                    Constant Estimate        -0.00003
                    Variance Estimate        0.000289
                    Std Error Estimate       0.016993
                    AIC                      -1334.68
                    SBC                      -1320.56
                    Number of Residuals           252
             * AIC and SBC do not include log determinant.

                      Autocorrelation Check of Residuals
  To      Chi-          Pr >
 Lag    Square   DF   ChiSq   -------------Autocorrelations------------

   6      9.31    3  0.0254  -0.000   0.005  -0.025   0.000  -0.007  -0.187
  12     16.72    9  0.0534  -0.025   0.021  -0.038  -0.008  -0.109   0.116
  18     19.68   15  0.1847   0.045  -0.052   0.006   0.069   0.007   0.037
  24     25.59   21  0.2224  -0.028  -0.010  -0.093   0.073   0.079   0.011
  30     34.54   27  0.1510  -0.004   0.109   0.033  -0.059  -0.122   0.004
```

4.7.4 The differencing operator

The *differencing operator* is another useful notation and is defined as $\Delta = 1 - B$, where B is the backwards operator so that

$$\Delta Y_t = Y_t - B\,Y_t = Y_t - Y_{t-1}.$$

Thus, differencing a time series produces a new time series consisting of the changes in the original series. For example, if $p_t = \log(P_t)$ is the log price, then the log return is

$$r_t = \Delta p_t.$$

Differencing can be iterated. For example,

$$\begin{aligned}\Delta^2 Y_t = \Delta(\Delta Y_t) &= \Delta(Y_t - Y_{t-1}) = (Y_t - Y_{t-1}) - (Y_{t-1} - Y_{t-2})\\ &= Y_t - 2Y_{t-1} + Y_{t-2}.\end{aligned}$$

4.7.5 From ARMA processes to ARIMA processes

Often the first or second differences of nonstationary time series are stationary. For example, the first differences of a random walk (nonstationary) are white noise (stationary).

A time series Y_t is said to be *ARIMA*(p, d, q) if $\Delta^d Y_t$ is ARMA(p, q). For example, if log returns (r_t) on an asset are ARMA(p, q), then the log prices (p_t) are ARIMA$(p, 1, q)$.

Notice that an ARIMA$(p, 0, q)$ model is the same as an ARMA(p, q) model. ARIMA$(p, 0, 0)$, ARMA$(p, 0)$, and AR(p) models are the same. Also, ARIMA$(0, 0, q)$, ARMA$(0, q)$, and MA(q) models are the same. A random walk is an ARIMA$(0, 1, 0)$ model. Why?

The inverse of differencing is "integrating." The integral of a process Y_t is the process w_t, where

$$w_t = w_{t_0} + Y_{t_0} + Y_{t_0+1} + \cdots + Y_t, \tag{4.22}$$

where t_0 is an arbitrary starting time point and w_{t_0} is the starting value of the w_t process. It is easy to check (see Problem 8) that

$$\Delta w_t = Y_t; \tag{4.23}$$

that is, integrating and differencing are inverse processes.[6]

Figure 4.6 shows an AR(1) process, its "integral," and its "second integral," meaning the integral of its integral. The three processes behave in entirely different ways. The AR(1) is stationary and varies randomly about

[6] An analogue is, of course, differentiation and integration in calculus which are inverses of each other.

its mean which is 0; one says that the process *reverts* to its mean. The integral of this process behaves much like a random walk in having no fixed level to which it reverts. The second integral has *momentum*. Once it starts moving upward or downward, it tends to continue in that direction. If data show momentum like this, this is an indication that $d = 2$.

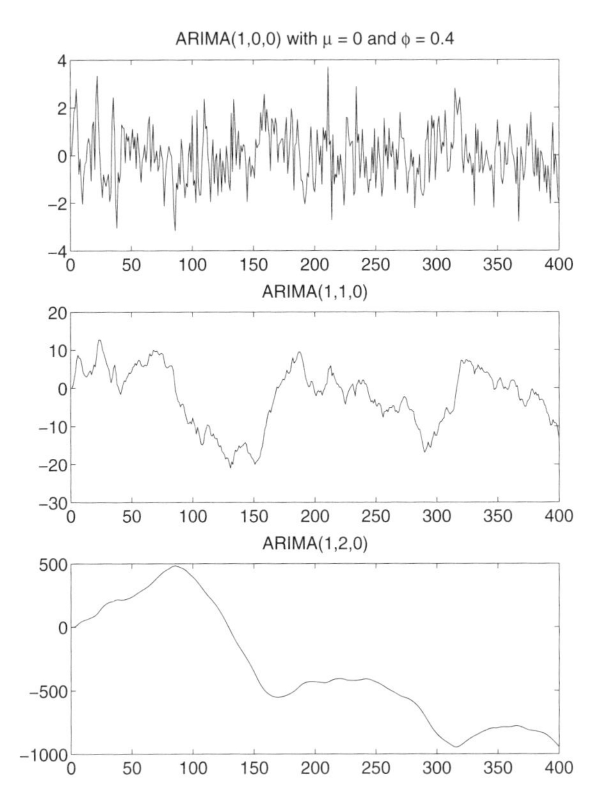

Fig. 4.6. *The top plot is of an $AR(1)$ process with $\mu = 0$ and $\phi = 0.4$. The middle and bottom plots are, respectively, the integral and second integral of this $AR(1)$ process.*

4.7.6 ARIMA(2,1,0) model for GE daily log prices

The following SAS program fits an ARIMA(2,1,0) process to the GE daily log *prices*. This is the same as fitting an ARIMA(2,0,0) process to the daily log *returns*. The statement `var = logP(1)` specifies `logP` (log prices) as the variable to be analyzed and set d equal to 1. In general, d is placed in parentheses after the name of the variable.

```
(data step omitted, but as in previous programs)
proc arima ;
identify var = logP(1) ;
estimate p=2 ;
run ;
```

4.8 Model Selection

The ARIMA procedure in SAS allows one to specify p, d, and q. Once the parameters p, d, and q of an ARIMA process have been selected, the AR and MA coefficients can be estimated by maximum likelihood. But how do we choose p, d, and q?

Generally, d is either 0, 1, or 2 and is chosen by looking at the SACF of Y_t, ΔY_t, and $\Delta^2 Y_t$. A sign that a process is nonstationary is that its SACF decays to zero very slowly and is consistently positive and well above the upper test bound for 30 or more lags, often many more. If this is true of Y_t then Y_t is nonstationary and should be differenced at least once.

If the SACF of ΔY_t looks stationary then use $d = 1$. Otherwise, look at the SACF of $\Delta^2 Y_t$; if this looks stationary use $d = 2$. As an example, see Figure 4.7 where the SACF of the original series looks nonstationary but the sample SACF of the differenced series looks stationary so that $d = 1$ is appropriate. I have never seen a real time series where $\Delta^2 Y_t$ did not look stationary, but if one were encountered then $d > 2$ would be used.

Once d has been chosen, we know that we will fit an ARMA(p, q) process to $\Delta^d Y_t$, but we still need to select p and q. This can be done by comparing various choices of p and q by some criterion that measures how well a model fits the data and, more important, how well it can be expected to predict new data from the same time series.

4.8.1 AIC and SBC

AIC and SBC are model selection criteria based on the log-likelihood and can be used to select p and q assuming that d has already been selected as described above.

AIC (Akaike's information criterion) is defined as

$$-2\log(L) + 2(p+q), \tag{4.24}$$

where L is the likelihood evaluated at the MLE. **Schwarz's Bayesian Criterion (SBC)** is also called the Bayesian Information Criterion (BIC) and is defined as

$$-2\log(L) + \log(n)(p+q), \tag{4.25}$$

where n is the length of the time series. The "best" model according to either criterion is the model that minimizes that criterion. Both criteria will tend to select models with large values of the likelihood. This makes perfect sense since a large value of L means that the observed data are likely under that model.

The term $2(p+q)$ in AIC or $\log(n)(p+q)$ in SBC is a penalty on having too many parameters (lack of parsimony). Therefore, AIC and SBC both try to trade off a good fit to the data measured by L with the desire to use as few parameters as possible.

Note that $\log(n) > 2$ if $n \geq 8$. Since most time series are much longer than 8, SBC penalizes $p+q$ more than AIC does. Therefore, AIC will tend to choose models with more parameters than SBC. Compared to SBC, with AIC the trade off is more in favor of a large value of L than a small value of $p+q$.

This difference between AIC and SBC is due to the way they were designed. AIC is designed to select the model that will predict best and is less concerned than SBC with having a few too many parameters. SBC is designed to select the true values of p and q exactly. In practice the best AIC model is usually close to the best SBC model and often they are the same model. Because AIC and SBC are based on the log-likelihood, use of either of these criteria is closely connected with likelihood ratio tests.

Unfortunately, not all software packages compute AIC and SBC though SAS computes both. Here's how you can calculate approximate AIC and SBC values using software that does not include AIC and SBC in its output. It can be shown that $\log(L) \approx (-n/2)\log(\widehat{\sigma}^2) + K$, where K is a constant that does not depend on the model or on the parameters. Since we only want to minimize AIC and SBC, the exact value of K is irrelevant and we drop K. Thus, one can use the approximations

$$\text{AIC} \approx n\log(\widehat{\sigma}^2) + 2(p+q), \tag{4.26}$$

and

$$\text{SBC} \approx n\log(\widehat{\sigma}^2) + \log(n)(p+q). \tag{4.27}$$

The estimator $\widehat{\sigma}^2$ is called MSE (mean squared error) or simply MS in computer output. These approximations to AIC and SBC have been simplified by dropping certain constant terms (the constant K above) that do not depend on the model and therefore do not affect which model minimizes each criterion. Because these terms have been dropped, AIC or SBC calculated from these approximations will differ from AIC or SBC in, for example, SAS output. However, the model with the smallest value of AIC or SBC should be the same when using these approximations as when using SAS or other software.

4.8.2 GE daily log returns: Choosing the AR order

We have fit a variety of models to the GE daily log returns. In this section we compare various AR models using AIC and SBC. Table 4.1 compares AR(p) models for $p = 1, \ldots, 8$ to an AR(1) model using AIC and SBC. AIC and SBC values are often large; e.g., -1331.75 is the AIC value for an AR(1) model in

Table 4.1. GE daily log return data. AIC and SBC for AR(p) compared to AR(1). The column AIC is AIC for an AR(p) model minus AIC for an AR(1) model. The column SBC is analogous.

p	AIC	SBC
1	0	0
2	−0.41	3.12
3	1.03	8.09
4	2.44	13.03
5	4.43	18.54
6	−7.04	10.61
7	-6.06	15.11
8	4.50	20.20

this example. For this reason, when we look for the model that minimizes AIC or SBC it is easier to look at the *changes* in AIC and SBC as p varies, as is done here. In this example, AIC and SBC disagree more than is typical. SBC is minimized at $p = 1$. AIC has a local minimum at $p = 2$, close but slightly larger than the value of p that minimizes SBC. This behavior is typical. What is atypical is that AIC has a minimum at a far larger value of p, $p = 6$. Thus, SBC chooses an AR(1) model, while AIC chooses a much more complex model.

My interpretation of these results is that AR(1) is our best guess at the "correct" model, but we cannot be sure of the correct model so a more complex AR(6) model is suggested by AIC for forecasting. Statisticians believe that one should never use criteria such as AIC and SBC blindly. I am skeptical about the need for the complexity of the AR(6) model and would probably use the AR(1) model even for forecasting.

4.9 Three-Month Treasury Bill Rates

The efficient market hypothesis predicts that log returns on stocks will be white noise, and our empirical results are that log returns have little autocorrelation even if they are not exactly white noise. For this reason, log returns are rather easy to model by ARMA processes. Other financial time series do have substantial autocorrelation and are more challenging to model, as is shown in this example.

The time series in this example is monthly interest rates on three-month U.S. Treasury bills from December 1950 until February 1996. The data are used as Example 16.1 of Pindyck and Rubinfeld (1998). The rates are plotted in Figure 4.7 and the SACF has behavior typical of a nonstationary series with values that are consistently positive and decaying very slowly to zero and still well above the test bound at a lag of 60. The first differences look stationary, and we fit ARMA models to them. The first differences do show nonconstant volatility. This is not necessarily a sign of nonstationarity and the stationary

GARCH processes studied in Chapter 12 show this type of behavior. In fact, a GARCH model would probably be recommended for this series, but as an illustration we use ARIMA models here.

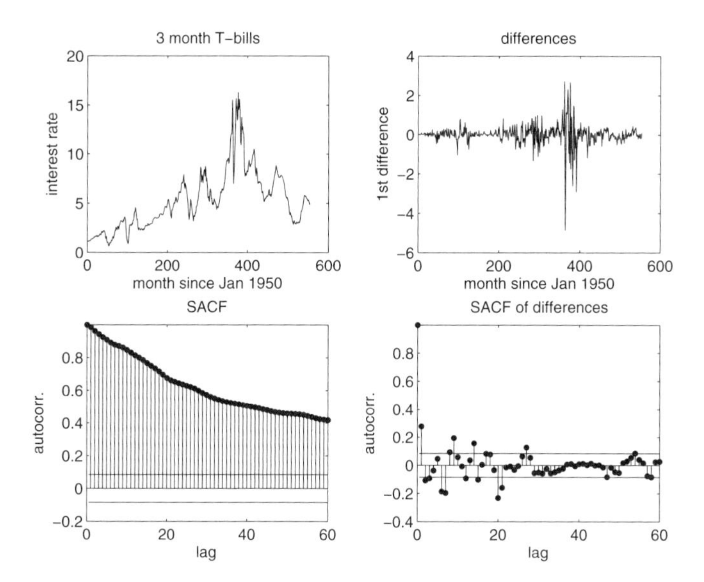

Fig. 4.7. *Time series plot of* 3 *month Treasury bill rates, plot of first differences, and sample autocorrelation function of first differences. The data set contains monthly values of the* 3 *month rates from January* 1950 *until March* 1996.

We fit a large model, specifically an AR(20) model, with SAS's PROC ARIMA to see which coefficients are significant. Here is the SAS program. The statement `identify var=z(1)` specifies that the model should be fit to the first differences of the variable z; z is the interest rate.

```
(data step omitted)
proc arima ;
identify var=z(1) ;
estimate p=20;
run ;
```

Here are selected portions of the SAS output.

Conditional Least Squares Estimation

Parameter	Estimate	Standard Error	t Value	Approx Pr > |t|	Lag
MU	0.0068268	0.02056	0.33	0.7400	0
AR1,1	0.36984	0.04316	8.57	<.0001	1
AR1,2	-0.18402	0.04584	-4.01	<.0001	2
AR1,3	0.01914	0.04647	0.41	0.6807	3
AR1,4	-0.07064	0.04649	-1.52	0.1293	4
AR1,5	0.11528	0.04616	2.50	0.0128	5
AR1,6	-0.21882	0.04576	-4.78	<.0001	6
AR1,7	-0.05070	0.04589	-1.10	0.2698	7

```
AR1,8          0.09568      0.04589      2.08     0.0376      8
AR1,9          0.06849      0.04592      1.49     0.1364      9
AR1,10        -0.01316      0.04582     -0.29     0.7740     10
AR1,11         0.10030      0.04583      2.19     0.0290     11
AR1,12        -0.08882      0.04593     -1.93     0.0537     12
AR1,13        -0.05032      0.04591     -1.10     0.2736     13
AR1,14         0.20361      0.04591      4.44     <.0001     14
AR1,15        -0.18200      0.04577     -3.98     <.0001     15
AR1,16         0.14535      0.04620      3.15     0.0017     16
AR1,17      -0.0079428      0.04652     -0.17     0.8645     17
AR1,18         0.05260      0.04654      1.13     0.2589     18
AR1,19        -0.10162      0.04590     -2.21     0.0272     19
AR1,20        -0.08467      0.04326     -1.96     0.0508     20

                  Constant Estimate      0.006024
                  Variance Estimate      0.181664
                  Std Error Estimate     0.426221
                  AIC                    647.8759
                  SBC                    738.5363
                  Number of Residuals         554
              * AIC and SBC do not include log determinant.

                  Autocorrelation Check of Residuals

 To       Chi-         Pr >
 Lag    Square  DF  ChiSq  ------------Autocorrelations------------
  24      4.14   4 0.3871 -0.007 -0.006 -0.051 -0.037 -0.022 -0.003
  30      6.17  10 0.8007 -0.030  0.007  0.012  0.014  0.028 -0.037
  36     11.46  16 0.7801 -0.029 -0.028 -0.028  0.029 -0.069 -0.032
  42     12.17  22 0.9538 -0.006  0.019  0.002 -0.023 -0.009 -0.013
  48     15.74  28 0.9696  0.003 -0.013 -0.007  0.007 -0.050  0.056
```

We see that some of the AR coefficients at high lags, e.g., ϕ_{14}, ϕ_{15}, ϕ_{16}, and ϕ_{19} are significant. This is a sign of rather complex dependencies. The residuals pass the white noise test.

4.10 Forecasting

ARIMA models are often used to forecast future values of a time series. Consider forecasting using an AR(1) process. Suppose that we have data $Y_1, \ldots, Y_n$ and estimates $\widehat{\mu}$ and $\widehat{\phi}$. We know that

$$Y_{n+1} = \mu + \phi(Y_n - \mu) + \epsilon_{n+1}. \tag{4.28}$$

Since ϵ_{n+1} is independent of the past and present, by Result 2.15.1 the best predictor of ϵ_{n+1} is its expected value, which is 0. On the other hand, we know or have estimates of all the other quantities in (4.28). Therefore we predict Y_{n+1} by

$$\widehat{Y}_{n+1} = \widehat{\mu} + \widehat{\phi}(Y_n - \widehat{\mu}).$$

By the same reasoning we forecast Y_{t+2} by

$$\widehat{Y}_{n+2} = \widehat{\mu} + \widehat{\phi}(\widehat{Y}_{n+1} - \widehat{\mu}) = \widehat{\mu} + \widehat{\phi}\{\widehat{\phi}(Y_n - \widehat{\mu})\}, \tag{4.29}$$

and so forth. Notice that in (4.29) we do not use Y_{n+1} which is unknown at time n but rather the forecast $\widehat{Y}_{n+1}$. Continuing in this way we find the general formula for the k-step ahead forecast:

$$\widehat{Y}_{n+k} = \widehat{\mu} + \widehat{\phi}^k (Y_n - \widehat{\mu}). \tag{4.30}$$

If $\widehat{\phi} < 1$ as is true for a stationary series, then as k increases the forecasts will converge exponentially fast to $\widehat{\mu}$.

Formula (4.30) is valid only for AR(1) processes, but forecasting other AR(p) processes is similar. For example, for an AR(2) process

$$\widehat{Y}_{n+1} = \widehat{\mu} + \widehat{\phi}_1 (Y_n - \widehat{\mu}) + \widehat{\phi}_2 (Y_{n-1} - \widehat{\mu})$$

and

$$\widehat{Y}_{n+2} = \widehat{\mu} + \widehat{\phi}_1 (\widehat{Y}_{n+1} - \widehat{\mu}) + \widehat{\phi}_2 (Y_n - \widehat{\mu}),$$

and so on.

Forecasting ARMA and ARIMA processes is similar to forecasting AR processes. For example, consider the MA(1) process, $Y_t - \mu = \epsilon_t - \theta \epsilon_{t-1}$. Then the next observation will be

$$Y_{n+1} = \mu + \epsilon_{n+1} - \theta \epsilon_n. \tag{4.31}$$

In the right-hand side of (4.31) we replace μ and θ by estimates and ϵ_n by the residual $\widehat{\epsilon}_n$. Also, since ϵ_{n+1} is independent of the observed data, it is replaced by 0. Then the forecast is

$$\widehat{Y}_{n+1} = \widehat{\mu} - \widehat{\theta}\,\widehat{\epsilon}_n.$$

The two-step ahead forecast of $Y_{n+2} = \mu + \epsilon_{n+2} - \theta \epsilon_{n+1}$ is simply $\widehat{Y}_{n+2} = \widehat{\mu}$, since ϵ_{n+1} and ϵ_{n+2} are independent of the observed data. Similarly, $\widehat{Y}_{n+k} = \widehat{\mu}$ for all $k > 2$.

To forecast the ARMA(1,1) process

$$Y_t - \mu = \phi(Y_{t-1} - \mu) + \epsilon_t - \theta \epsilon_{t-1}$$

we use

$$\widehat{Y}_{n+1} = \widehat{\mu} + \widehat{\phi}(Y_n - \widehat{\mu}) - \widehat{\theta}\,\widehat{\epsilon}_n$$

as the one-step ahead forecast and

$$\widehat{Y}_{n+k} = \widehat{\mu} + \widehat{\phi}(\widehat{Y}_{n+k-1} - \widehat{\mu}),\ k \geq 2$$

for forecasting two or more steps ahead.

As a final example, suppose that Y_t is ARIMA(1,1,0), so that ΔY_t is AR(1). To forecast Y_{n+k}, $k \geq 1$, one first fits an AR(1) model to the ΔY_t process and forecasts ΔY_{n+k}, $k \geq 1$. Let the forecasts be denoted by $\widehat{\Delta Y}_{n+k}$. Then since

$$Y_{n+1} = Y_n + \Delta Y_{n+1}$$

the forecast of Y_{n+1} is

$$\widehat{Y}_{n+1} = Y_n + \widehat{\Delta Y}_{n+1},$$

and similarly

$$\widehat{Y}_{n+2} = \widehat{Y}_{n+1} + \widehat{\Delta Y}_{n+2} = Y_n + \widehat{\Delta Y}_{n+1} + \widehat{\Delta Y}_{n+2},$$

and so on.

4.10.1 Forecasting GE daily log returns and log prices

We have learned that fitting an ARMA(1,0) (= ARIMA(1,0,0)) model to log returns is equivalent to fitting an ARIMA(1,1,0) model to the log prices. Here we fit both models to the GE daily price data. Figure 4.8 shows the forecasts of the log returns up to 24 days ahead. Both the forecasts and 95% confidence limits on the forecasts are shown.

Next we fit an ARIMA(1,1,0) model to the log prices. Although this model is equivalent to the last model, it generates forecasts of the log prices, not the log returns since SAS forecasts the *input* series. The forecasts are given in Figure 4.9. Notice that the forecasts predict that the price of GE will stay constant, but the confidence limits on the forecasts get wider as we forecast further ahead. This is exactly the type of behavior we would expect from a random walk [ARIMA(0,1,0)] model. The ARIMA(1,1,0) model for the log prices isn't quite a random walk model, but it is similar to a random walk model with zero drift ($\mu = 0$) since $\widehat{\phi}$ is close to 0 and $\widehat{\mu}$ is *extremely* close to 0.

The forecast limits suggest that accurately forecasting GE stock prices far into the future is pretty hopeless. For practical purposes the log prices behave as a random walk so that the prices behave as a geometric random walk.

Figures 4.8 and 4.9 show that forecasts of a stationary process behave very differently from forecasts of a nonstationary process. The forecasts of the stationary AR(1) process in Figure 4.8 converge to the mean of the observed series and the distance between the confidence limits converges to a constant muliple of $\widehat{\sigma}$. In contrast, the forecasts of the nonstationary ARIMA(1,1,0) process in Figure 4.9 are close to the average of the last few observations rather than equaling the average of all the observations. Also, for the nonstationary process the distance between the confidence limits increases as one forecasts farther into the future.

Example 4.1. Here is a SAS program to fit an ARIMA(1,1,0) model to the GE log prices and to plot the forecasts and the lower and upper 95% confidence limits on the forecasts. The program will produce a plot similar to that of Figure 4.9. The `forecast` statement in PROC ARIMA creates an output data set containing one-step ahead forecasts[7] for all observations and its specification `lead=20` also creates forecasts 1 to 20 days after the last observation. The variables `l95` and `u95` in the output data set are the lower and upper 95% confidence limits.

PROC GPLOT is used to plot the forecasts, lower and upper confidence limits, and the log prices. Note the use of `overlay` to put all four plots on the axes. There are three symbol statements; `symbol1` is used for the first plot, that of the forecasts, `symbol2` is used for the second plot, that of the lower and upper confidence limits, and `symbol3` for the plot of the log prices. In

[7] The one-step ahead forecast of Y_t is the best prediction of Y_t based on $Y_{t-1}, Y_{t-2}, \ldots$.

each symbol statement, `i` (meaning `interpolate`) specifies how the points will be connected, `v` (meaning `value`) is the plotting symbol, `l` is the line type, and `h` is the height. It is important to specify a different color for each plot, even if the graph will be printed in black-and-white, because SAS will cycle through the symbols only if each has a different color.

The specification `haxis=220 to 280 by 10` limits the range of the horizontal axis so that the first 219 observations are not plotted. This zooms in on the forecasts after the last observation. The statement `href=253` puts a vertical line through the last observation (the 253rd).

```
options linesize = 72 ;
data ge ;
infile 'C:\book\SAS\ge.dat' ;
input close ;
D_p = dif(close);
logP = log(close) ;
logR = dif(logP) ;
time=_N_ ;
run ;
title 'GE - Daily prices, Dec 17, 1999 to Dec 15, 2000' ;
proc arima ;
identify var=logP(1) ;
estimate p=1 ;
forecast id=time  lead=20 out=geout ;
run ;
proc gplot data=geout ;
symbol1 i=spline v=star l=1 h=1 color=red;
symbol2 i=spline v=none l=5 h=1 color=green;
symbol3 i=none v=circle h=1 color=blue;
plot  forecast*time=1 (l95 u95)*time=2 logP*time=3/overlay haxis=220 to 280 by 10 href=253;
run ;
quit ;
```

To obtain a figure like Figure 4.8 the `identify` statement in PROC ARIMA would be changed to

```
identify var=logR ;
```

so that the log returns rather than the log prices would be forecast and an ARIMA$(1,0,0)$ rather than an ARIMA$(1,1,0)$ model would be fit.

4.11 Summary

A time series is a sequence of observations taken in time. A process is *weakly stationary* if its mean, variance, and covariance are unchanged by time shifts. Thus, $Y_1, Y_2, \ldots$ is a weakly stationary process if $E(Y_i) = \mu$ (a constant) for all i, $\mathrm{Var}(Y_i) = \sigma^2$ (a constant) for all i, and $\mathrm{Corr}(Y_i, Y_j) = \rho(|i-j|)$ for all i and j for some function ρ.

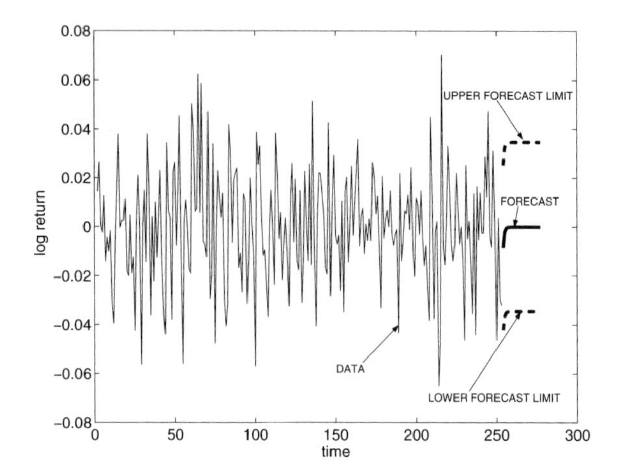

Fig. 4.8. *Time series plot of the daily GE log returns with forecasts from an AR(1) model.*

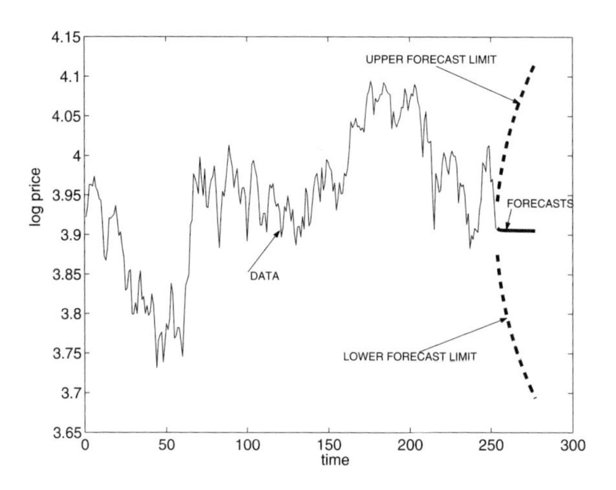

Fig. 4.9. *Time series plot of the daily GE log prices with forecasts from an ARIMA(1, 1, 0) model.*

The covariance between Y_t and Y_{t+h} is denoted by $\gamma(h)$. $\gamma(\cdot)$ is called the autocovariance function. Note that $\gamma(h) = \sigma^2\rho(h)$ and that $\gamma(0) = \sigma^2$ since $\rho(0) = 1$. Therefore $\rho(h) = \gamma(h)/\sigma^2$.

$Y_1, Y_2, \dots$ is WhiteNoise(μ, σ^2) if $E(Y_i) = \mu$ for all i, $\mathrm{Var}(Y_i) = \sigma^2$ (a constant) for all i, $\mathrm{Corr}(Y_i, Y_j) = 0$ for all $i \neq j$.

AR(1) is the simplest AR process. If $\epsilon_1, \epsilon_2, \dots$ are WhiteNoise($0,\sigma_\epsilon^2$), then $Y_1, Y_2, \dots$ is an AR(1) process if

$$Y_t - \mu = \phi(Y_{t-1} - \mu) + \epsilon_t \tag{4.32}$$

for all t. If $|\phi| < 1$, then $Y_1, \ldots$ is a weakly stationary process with mean μ and

$$\rho(h) = \text{Corr}(Y_t, Y_{t+h}) = \phi^{|h|} \quad \forall t.$$

The least squares estimators of ϕ and μ minimize

$$\sum_{t=2}^{n} \Big[\{Y_t - \mu\} - \{\phi(Y_{t-1} - \mu)\} \Big]^2.$$

The residuals are

$$\widehat{\epsilon}_t = Y_t - \widehat{\mu} - \widehat{\phi}(Y_{t-1} - \widehat{\mu}).$$

Autocorrelation in the residuals is evidence against AR(1) assumption. The process Y_t is AR(p) if

$$(Y_t - \mu) = \phi_1(Y_{t-1} - \mu) + \phi_2(Y_{t-2} - \mu) + \cdots + \phi_p(Y_{t-p} - \mu) + \epsilon_t.$$

The backwards operator B is defined by

$$B\,Y_t = Y_{t-1}.$$

An ARMA(p, q) process satisfies the equation

$$(1 - \phi\, B - \cdots - \phi_p\, B^p)(Y_t - \mu) = (1 - \theta_1\, B - \cdots - \theta_q\, B^q)\epsilon_t. \tag{4.33}$$

The differencing operator is $\Delta = 1 - B$ so that

$$\Delta Y_t = Y_t - B\,Y_t = Y_t - Y_{t-1}.$$

A time series Y_t is said to be ARIMA(p, d, q) if $\Delta^d Y_t$ is ARMA(p, q).

AIC and SBC are model selection criteria based on the log-likelihood. The "best" model by either criterion is the model that minimizes that criterion. AIC and SBC try to trade off good fit to the data measured by L and the desire to use few parameters. SBC penalizes the number of coefficient, $p + q$, more than AIC so that SBC tends to select fewer coefficients than AIC does.

ARIMA models can be used to forecast future values of a time series. Consider forecasting using an AR(1) process. We predict Y_{n+1} by

$$\widehat{Y}_{n+1} := \widehat{\mu} + \widehat{\phi}(Y_n - \widehat{\mu})$$

and Y_{n+2} by

$$\widehat{Y}_{n+2} := \widehat{\mu} + \widehat{\phi}(\widehat{Y}_{n+1} - \widehat{\mu}) = \widehat{\phi}\{\widehat{\phi}(Y_n - \widehat{\mu})\},$$

and so forth. Forecasting ARMA and ARIMA processes is slightly more complicated and is discussed in time series textbooks. The forecasts can be generated automatically by statistical software such as SAS.

4.12 Bibliographic Notes

Box, Jenkins, and Reinsel (1994) did so much to popularize ARIMA models that these are often called "Box-Jenkins models." (Reinsel's name was not attached since he was not a coauthor of the first two editions.) Brockwell and Davis (1991) is particularly recommended for those with a strong mathematical preparation wishing to understand the theory of time series analysis. Enders (1995) and Tsay (2002) are time series textbooks concentrating on economic and financial applications. A recent book by Gourieroux and Jasiak (2001) has a chapter on the applications of univariate time series in financial econometrics and another recent text by Alexander (2001) has a chapter on time series models including an introduction to vector AR models for multivariate time series.

SAS Institute (1993) discusses the time series software available in SAS and gives much useful information about using PROC AUTOREG and PROC ARIMA, though much of this information is available from SAS's OnlineDoc that is generally installed with the SAS software.

4.13 References

Alexander, C. (2001) *Market Models: A Guide to Financial Data Analysis*, Wiley, Chichester.

Box, G. E. P., Jenkins, G. M., and Reinsel, G. C. (1994) *Times Series Analysis: Forecasting and Control, 3rd Ed.*, Prentice-Hall, Englewood Cliffs, NJ.

Brockwell, P. J. and Davis, R. A. (1991) *Time Series: Theory and Methods, 2nd Ed.*, Springer, NJ.

Enders, W. (1995) *Applied Econometric Time Series*, Wiley, New York.

Gourieroux, C. and Jasiak, J. (2001) *Financial Econometrics*, Princeton University Press, Princeton, NJ.

Pindyck, R. S. and Rubinfeld, D. L. (1998) *Econometric Models and Economic Forecasts, 4th Ed.*, Irwin McGraw-Hill, Boston.

SAS Institute (1993) *SAS/ETS User's Guide, Version 6, 2nd Ed.*, SAS Institute, Cary, NC.

Tsay, R. S. (2002) *Analysis of Financial Time Series*, Wiley, New York.

4.14 Problems

1. Consider the AR(1) model

$$Y_t = 5 - 0.7Y_{t-1} + \epsilon_t$$

and assume that $\sigma_\epsilon^2 = 2$.

(a) Is this process stationary? Why or why not?
(b) What is the mean of this process?
(c) What is the variance of this process?
(d) What is the covariance function of this process?

2. Suppose that $Y_1, Y_2, \ldots$ is an AR(1) process with $\mu = 1$, $\phi = .3$, and $\sigma_\epsilon^2 = 2$.
(a) What is the variance of Y_1?
(b) What is the covariance between Y_1 and Y_3?
(c) What is the variance of $(Y_1 + Y_3)/2$?
3. An AR(3) model has been fit to a time series. The estimates are $\hat{\mu} = 102$, $\hat{\phi}_1 = 0.5$, $\hat{\phi}_2 = 0.2$, and $\hat{\phi}_3 = 0.1$. The last four observations were $Y_{n-3} = 104$, $Y_{n-2} = 101$, $Y_{n-1} = 102$, and $Y_n = 99$. Forecast Y_{n+1} and Y_{n+2} using these data and estimates.
4. Explain why a random walk is an ARIMA(0,1,0) process.
5. Let Y_t be an MA(2) process,

$$Y_t = \mu + \epsilon_t - \theta_1 \epsilon_{t-1} - \theta_2 \epsilon_{t-2}.$$

Find formulas for the autocovariance and autocorrelation functions of Y_t.
6. Let Y_t be a stationary AR(2) process,

$$(Y_t - \mu) = \phi_1(Y_{t-1} - \mu) + \phi_2(Y_{t-2} - \mu) + \epsilon_t.$$

(a) Show that the ACF of Y_t satisfies the equation

$$\rho(k) = \phi_1 \rho(k-1) + \phi_2 \rho(k-2)$$

for all values of $k > 0$. (These are called the Yule-Walker equations.) [*Hint:* $\gamma(k) = \text{Cov}(Y_t, Y_{t-k}) = \text{Cov}\{\phi_1(Y_{t-1} - \mu) + \phi_2(Y_{t-2} - \mu) + \epsilon_t, Y_{t-k}\}$ and ϵ_t and Y_{t-k} are independent if $k > 0$.]
(b) Use part (a) to show that (ϕ_1, ϕ_2) solves the following system of equations

$$\begin{pmatrix} \rho(1) \\ \rho(2) \end{pmatrix} = \begin{pmatrix} 1 & \rho(1) \\ \rho(1) & 1 \end{pmatrix} \begin{pmatrix} \phi_1 \\ \phi_2 \end{pmatrix}.$$

(c) Suppose that $\rho(1) = 0.4$ and $\rho(2) = 0.2$. Find ϕ_1, ϕ_2, and $\rho(3)$.
7. Use (4.11) to verify equation (4.12).
8. Show that if w_t is defined by (4.22) then (4.23) is true.
9. The times series in the middle and bottom panels of Figure 4.6 are both nonstationary, but they clearly behave in different manners. The time series in the bottom panel exhibits "momentum" in the sense that once it starts moving upwards or downwards it often moves consistently in that direction for a number of steps. In contrast, the series in the middle panel does not have this type of momentum and a step in one direction is quite likely to be followed by a step in the opposite direction.
(a) Explain why the bottom panel shows momentum but not the middle panel.

(b) Do you think the time series in the bottom panel would be a good model for the price of a stock? Why or why not?

5

Portfolio Theory

5.1 Trading Off Expected Return and Risk

How should we invest our wealth? Portfolio theory is based upon two principles:[1]

- We want to maximize the expected return; and
- We want to minimize the risk which we define in this chapter to be the standard deviation of the return, though we are ultimately concerned with the probabilities of large losses.

These goals are somewhat at odds because riskier assets generally have a higher expected return, since investors demand a reward for bearing risk. The difference between the expected return of a risky asset and the risk-free rate of return is called the **risk premium**. Without risk premiums, few investors would invest in risky assets.

Nonetheless, there are optimal compromises between expected return and risk. In this chapter we show how to maximize expected return subject to an upper bound on the risk, or to minimize the risk subject to a lower bound on the expected return. One key concept that we discuss is reduction of risk by diversifying the portfolio of assets held.

5.2 One Risky Asset and One Risk-Free Asset

We start with a simple example where we have one risky asset, which could be a portfolio, e.g., a mutual fund. Assume that the expected return is 0.15 and the standard deviation of the return is 0.25. Assume that there is also one **risk-free asset**, e.g., a 30-day T-bill and the return is 0.06. The standard deviation of the return is 0 by definition of "risk-free."

[1] Another approach would be to maximize expected utility; see Section 5.8.

We are faced with the problem of constructing an investment portfolio that we will hold for one time period which is called the **holding period** and which could be a day, a month, a quarter, a year, 10 years, and so forth. At the end of the holding period we might want to readjust the portfolio, so for now we are only looking at returns over one time period. Suppose that a fraction w of our wealth is invested in the risky asset and the remaining fraction $1 - w$ is invested in the risk-free asset. Then the expected return is

$$E(R) = w(0.15) + (1 - w)(0.06) = .06 + .09w, \tag{5.1}$$

the variance of the return is

$$\sigma_R^2 = w^2\,(.25)^2 + (1 - w)^2\,(0)^2 = w^2(0.25)^2,$$

and the standard deviation of the return is

$$\sigma_R = 0.25\,w. \tag{5.2}$$

To decide what proportion w of one's wealth to invest in the risky asset one either chooses the expected return $E(R)$ one wants or the amount of risk σ_R with which one is willing to live. Once either $E(R)$ or σ_R is chosen, w can be determined.

Example 5.1. Suppose you want an expected return of 0.10? What should w be?

Answer: By (5.1) w solves $0.1 = 0.06 + 0.09w$ so $w = 4/9$.

Example 5.2. Suppose you want $\sigma_R = 0.05$. What should w be?

Answer: By (5.2) $.05 = .25w$ so that $w = 0.2$.

Although σ is a measure of risk, a more direct measure of risk is actual monetary loss. In the next example, w is chosen to control the maximum size of the loss. This is a more realistic example of how a portfolio would be chosen in practice.

Example 5.3. Suppose that a firm is planning to invest \$1,000,000 and has capital reserves that could cover a loss of \$150,000 but no more. Therefore, the firm would like to be certain that if there is a loss then it is no more that 15%, that is, that R is greater than -0.15. Suppose that R is normally distributed. Then the only way to guarantee that R is greater than -0.15 with probability equal to 1 is to invest entirely in the risk-free asset. The firm might instead be more modest and require only that $P(R < -0.15)$ be small, for example, 0.01. Therefore, by (2.15) the firm should find the value of w such that

$$P(R < -0.15) = \Phi\left(\frac{-0.15 - (0.06 + 0.09w)}{.25w}\right) = 0.01.$$

The solution is

$$w = \frac{-0.21}{0.25\Phi^{-1}(0.01) + 0.9} = 0.4264.$$

In Chapter 11 we show that \$150,000 is called the value-at-risk and 0.01 is called the confidence coefficient. What we have done in this example is to find the portfolio that maximizes the expected return subject to having a VaR of \$150,000 with 0.01 confidence.

More generally, if the expected returns on the risky and risk-free assets are μ_1 and μ_f and if the standard deviation of the risky asset is σ_1, then the expected return on the portfolio is $w\mu_1 + (1-w)\mu_f$ while the standard deviation of the portfolio's return is $w\,\sigma_1$.

This model is simple but not as useless as it might seem at first. As discussed later in this course, finding an optimal portfolio can be achieved in two steps:

1. finding the "optimal" portfolio of risky assets, called the "tangency portfolio," and
2. finding the appropriate mix of the risk-free asset and the tangency portfolio.

So we now know how to do the second step, since that is exactly the type of problem we have solved in this section. What we need to learn is how to mix optimally a number of risky assets. We do this in the next sections.

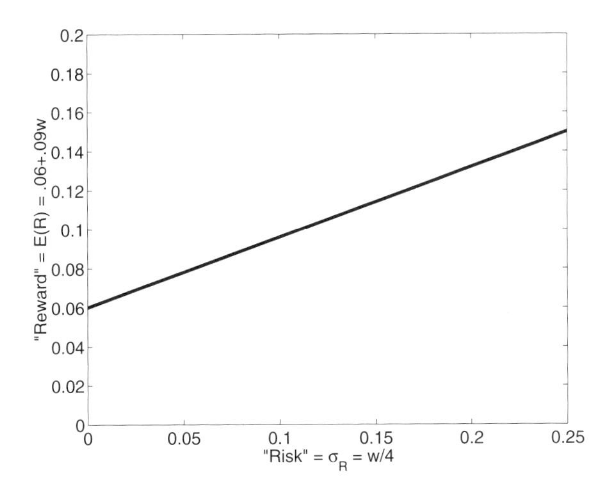

Fig. 5.1. *Expected return for a portfolio with allocation w to the risky asset with expected return* 0.15 *and allocation $1-w$ to the risk-free return with return* 0.06.

5.2.1 Estimating $E(R)$ and σ_R

The value of the risk-free rate, μ_f, will be known since Treasury bill rates are published in most newspapers and websites providing financial information.

What should we use as the values of $E(R)$ and σ_R? If returns on the asset are assumed to be stationary, then we can take a time series of past returns and use the sample mean and standard deviation. Whether the stationarity assumption is realistic is always debatable. If we think that $E(R)$ and σ_R will be different from the past, we could subjectively adjust these estimates upward or downward according to our opinions, but we must live with the consequences if our opinions prove to be incorrect.

Another question is how long a time series to use, that is, how far back in time one should gather data. A long series, say 10 or 20 years, will give much less variable estimates. However, if the series is not stationary but rather has slowly drifting parameters, then a shorter series (maybe 1 or 2 years) will be more representative of the future. Almost every time series of returns is nearly stationary over short enough time periods.

5.3 Two Risky Assets

5.3.1 Risk versus expected return

The mathematics of mixing risky assets is most easily understood when there are only two risky assets. This is where we start.

Suppose the two risky assets have returns R_1 and R_2 and that we mix them in proportions w and $1-w$, respectively. The return is $R = wR_1 + (1-w)R_2$.[2] The expected return on the portfolio is $E(R) = w\mu_1 + (1-w)\mu_2$. Let ρ_{12} be the correlation between the returns on the two risky assets. The variance of the return on the portfolio is

$$\sigma_R^2 = w^2\sigma_1^2 + (1-w)^2\sigma_2^2 + 2w(1-w)\rho_{12}\sigma_1\sigma_2. \tag{5.3}$$

Note that $\sigma_{R_1,R_2} = \rho_{12}\sigma_1\sigma_2$.

Example 5.4. If $\mu_1 = .14$, $\mu_2 = .08$, $\sigma_1 = .2$, $\sigma_2 = .15$, and $\rho_{12} = 0$, then

$$E(R) = 0.08 + 0.06w.$$

Also, because $\rho_{12} = 0$ in this example

$$\sigma_R^2 = (.2)^2\, w^2 + (.15)^2\, (1-w)^2.$$

[2] See Problem 4.

Using differential calculus, one can easily show that the portfolio with the minimum risk is $w = 0.045/0.125 = 0.36$. For this portfolio $E(R) = 0.08 + (0.06)(0.36) = 0.1016$ and $\sigma_R = \sqrt{(0.2)^2(0.36)^2 + (0.15)^2(0.64)^2} = 0.12$.

Here are values of $E(R)$ and σ_R for some other values of w:

w	$E(R)$	σ_R
0	0.080	0.150
1/4	0.095	0.123
1/2	0.110	0.125
3/4	0.125	0.155
1	0.140	0.200

The somewhat parabolic curve[3] in Figure 5.2 is the locus of values of $(\sigma_R, E(R))$ when $0 \le w \le 1$. The leftmost point on this locus achieves the minimum value of the risk and is called the minimum variance portfolio. The points on this locus that have an expected return at least as large as the minimum variance portfolio are called the **efficient frontier**. Portfolios on the efficient frontier are called **efficient portfolios** or, more precisely, **mean-variance efficient portfolios.**[4] The points labeled R_1 and R_2 correspond to $w = 1$ and $w = 0$, respectively. The other features of this figure are explained in Section 5.4.

5.3.2 Estimating means, standard deviations, and covariances

Estimates of μ_1 and σ_1 can be obtained from a univariate times series of past returns on the first risky asset. Denote this time series by $R_{1,1}, \ldots, R_{1,n}$ where the first subscript indicates the asset and the second subscript is for time. Let $\overline{R}_1$ and s_{R_1} be the sample mean and standard deviation of this series. Similarly, μ_2 and σ_2 can be estimated from a time series of past returns on the second risky asset denoted by $R_{2,1}, \ldots, R_{2,n}$. The covariance σ_{12} can be estimated by sample covariance

$$\widehat{\sigma}_{12} = n^{-1} \sum_{t=1}^{n} (R_{1,t} - \overline{R}_1)(R_{2,t} - \overline{R}_2).$$

The correlation ρ_{12} can be estimated by the sample correlation

$$\widehat{\rho}_{12} = \frac{\widehat{\sigma}_{12}}{s_{R_1} s_{R_2}}.$$

The estimator $\widehat{\rho}_{12}$, sometimes denoted by r_{12}, is called the sample cross-correlation coefficient between R_1 and R_2 at lag 0, since we are correlating

[3] In fact, the curve would be parabolic if σ_R^2 were plotted on the x-axis instead of σ_R. Why?

[4] When a risk-free asset is available then the efficient portfolios are no longer those on the efficient frontier but rather are characterized by Result 5.4.1 below.

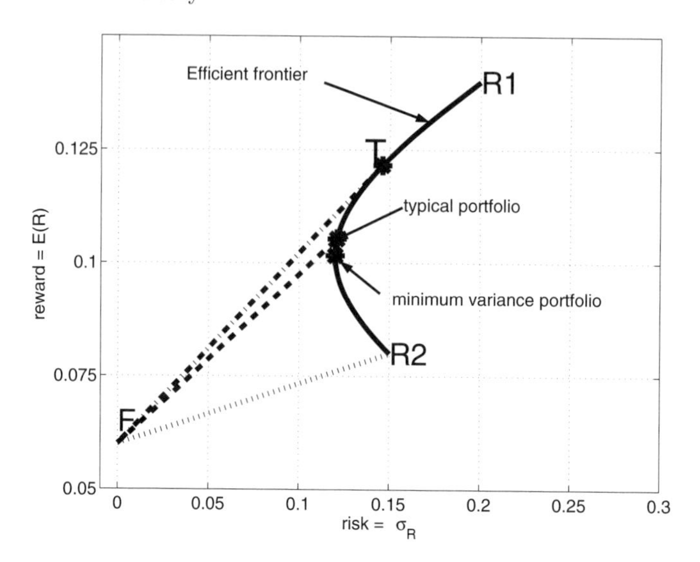

Fig. 5.2. *Expected return versus risk for Example 5.4. F = risk-free asset. T = tangency portfolio.* R_1 *is the first risky asset.* R_2 *is the second risky asset.*

the return on the first risky asset with the return on the second during the same time periods. Cross-correlations at other lags can be defined but are not needed here. They play an important role in the analysis of multivariate time series, a topic beyond the scope of this book. Sample correlations and covariances can be computed using SAS's PROC CORR.

5.4 Combining Two Risky Assets with a Risk-Free Asset

Our ultimate goal is to find optimal portfolios combining many risky assets with a risk-free asset. However, many of the concepts needed for this task can be mostly understood at first when there are only two risky assets, so we study that subject next.

5.4.1 Tangency portfolio with two risky assets

As mentioned in Section 5.3.1 each point on the efficient frontier in Figure 5.2 is $(\sigma_R, E(R))$ for some value of w between 0 and 1. If we fix w, then we have a fixed portfolio of the two risky assets. Now let us mix that portfolio of risky assets with the risk-free asset. The point F in Figure 5.2 gives $(\sigma_R, E(R))$ for the risk-free asset; of course, $\sigma_R = 0$ at F. The possible values of $(\sigma_R, E(R))$ for a portfolio consisting of the fixed portfolio of two risky assets and the risk-free asset is a line connecting the point F with a point on the efficient frontier, e.g., the dashed line. The dotted line connecting F with R_2 mixes the risk-free asset with the second risky asset.

Notice that the dashed and dotted line connecting F with the point labeled T lies above the dashed line connecting F and the typical portfolio. This means that for any value of σ_R, the dashed and dotted line gives a higher expected return than the dashed line. The slope of each line is called its **Sharpe's ratio**, named after William Sharpe whom we have met before in Section 3.4 and meet again in Chapter 7. Sharpe's ratio can be thought of as a "reward-to-risk" ratio. It is the ratio of the reward quantified by the "excess expected return" to the risk as measured by the standard deviation.

A line with a larger slope gives a higher expected return for a given level of risk, so the larger the Sharpe's ratio the better regardless of what level of risk one is willing to accept. The point T on the parabola represents the portfolio with the highest Sharpe's ratio. It is the optimal portfolio for the purpose of mixing with the risk-free asset. This portfolio is called the **tangency portfolio** since its line is tangent to the efficient frontier.

Result 5.4.1 *The optimal or **efficient** portfolios mix the tangency portfolio of two risky assets with the risk-free asset. Each efficient portfolio has two properties:*

- *It has a higher expected return than any other portfolio with the same (or smaller) risk, and*
- *It has a smaller risk than any other portfolio with the same (or smaller) expected return.*

Thus we can only improve (reduce) the risk of an efficient portfolio by accepting a worse (smaller) expected return, and we can only improve (increase) the expected return of an efficient portfolio by accepting worse (higher) risk.

Note that all efficient portfolios use the same mix of the two risky assets, namely, the tangency portfolio. Only the proportion allocated to the tangency portfolio and the proportion allocated to the risk-free asset vary.

Given the importance of the tangency portfolio, you may be wondering "how do we find it?" Again let μ_1, μ_2, and μ_f be the expected returns on the two risky assets and the return on the risk-free asset. Let σ_1 and σ_2 be the standard deviations of the returns on the two risky assets and let ρ_{12} be the correlation between the returns on the risky assets.

Define $V_1 = \mu_1 - \mu_f$ and $V_2 = \mu_2 - \mu_f$. V_1 and V_2 are called the **excess returns**, that is, the expected returns in excess of the risk-free rate. Then the tangency portfolio uses weight

$$w_T = \frac{V_1\sigma_2^2 - V_2\rho_{12}\sigma_1\sigma_2}{V_1\sigma_2^2 + V_2\sigma_1^2 - (V_1 + V_2)\rho_{12}\sigma_1\sigma_2} \tag{5.4}$$

for the the first risky asset and weight $(1 - w_T)$ for the second. Formula (5.4) is derived in Section 5.5.5.

Let R_T, $E(R_T)$, and σ_T be the return, expected return, and standard deviation of the return on the tangency portfolio. Then $E(R_T)$ and σ_T can be found by first finding w_T using (5.4) and then using the formulas

$$E(R_T) = w_T\mu_1 + (1 - w_T)\mu_2$$

and

$$\sigma_T = \sqrt{w_T^2\sigma_1^2 + (1 - w_T)^2\sigma_2^2 + 2w_T(1 - w_T)\rho_{12}\sigma_1\sigma_2}\,.$$

Example 5.5. Suppose as before that $\mu_1 = .14$, $\mu_2 = .08$, $\sigma_1 = .2$, $\sigma_2 = .15$, and $\rho_{12} = 0$. Suppose as well that $\mu_f = .06$. Then $V_1 = 0.14 - 0.06 = 0.08$ and $V_2 = 0.08 - 0.06 = 0.02$. Plugging these values into formula (5.4) we get $w_T = 0.693$ and $1 - w_t = 0.307$. Therefore,

$$E(R_T) = (0.693)(0.14) + (0.307)(0.08) = 0.122,$$

and

$$\sigma_T = \sqrt{(0.693)^2(0.2)^2 + (0.307)^2(0.15)^2} = 0.146.$$

5.4.2 Combining the tangency portfolio with the risk-free asset

Let R be the return on the portfolio that allocates a fraction ω of the investment to the tangency portfolio and $1 - \omega$ to the risk-free asset. Then $R = \omega R_T + (1 - \omega)\mu_f = \mu_f + \omega(R_T - R_f)$ so that

$$E(R) = \mu_f + \omega\{E(R_T) - \mu_f\} \quad \text{and} \quad \sigma_R = \omega\sigma_T.$$

Example 5.6. (Continuation of example 5.4) What is the optimal investment with $\sigma_R = 0.05$?

Answer: The maximum expected return with $\sigma_R = 0.05$ mixes the tangency portfolio and the risk-free asset such that $\sigma_R = 0.05$. Since $\sigma_T = 0.146$, we have that $0.05 = \sigma_R = \omega\,\sigma_T = 0.146\,\omega$, so that $\omega = 0.05/0.146 = 0.343$ and $1 - \omega = 0.657$.

So 65.7% of the portfolio should be in the risk-free asset. 34.3% should be in the tangency portfolio. Thus (0.343)(69.3%) = 23.7% should be in the first risky asset and (0.343)(30.7%) = 10.5% should be in the second risky asset. The total is not quite 100% because of rounding. The allocation is summarized in Table 5.1.

Example 5.7. (Continuation of example 5.4) Now suppose that you want a 10% expected return. Compare

- The best portfolio of only risky assets, and
- The best portfolio of the risky assets and the risk-free asset.

Table 5.1. Optimal allocation to two risky assets and the risk-free asset to achieve $\sigma_R = 0.05$.

Asset	Allocation (%)
risk-free	65.7
risky 1	23.7
risky 2	10.5
Total	99.9

Answer: The best portfolio of only risky assets uses w solving $0.1 = w(0.14) + (1 - w)(0.08)$ which implies that $w = 1/3$. This is the *only* portfolio of risky assets with $E(R) = 0.1$, so by default it is best. Then

$$\sigma_R = \sqrt{w^2(0.2)^2 + (1-w)^2(0.15)^2} = \sqrt{(1/9)(0.2)^2 + 4/9(0.15)^2} = 0.120.$$

The best portfolio of the two risky assets and the risk-free asset can be found as follows. First, $0.1 = E(R) = \mu_f + \omega\{E(R_T) - \mu_f\} = 0.06 + 0.062\omega = 0.06 + 0.425\sigma_R$, since $\sigma_R = \omega\sigma_T$ or $\omega = \sigma_R/\sigma_T = \sigma_R/0.146$. This implies that $\sigma_R = 0.04/0.425 = 0.094$ and $\omega = 0.04/0.062 = 0.645$. So combining the risk-free asset with the two risky assets reduces σ_R from 0.120 to 0.094 while maintaining $E(R)$ at 0.1. The reduction in risk is $(0.120 - 0.094)/0.094 = 28\%$ which is substantial.

Example 5.8. (More on Example 5.4*)* What is the best we can do combining the risk-free asset with only one risky asset? Assume that we still want to have $E(R) = 0.1$. Also, assume that short selling and buying on margin are not permitted. Selling short means selling an asset one does not yet own and purchasing it later at delivery time; see Section 5.5.3. Buying on margin means borrowing money at the risk-free rate and using the money to purchase stock. Buying on margin or short selling allows some weights to exceed 1 or to be negative, though the sum of the weights is still constrained to equal 1.

Second risky asset with the risk-free: Since $\mu_f = 0.06 < 0.1$ and $\mu_2 = 0.08 < 0.1$, no portfolio with only the second risky asset and the risk-free asset will have an expected return of 0.1 unless weights greater than 1 are permitted. Another way to appreciate this fact is to solve $0.1 = \omega(0.08) + (1 - \omega)(0.06) = 0.06 + 0.02\omega$ to get $\omega = 2$ and $1 - \omega = -1$. However, ω and $1 - \omega$ must both be between 0 and 1 unless one is permitted to sell short or buy stock on margin.

If buying on margin is permitted, then one can borrow money equal to one's assets so that $\omega = -1$ and then invest one's entire assets plus the borrowed money in the second risky asset so that $1 - \omega = 2$. Then $\sigma_R = 2(0.15) = 0.3$.

First risky asset with the risk-free: $0.1 = \omega(0.14) + (1 - \omega)(0.06) = 0.06 + \omega(0.08)$ implies that $\omega = 0.04/0.08 = 1/2$. Then $\sigma_R = \omega(0.20) = 0.10$ which is greater than 0.094, the smallest risk with two risky assets and the risk-free asset such that $E(R) = 0.1$.

The minimum values of σ_R under various combinations of available assets are given in Table 5.2.

Table 5.2. *Minimum value of σ_R as a function of the available assets. In all cases, the expected return is 0.1. When only the risk-free asset and the second risky asset are available, then a return of 0.1 is achievable only if buying on margin is permitted.*

Available Assets	Minimum σ_R
1st risky, risk-free	0.1
2nd risky, risk-free	0.3
Both riskies	0.12
All three	0.094

5.4.3 Effect of ρ_{12}

Positive correlation between the two risky assets is bad because it increases risk. With positive correlation, the two assets tend to move together which increases the volatility of the portfolio. Conversely, negative correlation is good. If the assets are negatively correlated, a negative return of one tends to occur with a positive return of the other so the volatility of the portfolio decreases. Figure 5.3 shows the efficient frontier and tangency portfolio when $\mu_1 = 0.14$, $\mu_2 = 0.09$, $\sigma_1 = 0.2$, $\sigma_2 = 0.15$, and $\mu_f = 0.03$. The value of ρ_{12} is varied from 0.7 to -0.7. Notice that the Sharpe's ratio of the tangency portfolio returns increases as ρ_{12} decreases. This means that when ρ_{12} is small then efficient portfolios have less risk for a given expected return compared to when ρ_{12} is large.

5.5 Risk-Efficient Portfolios with N Risky Assets*

5.5.1 Efficient-portfolio mathematics

Efficient-portfolio mathematics generalizes our previous analysis with two risky assets to the more realistic case of many risky assets.

Assume that we have N risky assets and that the return on the ith risky asset is R_i and has expected value μ_i. Define

$$\boldsymbol{R} = \begin{pmatrix} R_1 \\ \vdots \\ R_N \end{pmatrix}$$

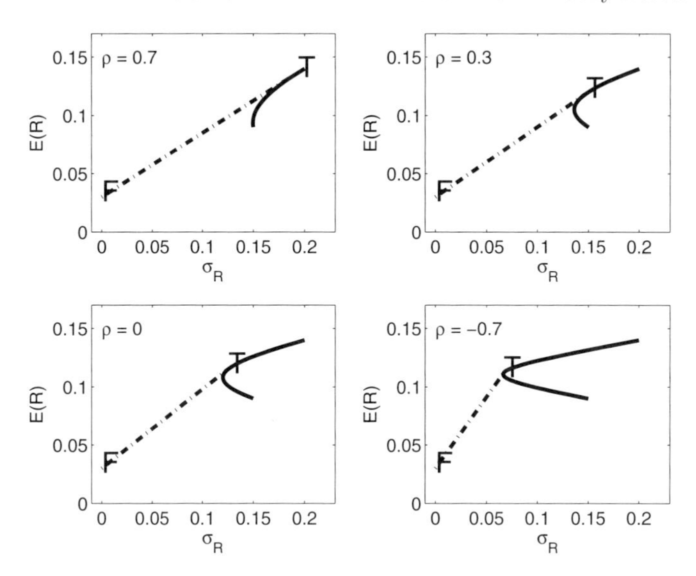

Fig. 5.3. *Efficient frontier and tangency portfolio when* $\mu_1 = 0.14$, $\mu_2 = 0.09$, $\sigma_1 = 0.2$, $\sigma_2 = 0.15$, *and* $\mu_f = 0.03$. *The value of* ρ_{12} *is varied from* 0.7 *to* −0.7.

to be the random vector of returns. Then

$$E(\boldsymbol{R}) = \boldsymbol{\mu} = \begin{pmatrix} \mu_1 \\ \vdots \\ \mu_N \end{pmatrix}.$$

Let Ω_{ij} be the covariance between R_i and R_j. Also, let $\sigma_i = \sqrt{\Omega_{ii}}$ be the standard deviation of R_i. Finally, let $\boldsymbol{\Omega}$ be the covariance matrix of $\boldsymbol{R}$, i.e.,

$$\boldsymbol{\Omega} = \text{COV}(\boldsymbol{R}),$$

so that the i, jth element of $\boldsymbol{\Omega}$ is $\Omega_{ij} = \text{Cov}(R_i, R_j)$.

Let

$$\boldsymbol{\omega} = \begin{pmatrix} \omega_1 \\ \vdots \\ \omega_N \end{pmatrix}$$

be a vector of portfolio weights and let

$$\mathbf{1} = \begin{pmatrix} 1 \\ \vdots \\ 1 \end{pmatrix}$$

be a column of N ones. We assume that[5] $\omega_1 + \cdots + \omega_N = \mathbf{1}^\mathsf{T}\boldsymbol{\omega} = 1$. The expected return on a portfolio with weights $\boldsymbol{\omega}$ is

[5] Recall that for any two vectors $\boldsymbol{a}$ and $\boldsymbol{b}$, their inner product is $\boldsymbol{a}^\mathsf{T}\boldsymbol{b} = \sum a_i b_i$. Therefore, $\boldsymbol{a}^\mathsf{T}\mathbf{1} = \sum a_i$.

$$\sum_{i=1}^{N} \omega_i \mu_i = \boldsymbol{\omega}^{\mathsf{T}} \boldsymbol{\mu}. \tag{5.5}$$

Suppose there is a target value, μ_P, of the expected return on the portfolio. When $N = 2$ the target, μ_P, is achieved by only one portfolio and its ω_1 value solves

$$\mu_P = \omega_1 \mu_1 + \omega_2 \mu_2 = \mu_2 + \omega_1(\mu_1 - \mu_2).$$

For $N \geq 3$, there will be an infinite number of portfolios achieving the target, μ_P. The one with the smallest variance is called the "efficient" portfolio. Our goal is to find the efficient portfolio.

By equation (2.53), the variance of the return on the portfolio with weights $\boldsymbol{\omega}$ is

$$\sum_{i=1}^{N} \sum_{j=1}^{N} \omega_i \, \omega_j \, \Omega_{i,j} = \boldsymbol{\omega}^{\mathsf{T}} \boldsymbol{\Omega} \boldsymbol{\omega}. \tag{5.6}$$

Thus, given a target μ_P, the efficient portfolio minimizes (5.6) subject to

$$\boldsymbol{\omega}^{\mathsf{T}} \boldsymbol{\mu} = \mu_P \tag{5.7}$$

and

$$\boldsymbol{\omega}^{\mathsf{T}} \mathbf{1} = 1. \tag{5.8}$$

We denote the weights of the efficient portfolio by $\boldsymbol{\omega}_{\mu_P}$. To find $\boldsymbol{\omega}_{\mu_P}$, form the Lagrangian[6]

$$L(\boldsymbol{\omega}, \delta_1, \delta_2) = \boldsymbol{\omega}^{\mathsf{T}} \boldsymbol{\Omega} \boldsymbol{\omega} + \delta_1(\mu_P - \boldsymbol{\omega}_{\mu_P}^{\mathsf{T}} \boldsymbol{\mu}) + \delta_2(1 - \boldsymbol{\omega}^{\mathsf{T}} \mathbf{1})$$

where δ_1 and δ_2 are Lagrange multipliers. Then we must solve

$$0 = \frac{\partial}{\partial \boldsymbol{\omega}} L(\boldsymbol{\omega}, \delta_1, \delta_2) = 2\boldsymbol{\Omega} \boldsymbol{\omega}_{\mu_P} - \delta_1 \boldsymbol{\mu} - \delta_2 \mathbf{1} \tag{5.9}$$

for $\boldsymbol{\omega}$ with δ_1 and δ_2. The solution will of course be a function of δ_1 and δ_2 but these will be determined from the constraints. Equation (5.9) uses the following definition and result.

Definition 5.9. *Here*

$$\frac{\partial}{\partial \boldsymbol{\omega}} L(\boldsymbol{\omega}, \delta_1, \delta_2) = \begin{pmatrix} \partial L(\boldsymbol{\omega}, \delta_1, \delta_2)/\partial \omega_1 \\ \vdots \\ \partial L(\boldsymbol{\omega}, \delta_1, \delta_2)/\partial \omega_N \end{pmatrix}$$

means the gradient of L with respect to $\boldsymbol{\omega}$ with the other variables in L held fixed.

---

[6] The method of Lagrange multipliers being used here is described in any textbook on multivariate calculus.

Result 5.5.1 *For an $n \times n$ matrix $\boldsymbol{A}$ and an n-dimensional vector $\boldsymbol{x}$,*

$$\frac{\partial}{\partial \boldsymbol{x}} \boldsymbol{x}^{\mathsf{T}} \boldsymbol{A} \boldsymbol{x} = (\boldsymbol{A} + \boldsymbol{A}^{\mathsf{T}})\boldsymbol{x}.$$

The solution to (5.9) is

$$\boldsymbol{\omega}_{\mu_P} = \frac{1}{2} \boldsymbol{\Omega}^{-1}(\delta_1 \boldsymbol{\mu} + \delta_2 \mathbf{1}) = \boldsymbol{\Omega}^{-1}(\lambda_1 \boldsymbol{\mu} + \lambda_2 \mathbf{1}), \tag{5.10}$$

where λ_1 and λ_2 are new Lagrange multipliers: $\lambda_1 = 1/2\, \delta_1$ and $\lambda_2 = 1/2\, \delta_2$. Thus, $\boldsymbol{\omega}_{\mu_P} = \lambda_1 \boldsymbol{\Omega}^{-1} \boldsymbol{\mu} + \lambda_2 \boldsymbol{\Omega}^{-1} \mathbf{1}$, where λ_1 and λ_2 are yet to be determined scalar quantities. We need to use the constraints (5.7) and (5.8) to find λ_1 and λ_2. Using (5.10), these constraints imply the equations

$$\mu_P = \boldsymbol{\mu}^{\mathsf{T}} \boldsymbol{\omega}_{\mu_P} = \lambda_1 \boldsymbol{\mu}^{\mathsf{T}} \boldsymbol{\Omega}^{-1} \boldsymbol{\mu} + \lambda_2 \boldsymbol{\mu}^{\mathsf{T}} \boldsymbol{\Omega}^{-1} \mathbf{1}, \tag{5.11}$$

and

$$1 = \mathbf{1}^{\mathsf{T}} \boldsymbol{\omega}_{\mu_P} = \lambda_1 \mathbf{1}^{\mathsf{T}} \boldsymbol{\Omega}^{-1} \boldsymbol{\mu} + \lambda_2 \mathbf{1}^{\mathsf{T}} \boldsymbol{\Omega}^{-1} \mathbf{1}. \tag{5.12}$$

These are equations in λ_1 and λ_2, since all other quantities in these equations are known. We introduce simpler notation for the coefficients:

$$A = \boldsymbol{\mu}^{\mathsf{T}} \boldsymbol{\Omega}^{-1} \mathbf{1} = \mathbf{1}^{\mathsf{T}} \boldsymbol{\Omega}^{-1} \boldsymbol{\mu}, \tag{5.13}$$
$$B = \boldsymbol{\mu}^{\mathsf{T}} \boldsymbol{\Omega}^{-1} \boldsymbol{\mu}, \tag{5.14}$$
$$C = \mathbf{1}^{\mathsf{T}} \boldsymbol{\Omega}^{-1} \mathbf{1}. \tag{5.15}$$

Then (5.11) and (5.12) can be rewritten as

$$\mu_P = B\lambda_1 + A\lambda_2 \tag{5.16}$$
$$1 = A\lambda_1 + C\lambda_2. \tag{5.17}$$

Let

$$D = BC - A^2 \tag{5.18}$$

be the determinant of this system of linear equations. The solution to (5.16) and (5.17) is

$$\lambda_1 = \frac{-A + C\,\mu_P}{D} \text{ and } \lambda_2 = \frac{B - A\,\mu_P}{D}.$$

It follows after some algebra that

$$\boldsymbol{\omega}_{\mu_P} = \boldsymbol{g} + \mu_P\, \boldsymbol{h}, \tag{5.19}$$

where

$$\boldsymbol{g} = \frac{B\,\boldsymbol{\Omega}^{-1}\mathbf{1} - A\,\boldsymbol{\Omega}^{-1}\boldsymbol{\mu}}{D} = \frac{B}{D}\boldsymbol{\Omega}^{-1}\mathbf{1} - \frac{A}{D}\boldsymbol{\Omega}^{-1}\boldsymbol{\mu}, \tag{5.20}$$

and

$$\boldsymbol{h} = \frac{C\boldsymbol{\Omega}^{-1}\boldsymbol{\mu} - A\boldsymbol{\Omega}^{-1}\mathbf{1}}{D} = \frac{C}{D}\boldsymbol{\Omega}^{-1}\boldsymbol{\mu} - \frac{A}{D}\boldsymbol{\Omega}^{-1}\mathbf{1}. \tag{5.21}$$

Notice that $\boldsymbol{g}$ and $\boldsymbol{h}$ are fixed vectors, since they depend on the fixed vector $\boldsymbol{\mu}$ and the fixed matrix $\boldsymbol{\Omega}$ but they do not depend on μ_P which will be varied. Also, the scalars A, C, and D are functions of $\boldsymbol{\mu}$ and $\boldsymbol{\Omega}$ so they are also fixed, that is, independent of μ_P. The target expected return, μ_P, can be varied over some range of values, e.g.,

$$\min_{i=1,\dots,N} \mu_i \le \mu_P \le \max_{i=1,\dots,N} \mu_i$$

or

$$\mu_{\min} \le \mu_P \le \max_{i=1,\dots,N} \mu_i, \tag{5.22}$$

where $\mu_{\min}$ is the expected return of minimum variance portfolio given in equation (5.25) in the next section.

As μ_P varies over the range (5.22), we get a locus $\boldsymbol{\omega}_{\mu_P}$ of efficient portfolios called the "efficient frontier." We can illustrate the efficient frontier by the following algorithm.

1. Vary μ_P along a grid. For each value of μ_P on this grid, compute σ_{μ_P} by:
   (a) computing $\boldsymbol{\omega}_{\mu_P} = \boldsymbol{g} + \mu_P \boldsymbol{h}$; and
   (b) then computing $\sigma_{\mu_P} = \sqrt{\boldsymbol{\omega}_{\mu_P}^{\mathsf{T}} \boldsymbol{\Omega} \boldsymbol{\omega}_{\mu_P}}$.
2. Plot the values (μ_P, σ_{μ_P}). The values (μ_P, σ_{μ_P}) with $\mu_P \ge \mu_{\min}$ are the efficient frontier. The other values of (μ_P, σ_{μ_P}) lie below the efficient frontier and are (very) inefficient portfolios.

This algorithm is implemented in the following MATLAB program. First the values of $\boldsymbol{\mu}$ and $\boldsymbol{\Omega}$ are input and $\mathbf{1}$ is created. In MATLAB, anything appearing after a percent sign is a comment. In this program, a variable name begins with a "b" if the same variable appears in a boldface font in the calculations above; e.g., "bmu" means "boldface mu," that is, $\boldsymbol{\mu}$, and "bone" means "boldface one," that is $\mathbf{1}$. The program begins as

```
bmu = [0.08;0.03;0.05] ;
bOmega = [ 0.3 0.02 0.01 ;
      0.02 0.15 0.03 ;
      0.01 0.03 0.18 ] ;
bone = ones(length(bmu),1) ;
```

Then $\boldsymbol{\Omega}$ is inverted and A, B, C, D, $\boldsymbol{g}$, and $\boldsymbol{h}$ are computed, so that the program continues as

```
ibOmega = inv(bOmega) ;
A = bone'*ibOmega*bmu ;
B = bmu'*ibOmega*bmu ;
C = bone'*ibOmega*bone ;
D = B*C - A^2 ;
bg = (B*ibOmega*bone - A*ibOmega*bmu)/D ;
bh = (C*ibOmega*bmu - A*ibOmega*bone)/D ;
```

Then the expected return and return standard deviation of the minimum variance portfolio, which are denoted by `mumin` and `sdmin`, are computed using equations (5.25) and (5.26) derived in Section 5.5.2.

```
gg = bg'*bOmega*bg ;
hh = bh'*bOmega*bh ;
gh = bg'*bOmega*bh ;
mumin = - gh/hh ;
sdmin = sqrt( gg * ( 1 - gh^2/(gg*hh)) ) ;
```

Next, a grid of 50 μ_P values is generated and a matrix for storing the corresponding 50 values of σ_{μ_P} is set up. Then ω_{μ_P} and σ_{μ_P} are computed for each value of μ_P on the grid.

```
muP = linspace(min(bmu),max(bmu),50) ;  %   muP grid
sigmaP = zeros(1,50) ;                  %   Storage
for i=1:50 ;
    omegaP = bg + muP(i)*bh ;
    sigmaP(i) = sqrt(omegaP'*bOmega*omegaP) ;
end ;
```

Then logical vectors, that is, vectors of 0s and 1s, are generated to indicate which of the 50 grid values are on the efficient frontier and which are not.

```
ind = (muP > mumin) ;   %   Indicates efficient frontier
ind2 = (muP < mumin) ;  %   Indicates locus below efficient frontier
```

Now the results are plotted. Notice that the efficient frontier, which has expected returns greater than the expected return of the minimum variance portfolio, is plotted with a solid line and points with an expected return less than that of the minimum variance return are on a locus plotted with a dashed curve.

```
%   Create plot - efficient frontier is shown as a solid curve
%         - the inefficient part of the locus is dashed
figure(1)
p1 = plot(sigmaP(ind),muP(ind),'-',sigmaP(ind2),muP(ind2),'--' , ...
    sdmin,mumin,'.') ;
%   Change line widths, marker sizes, and colors for better appearance
set(p1(1:2),'linewidth',4) ;
set(p1(1:2),'color','blue') ;
set(p1(3),'markersize',40) ;
set(p1(3),'color','red') ;

%   Label axes
fsize = 16 ;
xlabel('standard deviation of return','fontsize',fsize) ;
ylabel('expected return','fontsize',fsize) ;
set(gca,'xlim',[0, 0.5]) ;
set(gca,'ylim',[0, 0.08]) ;
grid ;
```

To use this program on other problems replace `bmu` and `bOmega` in the program by the vector of expected returns and covariance matrix of returns for the assets you wish to analyze. Figure 5.4 is the output from this program. The portfolio weights as functions of μ_P are plotted in Figure 5.5. The weights can be negative. Negative weights can be obtained by the technique of selling short which is described in Section 5.5.3.

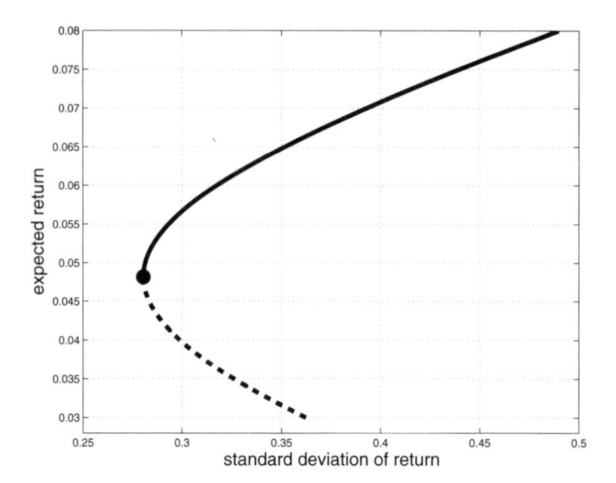

Fig. 5.4. *Efficient frontier (solid) plotted for $N = 3$ assets by the program in the text.*

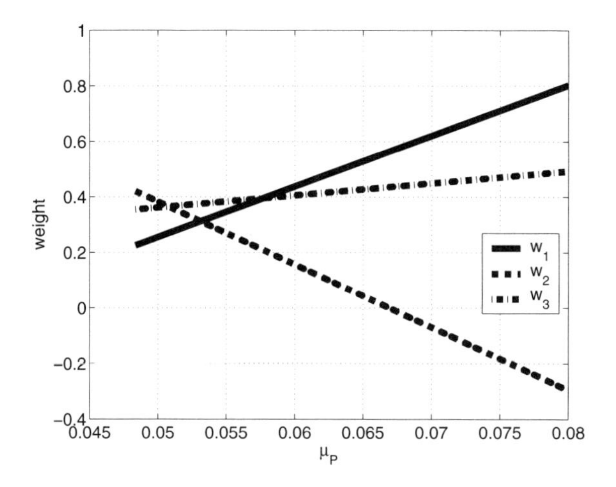

Fig. 5.5. *Weights for assets 1, 2, and 3 as functions of $\mu_p \geq \mu_{\min}$. Note that the weights for assets 1 and 2 can be negative, so that short selling would be required.*

5.5.2 The minimum variance portfolio

We just showed that the efficient portfolio with expected return equal to μ_P has weights

$$\boldsymbol{\omega}_{\mu_P} = \boldsymbol{g} + \boldsymbol{h}\,\mu_P. \tag{5.23}$$

The variance of the return on the portfolio $R_P = \boldsymbol{\omega}_{\mu_P}^{\mathsf{T}} \boldsymbol{R}$ is

$$\begin{aligned}\mathrm{Var}(R_P) &= (\boldsymbol{g} + \boldsymbol{h}\,\mu_P)^{\mathsf{T}} \boldsymbol{\Omega} (\boldsymbol{g} + \boldsymbol{h}\,\mu_P) \\ &= \boldsymbol{g}^{\mathsf{T}} \boldsymbol{\Omega} \boldsymbol{g} + 2\boldsymbol{g}^{\mathsf{T}} \boldsymbol{\Omega} \boldsymbol{h} \mu_P + \boldsymbol{h}^{\mathsf{T}} \boldsymbol{\Omega} \boldsymbol{h} \mu_P^2. \end{aligned} \tag{5.24}$$

To find the **minimum variance portfolio** we minimize this quantity over μ_P by solving

$$0 = \frac{d}{d\mu_p} \mathrm{Var}(R_P) = 2\boldsymbol{g}^{\mathsf{T}} \boldsymbol{\Omega} \boldsymbol{h} + 2\boldsymbol{h}^{\mathsf{T}} \boldsymbol{\Omega} \boldsymbol{h} \mu_P.$$

The solution is the expected return of the minimum variance portfolio given by

$$\mu_{\min} = -\frac{\boldsymbol{g}^{\mathsf{T}} \boldsymbol{\Omega} \boldsymbol{h}}{\boldsymbol{h}^{\mathsf{T}} \boldsymbol{\Omega} \boldsymbol{h}}. \tag{5.25}$$

Plugging $\mu_{\min}$ into (5.24), and calling the portfolio $R_{\min}$, we find that the smallest possible variance of a portfolio is

$$\mathrm{Var}(R_{\min}) = \boldsymbol{g}^{\mathsf{T}} \boldsymbol{\Omega} \boldsymbol{g} - \frac{(\boldsymbol{g}^{\mathsf{T}} \boldsymbol{\Omega} \boldsymbol{h})^2}{\boldsymbol{h}^{\mathsf{T}} \boldsymbol{\Omega} \boldsymbol{h}}. \tag{5.26}$$

If one wants to avoid short selling, then one must impose the additional constraints that $w_i \geq 0$ for $i = 1, \ldots, N$. Minimization of portfolio risk subject to $\boldsymbol{\omega}^{\mathsf{T}} \boldsymbol{\mu} = \mu_P$, $\boldsymbol{\omega}^{\mathsf{T}} \boldsymbol{1} = 1$, and these additional nonnegativity constraints is a quadratic programming problem. (This minimization problem cannot be solved by the method of Lagrange multipliers because of the inequality constraints.) Quadratic programming algorithms are not hard to find. For example, the program "quadprog" in MATLAB's Optimization Toolbox does quadratic programming.

Figure 5.6 was produced by the program given in Section 5.6 that uses "quadprog" in MATLAB. Figure 5.6 shows two efficient frontiers. The dashed curve is the nonconstrained efficient frontier and is the same as the efficient frontier in Figure 5.4. The second is the efficient frontier with weights constrained to be nonnegative using quadratic programming. The portfolio weights with the nonnegativity constraints are plotted as functions of μ_P in Figure 5.7.

Now suppose that we have a risk-free asset and we want to mix the risk-free asset with *some* efficient portfolio. One can see geometrically in Figure 5.8 that there is a tangency portfolio with the following remarkable property that leads to significant simplifications.

Result 5.5.2 *An optimal portfolio is* always *a mixture of the risk-free asset with the tangency portfolio.*

An old view of investing is that investors should select stocks that match their tolerance for risk. Investors with a low tolerance for risk should select stocks with low volatility, while aggressive investors should select risky stocks. Bernstein (1993) calls this view the "interior decorator fallacy" in analogy with a decorator who designs a client's interior according to the client's taste. The newer view derived from portfolio theory is that *all* investors should invest in the same mixture of low, moderate, and high risk stocks given by the tangency portfolio. An individual investor's risk tolerance should determine only that investor's allocation between the tangency portfolio and the risk-free asset.

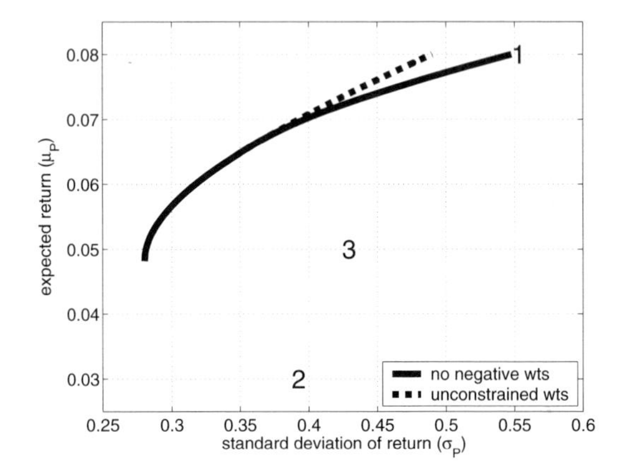

Fig. 5.6. *Efficient frontier for $N = 3$ assets plotted by the program described in Section 5.6. "1," "2," and "3" are the three single assets. The efficient frontiers are found with and without the constraint of no negative weights. The constrained efficient frontier is computed using MATLAB's quadratic programming algorithm.*

5.5.3 Selling short

Selling short is a way to profit if a stock price goes *down*. To sell a stock short, one sells the stock without owning it. The stock must be borrowed from a broker or another customer of the broker. At a later point in time, one buys the stock and gives it back to the lender. This closes the short position.

Suppose a stock is selling at \$25/share and you sell 100 shares short. This gives you \$2500. If the stock goes down to \$17/share, you can buy the 100 shares for \$1700 and close out your short position. You made a profit of \$800 (net transaction costs) because the stock went down 8 points. If the stock had gone up, then you would have had a loss.

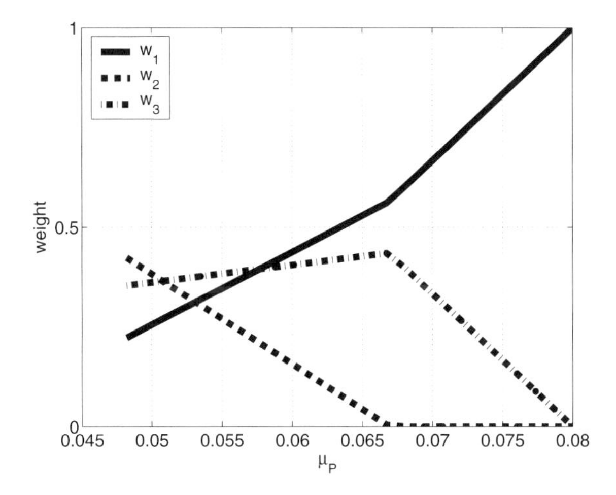

Fig. 5.7. *Weights for assets 1, 2, and 3 as functions of $\mu_p \geq \mu_{\min}$. The weights for all three assets are constrained to be nonnegative.*

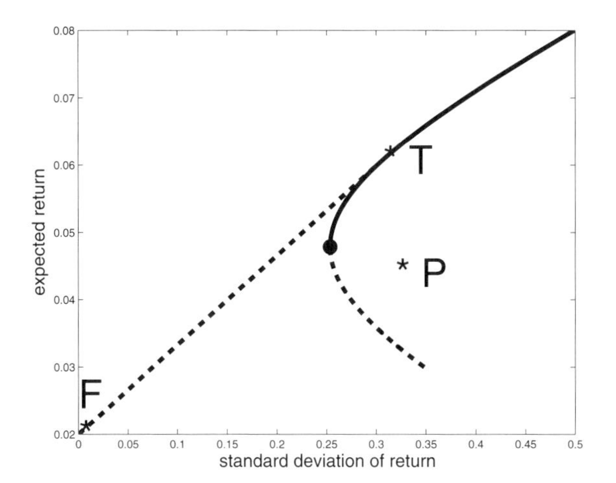

Fig. 5.8. *Efficient frontier and line of optimal combinations of $N = 3$ risky assets and the risk-free asset. "P" is the portfolio with weights (0 0.3 0.7) that is, an example of a portfolio not on the efficient frontier. "T" is the tangency portfolio and "F" is the risk-free asset.*

Suppose now that you have \$100 and there are two risky assets. With your money you could buy \$150 worth of risky asset 1 and sell \$50 short of risky asset 2. The net cost would be exactly \$100. If R_1 and R_2 are the returns on risky assets 1 and 2, then the return on your portfolio would be

$$\frac{3}{2}R_1 + \left(-\frac{1}{2}\right)R_2.$$

Your portfolio weights are $w_1 = 3/2$ and $w_2 = -1/2$. Thus, you hope that risky asset 1 rises in price and risky asset 2 falls in price. Here, as elsewhere, we have ignored transaction costs.

5.5.4 Back to the math — Finding the tangency portfolio

Here is the mathematics behind Figure 5.8. We now remove the assumption that $\boldsymbol{\omega}^\mathsf{T}\mathbf{1} = 1$. The quantity $1 - \boldsymbol{\omega}^\mathsf{T}\mathbf{1}$ is invested in the risk-free asset. (Does it make sense to have $1 - \boldsymbol{\omega}^\mathsf{T}\mathbf{1} < 0$?). The expected return is

$$\boldsymbol{\omega}^\mathsf{T}\boldsymbol{\mu} + (1 - \boldsymbol{\omega}^\mathsf{T}\mathbf{1})\mu_f, \tag{5.27}$$

where μ_f is the return on the risk-free asset. Suppose the target expected return is μ_P. Then the constraint to be satisfied is that (5.27) is equal to μ_P. Thus, the Lagrangian function is

$$L = \boldsymbol{\omega}^\mathsf{T}\boldsymbol{\Omega}\boldsymbol{\omega} + \delta\{\mu_P - \boldsymbol{\omega}^\mathsf{T}\boldsymbol{\mu} - (1 - \boldsymbol{\omega}^\mathsf{T}\mathbf{1})\mu_f\}.$$

Here δ is a Lagrange multiplier. Since

$$0 = \frac{\partial}{\partial\,\boldsymbol{\omega}}L = 2\boldsymbol{\Omega}\boldsymbol{\omega} + \delta(-\boldsymbol{\mu} + \mathbf{1}\mu_f),$$

the optimal weight vector, i.e., the vector of weights that minimizes risk subject to the constraint on the expected return, is

$$\boldsymbol{\omega}_{\mu_P} = \lambda\boldsymbol{\Omega}^{-1}(\boldsymbol{\mu} - \mu_f\mathbf{1}), \tag{5.28}$$

where $\lambda = \delta/2$. To find λ, we use our constraint:

$$\boldsymbol{\omega}_{\mu_P}^\mathsf{T}\boldsymbol{\mu} + (1 - \boldsymbol{\omega}_{\mu_P}^\mathsf{T}\mathbf{1})\mu_f = \mu_P. \tag{5.29}$$

Rearranging (5.29), we get

$$\boldsymbol{\omega}_{\mu_P}^\mathsf{T}(\boldsymbol{\mu} - \mu_f\mathbf{1}) = \mu_P - \mu_f. \tag{5.30}$$

Therefore, substituting (5.28) into (5.30) we have

$$\lambda(\boldsymbol{\mu} - \mu_f\mathbf{1})^\mathsf{T}\boldsymbol{\Omega}^{-1}(\boldsymbol{\mu} - \mu_f\mathbf{1}) = \mu_P - \mu_f,$$

or

$$\lambda = \frac{\mu_P - \mu_f}{(\boldsymbol{\mu} - \mu_f\mathbf{1})^\mathsf{T}\boldsymbol{\Omega}^{-1}(\boldsymbol{\mu} - \mu_f\mathbf{1})}. \tag{5.31}$$

Then substituting (5.31) into (5.28) gives

$$\boldsymbol{\omega}_{\mu_P} = c_P\,\overline{\boldsymbol{\omega}},$$

where

$$c_P = \frac{\mu_P - \mu_f}{(\boldsymbol{\mu} - \mu_f \mathbf{1})^{\mathsf{T}} \boldsymbol{\Omega}^{-1} (\boldsymbol{\mu} - \mu_f \mathbf{1})}$$

and

$$\overline{\boldsymbol{\omega}} = \boldsymbol{\Omega}^{-1} (\boldsymbol{\mu} - \mu_f \mathbf{1}). \tag{5.32}$$

Note that $(\boldsymbol{\mu} - \mu_f \mathbf{1})$ is the vector of **excess returns**, that is, the amount by which the expected returns on the risky assets exceed the risk-free return. The excess returns measure how much the market pays for assuming risk.

$\overline{\boldsymbol{\omega}}$ is not quite a portfolio because these weights do not necessarily sum to one. The tangency portfolio is a scalar multiple of $\overline{\boldsymbol{\omega}}$:

$$\boldsymbol{\omega}_T = \frac{\overline{\boldsymbol{\omega}}}{\mathbf{1}^{\mathsf{T}} \overline{\boldsymbol{\omega}}}. \tag{5.33}$$

The optimal weight vector $\boldsymbol{\omega}_{\mu_P}$ can be expressed in terms of the tangency portfolio as $\boldsymbol{\omega}_{\mu_P} = c_p \overline{\boldsymbol{\omega}} = c_p(\mathbf{1}^{\mathsf{T}} \overline{\boldsymbol{\omega}}) \boldsymbol{\omega}_T$. Therefore, $c_P\,(\mathbf{1}^{\mathsf{T}} \overline{\boldsymbol{\omega}})$ tells us how much weight to put on the tangency portfolio, $\boldsymbol{\omega}_T$. The amount of weight to put on the risk-free asset is $= 1 - c_p\,(\overline{\boldsymbol{\omega}}^{\mathsf{T}} \mathbf{1})$.

Note that $\overline{\boldsymbol{\omega}}$ and $\boldsymbol{\omega}_T$ do not depend on μ_P. To compute the tangency portfolio, the following MATLAB code can be added to the program given in Section 5.5.1.

```
bomegabar = ibOmega*(bmu - muf*bone) ;
bomegaT = bomegabar/(bone'*bomegabar) ;
sigmaT = sqrt(bomegaT'*bOmega*bomegaT) ;
muT = bmu'*bomegaT ;
```

Then Figure 5.8 can be produced with the following MATLAB statements.

```
fsize = 16 ;
fsize2 = 35 ;
bomegaP2 = [0;.3;.7] ;  % "arbitrary" portfolio weights
sigmaP2 = sqrt(bomegaP2'*bOmega*bomegaP2) ;
muP2 = bmu'*bomegaP2 ;
ind = (muP > mumin) ;   %   Indicates efficient frontier
ind2 = (muP < mumin) ;  %   Indicates locus below efficient frontier
p1 = plot(sigmaP(ind),muP(ind),'-',sigmaP(ind2),muP(ind2), ...
    '--' ,sdmin,mumin,'.') ;
l1 = line([0,sigmaT],[muf,muT]) ;
t1= text(sigmaP2,muP2,'* P','fontsize',fsize2) ;
t2= text(sigmaT,muT,'* T','fontsize',fsize2) ;
t3=text(0,muf+.006,'F','fontsize',fsize2) ;
t3B= text(0,muf,'*','fontsize',fsize2) ;
set(p1(1:2),'linewidth',4) ;
set(p1(3),'markersize',40) ;
set(l1,'linewidth',4) ;
set(l1,'linestyle','--') ;
xlabel('standard deviation of return','fontsize',fsize) ;
ylabel('expected return','fontsize',fsize) ;
grid ;
```

5.5.5 Examples

Example 5.10. (Deriving formulas for $N = 2$) If $N = 2$, then

$$\boldsymbol{\Omega} = \begin{pmatrix} \sigma_1^2 & \rho_{12}\sigma_1\sigma_2 \\ \rho_{12}\sigma_1\sigma_2 & \sigma_2^2 \end{pmatrix}.$$

You should check that

$$\boldsymbol{\Omega}^{-1} = \frac{1}{1-\rho_{12}^2} \begin{pmatrix} \sigma_1^{-2} & -\rho_{12}\sigma_1^{-1}\sigma_2^{-1} \\ -\rho_{12}\sigma_1^{-1}\sigma_2^{-1} & \sigma_2^{-2} \end{pmatrix}. \tag{5.34}$$

Also,

$$\boldsymbol{\mu} - \mu_f \mathbf{1} = \begin{pmatrix} \mu_1 - \mu_f \\ \mu_2 - \mu_f \end{pmatrix}.$$

Therefore,

$$\overline{\boldsymbol{\omega}} = \boldsymbol{\Omega}^{-1}(\boldsymbol{\mu} - \mu_f \mathbf{1}) = \frac{1}{1-\rho_{12}^2} \begin{pmatrix} \frac{\mu_1-\mu_f}{\sigma_1^2} - \frac{\rho_{12}(\mu_2-\mu_f)}{\sigma_1\sigma_2} \\ -\frac{\rho_{12}(\mu_1-\mu_f)}{\sigma_1\sigma_2} + \frac{\mu_2-\mu_f}{\sigma_2^2} \end{pmatrix}.$$

Next, let $V_1 = \mu_1 - \mu_f$ and $V_2 = \mu_2 - \mu_f$. Then,

$$\mathbf{1}^{\mathsf{T}}\overline{\boldsymbol{\omega}} = \frac{V_1\,\sigma_2^2 + V_2\,\sigma_1^2 - (V_1 + V_2)\,\rho_{12}\sigma_1\sigma_2}{\sigma_1^2\sigma_2^2(1-\rho_{12}^2)}.$$

It follows that

$$\boldsymbol{\omega}_T = \frac{\overline{\boldsymbol{\omega}}}{\mathbf{1}^{\mathsf{T}}\overline{\boldsymbol{\omega}}} = \frac{1}{V_2\,\sigma_2 + V_2\,\sigma_1^2 - (V_1+V_2)\,\rho_{12}\sigma_1\sigma_2} \begin{pmatrix} V_1\sigma_2^2 - V_2\,\rho_{12}\sigma_1\sigma_2 \\ V_2\,\sigma_1^2 - V_1\,\rho_{12}\sigma_1\sigma_2 \end{pmatrix}.$$

Compare the first element of this vector with (5.4), the formula that gives the weight of the first of two risky assets in the tangency portfolio.

Example 5.11. (A numerical example with $N = 3$) In principle, one could find formulas analogous to (5.4) for cases where N is bigger than 2, but they would be cumbersome. It is better to use matrix-based formulas such as (5.33).

In this section we work through a numerical example with $N = 3$. In practice, we would want to do these calculations using a computer program and, in fact, they really were done with a program. However, I go through the calculations as if they were done "by hand" as an illustration of the mathematics.

Assume that the mean returns are 0.07, 0.12, and 0.09. Assume as well that the variances of the returns are 0.2, 0.3, and 0.25. Finally, to make the calculations simple, assume that the returns are uncorrelated. The mean vector is

$$\boldsymbol{\mu} = \begin{pmatrix} 0.07 \\ 0.12 \\ 0.09 \end{pmatrix}.$$

Because the returns are uncorrelated the covariance matrix is the diagonal matrix

$$\boldsymbol{\Omega} = \begin{pmatrix} 0.2 & 0 & 0 \\ 0 & 0.3 & 0 \\ 0 & 0 & 0.25 \end{pmatrix}.$$

Diagonal matrices are easy to invert and the inverse covariance matrix is

$$\boldsymbol{\Omega}^{-1} = \begin{pmatrix} 5 & 0 & 0 \\ 0 & 10/3 & 0 \\ 0 & 0 & 4 \end{pmatrix}.$$

Then using equations (5.13) to (5.18),

$$A = \boldsymbol{\mu}^{\mathsf{T}}\boldsymbol{\Omega}^{-1}\mathbf{1} = (\,0.07 \quad 0.12 \quad 0.09\,) \begin{pmatrix} 5 & 0 & 0 \\ 0 & 10/3 & 0 \\ 0 & 0 & 4 \end{pmatrix} \begin{pmatrix} 1 \\ 1 \\ 1 \end{pmatrix} = 1.1100,$$

$$B = \boldsymbol{\mu}^{\mathsf{T}}\boldsymbol{\Omega}^{-1}\boldsymbol{\mu} = (\,0.07 \quad 0.12 \quad 0.09\,) \begin{pmatrix} 5 & 0 & 0 \\ 0 & 10/3 & 0 \\ 0 & 0 & 4 \end{pmatrix} \begin{pmatrix} 0.07 \\ 0.12 \\ 0.09 \end{pmatrix} = 0.1049,$$

and

$$C = \mathbf{1}^{\mathsf{T}}\boldsymbol{\Omega}^{-1}\mathbf{1} = (\,1 \quad 1 \quad 1\,) \begin{pmatrix} 5 & 0 & 0 \\ 0 & 10/3 & 0 \\ 0 & 0 & 4 \end{pmatrix} \begin{pmatrix} 1 \\ 1 \\ 1 \end{pmatrix} = 12.3333.$$

Also $D = BC - A^2 = 0.0617$. Then by (5.20),

$$\begin{aligned} \boldsymbol{g} = {} & \frac{.1049}{.0617} \begin{pmatrix} 5 & 0 & 0 \\ 0 & 10/3 & 0 \\ 0 & 0 & 4 \end{pmatrix} \begin{pmatrix} 1 \\ 1 \\ 1 \end{pmatrix} \\ & - \frac{1.1100}{.0617} \begin{pmatrix} 5 & 0 & 0 \\ 0 & 10/3 & 0 \\ 0 & 0 & 4 \end{pmatrix} \begin{pmatrix} 0.07 \\ 0.12 \\ 0.09 \end{pmatrix} = \begin{pmatrix} 2.2054 \\ -1.5297 \\ 0.3243 \end{pmatrix}, \end{aligned}$$

and by equation (5.21)

$$\begin{aligned} \boldsymbol{h} = {} & \frac{12.3333}{.0617} \begin{pmatrix} 5 & 0 & 0 \\ 0 & 10/3 & 0 \\ 0 & 0 & 4 \end{pmatrix} \begin{pmatrix} 0.07 \\ 0.12 \\ 0.09 \end{pmatrix} \\ & - \frac{1.1100}{.0617} \begin{pmatrix} 5 & 0 & 0 \\ 0 & 10/3 & 0 \\ 0 & 0 & 4 \end{pmatrix} \begin{pmatrix} 1 \\ 1 \\ 1 \end{pmatrix} = \begin{pmatrix} -20 \\ 20 \\ 0 \end{pmatrix}. \end{aligned}$$

Then by equation (5.19) the efficient weight vector for a portfolio with expected return equal to μ_P is

$$\boldsymbol{\omega}_{\mu_P} = \boldsymbol{g} + \mu_P\,\boldsymbol{h} = \begin{pmatrix} 2.2054 - 20\mu_P \\ -1.5297 + 20\mu_P \\ 0.3243 \end{pmatrix}.$$

In other words, if we select a target expected return μ_P, these are the weights that give us that expected return with the smallest possible risk. As μ_P varies, the first and second components of $\boldsymbol{\omega}_{\mu_P}$ change. The third component of $\boldsymbol{\omega}_{\mu_P}$ does not change in this example because this is a somewhat unusual case where one of the components of $\boldsymbol{h}$ is exactly zero. Notice what happens as we *increase* μ_P: the first component of $\boldsymbol{\omega}_{\mu_P}$ *decreases*, the second component *increases*, and the third component *stays the same.* This makes perfect sense. The first risky asset has the smallest expected return and the second risky asset has the greatest expected return. Thus, to increase the portfolio's expected return we move some of our allocation from the first risky asset to the second.

To find the minimum variance portfolio, note that

$$\boldsymbol{g}^{\mathsf{T}}\boldsymbol{\Omega}\boldsymbol{h} = -18$$

and

$$\boldsymbol{h}^{\mathsf{T}}\boldsymbol{\Omega}\boldsymbol{h} = 200.$$

Then by (5.25)

$$\mu_{\min} = -\frac{\boldsymbol{g}^{\mathsf{T}}\boldsymbol{\Omega}\boldsymbol{g}}{\boldsymbol{h}^{\mathsf{T}}\boldsymbol{\Omega}\boldsymbol{h}} = 0.0900.$$

The weights for the minimum variance portfolio are

$$\boldsymbol{g} + \mu_{\min}\boldsymbol{h} = \begin{pmatrix} 0.4054 \\ 0.2703 \\ 0.3243 \end{pmatrix}.$$

Now let's find the tangency portfolio when $\mu_f = 0.05$. We have by (5.32),

$$\overline{\boldsymbol{\omega}} = \begin{pmatrix} 5 & 0 & 0 \\ 0 & 10/3 & 0 \\ 0 & 0 & 4 \end{pmatrix} \begin{pmatrix} .07 - 0.05 \\ 0.12 - 0.05 \\ 0.09 - 0.05 \end{pmatrix} = \begin{pmatrix} 0.1000 \\ 0.2333 \\ 0.1600 \end{pmatrix}$$

and then by (5.33),

$$\boldsymbol{\omega}_T = \frac{1}{0.1000 + 0.2333 + 0.1600} \begin{pmatrix} .1000 \\ 0.2333 \\ 0.1600 \end{pmatrix} = \begin{pmatrix} 0.2027 \\ 0.4730 \\ 0.3243 \end{pmatrix}.$$

5.6 Quadratic Programming*

Quadratic programming is used to minimize a quadratic objective function subject to linear constraints. In applications to portfolio optimization, the objective function is the variance of the portfolio return and the linear constraints could be that the portfolio weights are nonnegative so that selling short is precluded.

The objective function is a function of N variables, e.g., the weights of N assets, that are denoted by an $N \times 1$ vector $\boldsymbol{x}$. Suppose that the quadratic objective function to be minimized is

$$\frac{1}{2}\boldsymbol{x}^{\mathsf{T}}\boldsymbol{H}\boldsymbol{x} + \boldsymbol{f}^{\mathsf{T}}\boldsymbol{x}, \tag{5.35}$$

where $\boldsymbol{H}$ is an $N \times N$ matrix and $\boldsymbol{f}$ is an $N \times 1$ vector. The factor of 1/2 is not essential but is used here to keep our notation consistent with MATLAB. There are two types of linear constraints on $\boldsymbol{x}$, inequality and equality constraints. The linear inequality constraints are

$$\boldsymbol{A}\boldsymbol{x} \leq \boldsymbol{b}, \tag{5.36}$$

where $\boldsymbol{A}$ is an $m \times N$ matrix and $\boldsymbol{b}$ is an $m \times 1$ vector where m is the number of inequality constraints. The equality constraints are

$$\boldsymbol{A}_{eq}\boldsymbol{x} = \boldsymbol{b}_{eq}, \tag{5.37}$$

where $\boldsymbol{A}_{eq}$ is an $n \times N$ matrix and $\boldsymbol{b}_{eq}$ is an $n \times 1$ vector where n is the number of equality constraints. In other words, quadratic programming minimizes the quadratic objective function (5.35) subject to linear inequality constraints (5.36) and linear equality constraints (5.37).

We can impose nonnegativity constraints on the weights of a portfolio by using quadratic programming. The variables are the portfolio weights so that $\boldsymbol{x} = \boldsymbol{\omega}$. The objective function is one-half the variance of the portfolio's return; again the "half" is not essential. Therefore, $\boldsymbol{H} = \boldsymbol{\Omega}$ and $\boldsymbol{f}$ is equal to an $N \times 1$ vector of zeros so that (5.35) is half the variance of the portfolio's return, $\boldsymbol{\omega}^{\mathsf{T}}\boldsymbol{\Omega}\boldsymbol{\omega}/2$, that is, one-half the objective function (5.6). There are two equality constraints, one that the weights sum to 1 and the other that the portfolio return is a specified target μ_P. Therefore, we define

$$\boldsymbol{A}_{eq} = \begin{pmatrix} \boldsymbol{1}^{\mathsf{T}} \\ \boldsymbol{\mu}^{\mathsf{T}} \end{pmatrix},$$

and

$$\boldsymbol{b}_{eq} = \begin{pmatrix} 1 \\ \mu_P \end{pmatrix},$$

so that (5.37) becomes

$$\begin{pmatrix} \boldsymbol{1}^{\mathsf{T}}\boldsymbol{\omega} \\ \mu^{\mathsf{T}}\boldsymbol{\omega} \end{pmatrix} = \begin{pmatrix} 1 \\ \mu_P \end{pmatrix},$$

which is the same as constraints (5.7) and (5.8). Thus, we are solving the same minimization problem as in Section 5.5.1. However, we can now add the constraints that all of the ω_i be nonnegative. This can be done by defining $\boldsymbol{A}$ to be $-\boldsymbol{I}$ and $\boldsymbol{b} = 0$ so that (5.36) is

$$-\boldsymbol{\omega} \leq \boldsymbol{0},$$

which is vector shorthand for the inequalities $-\omega_i \le 0$ (or $\omega_i \ge 0$) for $i = 1, \ldots, N$.

In MATLAB, the function "quadprog" does quadratic programming. The MATLAB statement `x=quadprog(H,f,A,b,Aeq,beq,LB,UB)` minimizes (5.35) subject to the inequality and equality constraints (5.36) and (5.37) and the additional inequality constraints that $LB_i \le X_i \le UB_i$ for all $i = 1, \ldots, N$. Here "LB" means "lower bound" and "UB" means "upper bound." The constraint that the portfolio weights be nonnegative can be imposed in either of two ways. One way is to let $\boldsymbol{A} = -\boldsymbol{I}$ and $\boldsymbol{b} = \boldsymbol{0}$ as discussed above. In this case, we use no lower or upper bounds. The other way is to omit $\boldsymbol{A}$ and $\boldsymbol{b}$ and to use lower bounds that are all equal to 0 and no upper bounds. Here is a program that uses the second method. First we enter $\boldsymbol{\mu}$ and $\boldsymbol{\Omega}$ and the risk-free rate.

```
bmu = [0.08;0.03;0.05] ;
bOmega = [ 0.3 0.02 0.01 ;
    0.02 0.15 0.03 ; 0.01 0.03 0.18 ] ;
    rf = .04 ;  % Risk-free rate
```

Then $\boldsymbol{A}_{eq}$ is defined.

```
Aeq = [ones(1,length(bmu));bmu'] ;
```

Then we create a grid of 50 μ_P values from 0.03 to 0.08.

```
ngrid = 50 ;
muP = linspace(.03,.08,ngrid) ; %   muP grid
```

Then we set up matrices for storage.

```
sigmaP = muP ;     %   Set up storage.  These matrices will be overwritten
sigmaP2 = sigmaP ;
omegaP = zeros(3,ngrid) ;
omegaP2 = omegaP ;
```

Finally, we find the portfolio weights for each value of μ_P.

```
for i = 1:ngrid ;
    omegaP(:,i)=quadprog(bOmega,zeros(length(bmu),1),'','', ...
    Aeq,[1;muP(i)],zeros(3,1)) ;
end ;
```

We can also find the minimum variance portfolio and the tangency portfolio. In the following lines of the program, `imin` indicates the minimum variance portfolio and `Ieff` indicates which points have a return larger than or equal to the return of the minimum variance portfolio and therefore are on the efficient frontier. Also, `sharpratio` is the vector of Sharpe's ratio of the portfolios and `Itangenc` indicates which portfolio has the largest Sharpe's ratio, that is, which portfolio is the tangency portfolio.

```
imin=find(sigmaP==min(sigmaP)) ;
Ieff = (muP >= muP(imin)) ;
sharperatio = (muP-rf) ./ sigmaP ;
Itangency = find(sharperatio == max(sharperatio))
```

Now we plot the efficient frontier indicating the tangency portfolio with an "*" and the minimum variance portfolio with an "o". Also, the risk-free asset is indicated by an "x".

```
plot(sigmaP(Ieff),muP(Ieff),sigmaP(Itangency),...
    muP(Itangency),'*',sigmaP(imin),muP(imin),'o', ...
    0,rf,'x') ;
```

Finally, a line is drawn from the risk-free asset to the tangency portfolio.

```
line([0 sigmaP(Itangency)],[rf,muP(Itangency)]) ;
```

The plot produced by this program is Figure 5.9.

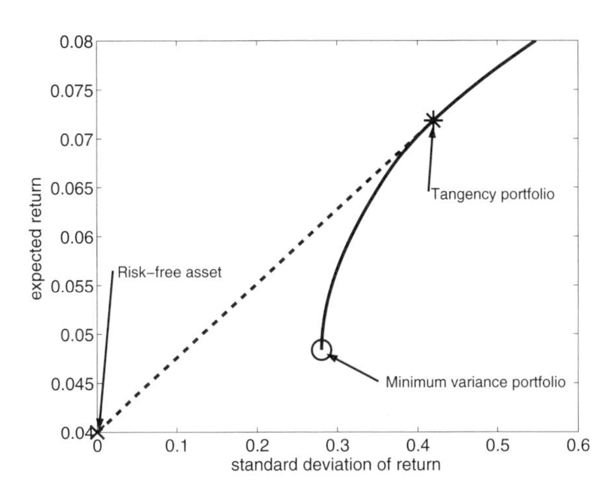

Fig. 5.9. *Efficient frontier and tangency portfolio with the constraints that the weights are nonnegative.*

5.7 Is the Theory Useful?

This theory of portfolio optimization could be used if N were small. We would need estimates of $\boldsymbol{\mu}$ and $\boldsymbol{\Omega}$. These would be obtained from recent returns data as described in Section 5.3.2. Of course, there is no guarantee that future returns will behave as returns in the past, but this is the working assumption.

Michaud (1998) gives an example of using portfolio theory to allocate capital among eight international markets. With a total of only eight assets, implementing the theory is computationally feasible, but estimation error in $\widehat{\mu}$ and $\widehat{\boldsymbol{\Omega}}$ causes substantial error in estimation of optimal portfolios; see Chapter 10 where the effects of estimation error are investigated.

However, suppose that we were considering selecting a portfolio from all 500 stocks on the S&P 500 index. Or, even worse, consider all 3000 stocks on the Russell index. In the latter case, there would be $(3000)(2999)/2 \approx 4.5$ million covariances to estimate. A tremendous amount of data would need to be collected and, after that effort, the effects of estimation error in 3000 means and 4.5 million covariances would be overwhelming.

Portfolio theory was considered such an important theoretical development that Harry Markowitz was awarded the Nobel Prize in Economic Sciences for this work. However, a more practical version of this theory awaited further work such as that of William Sharpe and John Lintner. Sharpe, who was Markowitz's PhD student, shared the Nobel Prize with Markowitz. As we show in Chapter 7, Sharpe's CAPM (capital asset pricing model) asserts that the tangency portfolio is also the market portfolio. This is a tremendous simplification. The CAPM also provides a simple way to calculate all possible correlations between assets; see Section 7.7. Moreover, although CAPM is now considered too simple an economic theory to be entirely realistic, factor models, which include CAPM as a simple case, are more realistic and yet simple enough to make portfolio optimization feasible; see Section 7.8.

5.8 Utility Theory*

Economists generally do not model economic decisions in terms of the mean and variance of the return but rather by using a *utility function.* The utility of an amount X of money is said to be $U(X)$ where the utility function U generally has the properties:

1. $U(0) = 0$;
2. U is strictly increasing;
3. the first derivative $U'(X)$ is strictly decreasing.

Assumption 1 is not necessary but is reasonable. Assumption 2 says simply that more money is better than less. Assumption 3 implies that the more money we have the less we value an extra dollar. Assumption 3. is called *risk aversion.* Assumption 3 implies that we would decline a bet that pays $\pm\Delta$ with equal probabilities. In fact, Assumption 3 implies that we would decline any bet that is symmetrically distributed about 0, because the expected utility of our wealth would be reduced if we accepted the bet. If the second derivative U'' exists then 3 is equivalent to the assumption that $U''(X) < 0$ for all X.

It is assumed that a rational person will make decisions so as to maximize $E\{U(X)\}$ where X is that person's final wealth. In economics this is almost a part of the definition of a rational person, with another component of the definition being that a rational person will update probabilities using Bayes' law. Each individual is assumed to have his or her own utility function and two different rational people may make different decisions because they have different utility functions.

Utility theory is generally accepted by economists as the way people *should* make decisions in an ideal world. How people actually do go about making decisions is another issue. In business and finance, utility functions are generally avoided. One problem is the difficulty of actually eliciting a person's utility function. In portfolio theory, another problem is that expected utilities typically depend on the distribution of the return, not just the mean and variance of the return and the distribution of the return is sensitive to modeling assumptions.

How different are mean-variance efficient portfolios and portfolios that maximize expected utility? In the case that returns are normally distributed, this question can be answered.

Result 5.8.1 *If returns on all portfolios are normally distributed and if U satisfies Assumptions 1 to 3, then the portfolio that maximizes expected utility is on the efficient frontier.*

So, if one chose a portfolio to maximize expected utility, then a mean-variance efficient portfolio would be selected. Exactly which portfolio on the efficient frontier one chose would depend on one's utility function. This result can be proven by first proving the following fact: if R_1 and R_2 are normally distributed with the same means and with standard deviations σ_1 and σ_2 such that $\sigma_1 < \sigma_2$, then $E\{U(R_1)\} > E\{U(R_2)\}$.

5.9 Summary

The efficient market hypothesis suggests that trying to pick a few great stocks is not an effective investment strategy. Instead of trying to buy a few "sure winners" it is better to diversify one's holdings. Riskier assets have higher expected returns because investors demand a reward for assuming risk. This reward is called the risk premium.

If there is one risky asset and one risk-free asset, then as one increases one's percentage of holdings in the risky asset, expected return and risk both increase. If there are two or more risky assets, then one can remove some risk without a cost in expected returns by diversifying one's holdings of risky assets. The efficient frontier is a locus of points in reward-risk space that maximize expected return for a given level of risk or minimize risk for a given expected return. The efficient frontier can be found using "efficient-set mathematics" that is based on Lagrange multipliers

When one can invest in two or more risky assets and in a risk-free asset, then the optimal portfolios are combinations of the tangency portfolio of risky assets and the risk-free asset. The tangency portfolio is on the efficient frontier and can be found by efficient-set mathematics. Sharpe's ratio is the reward to risk ratio and is maximized by the tangency portfolio.

Efficient-set mathematics does not allow one to constrain the portfolio weights to be nonnegative. Thus, short selling is needed to implement the optimal portfolios on the efficient frontier. To avoid negative portfolio weights, one can use quadratic programming. Quadratic programming results in a new and different efficient frontier with all portfolio weights nonnegative. These portfolios do not allow short selling.

5.10 Bibliographic Notes

Markowitz (1952) was the original paper on portfolio theory and was expanded into the book Markowitz (1959). Bernstein (1993) calls a chapter discussing Markowitz and his work "Fourteen Pages To Fame" in reference to this paper. Bodie and Merton (2000) provide an elementary introduction to portfolio selection theory. Bodie, Kane, and Marcus (1999) and Sharpe, Alexander, and Bailey (1999) give a more comprehensive treatment. I learned much of the material in Section 5.5 from Section 5.2 of Campbell, Lo, and MacKinlay (1997). See also Merton (1972). Michaud (1998) discusses the sensitivity of portfolio optimization to estimation error in the expectation vector and covariance matrix; see also Chapter 10.

5.11 References

Bernstein, P., (1993) *Capital Ideas: The Improbable Origins of Modern Wall Street*, Free Press, New York.

Bodie, Z. and Merton, R. C. (2000) *Finance*, Prentice-Hall, Upper Saddle River, NJ.

Bodie, Z., Kane, A., and Marcus, A. (1999) *Investments, 4th Ed.*, Irwin/McGraw-Hill, Boston.

Campbell, J. Y., Lo, A. W., and MacKinlay, A. C. (1997) *The Econometrics of Financial Markets*, Princeton University Press, Princeton, NJ.

Markowitz, H. (1952) Portfolio Selection, *Journal of Finance*, **7**, 77-91.

Markowitz, H. (1959) *Portfolio Selection: Efficient Diversification of Investment*, Wiley, New York.

Merton, R.C. (1972) An analytic derivation of the efficient portfolio frontier, *Journal of Financial and Quantitative Analysis*, **7**, 1851–1872.

Michaud, R. O. (1998) *Efficient Asset Management: A Practical Guide to Stock Portfolio Optimization and Asset Allocation*, Harvard Business School Press, Boston.

Sharpe, W. F., Alexander, G. J., and Bailey, J. V. (1999) *Investments, 6th Ed.*, Prentice-Hall, Upper Saddle River, NJ.

Williams, J. B. (1938) *The Theory of Investment Value*, Harvard University Press, Cambridge, MA.

5.12 Problems

1. Suppose that there are two risky assets, A and B, with expected returns equal to 2% and 5%, respectively. Suppose that the standard deviations

of the returns are $\sqrt{6}\%$ and $\sqrt{11}\%$ and that the returns on the assets have a correlation of 0.1.

(a) What portfolio of A and B achieves a 3% rate of expected return?

(b) What portfolios of A and B achieve an $\sqrt{5}\%$ standard deviation of return? Among these, which has the largest expected return?

2. Suppose there are two risky assets, C and D, the tangency portfolio is 60% C and 40% D, and the expected return and standard deviation of the return on the tangency portfolio are 5% and 7%, respectively. Suppose also that the risk-free rate of return is 2%.
   If you want the standard deviation of your return to be 5%, what proportions of your capital should be in the risk-free asset, asset C, and asset D?
3. (a) Suppose that stock A shares sell at \$85 and Stock B shares at \$35. A portfolio has 300 shares of stock A and 100 of stock B. What are the weight w and $1 - w$ of stocks A and B in this portfolio?
   (b) More generally, if a portfolio has N stocks, if the price per share of the jth stock is P_j, and if the portfolio has n_j shares of stock j, then find a formula for w_j as a function of $n_1, \ldots, n_N$ and $P_1, \ldots, P_N$.
4. Let $\mathcal{R}_P$ be a return of some type on a portfolio and let $\mathcal{R}_1, \ldots, \mathcal{R}_N$ be the same type of returns on the assets in this portfolio. Is

$$\mathcal{R}_P = w_1\mathcal{R}_1 + \cdots + w_N\mathcal{R}_N$$

   true if $\mathcal{R}_P$ is a net return? Is this equation true if $\mathcal{R}_P$ is a gross return? Is it true if $\mathcal{R}_P$ is a log return? Justify your answers.
5. Verify equation (5.34).
6. Write a MATLAB program to find the efficient frontier, minimum variance portfolio, and tangency portfolio subject to the constraints that there is no short selling and that no asset is more that 33% of the portfolio. Assume that there are five assets, that the expected returns are

$$\boldsymbol{\mu} = \begin{pmatrix} .1 \\ 0.04 \\ 0.07 \\ 0.11 \\ 0.05 \end{pmatrix},$$

   and that the covariance matrix of the returns is

$$\boldsymbol{\Omega} = \begin{pmatrix} 0.38 & 0.02 & 0.01 & 0 & 0 \\ 0.02 & 0.17 & 0.03 & 0.01 & 0.02 \\ 0.01 & 0.03 & 0.38 & 0.04 & 0.05 \\ 0 & 0.01 & 0.04 & 0.47 & 0.03 \\ 0 & 0.02 & 0.05 & 0.03 & 0.26 \end{pmatrix}.$$

   Suppose that the risk-free rate is 0.028. Plot the efficient frontier, risk-free asset, minimum variance portfolio, and tangency portfolio as in Figure 5.9.

6

Regression

6.1 Introduction

Regression is one of the most widely used of all statistical methods. The available data are one response variable and p predictor variables, all measured on each of n observations. We let Y_i be the value of the response variable for the ith observation and $X_{i,1}, \ldots, X_{i,p}$ be the values of predictor variables 1 through p for the ith observation. The goals of regression modeling include investigation of how Y is related to $X_1, \ldots, X_p$, estimation of the conditional expectation of Y given $X_1, \ldots, X_p$, and prediction of future Y values when the corresponding values of $X_1, \ldots, X_p$ are already available. These goals are closely connected.

The *multiple linear regression* model relating Y to the predictor variables is

$$Y_i = \beta_0 + \beta_1 X_{i,1} + \cdots + \beta_p X_{i,p} + \epsilon_i,$$

where ϵ_i is random noise. The ϵ_i are often called "errors" but that term suggests that they are caused by mistakes, which is only sometimes the case. The parameter β_0 is the intercept. It is the expected value of Y_i when all the $X_{i,j}$s are zero. The regression coefficients $\beta_1, \ldots, \beta_p$ are the slopes. More precisely, β_j is the partial derivative of the expected response with respect to the jth predictor:

$$\beta_j = \frac{\partial\, E(Y_i)}{\partial\, X_{i,j}}.$$

In other words, β_j is the change in the expected value of Y_i when $X_{i,j}$ changes one unit. It is assumed that the noise is white so that

$$\epsilon_1, \ldots, \epsilon_n \text{ are i.i.d. with mean 0 and constant variance } \sigma_\epsilon^2. \tag{6.1}$$

If $X_1, \ldots, X_n$ are random,[1] then it is assumed that

[1] In applications in economics and finance the predictor variables are usually from a random sample and therefore considered random, but in experimental studies the

$$\epsilon_1, \dots, \epsilon_n \text{ are independent of } X_1, \dots, X_n. \tag{6.2}$$

Often the ϵ_is are assumed to be normally distributed, which with (6.1) implies Gaussian white noise.

6.1.1 Straight line regression

Straight line regression is regression with only one predictor variable. The model is

$$Y_i = \beta_0 + \beta_1 X_i + \epsilon_i,$$

where β_0 and β_1 are the unknown intercept and slope of the line.

6.2 Least Squares Estimation

The regression coefficients $\beta_0, \dots, \beta_p$ can be estimated by the *method of least squares.* This method is first explained for straight line regression.

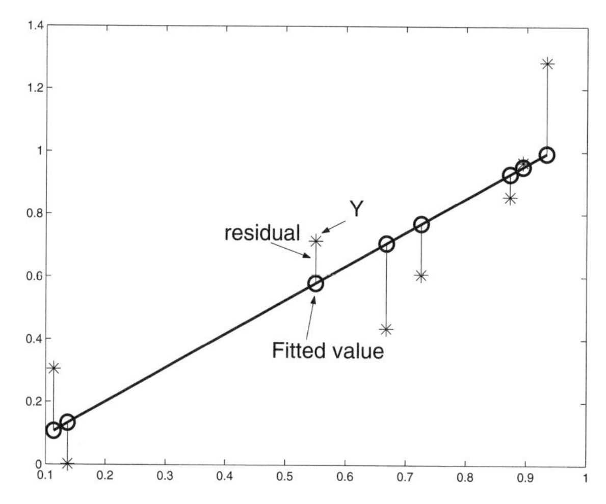

Fig. 6.1. *Least squares estimation. The vertical lines are the residuals. The least squares line is defined as the line making the sum of the squared residuals as small as possible.*

---

predictors are often chosen by an investigator and considered fixed, not random. Fortunately, the regression analysis is the same whether the predictor variables are random or fixed and we need not worry about the distinction.

6.2.1 Estimation in straight line regression

The least squares estimates are the values of $\widehat{\beta}_0$ and $\widehat{\beta}_1$ that minimize

$$\sum_{i=1}^{n} \left\{ Y_i - (\widehat{\beta}_0 + \widehat{\beta}_1 X_i) \right\}^2. \tag{6.3}$$

Geometrically, we are minimizing the sum of the squared lengths of the vertical lines in Figure 6.1. The data points are shown as asterisks. The vertical lines represent the distances between the data points and the predictions using the linear equation. The predictions themselves are called the **fitted values** or "y-hats" and shown as open circles and the differences between the Y values and the fitted values are called the **residuals**. Using calculus to minimize (6.3) one can show that

$$\widehat{\beta}_1 = \frac{\sum_{i=1}^n (Y_i - \overline{Y})(X_i - \overline{X})}{\sum_{i=1}^n (X_i - \overline{X})^2} = \frac{\sum_{i=1}^n Y_i (X_i - \overline{X})}{\sum_{i=1}^n (X_i - \overline{X})^2}. \tag{6.4}$$

and

$$\widehat{\beta}_0 = \overline{Y} - \widehat{\beta}_1 \overline{X}.$$

The *least squares line* is

$$\begin{aligned} \widehat{Y} &= \widehat{\beta}_0 + \widehat{\beta}_1 X = \overline{Y} + \widehat{\beta}_1 (X - \overline{X}) \\ &= \overline{Y} + \left\{ \frac{\sum_{i=1}^n (Y_i - \overline{Y})(X_i - \overline{X})}{\sum_{i=1}^n (X_i - \overline{X})^2} \right\} (X - \overline{X}) \\ &= \overline{Y} + \frac{s_{XY}}{s_X^2} (X - \overline{X}), \end{aligned}$$

where s_{XY} is the sample covariance between X and Y and s_X^2 is the sample variance of the X_is.

Example 6.1. Data on weekly interest rates from Jan 1, 1970 to Dec 31, 1993 were obtained from the Federal Reserve Bank of Chicago. The URL is:

http://www.chicagofed.org/economicresearchanddata/data/index.cfm

Figure 6.2 is a plot of changes in the 10-year Treasury rate and changes in the corporate AAA bond yield. The plot looks linear, so we try linear regression using PROC REG in SAS. In the following program, the variables `aaa_dif` and `cm10_dif` are the changes (differences) in the corporate AAA rate and 10-year Treasury rate time series and were produced by SAS's differencing command "`dif`."

```
options linesize = 72 ;
data WeeklyInterest ;
infile 'C:\book\SAS\WeeklyInterest.dat' ;
input month day year ff tb03 cm10 cm30 discount prime aaa ;
aaa_dif = dif(aaa) ;
cm10_dif = dif(cm10) ;
```

```
run ;
title 'Weekly Interest Rates' ;
proc reg ;
model aaa_dif = cm10_dif ;
run ;
```

Here is the SAS output.

```
                         Weekly Interest Rates                              1

                          The REG Procedure
                           Model: MODEL1
                     Dependent Variable: aaa_dif

                         Analysis of Variance

                                     Sum of           Mean
Source                  DF          Squares         Square   F Value   Pr > F

Model                    1         10.89497       10.89497   2716.14   <.0001
Error                 1249          5.00998        0.00401
Corrected Total       1250         15.90495


        Root MSE              0.06333    R-Square     0.6850
        Dependent Mean    -0.00077538    Adj R-Sq     0.6848
        Coeff Var          -8168.12675


                         Parameter Estimates

                          Parameter       Standard
Variable        DF         Estimate          Error    t Value    Pr > |t|

Intercept        1       0.00022661        0.00179       0.13      0.8993
cm10_dif         1          0.57237        0.01098      52.12      <.0001
```

From the output we see that the least squares estimates of the intercept and slope are 0.000227 and 0.572. The `Root MSE` is 0.06333; this is what we call $\widehat{\sigma}_\epsilon$ or s, the estimate of σ_ϵ; see Section 6.2.3. The remaining items of the output are also explained shortly in Sections 6.3 and 6.4.

6.2.2 Variance of $\widehat{\beta}_1$

It is useful to have a formula for the variance of an estimator to tell us how the estimator's precision depends on various aspects of the data such as the sample size and the values of the predictor variables. Fortunately, it is easy to derive a formula for the variance of $\widehat{\beta}_1$. By (6.4), we can write $\widehat{\beta}_1$ as a weighted average of the responses

$$\widehat{\beta}_1 = \sum_{i=1}^{n} w_i Y_i,$$

where w_i is the weight given by

$$w_i = \frac{X_i - \overline{X}}{\sum_{i=1}^{n}(X_i - \overline{X})^2}.$$

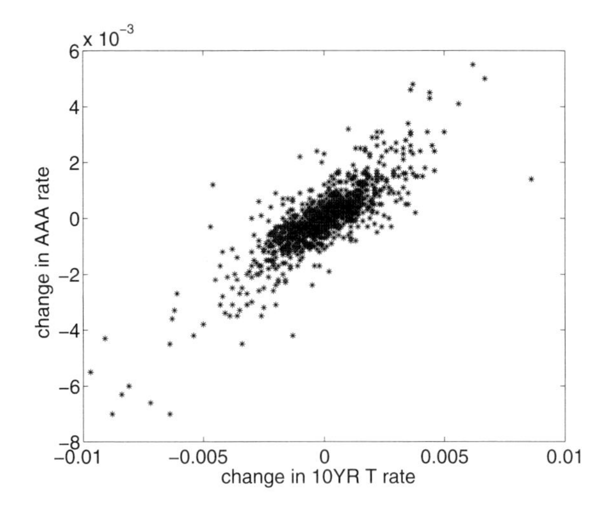

Fig. 6.2. *Change in "CM10 =* 10*-YEAR TREASURY CONSTANT MATURITY RATE (AVERAGE, NSA)" plotted against "AAA = MOODYS SEASONED CORPORATE AAA BOND YIELDS." Data from Federal Reserve Statistical Release H.*15 *and were taken from the Chicago Federal Bank's website.*

We consider $X_1, \ldots, X_n$ as fixed, so that if they are random we are conditioning upon their values. It then follows (see Problem 3) that, conditionally on $X_1, \ldots, X_n$,

$$\mathrm{Var}(\widehat{\beta}_1) = \sigma_\epsilon^2 \sum_{i=1}^n w_i^2 = \frac{\sigma_\epsilon^2}{\sum_{i=1}^n (X_i - \overline{X})^2} = \frac{\sigma_\epsilon^2}{(n-1)s_X^2}. \tag{6.5}$$

It is worth the trouble to think about this formula. First, the numerator σ_ϵ^2 is simply the variance of the ϵ_i which is also the conditional variance of Y_i given the $X_{i,j}$. So, we see that the variance of $\widehat{\beta}_1$ is proportional to the conditional variance of the Y_i. This is not surprising. More variable data mean more variable estimates. The denominator shows us that the variance of $\widehat{\beta}_1$ is inversely proportional to $(n-1)$ and to s_X^2. So the precision of $\widehat{\beta}_1$ increases as σ_ϵ^2 is reduced, n is increased, or s_X^2 is increased. Why does increasing s_X^2 decrease $\mathrm{Var}(\widehat{\beta}_1)$? The reason is that increasing s_X^2 means that the X_i are spread farther apart which makes the slope of the line easier to estimate.

Example 6.2. Here is an important application of (6.5). Suppose that we have two time series, X_t and Y_t, and we wish to regress Y_t on X_t. An example is given in Chapter 7 where X_t is the log return on a market index such as the S&P 500 and Y_t is the log return on some stock. A significant practical question is whether one should use daily or weekly data, or perhaps even monthly or quarterly data. Does it matter which sampling frequency we use? To answer this question, we compare daily and weekly data. Assume that the X_t and the Y_t are white noise sequences. Since a weekly log return is simply the sum of the five daily log returns within a week, σ_ϵ^2 and s_X^2 will each increase by a factor of five if we change from daily to weekly log returns, so the ratio σ_ϵ^2/s_X^2 will not change. However, by changing from daily to weekly log returns, $(n-1)$ is reduced by approximately a factor of five. The result is that $\mathrm{Var}(\widehat{\beta}_1)$ is approximately five times smaller using daily rather than weekly log returns. Similarly $\mathrm{Var}(\widehat{\beta}_1)$ is about four times larger using monthly rather than weekly returns. The obvious conclusion is that one should use the highest sampling frequency available, which is often daily returns.

6.2.3 Estimation in multiple linear regression

The least squares estimates for the multiple linear regression model

$$Y_i = \beta_0 + \beta_1 X_{i,1} + \cdots + \beta_p X_{i,p} + \epsilon_i$$

are the values $\widehat{\beta}_0, \widehat{\beta}_1, \ldots, \widehat{\beta}_p$ that minimize

$$\sum_{i=1}^{n} \{Y_i - (\widehat{\beta}_0 + \widehat{\beta}_1 X_{i,1} + \cdots + \widehat{\beta}_p X_{i,p})\}^2. \tag{6.6}$$

Calculation of the least squares estimates is discussed in regression textbooks. The details do not concern us since software for least squares estimation is readily available.

The estimate of σ_ϵ^2 is (6.6) divided by $(n-p-1)$; see Section 6.4.4.

6.3 Standard Errors, *T*-Values, and *P*-Values

In this section we explain useful statistics included in the regression output of SAS and other statistical software packages. We start with the straight line regression example used earlier and use SAS output as an illustration. The SAS output includes certain important information about β_0 and β_1 which for convenience is listed again:

```
                       Parameter Estimates

                         Parameter      Standard
Variable       DF         Estimate         Error    t Value    Pr > |t|

Intercept       1       0.00022661       0.00179       0.13      0.8993
cm10_dif        1          0.57237       0.01098      52.12      <.0001
```

As noted before, the estimated coefficients are $\widehat{\beta}_0 = 0.000227$ and $\widehat{\beta_1} = 0.57237$. Each of these coefficients has three other statistics associated with it.

- The standard error (SE) which is the estimated standard deviation of the least squares estimator and tells us the precision of that estimator. The standard errors are 0.001791 for β_0 and 0.01098 for β_1.

- The t-value which is the t-statistic for testing that the coefficient is 0. The t-values are 0.13 for β_0 and 52.12 for β_1. The t-value is the ratio of the estimate to its standard error; e.g., $0.13 = 0.000227/0.001791$ for β_0.
- The p-value (`Pr > |t|` in the SAS output) for testing the null hypothesis that the coefficient is 0 versus the alternative that it is not 0. The p-value for β_1 is *very* small, less than 0.0001, so we can conclude that the slope is *not* 0. The p-value is large (0.899) for β_0 so we would not reject the hypothesis that the intercept is 0. However, the null hypothesis that the intercept is 0 is generally not interesting, because it is the slope, not the intercept, that tells us how Y and X are related.

In general, if a p-value for a slope parameter is small as it is here for β_1, then this is evidence that the corresponding coefficient is *not* 0 which means that the predictor has some relationship with the response.

For regression with one predictor variable, by (6.5) the standard error of $\widehat{\beta}_1$ is $\widehat{\sigma}_\epsilon / \sqrt{\sum_{i=1}^n (X_i - \overline{X})^2}$. When there are more than two predictor variables, formulas of standard errors are similar but somewhat more complex and are facilitated by the use of matrix notation. This topic is not discussed because standard errors can be computed with standard software such as SAS, and it is treated in most regression textbooks.

Example 6.3. (Multiple Linear Regression) As an example, we continue the analysis of the weekly interest rates data but now with the 30-year Treasury rates as a second predictor. The changes in the 30-year rate are the variable `cm30_dif`. Thus $p = 2$.

Here is the SAS program. In the data step, the variable `cm30` has missing values at the beginning of the data set. These are coded as zeros. The statement "`if lag(cm30)` $>$ 0" deletes the observations where the lagged value of `cm30` is missing. The statement "`id = _N_`" creates a variable `id` which is the observation number, that is, "1" for the first observation and so forth.[2] This variable is used in plotting.

In the proc step, the model statement in PROC REG specifies `aaa_dif` as the response and `cm10_dif` and `cm30_dif` as the predictors. Everything after the slash (/) in the model statement is an *option*. The options request that Type I and Type II sums of squares (ss1 and ss2) and variance inflation factors (vif) be printed in the output; these are explained in later sections. The output statement of PROC REG requests an output data set, which is named `WeeklyInterest` and so overwrites the data set created in the data step. The output data set contains all variables originally in `WeeklyInterest` as well as the predicted values (predicted), RSTUDENT, Cook's D, and leverage values;

[2] The variable `_N_` is created by SAS as the data are read in and is the observation number. However, `_N_` is *not* retained in the data set and therefore cannot be used in subsequent proc steps, so the program creates a variable `id` that equals `_N_`. The variable `id` is retained.

these are explained shortly. PROC GPLOT, which does plotting, is used to produce diagnostic plots that are explained in later sections.

```
options linesize = 72 ;
data WeeklyInterest ;
infile 'C:\book\SAS\WeeklyInterest.dat' ;
input month day year ff tb03 cm10 cm30 discount prime aaa ;
if lag(cm30) > 0 ;
aaa_dif = dif(aaa) ;
cm10_dif = dif(cm10) ;
cm30_dif = dif(cm30) ;
id = _N_ ;
run ;
title 'Weekly Interest Rates' ;
proc reg ;
model aaa_dif = cm10_dif cm30_dif / ss1 ss2 vif ;
output out=WeeklyInterest predicted=predicted rstudent=rstudent cookd=cookd h=leverage ;
run ;
proc gplot ;
plot rstudent*predicted ;
plot (rstudent cookd leverage cm10_dif cm30_dif)*id ;
plot cm10_dif*cm30_dif ;
run ;
```

Here is the output from this program.

```
                          The REG Procedure
                     Dependent Variable: aaa_dif

                         Analysis of Variance
                                Sum of           Mean
Source                   DF    Squares         Square  F Value  Pr > F

Model                     2   11.35366        5.67683  1357.95  <.0001
Error                   876    3.66206        0.00418
Corrected Total         878   15.01572

        Root MSE              0.06466    R-Square     0.7561
        Dependent Mean       -0.00130    Adj R-Sq     0.7556
        Coeff Var         -4985.33904

                         Parameter Estimates
                   Parameter      Standard
Variable     DF     Estimate         Error  t Value  Pr > |t|    Type I SS

Intercept     1  -0.00010686       0.00218    -0.05    0.9609      0.00148
cm10_dif      1      0.36041       0.04456     8.09    <.0001     11.20585
cm30_dif      1      0.29655       0.04987     5.95    <.0001      0.14781
                         Parameter Estimates
                                                Variance
               Variable     DF   Type II SS     Inflation

               Intercept     1   0.00001004             0
               cm10_dif      1      0.27353      14.03581
               cm30_dif      1      0.14781      14.03581
```

We see that $\widehat{\beta}_0 = -0.00010686$, $\widehat{\beta}_1 = 0.36041$, and $\widehat{\beta}_2 = 0.29655$. The standard error of $\widehat{\beta}_0$ is 0.00218, the standard error of $\widehat{\beta}_1$ is 0.04456, and the standard error of $\widehat{\beta}_2$ is 0.04987. The p-values of both slopes are small, less than 0.0001, indicating that both predictors are related to the response. The estimate of σ_ϵ is 0.06466.

6.4 Analysis Of Variance, R^2, and F-Tests

6.4.1 AOV table

The SAS output also contains an *analysis of variance table*. In this section, the entries in such tables are explained. The idea behind the AOV table is to describe how much of the variation in Y is predictable if one knows X. Here is the AOV table for the weekly interest rates example with only `cm10_dif` as the predictor.

```
                 Dependent Variable: aaa_dif
                    Analysis of Variance

                                 Sum of           Mean
Source                  DF      Squares         Square  F Value  Pr > F

Model                    1     10.89497       10.89497  2716.14  <.0001
Error                 1249      5.00998        0.00401
Corrected Total       1250     15.90495
```

6.4.2 Sums of squares (SS) and R^2

The total variation in Y can be partitioned into two parts, the variation that can be predicted by X and the part that cannot be predicted. The total variation is measured by the total sum of squares (total SS) which is the sum of the squared deviations of Y from its mean; that is,

$$\text{total SS} = \sum_{i=1}^{n} (Y_i - \overline{Y})^2.$$

The total SS is called "Corrected Total" in the table and is 15.905.[3] The variation that can be predicted is measured by the regression sum of squares which is

$$\text{regression SS} = \sum_{i=1}^{n} (\widehat{Y}_i - \overline{Y})^2.$$

The regression sum of squares in the table is called the model sum of squares and is 10.895. Finally, the amount of variation in Y that cannot be predicted by a linear function of X is measured by the residual error sum of squares which is the sum of the squared residual; i.e.,

$$\text{residual error SS} = \sum_{i=1}^{n} (Y_i - \widehat{Y}_i)^2.$$

In the table, the residual error sum of squares is called the error sum of squares and is 5.010. It can be shown algebraically that

[3] SAS used two types of total sums of squares, the corrected where $\overline{Y}$ is subtracted from Y_i before squaring and the uncorrected where $\overline{Y}$ is not subtracted. The latter has some specialized uses.

$$\text{total SS} = \text{regression SS} + \text{residual error SS}.$$

This fact can be verified in the table: 15.905 = 10.895 + 5.010. R-squared, denoted by R^2, is

$$R^2 = \frac{\text{regression SS}}{\text{total SS}} = 1 - \frac{\text{residual error SS}}{\text{total SS}}$$

and measures the proportion of the total variation in Y that can be linearly predicted by X. In the table, R^2 is 68.5% or 0.685 = 10.895/15.905. When there is only a single X variable, then $R^2 = r_{XY}^2 = r_{\widehat{Y}Y}^2$, where r_{XY} and $r_{\widehat{Y}Y}$ are the sample correlations between Y and X and between Y and the predicted values, respectively. Put differently, R^2 is the squared correlation between Y and X and also between Y and $\widehat{Y}$. When there are multiple predictors, then we still have $R^2 = r_{\widehat{Y}Y}^2$. For this reason, R can be viewed as the "multiple" correlation between Y and many Xs. The residual error sum of squares is also called the error sum of squares or sum of squared errors and is denoted by SSE.

6.4.3 Degrees of freedom (DF)

There are degrees of freedom (DF) associated with each of these "sources" of variation. The degrees of freedom for regression is p, which is the number of predictor variables. Note that p is 1 for straight-line regression. The total degrees of freedom is $n-1$. The residual error degrees of freedom is $n-p-1$ which is the total degrees of freedom minus the regression degrees of freedom. Here is a way to think of degrees of freedom. Initially, there are n degrees of freedom, one for each observation. Then one degree of freedom is allocated to estimation of the "constant," that is, the intercept, of the model. This leaves a total of $n-1$ degrees of freedom left for modeling the effects of the X variables. Each regression parameter uses one degree of freedom for estimation. Then there are $(n-1)-p$ degrees of freedom remaining for estimation of σ_ϵ^2 using the residuals. There is an elegant geometrical theory of regression where the responses are viewed as lying in an n-dimensional vector space and degrees of freedom are the dimensions of various subspaces. However, there is not sufficient space to pursue this subject here.

6.4.4 Mean sums of squares (MS) and testing

The mean sum of squares (MS) for any source is its sum of squares divided by its degrees of freedom. The total MS is $(n-1)^{-1}\sum_{i=1}^n (Y_i - \overline{Y})^2$ which is the sample variance of the Y_is. The residual error MS is an unbiased estimator of σ_ϵ^2 and is often denoted by s^2 or $\widehat{\sigma}_\epsilon^2$. To understand why this is so, recall that σ_ϵ^2 is the variance of the ϵ_i and notice that

$$\text{residual error MS} = \frac{\sum_{i=1}^n (Y_i - \widehat{Y}_i)^2}{n-p-1}. \tag{6.7}$$

If there were no estimation errors in the regression coefficients, then the numerator would be the sum of the squared ϵ_is which when divided by n gives an unbiased estimator of σ_ϵ^2. Of course, there is estimation error in the regression coefficients and this causes bias. However, the size of the bias can be computed and it is known that the bias is removed if $n-p-1$ rather than n is used in the denominator. This method of bias correction is analogous to using $n-1$ rather than n in the denominator of the sample variance. The estimate `Root MSE` in the SAS output is the square root of the residual error MS.

The other mean sums of squares are used for testing. The regression MS divided by the residual error MS is called the F-statistic, or just F. The F-statistic is used to perform the so-called F-test of the null hypothesis that there is no linear relationship between any of the predictors and Y. This null hypothesis can be stated succinctly as $\mathrm{H_0}$: $\beta_1 = 0, \ldots, \beta_p = 0$; that is, *all* of the slopes are zero. The alternative is that one or more of the slopes is not zero so that at least one predictor has an effect on the response.[4]

The entry in the column labelled "P" is the p-value of this test. In our example, the p-value is 0.000 which is very strong evidence against the null hypothesis. We conclude that there *is* a relationship between changes in `cm10` and changes in `aaa`.

6.4.5 Adjusted R^2

We have seen that

$$R^2 = 1 - \frac{\text{residual error SS}}{\text{total SS}} = 1 - \frac{n^{-1}\text{residual error SS}}{n^{-1}\text{total SS}}.$$

As just discussed, the quantity (n^{-1}residual error SS) is a biased estimate of the variance of ϵ_i. Moreover, the variance of ϵ_i is the expected squared prediction error when Y is predicted by $\beta_0 + \beta_1 X_1 + \cdots + \beta_p X_p$. The quantity ($n^{-1}$total SS) is a biased estimate of the expected squared prediction error when Y is predicted by $E(Y)$. The fact that both estimators are biased is what biases R^2 towards larger models. The biases can be removed by replacing n^{-1} by $(n-p-1)^{-1}$ and by $(n-1)^{-1}$ in (n^{-1}residual error SS) and (n^{-1}total SS), respectively. The new value of R^2 after these bias corrections is called the adjusted R^2 and is given by

$$\text{adjusted } R^2 = 1 - \frac{(n-p-1)^{-1}\text{residual error SS}}{(n-1)^{-1}\text{total SS}} = 1 - \frac{\text{residual error MS}}{\text{total MS}}.$$

Example 6.4. (Multiple Regression, Example 6.3, *Continued)* For the example of regressing the AAA rate change on `cm10_dif` and `cm30_dif` the analysis of variance table was:

---

[4] By "has an effect on the response" we do not necessarily mean a causal relationship but only that a predictor is related to the response.

```
                      Analysis of Variance

                               Sum of           Mean
Source                   DF     Squares        Square   F Value   Pr > F

Model                     2    11.35366       5.67683   1357.95   <.0001
Error                   876     3.66206       0.00418
Corrected Total         878    15.01572


       Root MSE              0.06466    R-Square     0.7561
       Dependent Mean       -0.00130    Adj R-Sq     0.7556
       Coeff Var         -4985.33904
```

6.4.6 Sequential and partial sums of squares

When there are two or more predictor variables, the regression sum of squares can be divided into the portion due to X_1, the portion due to X_2, and so forth. This is done by the SAS's PROC REG. There are two different sums of squares associated with each predictor, a sequential sum of squares (called Type I SS in SAS) and a partial sum of squares (called Type II SS in SAS).[5] Sequential sums of squares show how prediction is improved as we add the predictors in the order listed in the program, e.g., in our example first `cm10_dif` and then `cm30_dif`. More precisely, the sequential sum of squares for a predictor is the increase in regression sum of squares when that predictor is added to the model that contains all other predictors that precede it in the program. Clearly, some thought on the order in which to list the predictors is needed if one plans to analyze the sequential sums of squares.

Using sequential sums of squares is particularly useful when there is a natural order to the predictor variables. For example, suppose we are building a polynomial regression model with a predictor X. Then, we would add X^2, X^3, and possibly even higher powers of X to the set of possible predictors. The predictors could be ordered by their exponents, i.e., X, X^2, and so forth. The sequential sums of squares would then show the improvements as the degree of the polynomial model increased. If, for example, the sequential sums of squares for X^3 and all higher powers of X were small, then a quadratic model would be chosen.

A partial sum of squares shows how much a predictor improves over all the other predictors and is defined as the increase in the regression sum of squares if that predictor is added to the model containing *all* of the other predictors. The order of the predictors in the program is irrelevant to the values of the partial sums of squares because of the way they are defined. By definition, the Type I and Type II sums of squares must be equal for the predictor that is last in the list, e.g., `cm30_dif` in this example. Any variable that has an

[5] SAS also has Type III and IV sums of squares, but in regression these are equivalent to the Type II sum of squares.

insignificant Type II SS could be dropped from the model, provided that *all* the other predictors are retained.[6]

Example 6.5. In Example 6.3 the model statement

```
model aaa_dif = cm10_dif cm30_dif / ss1 ss2 vif ;
```

requests that Type I and Type II sums of squares be printed in the output. The following output is the result of requesting Type I and II sums of squares.

Parameter Estimates

Variable	DF	Parameter Estimate	Standard Error	t Value	Pr > |t|	Type I SS
Intercept	1	-0.00010686	0.00218	-0.05	0.9609	0.00148
cm10_dif	1	0.36041	0.04456	8.09	<.0001	11.20585
cm30_dif	1	0.29655	0.04987	5.95	<.0001	0.14781

Parameter Estimates

Variable	DF	Type II SS	Variance Inflation
Intercept	1	0.00001004	0
cm10_dif	1	0.27353	14.03581
cm30_dif	1	0.14781	14.03581

Here we see that the regression sum of squares has been divided into two *Type I SS*, 11.20585 due to `cm10_dif` and 0.14781 due to `cm30_dif`. The first Type I SS, 11.20585, is the regression sum of squares when only `cm10_dif` is in the model. The second sequential sum of squares, 0.14781, is the increase in regression sum of squares when `cm30_dif` is added to the model already containing `cm10_dif`. The small size of 0.14781 compared to 11.20585 should not be interpreted as `cm30_dif` being less important than `cm10_dif`. The variables that enter the model first generally have larger values of Type I SS. If we had made `cm30_dif` the first rather than the second predictor variable, then `cm30_dif`, not `cm10_dif`, would have had the larger Type I SS. The point here is that adding a second predictor variable does not improve prediction very much, not necessarily that `cm10_dif` is "better" than `cm30_dif`. This can be seen in the partial (Type II) sums of squares which are small for both variables.

Both predictors have very small p-values indicating that they are statistically significant, but since there are 879 data points, a very small effect of no practical significance could be statistically significant. This may very well be what is happening here since both partial sums of squares are small.

6.5 Regression Hedging*

An important application of regression is determining the optimal hedge of a bond position. Market makers buy securities at a **bid price** and make a

[6] This means if two predictors each have an insignificant Type II SS, then either of them could be deleted, but not necessarily both of them.

profit by selling them at lower **ask price**. Suppose a market maker has just purchased a bond from a pension fund. Ideally, the market maker would sell the bond immediately after purchasing it. However, many bonds are *illiquid*, meaning that they trade infrequently, so it may take some time before the bond could be sold. During the period that a market maker is holding a bond, the market maker is at risk that the bond price could drop due to a change in interest rates. The change could wipe out the profit due to the small spread between the bid and ask prices. The market maker would prefer to hedge this risk. **Hedging** a risk means assuming another risk which is likely to be in the opposite direction, so that the two risks nearly cancel. To hedge the interest-rate risk of the bond being held, the market maker can sell other, more liquid, bonds short. Suppose that the market maker decides to sell short a 30-year Treasury bond, which is quite liquid and could be sold short immediately.

Regression hedging determines the optimal amount of the 30-year Treasury to sell short in order to hedge the risk of the bond just purchased so that the price of the portfolio of the long position in the first bond and the short position in the 30-year Treasury changes little as yields change. It is also necessary to determine the sensitivity of each bond price to changes in interest rates. This is discussed in more detail in Section 9.8. For now, only the essential ideas are outlined. Suppose the first bond has a maturity of 25 years. Let y_{30} be the yield on 30-year bonds and let P_{30} be the price of \$1 in face amount (par value) of 30-year bonds.[7] Then there is a quantity called the duration of the bond and denoted by DUR_{30} such that the change in price, ΔP_{30}, and the change in yield, Δy_{30}, are related by

$$\Delta P_{30} \approx -P_{30}\mathrm{DUR}_{30}\Delta y_{30}$$

for small values of Δy_{30}. A similar result holds for the 25-year bonds. Consider a portfolio that holds F_{25} in 25-year bonds and is short F_{30} in 30-year bonds. The value of the portfolio is

$$F_{25}P_{25} - F_{30}P_{30}.$$

If Δy_{25} and Δy_{30} are the changes in the yields, then the change in value of the portfolio is approximately

$$\{F_{30}P_{30}\mathrm{DUR}_{30}\Delta y_{30} - F_{25}P_{25}\mathrm{DUR}_{25}\Delta y_{25}\}. \tag{6.8}$$

Suppose that the regression of Δy_{30} on Δy_{25} is

$$\Delta y_{30} = \widehat{\beta}_0 + \widehat{\beta}_1 \Delta y_{25} \tag{6.9}$$

and $\widehat{\beta}_0 \approx 0$, as is usually the case for regression of changes in interest rates, e.g., in Example 6.1. Then substituting (6.9) into (6.8), the change in price of the portfolio is approximately

[7] The yield, also called the yield to maturity, is the interest rate. The face or par amount is the amount paid to the bond holder at maturity, that is, 30 years from now. See Chapter 9.

$$\{F_{30}P_{30}\mathrm{DUR}_{30}\widehat{\beta}_1 - F_{25}P_{25}\mathrm{DUR}_{25}\}\Delta y_{25}. \tag{6.10}$$

This change is approximately zero for all values of Δy_{25} if

$$F_{30} = F_{25}\frac{P_{25}\mathrm{DUR}_{25}}{P_{30}\mathrm{DUR}_{30}\widehat{\beta}_1}. \tag{6.11}$$

Equation (6.11) tells us how much face value of the 30-year bond to sell short in order to hedge F_{25} face value of the 25-year bond. The higher the R^2 of the regression, the better the hedge works. Hedging with two or more liquid bonds, say a 30-year and a 10-year can be done by multiple regression; see Problem 11.

6.6 Regression and Best Linear Prediction

Note the similarity between the best linear predictor

$$\widehat{Y} = E(Y) + \frac{\sigma_{XY}}{\sigma_X^2}\{X - E(X)\},$$

and the least squares line

$$\widehat{Y} = \overline{Y} + \frac{s_{XY}}{s_X^2}(X - \overline{X})$$

in the case of a single predictor. The least squares line is a sample version of the best linear predictor. Also ρ_{XY}^2, the squared correlation between X and Y, is the fraction of variation in Y that can be predicted using the linear predictor and the sample version of ρ_{XY}^2 is $R^2 = r_{XY}^2 = r_{\widehat{Y}Y}^2$.

6.7 Model Selection

Model selection means, among other things, the selection of the predictor variables to use in the prediction equation. Often, a large number of potential predictors are available, but we do not necessarily want to use all of them. As discussed in Chapter 4, there are two principles to balance:

- Larger models, that is, models with more predictor variables, have less bias and they *would* give the best predictions *if* all coefficients could be estimated without error;
- When unknown coefficients are replaced by estimates, then the prediction becomes less accurate, and this effect is worse when there are more coefficients to estimate.

Thus, larger models have less bias but more variability. Models with too few parameters and sizeable bias are said to **underfit** while models with too many parameters are said to **overfit**.

Selection criteria, such as AIC and SBC discussed in Chapter 4 can help us select a parsimonious model to achieve a good bias–variance trade-off. However, one should not use automatic model selection software blindly. Such software is not guaranteed to produce good results in all situations, and to obtain good models one must also use common sense and knowledge of the subject matter. Nonetheless, model selection software can be used as a guide to what models one might consider.

R^2 is not a useful statistic for comparing models of different sizes. It is biased towards large models and it always chooses the largest model. As we have seen, the adjusted R^2 statistic is adjusted to remove biases and *can* be used to select models.

Another useful model selection criterion is C_p. C_p is a statistic that estimates how well a model predicts new responses. C_p is closely related to the AIC statistic used in Chapter 4 to compare time series models. In particular, both AIC and C_p consist of a term measuring how well the model fits the data and another term that penalizes the number of parameters.

Here is the definition of C_p. Suppose there are M predictor variables. Let $\widehat{\sigma}^2_{\epsilon,M}$ be the estimate of σ^2_ϵ using all of them and let SSE(p) be the sum of squares for residual error for a model with only $p \leq M$ predictors. As usual, n is the sample size. Then C_p is

$$C_p = \frac{SSE(p)}{\widehat{\sigma}^2_{\epsilon,M}} - n + 2(p+1). \tag{6.12}$$

As we look at different subsets of the M predictors, $\widehat{\sigma}^2_{\epsilon,M}$ which is based on *all* of the predictors stays fixed. The two quantities in (6.12) that vary are SSE(p) and p. SSE(p) measures the size of the squared residuals, and the smaller the better so we want SSE(p) to be as small as possible. The quantity $2(p+1)$ is simply twice the number of regression coefficients to estimate, since there are p slope parameters plus the intercept. Thus, SSE(p) expresses goodness-of-fit of the model to the data while $2(p+1)$ is a penalty on the number parameters. For models with little bias, C_p should be equal to $p+1$, the number of parameters. Therefore, any model with C_p substantially greater than $p+1$ should be considered as an underfit. For models with many parameters, typically C_p is close to $p+1$ but both of these quantities are large, in particular substantially larger than the minimum value of C_p. Such models should be considered overfits, that is, not biased but with too many parameters.

Recall the AIC approximation

$$\text{AIC} \approx n\log(\widehat{\sigma}_\epsilon^{\,2}) + 2(p+q),$$

used for ARMA models in Chapter 4. The term $p+q$ is simply the number of ARMA coefficients and is analogous here to $p+1$, the number of regression parameters. Also, for a regression model

$$\widehat{\sigma_\epsilon}^2 = \text{residual error MS} = \frac{\text{SSE}(p)}{n-p-1},$$

so AIC and C_p are both using SSE(p) to measure goodness-of-fit. The ability just discussed of C_p to judge whether a model is biased is an advantage of C_p over AIC.

SAS's PROC REG command computes all possible models and compare models of the same size by R^2, C_p, AIC, SBC, and adjusted R^2.[8]

Example 6.6. We return to the interest rates example. Here is a SAS program to compare the possible models by R^2, C_p, SBC, and AIC.

```
proc reg ;
model aaa_dif = cm10_dif cm30_dif/selection=rsquare cp sbc aic ;
run ;
```

Here is the output.

```
                        The REG Procedure
                          Model: MODEL1
                   Dependent Variable: aaa_dif

                    R-Square Selection Method

Number in              Adjusted
  Model     R-Square   R-Square       C(p)          AIC            SBC

        1     0.7463     0.7460    36.3579   -4778.8055   -4769.24795
        1     0.7379     0.7376    66.4320   -4750.2675   -4740.70994
-------------------------------------------------------------------------
        2     0.7561     0.7556     3.0000   -4811.5872   -4797.25086

Number in
  Model     R-Square   Variables in Model

        1     0.7463   cm10_dif
        1     0.7379   cm30_dif
----------------------------------------
        2     0.7561   cm10_dif cm30_dif
```

Looking at the adjusted R^2 values, the model with just `cm10_dif` is a very close second to the two-variable model and in practical terms is probably just as good. Also, the model using only `cm30_dif` is nearly not much inferior to the model with just `cm10_dif`.

Example 6.7. Now let us add more potential predictor variables to make things a bit more interesting. In particular, we add `ff_dif` which is the weekly change in the Federal funds rate and `prime_dif` which is the weekly change in the prime rate. `cm10_dif` and `cm30_dif` continue to be used as predictors as well. Here is the SAS program.

[8] When comparing models of the *same* size, the rankings by R^2, adjusted R^2, AIC, SBC, and C_p are identical. Remember that "best" means largest R^2 or adjusted R^2 value or smallest C_p, AIC, or SBC value.

```
options linesize = 72 ;
data WeeklyInterest ;
infile 'C:\book\SAS\WeeklyInterest.dat' ;
input month day year ff tb03 cm10 cm30 discount prime aaa ;
if lag(cm30) > 0 ;
aaa_dif = dif(aaa) ;
cm10_dif = dif(cm10) ;
cm30_dif = dif(cm30) ;
ff_dif = dif(ff) ;
prime_dif = dif(prime) ;
run ;
title 'Weekly Interest Rates' ;
proc reg ;
model aaa_dif = cm10_dif cm30_dif ff_dif prime_dif/selection=rsquare adjrsq cp sbc aic ;
run ;
```

Here is SAS's output. The best model according to adjusted R^2, C_p, AIC, and SBC uses two predictors, `cm10_dif` and `cm30_dif`.[9] The variable `cm10_dif` and `cm30_dif` are both contained in the best two-variable model, the two best three-variable models, and of course the four-variable model. All other models are missing one or both of `cm10_dif` and `cm30_dif`. Thus, one can see that if we are with `cm10_dif` and `cm30_dif` in the model, then adding either or both of the other predictors (`ff_dif` and `prime_dif`) does not increase C_p by much (to no more than 5), but deleting either or both of `cm10_dif` or `cm30_dif` increases C_p to at least 37.2491.

```
                         The REG Procedure
                          Model: MODEL1
                   Dependent Variable: aaa_dif

                     R-Square Selection Method
```

Number in Model	R-Square	Adjusted R-Square	C(p)	AIC	SBC
1	0.7463	0.7460	35.4718	-4778.8055	-4769.24795
1	0.7379	0.7376	65.5166	-4750.2675	-4740.70994
1	0.0625	0.0615	2489.033	-3630.0113	-3620.45378
1	0.0320	0.0309	2598.720	-3601.8083	-3592.25074
2	0.7561	0.7556	2.1482	-4811.5872	-4797.25086
2	0.7463	0.7458	37.2491	-4777.0205	-4762.68417
2	0.7463	0.7457	37.4036	-4776.8714	-4762.53501
2	0.7391	0.7385	63.2227	-4752.2898	-4737.95341
2	0.7379	0.7373	67.5166	-4748.2675	-4733.93116
2	0.0727	0.0706	2454.497	-3637.6104	-3623.27404
3	0.7563	0.7555	3.5415	-4810.1968	-4791.08170
3	0.7562	0.7553	3.9224	-4809.8141	-4790.69896
3	0.7464	0.7455	39.0751	-4775.1885	-4756.07337
3	0.7392	0.7383	64.8002	-4750.6865	-4731.57137
4	0.7564	0.7553	5.0000	-4808.7412	-4784.84732

Number in Model	R-Square	Variables in Model
1	0.7463	cm10_dif
1	0.7379	cm30_dif

[9] Actually, the best two- and three- variable models are tied by adjusted R^2, but the best two-variable model is better according to the other criteria.

```
       1      0.0625    ff_dif
       1      0.0320    prime_dif
----------------------------------------------------------
       2      0.7561    cm10_dif cm30_dif
       2      0.7463    cm10_dif prime_dif
       2      0.7463    cm10_dif ff_dif
       2      0.7391    cm30_dif ff_dif
       2      0.7379    cm30_dif prime_dif
       2      0.0727    ff_dif prime_dif
----------------------------------------------------------
       3      0.7563    cm10_dif cm30_dif ff_dif
       3      0.7562    cm10_dif cm30_dif prime_dif
       3      0.7464    cm10_dif ff_dif prime_dif
       3      0.7392    cm30_dif ff_dif prime_dif
----------------------------------------------------------
       4      0.7564    cm10_dif cm30_dif ff_dif prime_dif
```

6.8 Collinearity and Variance Inflation

When two or more predictor variables are highly correlated with each other, then it is difficult to separate their effects on the response. For example, `cm10_dif` and `cm30_dif` have a correlation of 0.96 and a scatterplot shows that they are nearly equal to each other. If we regress `aaa_dif` on `cm10_dif` then the adjusted R^2 is 0.7460 but adjusted R^2 only increases to 0.7556 if we add `cm30_dif` as a second predictor. This suggests that `cm30_dif` might not be related to `aaa_dif`, but in fact the adjusted R^2 is 0.7376 when `cm30_dif` is the only predictor. The problem here is that `cm10_dif` and `cm30_dif` provide redundant information because of their high correlation. This problem is called *collinearity* or sometimes *multicollinearity*. Collinearity increases standard errors. The standard error of the β of `cm10_dif` is 0.01212 when only `cm10_dif` is in the model, but increases to 0.04456, a 368% increase, if `cm30_dif` is added to the model.

The *variance inflation factor* (*VIF*) of a variable tells us how much the squared standard error, i.e., the variance of $\widehat{\beta}$, of that variable is increased by having the other predictor variables in the model. For example, if a variable has a VIF of 4, then the variance of its $\widehat{\beta}$ is four times larger than it would be if the other predictors were either deleted or were not correlated with it. The standard error is increased by a factor of 2.

Suppose we have predictor variables $X_1, \ldots, X_p$. Then the VIF of X_j is found as follows. Regress X_j on the $p-1$ other predictors and let R_j^2 be the R^2 value of this regression, so that R_j^2 measures how well X_j can be predicted from the other Xs. Then the VIF of X_j is

$$\mathrm{VIF}_j = \frac{1}{1 - R_j^2},$$

so a value of R_j^2 close to 1 implies a large VIF. The more accurately that X_j can be predicted from the other Xs, the more redundant it is and the higher its VIF. The minimum value of VIF_j is 1 and occurs when R_j^2 is 0. There is, unfortunately, no upper bound to VIF_j. Variance inflation becomes infinite as R_j^2 approaches 1.

Example 6.8. VIFs can be requested in PROC REG by including "vif" among the options. This was done in the program in Example 6.3. Part of that output is shown again here.

Parameter Estimates

Variable	DF	Parameter Estimate	Standard Error	t Value	Pr > |t|	Type I SS
Intercept	1	-0.00010686	0.00218	-0.05	0.9609	0.00148
cm10_dif	1	0.36041	0.04456	8.09	<.0001	11.20585
cm30_dif	1	0.29655	0.04987	5.95	<.0001	0.14781

Parameter Estimates

Variable	DF	Type II SS	Variance Inflation
Intercept	1	0.00001004	0
cm10_dif	1	0.27353	14.03581
cm30_dif	1	0.14781	14.03581

The VIFs of `cm10_dif` and `cm30_dif` are 14.03 showing that their standard errors are increased by a factor of $3.75 = \sqrt{14.0358}$.

When there are exactly two predictors, then their VIFs are always equal, as seen in this example.

Example 6.9. Here are the VIFs when all four predictors in Example 6.7 are used. The VIFs for `cm10_dif` and `cm30_dif` are not changed much by the inclusion of two more predictors, because `cm10_dif` and `cm30_dif` are not correlated much with `ff_dif` and `prime_dif`. In addition, `ff_dif` and `prime_dif` are not highly correlated with each other so they have low VIFs.

Parameter Estimates

Variable	DF	Parameter Estimate	Standard Error	t Value	Pr > |t|	Type I SS
Intercept	1	-0.00010103	0.00218	-0.05	0.9631	0.00148
cm10_dif	1	0.35510	0.04517	7.86	<.0001	11.20585
cm30_dif	1	0.30093	0.05010	6.01	<.0001	0.14781
ff_dif	1	0.00531	0.00553	0.96	0.3371	0.00254
prime_dif	1	-0.00788	0.01071	-0.74	0.4620	0.00227

Parameter Estimates

Variable	DF	Type II SS	Variance Inflation
Intercept	1	0.00000897	0
cm10_dif	1	0.25860	14.41205
cm30_dif	1	0.15096	14.15236
ff_dif	1	0.00386	1.19941
prime_dif	1	0.00227	1.14743

6.9 Centering the Predictors

Centering or, more precisely, *mean-centering* a variable means expressing it as a deviation from its mean. Thus, if $X_{1,k}, \ldots, X_{n,k}$ and $\overline{X}_k$ are the values and mean of the kth predictor, then $(X_{1,k} - \overline{X}_k), \ldots, (X_{n,k} - \overline{X}_k)$ are values of the centered predictor.

Centering is useful for two reasons:

- Centering reduces collinearity (see Problem 4);
- If all predictors are centered, then β_0 is the expected value of Y when all of the predictors are equal to their mean. This gives β_0 a more interpretable meaning. In contrast, if the variables are not centered then β_0 is the expected value of Y when all of the predictors are equal to 0. Frequently, 0 is out of the range of some predictors, making the interpretation of β_0 of little real interest, unless the variables are centered.

6.10 Nonlinear Regression

Often we can derive a theoretical model relating predictor variables and a response, but the model we derive is not linear. In particular, models derived from economic theory are common in finance and many are not linear.

For example, consider the price of par $1000 zero-coupon bonds issued by a particular borrower, perhaps the Federal government or a corporation. "Zero-coupon" means that the owner of such a bond is paid $1000 at the time of maturity of that bond but receives no payments prior to maturity. The price of a zero-coupon bond always is less than par value, since the par value equals the repayment of principal plus interest.

Suppose that there is a variety of bonds with different maturities and that the ith type of bond has maturity T_i. Suppose also that all market participants of these zero-coupon bonds agree that the bonds should pay interest at a continuously compounded rate r that is constant across time from now until the maturity of all bonds and also constant across bonds.[10] We say that the market participants "agree" in that bonds trade only if the buyers and sellers agree on prices. The rate r is not publicly announced but rather is implied by prices. We are interested in the estimation of r from prices.

Under this assumption, the present price of a bond with maturity T_i is

$$P_i = 1000 \exp(-rT_i). \tag{6.13}$$

[10] The assumption that the rate is constant across time is not particularly realistic and is used only to keep this example simple. In Chapter 9 we consider more realistic models. The assumption that r is the same for all bonds is realistic if they are all subject to the same risk of default, which is true for U.S. Treasury bonds that have zero default risk.

There is some random variation in the observed prices. One reason for this variation is that the price of a bond can only be determined by the sale of the bond, so the observed prices have not been determined simultaneously. Instead, each bond's price was determined at the time of the last trade of a bond of that maturity. Thus, we add noise to (6.13) to obtain the regression model

$$P_i = 1000 \exp(-rT_i) + \epsilon_i. \tag{6.14}$$

An estimate of r can be determined by least squares, that is, by minimizing over r the sum of squares

$$\sum_{i=1}^{n} \left\{ P_i - 1{,}000 \exp(-rT_i) \right\}^2.$$

The least squares estimator is denoted by $\widehat{r}$.

Because the model is nonlinear, finding the least squares estimate requires solving nonlinear equations. Fortunately, most statistical software packages such as SAS have routines for nonlinear least squares estimation. This means that much of the difficult work has already been done for us. However, we do need to write an equation that tells the software what model we are using. In contrast, when using linear regression only the predictor variables need to be specified since the software knows that the model is linear.

Example 6.10. (Simulated Bond Prices) This example uses simulated data with $r = .06$. The data are listed in Table 6.1. The data and the predicted price curve using nonlinear regression are shown in Figure 6.3.

The following is the SAS program for the nonlinear regression. The "`model`" statement specifies the nonlinear equation that defines the model. Notice that `r`, which is the interest rate, appears in the models statement. In the previous "`parm`" statement, `r` is specified to be a parameter. The values "`r=0.02 to 0.09 by 0.005`" are trial values of `r`. SAS computes the sum of squares at each trial value. The trial value that minimizes the sum of squares is the starting value for an iterative minimization method called Gauss-Newton that is much like Newton's method.

```
data bondprices ;
infile 'c:\book\sas\bondprices.txt' ;
input maturity price ;
run ;
title 'Nonlinear regression using simulated zero-coupon bond data';
proc nlin ;
parm r=0.02 to 0.09 by 0.005 ;
model price = 1000*exp(-r*maturity) ;
run ;
```

Here is the SAS output. The first part of the SAS output shows the initial search of a minimum on the grid "`r=0.02 to 0.09 by 0.005`." The sum of squares is minimized on this grid at `r`=.06. This is the starting value of the Gauss-Newton search.

```
Nonlinear regression using simulated zero-coupon bond data

                    The NLIN Procedure
                       Grid Search
                Dependent Variable price

                                  Sum of
                          r      Squares

                     0.0200       390066
                     0.0250       279853
                     0.0300       192505
                     0.0350       124990
                     0.0400      74665.1
                     0.0450      39230.4
                     0.0500      16679.7
                     0.0550       5263.9
                     0.0600       3456.9
                     0.0650       9926.6
                     0.0700      23509.9
                     0.0750      43191.1
                     0.0800      68082.5
                     0.0850      97408.6
                     0.0900       130491
```

The next section of the output shows the results of the Gauss-Newton search for the least squares estimate. The Gauss-Newton method converged very quickly.

```
                    The NLIN Procedure
                     Iterative Phase
                Dependent Variable price
                 Method: Gauss-Newton

                                      Sum of
              Iter            r      Squares

                 0       0.0600       3456.9
                 1       0.0585       3072.8
                 2       0.0585       3072.8

NOTE: Convergence criterion met.
```

Next in the output, there is an AOV table including an F-test that r is zero. In this context, the null hypothesis that the rate of interest is 0 is a little silly and we shouldn't be surprised that the p-value is < 0.0001.

Source	DF	Sum of Squares	Mean Square	F Value	Approx Pr > F
Regression	1	4490011	4490011	11689.7	<.0001
Residual	8	3072.8	384.1		
Uncorrected Total	9	4493084			
Corrected Total	8	252587			

Finally, we find the estimation results. The least squares estimate is $\widehat{r} = .0585$ with a standard error of .00149. An approximate 95% confidence interval is (0.0551, 0.0619). The confidence interval does contain the true value of r, 0.06.

Parameter	Estimate	Approx Std Error	Approximate 95% Confidence Limits	
r	0.0585	0.00149	0.0551	0.0619

Table 6.1. Data for nonlinear regression example of zero-coupon bond prices.

Maturity	Price
0.75	967.26
2.50	834.21
3.00	810.52
5.00	769.30
7.50	656.64
7.90	639.71
8.00	604.61
12.00	502.11
16.00	393.38

In **nonlinear regression**, the form of the regression function is nonlinear but *known* up to a few unknown parameters. For example, the regression function has an exponential form in model (6.14). For this reason nonlinear regression would best be called *nonlinear parametric regression* to distinguish it from nonparametric regression where the regression function is nonlinear but not of a known parametric form.

Polynomial regression may appear to be nonlinear since polynomials are nonlinear functions. For example, the quadratic regression model

$$Y_i = \beta_0 + \beta_1 X_i + \beta_2 X_i^2 + \epsilon_i \tag{6.15}$$

is nonlinear in X_i. However, by defining X_i^2 as a second predictor variable, this model is linear in (X_i, X_i^2) and therefore is an example of multiple *linear* regression. What makes model (6.15) linear is that the right-hand side is a linear function of the parameters β_0, β_1, and β_2, and therefore can be interpreted as a linear regression with the appropriate definition of the variables. In contrast, the exponential model

$$Y_i = \beta_0 e^{\beta_1 X_i} + \epsilon_i$$

is nonlinear in the parameter β_1 so it cannot be made into a linear model by redefining the predictor variable.

6.11 The General Regression Model

Linear and nonlinear regression can be combined into a general regression model

$$Y_i = f(\boldsymbol{X}_i; \boldsymbol{\beta}) + \epsilon_i, \tag{6.16}$$

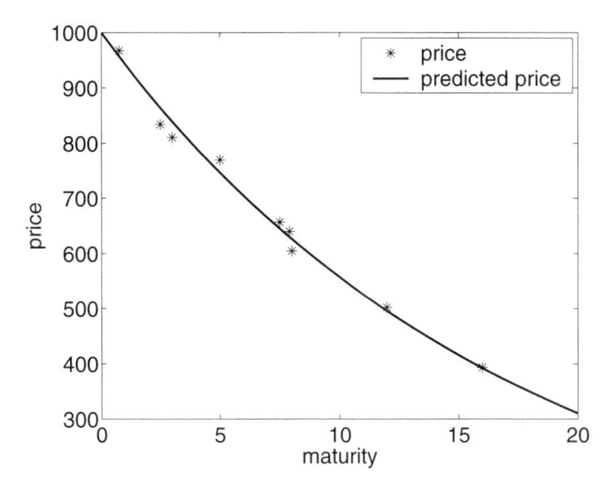

Fig. 6.3. *Plot of bond prices against maturities with the predicted price from the nonlinear least squares fit.*

where Y_i is the response measured on the ith observation, $f(\boldsymbol{X};\boldsymbol{\beta})$ is a *known* function, $\boldsymbol{\beta}$ is an unknown parameter vector, $\boldsymbol{X}_i$ is a vector of observed predictor variables for the ith observation, and $\epsilon_1,\ldots,\epsilon_n$ are i.i.d. with mean 0 and variance σ_ϵ^2. The least squares estimate $\widehat{\boldsymbol{\beta}}$ minimizes

$$\sum_{i=1}^{n}\{Y_i - f(\boldsymbol{X}_i;\widehat{\boldsymbol{\beta}})\}^2.$$

The predicted values are $\widehat{Y}_i = f(\boldsymbol{X}_i;\widehat{\boldsymbol{\beta}})$ and the residuals are $\widehat{\epsilon}_i = Y_i - \widehat{Y}_i$.

6.12 Troubleshooting

Many things can, and often do, go wrong when data are analyzed. There may be data that were entered incorrectly, one might not be analyzing the data set one thinks, the variables may have been mislabeled, and so forth. In Example 6.13, presented shortly, one of the weekly time series of interest rates began with 371 weeks of zeros indicating missing data. However, I was unaware of this problem when I first analyzed the data. The lesson here is that I should have plotted each of the data series first before starting to analyze them, but I hadn't. Fortunately, the diagnostics presented in this section showed quickly that there was some type of serious problem, and then after plotting each of the time series I easily discovered what the problem was. Besides problems with the data, the assumed model may not be a good approximation to reality. The usual estimation methods, such as least squares in regression, are highly nonrobust and therefore particularly sensitive to problems with the data or the model.

The usefulness of robust estimators was discussed in Section 2.18.5. One might replace least squares estimation with robust regression methods that are less sensitive to outlying data and to mild violations of the assumptions, but robust regression is not as widely used as it should be and robust regression programs are not available in some standard statistical software packages, including SAS. Instead of using robust regression methods, most practitioners use regression diagnostics to detect problems and then attempt to remedy any problems that were detected. Personally, I have found robust regression very useful and I wish that robust regression software were more widely available, but since it is not I emphasize regression diagnostics and only discuss robust regression briefly in Section 6.15.

Experienced data analysts know that they should always look at the raw data. Graphical analysis often reveals any problems that exist, especially the types of gross errors that can seriously degrade the analysis.

Figure 6.4 shows data simulated to illustrate some of the problems that can arise in regression. There are 11 observations. The predictor variable takes on values 0,1, ...,9, and 75. The last value is clearly an extreme value in X. Such a point is said to have *high leverage*. However, a high leverage point is not necessarily a problem, only a potential problem. In the top plot, Y is linearly related to X and the extreme X value is, in fact, helpful as it increases the precision of the estimated slope. In the bottom plot, the high leverage point is associated with a Y value that does *not* follow the linear relationship between Y and X seen in the other data. This data point is called a *residual outlier*. Clearly, the high leverage point has an extreme influence on the estimated slope *when it is also a residual outlier*.

In regression and time series analysis, one should also look at the residuals after the model has been fit, because the residual may indicate problems not visible in plots of the raw data. However, there are several types of residuals and as explained soon, one type, called RSTUDENT, is best for diagnosing problems. Ordinary (or raw) residuals are not necessarily useful for diagnosing problems. For example, in the bottom plot in Figure 6.4, none of the residuals is large, not even the one associated with the residual outlier. In fact, residual outliers only appear outlying if one uses the right type of residuals; see the next section.

6.12.1 Influence diagnostics and residuals

There are three important tools that are discussed for diagnosing problems with the model or the data:

- leverages;
- residuals; and
- Cook's D.

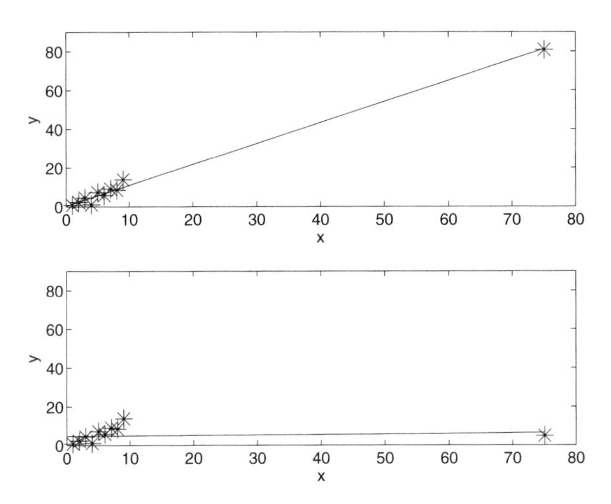

Fig. 6.4. *Linear regression example with a high leverage point.* **Top**: *no residual outliers.* **Bottom**: *residual outlier at high leverage point.*

Leverages

The *leverage* of the ith observations, denoted by H_{ii}, is a measure of how much influence Y_i has on its own fitted value $\widehat{Y}_i$. We do not go into the algebraic details. The end result is that there are weights H_{ij} depending on the values of the predictor variables but *not* on $Y_1, \ldots, Y_n$ such that

$$\widehat{Y}_i = \sum_{j=1}^{n} H_{ij} Y_j.$$

In other words, H_{ii} is the weight of Y_i in the determination of $\widehat{Y}_i$. It is bad if H_{ii} is large since that means that $\widehat{Y}_i$ is too much determined by Y_i itself with the result that the residual $\widehat{\epsilon}_i = Y_i - \widehat{Y}_i$ will be small and not a good estimate of ϵ_i. Also, the standard error of $\widehat{Y}_i$ is $\sigma_\epsilon \sqrt{H_{ii}}$, so a high value of H_{ii} means a fitted value with low accuracy. The leverage value H_{ii} is large when the predictor variables for the ith case are atypical of those values in the data, e.g., because one of the predictor variables for that case is extremely outlying. It can be shown by some elegant algebra that the average of $H_{11}, \ldots, H_{nn}$ is $(p + 1)/n$, where $p + 1$ is the number of parameters (one intercept and p slopes). A value of H_{ii} exceeding $2(p + 1)/n$, that is, over twice the average value, is generally considered to be too large and therefore a cause for concern (Belsley, Kuh, and Welsch, 1980). In Figure 6.4, the high leverage point on the right has a leverage of 0.984 and $2(p + 1)/n = (2)(2)/11 = 0.364$, so the leverage of this point is extreme. The H_{ii}s are sometimes called the *hat diagonals*.

Example 6.11. The output statement in the program of Example 6.3 caused the leverages to be included in the output data set and named `leverage`. The

PROC GPLOT step plotted the leverages against the case number (id). For convenience, these statements are printed again here.

```
proc reg ;
model aaa_dif = cm10_dif cm30_dif / ss1 ss2 vif ;
output out=WeeklyInterest predicted=predicted rstudent=rstudent cookd=cookd h=leverage ;
run ;
proc gplot ;
plot rstudent*predicted ;
plot (rstudent cookd leverage cm10_dif cm30_dif)*id ;
plot cm10_dif*cm30_dif ;
run ;
```

In general, `h=variable-name` outputs the leverages and names them.

Residuals

The *raw residual* is just $\widehat{\epsilon}_i = Y_i - \widehat{Y}_i$. The size of the raw residuals depends on σ_ϵ so we do not know how large a residual to consider unusually large. This problem is to some extent solved by using *standardized residuals*. The ith standardized residual is $\widehat{\epsilon}_i/s$, where s is the estimate of σ_ϵ. Under ideal circumstances such as a reasonably large sample and no outliers or high leverage points, the standardized residuals are approximately $N(0,1)$, so absolute values greater than 2 are outlying and greater than 3 are extremely outlying. However, circumstances are often not ideal. In the bottom plot of Figure 6.4, the standardized residual of the residual outlier/high leverage point is -0.36, not at all outlying. One problem with standardization is that the residuals do not have the same standard errors. They have all been standardized by the same value, s, but they should be standardized by their standard errors. The standard error of $\widehat{\epsilon}_i$ is $s\sqrt{1 - H_{ii}}$. The *studentized residual*,[11] sometimes called the *internally studentized residual*, is $\widehat{\epsilon}_i$ divided by its standard error; that is, $\widehat{\epsilon}_i/(s\sqrt{1 - H_{ii}})$. There is still one problem with studentized residuals. An extreme residual outlier can inflate s causing its studentized residual to appear too small. The solution is to redefine the ith studentized residual with an estimate of σ_ϵ that does not use the ith data point. Thus, the *externally studentized residual*, often denoted by *RSTUDENT*, is defined to be $\widehat{\epsilon}_i/\{s_{(-i)}\sqrt{1 - H_{ii}}\}$ where $s_{(-i)}$ is the estimate of σ_ϵ computed by fitting the model to the data with the ith observation deleted.[12]

In the bottom plot of Figure 6.4, the high leverage point has a standardized residual of -0.36, an internally studentized residual of -2.79, and an RSTUDENT of -7.18. A rule of thumb is that a standardized or studentized residual is outlying if its absolute value exceeds 2 and extremely outlying if it exceeds 3. Thus, we see that the standardized residual completely failed to indicate a problem, the internally studentized residual did show a problem but did not indicate just how extreme the problem was, while the RSTUDENT

[11] *Studentization* means dividing a statistic by its standard error.
[12] The notation $(-i)$ signifies the deletion of the ith observation.

both revealed the problem and indicated its true magnitude. I recommend that RSTUDENT be used for all diagnostics.

See Example 6.11 for an illustration of how to output and plot RSTUDENT.

Cook's D

A high leverage value or a large absolute value of RSTUDENT indicates a potential problem with a data point, but not how much influence a data point has on the estimates. *Cook's distance*, often called *Cook's D*, tells us how much the fitted values change if the ith observations is deleted. Let $\widehat{Y}_j(-i)$ be the jth fitted value using estimates of the $\widehat{\beta}$s obtained with the ith observation deleted. Then Cook's D is

$$\frac{\sum_{j=1}^{n}\{\widehat{Y}_j - \widehat{Y}_j(-i)\}^2}{(p+1)s^2}. \tag{6.17}$$

The numerator in (6.17) is the sum of squared changes in the fitted values when the ith observation is deleted. A large value of this sum indicates a large influence of the ith case on the $\widehat{\beta}$s. The denominator standardizes this sum by dividing by the number of estimated parameters and an estimate of σ_ϵ^2.

The easiest way to use Cook's D is to plot the values of Cook's D against case number and look for unusually large values. See Example 6.11 for an illustration of plotting Cook's D.

Example 6.12. Figure 6.5 shows the values of Cook's D in the simulated data example in Figure 6.4. Clearly, Cook's D for the high leverage point is very large both when that point is a residual outlier and when it is not, but Cook's D is even more extreme in the latter case. In both cases, the estimated slope is almost entirely determined by this one observation, and that is worrisome because the X value of this data point is extreme. There is no way of knowing for sure that the linear regression model holds all the way out to $X = 80$ even if it is a good fit for the other data that are between $X = 0$ and $X = 9$.

Example 6.13. It was mentioned in Example 6.3 that there were missing values of `cm30` at the beginning of the data set that were coded as zeros. In fact, there were 371 weeks of missing data for `cm30`. I started to analyze the data without realizing this problem. This created a huge outlying value of `cm30_dif` at observation number 372 when `cm30` jumps from 0 to the first nonmissing value. Fortunately, plots of RSTUDENT, leverages, and Cook's D all reveal a serious problem somewhere between the 300th and 400th observation; see Figure 6.6. The nature of the problem is not evident from these plots, so I plotted each of the series `aaa`, `cm10`, and `cm30`. When I saw the initial zero values of the latter series, the problem was then obvious. Please remember this lesson: *ALWAYS look at the data.*

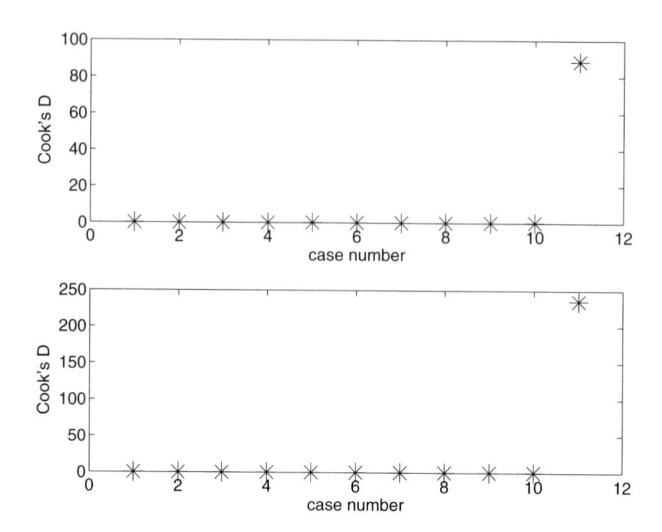

Fig. 6.5. *Cook's D in the linear regression example with a high leverage point.* ***Top:*** *no residual outliers.* ***Bottom:*** *residual outlier at high leverage point. Notice the very different vertical scales on the top and bottom. Cook's D of the leverage point is much larger in the bottom plot.*

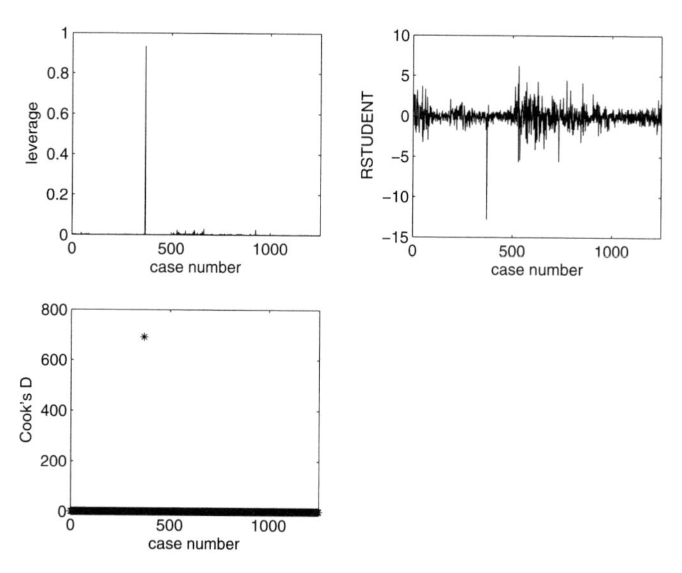

Fig. 6.6. *Weekly interest data. Full data set including the first* 371 *weeks of data where* cm30 *was missing and assigned a value of* 0. *This caused severe problems at case number* 372 *which are detected by Cook's D, the leverages, and RSTUDENT.*

6.12.2 Residual analysis

Because the ith residual $\widehat{\epsilon}_i$ estimates the "noise" ϵ_i, the residuals can be used to check the assumptions behind regression. Residual analysis generally consists of various plots of the residuals, each plot being designed to check one or more of the regression assumptions. Regression software will output the several types of residuals discussed in Section 6.12.1. I recommend using RSTUDENT.

Problems to look for include:

- Nonnormality of the residuals;
- Nonconstant variance of the residuals;
- Correlation of the residuals; and
- Nonlinearity of the effects of the predictor variables on the response.

Nonnormality

Nonnormality can be detected by a normal probability plot, boxplot, and histogram of the residuals. Not all three are needed, but I strongly recommend looking at a normal plot. Moreover, inexperienced data analysts have trouble with the interpretation of normal plots. Looking at a normal plot and a histogram, side by side, is very helpful when learning to use normal probability plots.

The residuals often appear nonnormal because there is an excess of outliers relative to the normal distribution. We have defined a value of RSTUDENT to be outlying if the absolute value of the raw residual exceeds 2 and extremely outlying if it exceeds 3. Of course these cutoffs of 2 and 3 are arbitrary and only intended to give rough guidelines.

It is the presence of outliers, particularly extreme outliers, that is a concern when we have nonnormality. A deficiency of outliers relative to the normal distribution is not a problem. Sometimes outliers are due to errors, such as mistakes in the entry of the data or, as in Example 6.13, misinterpreting a zero as a true data value rather than the indicator of a missing value. If possible, outliers due to mistakes should be corrected, of course. However, in financial time series, outliers are often "good observations" due to excess volatility in the markets on certain days. An excess of only positive outlying residuals occurs when the response is right skewed and similarly left skewness is associated with an excess of negative outlying residuals. If the residuals appear to be roughly symmetric with an excess of both negative and positive outlying residuals, then the noise distribution, that is, the distribution of $\epsilon_1, \ldots, \epsilon_n$, has **heavy tails**.

Another possible reason for an excess of both positive and negative outlying residuals in a normal probability plot is nonconstant residual variance, a problem which is explained shortly. Normal probability plots assume that all observations come from the same distribution, in particular, that they have the same variance. The purpose of that plot is to determine if the common distribution is normal or not. If there is not a common distribution, for example, because of nonconstant variance, then the normal plot is not readily interpretable. Therefore, one should check for a constant variance before attempting to interpret a normal plot.

Outliers can be a problem because they have an unduly large influence on the estimation results. A common solution to the problem of outliers is transformation of the response. It is always wise to check whether outliers are due to erroneous data, e.g., typing errors or other mistakes in data collection and entry. Of course, errors should be corrected if possible, but if this is not possible then erroneous data should be removed. Removal of outliers that are not known to be erroneous is dangerous and not recommended as routine statistical practice. However, reanalyzing the data with outliers removed is a sound practice. If the analysis changes drastically when the outliers are deleted, then one knows there is something about which to worry. On the other hand, if deletion of the outliers does not change the conclusions of the analysis then there is less reason to be concerned with whether the outliers were erroneous data.

Nonconstant variance

Nonconstant residual variance means that the conditional variance of the response given the predictor variables is not constant as assumed by standard regression models. Nonconstant variance is also called **heteroscedasticity**. Nonconstant variance can be detected by an absolute residual plot, that is, by plotting the absolute residuals against the predicted values ($\widehat{Y}_i$s) and, perhaps, also against the predictor variables. If the absolute residuals show a systematic trend, then this is an indication of nonconstant variance. Economic data often have the property that bigger responses are more variable. A more technical way of stating this is that the conditional variance of the response (given the predictor variables) is an increasing function of the conditional mean of the response. This type of behavior can be detected by plotting the absolute residuals versus the predicted values and looking for an increasing trend.

Often, trends are difficult to detect just by looking at the plotted points and adding a so-called *scatterplot smoother* is very helpful. A scatterplot smoother fits a smooth curve to a scatterplot. Nonparametric regression estimators such as loess, lowess,[13] and smoothing splines are commonly used scatterplot smoothers available in statistical software packages.

A potentially serious problem caused by nonconstant variance is inefficiency, that is, too-variable estimates. Here is a very simple example to illustrate the problem of inefficiency. We assume the simplest possible regression model with only an intercept and no predictor variables; that is, $Y_i = \beta_0 + \epsilon_i$. Not surprisingly, the estimate of the intercept is the sample

[13] Loess and lowess are different, but closely related, algorithms.

mean of the responses. Assume Y_1, Y_2, Y_3 are independent, all with mean β_0 and $\text{Var}(Y_1) = \text{Var}(Y_2) = 1$ and $\text{Var}(Y_3) = 10$. If we use all the data then

$$\text{Var}(\widehat{\beta}_0) = \text{Var}\{(Y_1 + Y_2 + Y_3)/3\} = (1 + 1 + 10)/9 = 12/9 = \frac{4}{3}.$$

If we delete Y_3 because it is less accurate than the other observations, then

$$\text{Var}(\widehat{\beta}_0) = \text{Var}\{(Y_1 + Y_2)/2\} = (1 + 1)/4 = \frac{1}{2} < \frac{4}{3}.$$

Our estimator is improved by deleting the inaccurate data point. However, deleting Y_3 is *not* the best thing we can do here. Although Y_3 contains less information about β_0 than either Y_1 or Y_2, it still contains some information. The most efficient (least variable) estimate is obtaining by *downweighting* Y_3 rather than deleting it altogether. The optimal weights are the reciprocal variances. Using these weights give us

$$\text{Var}(\widehat{\beta}_0) = \text{Var}\{(Y_1 + Y_2 + 0.1Y_3)/2.1\} = \{1 + 1 + (.1)^2 10\}/(2.1)^2 = \frac{1}{2.1} < \frac{1}{2}.$$

Another serious problem is that standard errors and confidence intervals assume a constant variance and can be seriously wrong if there is substantial nonconstant variance.

Transformation of the response and weighting are common solutions to the problem of nonconstant variance. Response transformations are presented in Section 6.13. Weighted least squares estimates $\boldsymbol{\beta}$ by minimizing

$$\sum_{i=1}^{n} w_i \{Y_i - f(\boldsymbol{X}_i; \widehat{\boldsymbol{\beta}})\}^2, \tag{6.18}$$

with w_i an estimate of the inverse (i.e., reciprocal) conditional variance of Y_i given $\boldsymbol{X}_i$. Estimation of the conditional variance function to determine the w_is is discussed in more advanced textbooks such as Carroll and Ruppert (1988).

Nonlinearity

If a plot of the residuals versus a predictor variable shows a systematic nonlinear trend, then this is an indication that the effect of that predictor on the response is nonlinear. Nonlinearity causes biased estimates. Confidence intervals, which assume unbiasedness, can be seriously incorrect if there is nonlinearity. The value $100(1 - \alpha)\%$ is called the *nominal value* of the coverage probability of a confidence interval and is guaranteed to be the actual coverage probability only if all modeling assumptions are met. For example, the probability that a 95% confidence interval contains its parameter might be far less than 0.95 if the estimate is very biased.

Response transformation, polynomial regression, and nonparametric regression (splines, loess) are common solutions to the problem of nonlinearity when it exists.

Residual correlation

If the data $(\boldsymbol{X}_i, Y_i)$ are a multivariate times series, then it is likely that the ϵ_i noises are correlated. This problem can be detected by looking at the SACF of the residuals. Similarly, in SAS, one can save the residuals in an output SAS data set, and then run, for example, PROC ARIMA on them. In PROC ARIMA, one can run the "identify" statement without an "estimate" statement if one wants only the SACF without the fitting of an ARMA model. Autocorrelation of the residuals is called *serial correlation* by some authors.

Residual correlation causes standard errors and confidence intervals to be incorrect. In particular, the coverage probability of confidence intervals can be much less than the nominal value. A solution to this problem is to model the ϵ_i as an AR process. This can be done using SAS's PROC AUTOREG.

Example 6.14. Data were simulated to get an example illustrating many of the ways to diagnose problems. In the example there are two predictor variables, X_1 and X_2. The assumed model is multiple linear regression.

Figure 6.7, which shows the responses plotted against each of the predictors, does not indicate any problems with this example, except the heteroscedasticity is suggested because the data are more scattered on the right sides of the plots. The point is that plots of the raw data often fail to reveal problems. Rather, it is plots of the residuals that can more reliably detect heteroscedasticity, nonnormality, and other difficulties.

Figure 6.8 is a normal plot and a histogram of the residuals. Notice the right skewness which suggests that a response transformation to remove right skewness, e.g., a square root or log transformation, should be investigated.

Figure 6.9 is a plot of the residuals versus X_1. The residuals appear to have a nonlinear trend. This is better revealed by adding a spline curve fit to the residuals. The curvature of the spline is evident. This pattern suggests that Y is not linear in X_1. A possible remedy is to add X_1^2 as a third predictor so that y is modeled as quadratic in X_1. Figure 6.10, a plot of the residuals against X_2, shows somewhat random scatter, indicating that Y appears to be linear in X_2. The spline fit does dip downward at the right side of Figure 6.10, but spline estimates are highly variable near the boundaries of the data and this feature is likely to be just random variation.

Figure 6.11 is a plot of the absolute residuals versus the predicted values. Note that the absolute residuals are largest where the fitted values are also largest, which is a clear sign of heteroscedasticity. A spline smooth has been added to make the heteroscedasticity clearer.

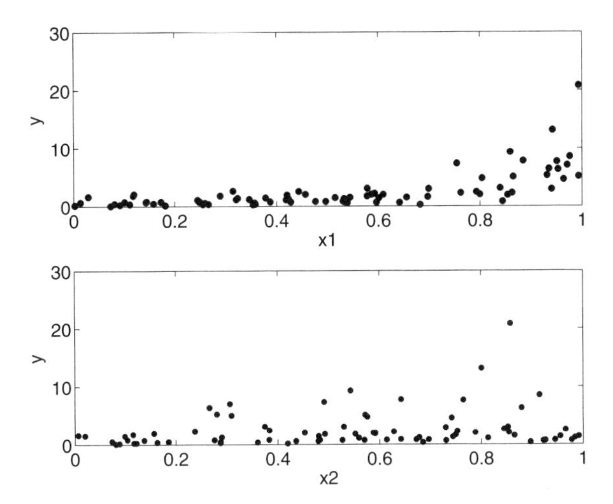

Fig. 6.7. *Simulated data. Responses plotted against the two predictor variables.*

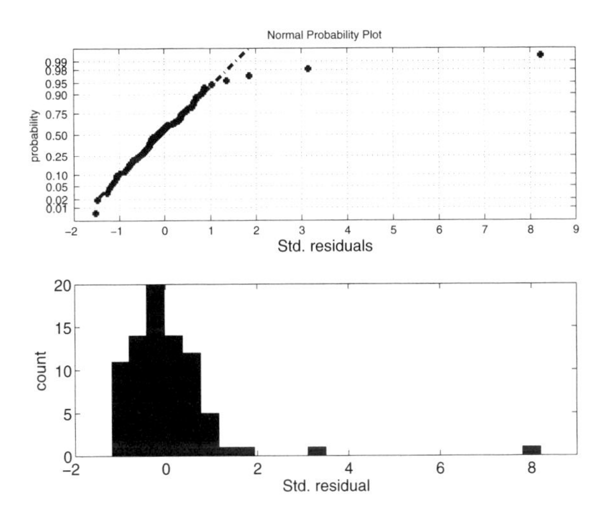

Fig. 6.8. *Simulated data. Normal plot and histogram of the studentized residuals. Right skewness is enough and perhaps a square root or log transformation of* Y *would be helpful.*

Example 6.15. (Estimating Default Probabilities) This example illustrates both nonlinear regression and the detection of heteroscedasticity by residual plotting.

Credit risk is the risk to a lender that a borrower will default on contractual obligations, in short, that a loan will not be repaid in full. A key parameter in the determination of credit risk is the probability of default. Bluhm, Overbeck,

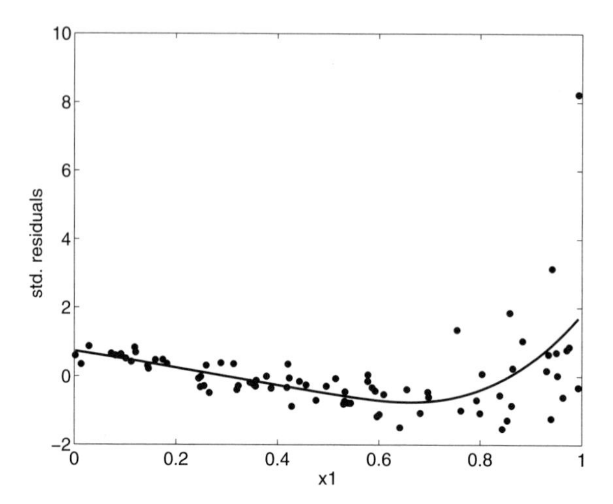

Fig. 6.9. *Simulated data. Plot of studentized residuals versus the X_1 with a spline smooth. This plot suggests that Y is not linearly related to X_1 and perhaps a model quadratic in X_1 is needed.*

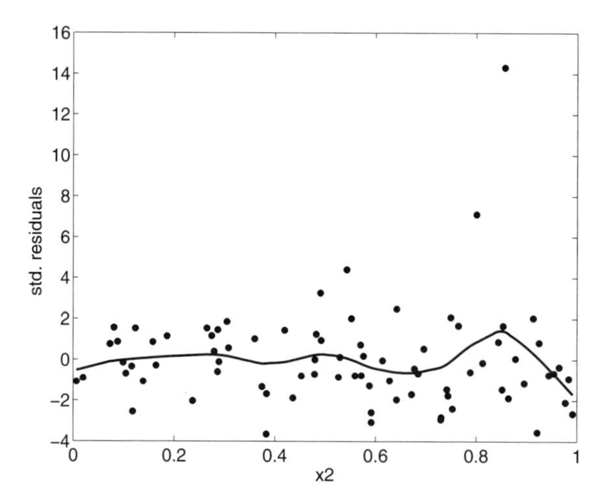

Fig. 6.10. *Simulated data. Plot of studentized residuals versus the X_2 with a spline smooth. This plot suggests that Y is linearly related to X_2 so that the component of the model relating Y to X_2 is satisfactory.*

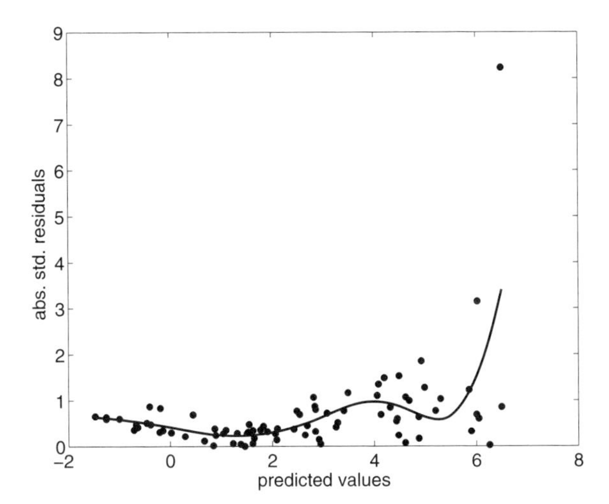

Fig. 6.11. *Simulated data. Plot of the absolute studentized residuals versus the predicted values with a spline smooth. This plot reveals heteroscedasticity.*

and Wagner (2003) illustrate how one can calibrate Moody's credit rating to estimate default probabilities. These authors use observed default frequencies for bonds in each of 16 Moody's ratings from Aaa (best credit rating) to B3 (worse rating). They convert the credit ratings to a 1 to 16 scale (Aaa = 1,..., B3 = 16). Figure 6.12 shows default frequencies plotted against the ratings. The data are from Bluhm, Overbeck, and Wagner (2003). The relationship is clearly nonlinear and the linear regression fit labelled "linear" in the figure has two obvious and serious problems. First, it does not follow the data closely, and, second, it gives negative estimated default probabilities at the better ratings. Unsurprisingly, Bluhm, Overbeck, and Wagner do not consider a linear fit to the default frequencies. Instead they assume that the default probability is an exponential function of the rating; that is,

$$Pr\{\text{default}|\text{rating}\} = \exp\{\beta_0 + \beta_1 \text{rating}\}. \tag{6.19}$$

To use this model they fit a linear function to the logarithms of the default frequencies. One difficulty with doing this is that many of the default frequencies are zero giving a log transformation of $-\infty$.

Bluhm, Overbeck, and Wagner address this issue by labelling default frequencies equal to zero as "unobserved" and not using them in the estimation process. The problem with their technique is that they have deleted the data with the lowest observed default frequencies. This biases their estimates of default probabilities in an upward direction. As we show, the bias is sizeable. Bluhm, Overbeck, and Wagner argue that an observed default frequency of zero does not imply that the true default probability is zero. This is certainly true. However, the default frequencies, even when they are zero, are unbiased estimates of the true default probabilities. I do not wish to seem critical of their book, which I find very interesting and useful. However, one can avoid the bias of their method by using nonlinear regression[14] with model (6.19). The advantage of fitting (6.19) by nonlinear regression is that it avoids the use of a logarithm transformation thus allowing the use of all the data, even data with a default frequency of zero. The fits by the Bluhm, Overbeck, and Wagner method and by nonlinear regression using model (6.19) are shown in Figure 6.13 with a log scale on the vertical axis so that the fitted functions are linear. Notice that at good credit ratings the estimated default probabilities are lower using nonlinear regression compared to Bluhm, Overbeck, and Wagner's biased method. The differences between the two sets of estimated default probabilities can be substantial. Bluhm, Overbeck, and Wagner estimate the default probability of an Aaa bond as 0.005%. In contrast, the unbiased estimate by nonlinear regression is only 40% of that figure, specifically, 0.0020%. Thus, the bias in the Bluhm, Overbeck, and Wagner estimate leads to a substantial overestimate of the credit risk of Aaa bonds and similar overestimation at other good credit ratings.

Figure 6.14 is an absolute residual plot, that is, a plot of the absolute residuals versus the fitted values. We see a clear indication of heteroscedasticity. Heteroscedasticity does not cause bias but it does cause inefficient, that is, unnecessarily variable, estimates. This problem is fixed in the next section. Figure 6.15 is a normal probability plot of the residuals. Outliers with both negative and positive values can be seen. These are due to the nonconstant variance and are not necessarily a sign of nonnormality. This plot illustrates the danger of attempting to interpret a normal plot when the data have a nonconstant variance.

6.13 Transform-Both-Sides Regression*

Suppose we have a theoretical model which states that in the absence of any noise

$$Y_i = f(\boldsymbol{X}_i; \boldsymbol{\beta}). \tag{6.20}$$

Model (6.20) is identical to the model

$$h\{Y_i\} = h\{f(\boldsymbol{X}_i; \boldsymbol{\beta})\}, \tag{6.21}$$

[14] Strictly speaking, nonlinear least squares can be biased, but the bias is usually small.

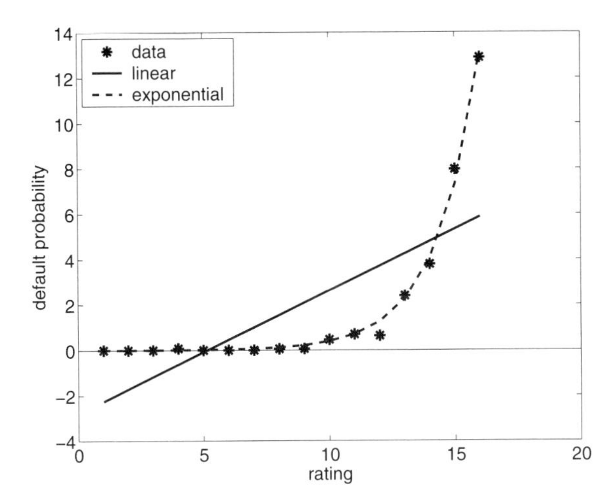

Fig. 6.12. *Default frequencies versus rating with a linear regression fit. The default frequency is expressed as a percentage. "Rating" is a conversion of the Moody's Rating to a* 1 *to* 16 *point scale as follows:* 1 = *Aaa,* 2 = *Aa*1, 3 = *Aa*3, 4 = *A*1, ..., 16 = *B*3. *Default probabilities are expressed as percentages.*

where h is *any* one-to-one function.[15] When we have noisy data, equation (6.21) can be converted to the nonlinear regression model

$$h\{Y_i\} = h\{f(\boldsymbol{X}_i; \boldsymbol{\beta})\} + \epsilon_i. \tag{6.22}$$

Model (6.22) is called the transform-both-sides (TBS) regression model because both sides of equation (6.21) have been transformed by the same function h.

Here is what is particularly interesting about using the transformation h. Although model (6.21) with no noise is the same model for all one-to-one functions h, when we do have noise then the ϵ_is in (6.22) are $N(0, \sigma_\epsilon^2)$ for *at most one* function h. The "best" h for our purposes is the one that makes the ϵ_is as close to $N(0, \sigma_\epsilon^2)$ as possible. In other words, we try to find a function h that induces normal residuals with a constant variance. This can be done by trying different functions h to find one with residuals plots showing little sign of nonnormality or nonconstant variance. Commonly used h functions for nonnegative responses are the square root, log, reciprocal, and identity transformations, the last-named giving the ordinary nonlinear regression as a special case of TBS regression. The log function can, of course, only be used for positive response, and if some response is equal to zero, one often uses $h(y) = \log(y + c)$ where c is a positive constant. The transformation

[15] The function h is one-to-one if $h(x_1) = h(x_2)$ only if $x_1 = x_2$ so that h has an inverse function. To appreciate why h should be one-to-one, consider the non-one-to-one function $h(x) = 1$ for all x. Then (6.21) becomes $1 = 1$ which, though true, is certainly not identical to model (6.20).

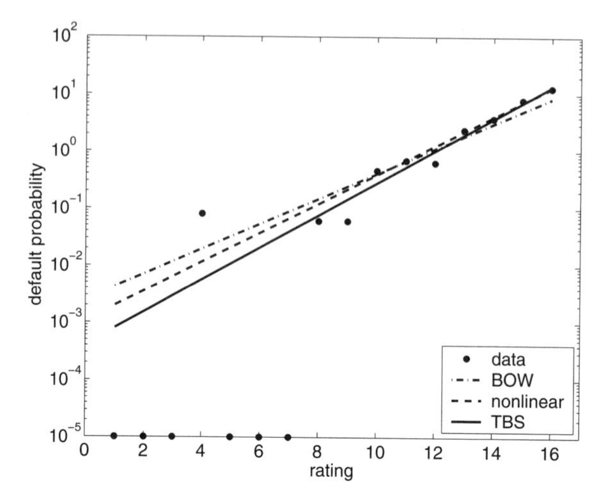

Fig. 6.13. *Estimation of default probabilities by Bluhm, Overbeck, and Wagner's (2003) linear regression with ratings removed that have no observed defaults (biased estimates) and by nonlinear regression with all data (unbiased). The default probability is expressed as a percentage. "Rating" is a conversion of the Moody's Rating to a 1 to 16 point scale as follows: 1 = Aaa, 2 = Aa1, 3 = Aa3, 4 = A1, ..., 16 = B3. Because some default frequencies are zero, when plotting the data on a semilog plot 10^{-5} was added to the default frequencies. This constant was* not *added when estimating default frequencies, only for plotting the raw data. The six observations along the bottom of the plot are the ones removed by Bluhm, Overbeck, and Wagner. The fit labeled "nonlinear" uses model (6.19) and "BOW" is the Bluhm, Overbeck, and Wagner estimate. Compare with Figure 1.1 of Bluhm, Overbeck, and Wagner which shows their fit and does not include the zero default frequency data. "TBS" is the transform-both-sides estimate.*

$\log(y + c)$ is rather sensitive to the value of c, and if the log transformation is to be used then c should be chosen carefully. If c is too small, then the observations with a response of 0 appears as negative outliers and if c is very large then the transformation will have only a weak effect on nonconstant variance and nonnormality. One should try a range of values of c and select a value giving residual plots that indicate normally distributed residuals with little heteroscedasticity.

To estimate $\boldsymbol{\beta}$ by TBS regression, after h has been selected one minimizes

$$\sum_{i=1}^{n}[h\{Y_i\} - h\{f(\boldsymbol{X}_i; \widehat{\boldsymbol{\beta}})\}]^2. \tag{6.23}$$

Example 6.16. TBS regression was applied to the default frequency data. The transformation $h(y) = y^{\alpha}$ was tried with various positive values of α. It was found that $\alpha = 1/2$ gave residuals that appeared normally distributed with

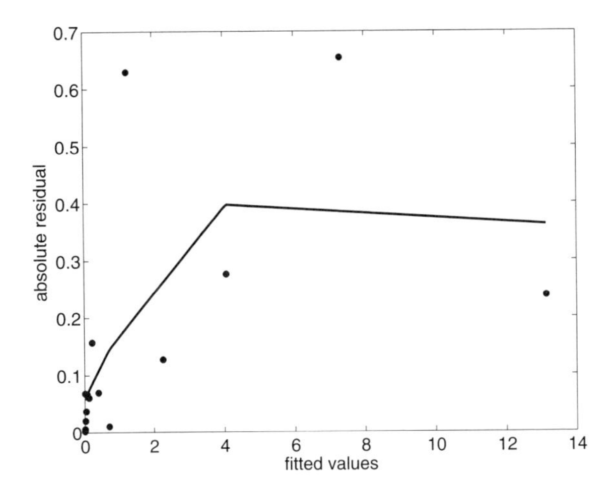

Fig. 6.14. *Estimation of default probabilities by nonlinear regression. Absolute studentized residuals plotted against fitted values with a spline smooth. Substantial heteroscedasticity is indicated because the data on the left side are much more scattered than elsewhere.*

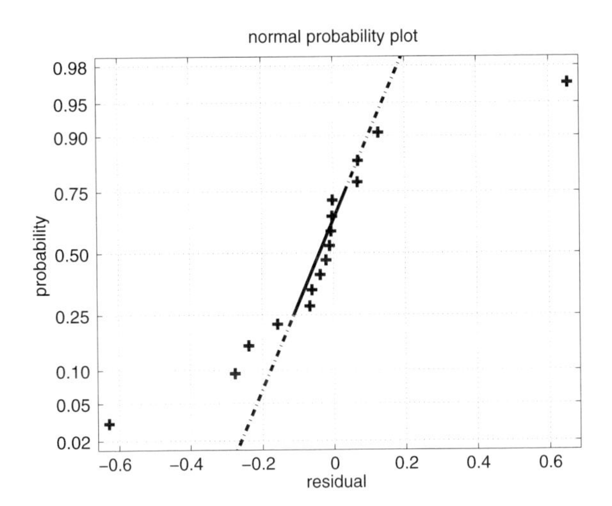

Fig. 6.15. *Estimation of default probabilities by nonlinear regression. Normal probability plot of the studentized residuals. Notice that there are both negative and positive outliers caused by the nonconstant variance.*

a constant variance, so the square root transformation was used for estimation; see Figures 6.16 and 6.17. With this transformation, $\boldsymbol{\beta}$ is estimated by minimizing

$$\sqrt{Y_i} = \exp\{\beta_0/2 + \beta_1/2X_i\}, \tag{6.24}$$

where Y_i is the ith default frequency and X_i is the ith rating. The square root transformation of the model is accomplishing by dividing β_0 and β_1 by 2. Using TBS regression, the estimated default probability of Aaa bonds is 0.0008%, only 16% of the estimate given by Bluhm, Overbeck, and Wagner and only 40% of the estimate given by nonlinear regression without a transformation. Of course, a reduction in estimated risk by 84% is a huge change. This shows how proper statistical modeling, e.g., using all the data and an appropriate transformation, can have a major impact on financial analysis.

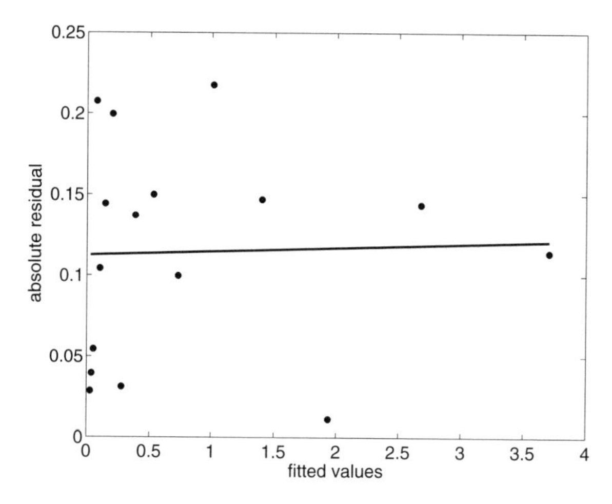

Fig. 6.16. *Transform-both-sides regression (TBS) with $h(y) = \sqrt{y}$. Absolute studentized residuals plotted against fitted values with a spline smooth.*

6.13.1 How TBS works

TBS in effect weights the data. To appreciation this, we use a Taylor series linearization[16] to obtain

[16] A Taylor series linearization of the function h about the point x is $h(y) \approx h(x) + h^{(1)}(x)(y - x)$ where $h^{(1)}$ is the first derivative of h. See any calculus textbook for further discussion of Taylor series.

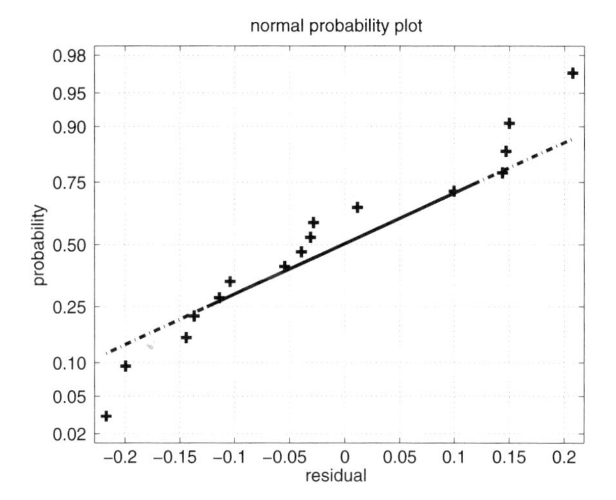

Fig. 6.17. *Transform-both-sides regression (TBS) with $h(y) = \sqrt{y}$. Normal plot of the studentized residuals.*

$$\begin{aligned} &\sum_{i=1}^{n}[h(Y_i) - h\{f(\boldsymbol{X}_i;\widehat{\boldsymbol{\beta}})\}]^2 \\ &\approx \sum_{i=1}^{n}\left[h^{(1)}\{f(\boldsymbol{X}_i;\widehat{\boldsymbol{\beta}})\}\{Y_i - f(\boldsymbol{X}_i;\widehat{\boldsymbol{\beta}})\}\right]^2 \\ &= \sum_{i=1}^{n}\left[h^{(1)}\{f(\boldsymbol{X}_i;\widehat{\boldsymbol{\beta}})\}\right]^2\{Y_i - f(\boldsymbol{X}_i;\widehat{\boldsymbol{\beta}})\}^2. \end{aligned}$$

Since the best weights are inverse variances, the most appropriate transformation h solves

$$\mathrm{Var}(Y_i|\boldsymbol{X}_i) \propto [h^{(1)}\{f(\boldsymbol{X}_i;\widehat{\boldsymbol{\beta}})\}]^{-2}. \tag{6.25}$$

For example, if $h(y) = \log(y)$, then $h^{(1)}(y) = 1/y$ and (6.25) becomes

$$\mathrm{Var}(Y_i|\boldsymbol{X}_i) \propto \{f(\boldsymbol{X}_i;\widehat{\boldsymbol{\beta}})\}^2; \tag{6.26}$$

that is, the conditional variance of the response is proportional to its conditional mean squared. This occurs frequently. For example, if the response is exponentially distributed then (6.26) must hold. Equation (6.26) holds also if the response is lognormally distributed and the variance of the log response is constant. In this case it is not surprising that the log transformation is best since the log transforms to i.i.d. normal noise. When (6.26) holds, the response has a constant coefficient of variation.[17]

[17] The *coefficient of variation* of a random variable is the ratio of its standard deviation to its expected value.

A transformation which causes that conditional variance to be constant is called the *variance-stabilizing transformation.* We have just shown that when the coefficient of variation is constant, that is, when the conditional response variance is proportional to the conditional response mean squared, then the variance-stabilizing transformation is the logarithm.

Example 6.17. (Poisson responses) Assume $Y_i|\boldsymbol{X}_i$ is Poisson distributed with mean $f(\boldsymbol{X}_i;\widehat{\boldsymbol{\beta}})$, as might, for example, happen if Y_i were of the number of companies declaring bankruptcy in a year with $f(\boldsymbol{X}_i;\boldsymbol{\beta})$ modeling how that expected number depends on macroeconomic variables in $\boldsymbol{X}_i$. The variance equals the mean for the Poisson distribution, so

$$\mathrm{Var}(Y_i|\boldsymbol{X}_i) = f(\boldsymbol{X}_i;\widehat{\boldsymbol{\beta}}).$$

Thus, we should use $\alpha = 1/2$; that is, the square root transformation is the variance-stabilizing transformation for Poisson distributed responses.

Example 6.18. (TBS regression in SAS) Here is a SAS program to do TBS regression on the default probability data of Bluhm, Overbeck, and Wagner (2003). The first section of the program reads in the data and raises the default frequency (called `prob`) to the α power (called `alpha` in the program). The program was run several times for various values of `alpha`, and `alpha` equal to 0.5 gave "good" residual plots showing little heteroscedasticity or nonnormality of the residuals. Therefore, `alpha` was set to 0.5 as seen here.

```
data DefaultProb ;
infile 'C:\book\sas\DefaultData.txt' ;
input rating prob;
alpha = .5 ;
transprob = prob**alpha ;
run ;
```

Next the TBS regression model is fit. Note that the `alpha` power of the model is taken by multiplying `beta0 + beta1*rating` by `alpha`. The predicted values and residuals are saved in an output file named "`outdata`" and named `PredValue` and `Residual`, respectively.

```
proc nlin ;
parm beta0=-8 to -1 by 1 beta1=0 to 3 by .5 ;
model transprob = exp(alpha*(beta0 + beta1*rating)) ;
output out = outdata p=PredValue r=Residual ;
run ;
```

The output data set is updated to include the absolute residuals. Then the absolute residuals are plotted against the predicted values.

```
data outdata;
set outdata ;
absresid = abs(Residual) ;
run ;
proc gplot;
plot absresid*PredValue ;
run ;
```

Finally PROC UNIVARIATE is used to test normality of the residuals and to generate a normal plot of the residuals. The "normal" option in the proc statement specifies tests of normality. The statement "probplot" specifies a probability plot.

```
proc univariate normal;
var Residual ;
probplot;
run ;
```

Here are selected parts of the output. The plots are not shown here but are similar to Figures 6.16 and 6.17. First iteration of the Gauss-Newton procedure locates the nonlinear least squares estimator. Notice that the convergence criterion is met. This is not a guarantee that the procedure has converged to the least squares estimator, but it is a good sign.

```
                    The NLIN Procedure
               Dependent Variable SqrtProb
                   Method: Gauss-Newton
                     Iterative Phase
                                                 Sum of
          Iter        beta0        beta1        Squares

             0      -6.0000       0.5000         1.7369
             1      -8.0208       0.6710         0.3230
             2      -7.7480       0.6480         0.2748
             3      -7.7690       0.6493         0.2747
             4      -7.7676       0.6492         0.2747
             5      -7.7677       0.6492         0.2747
NOTE: Convergence criterion met.
```

Next we see the least squares estimates, their standard errors, and confidence intervals.

```
                              Approx        Approximate 95%
Parameter      Estimate    Std Error       Confidence Limits

beta0           -7.7677       0.5182     -8.8791     -6.6563
beta1            0.6492       0.0346      0.5750      0.7234
```

After this we have the output from PROC UNIVARIATE. The skewness is relatively small (0.271) and the excess kurtosis is negative (-1.128) so the residuals appear close to being normally distributed.[18]

```
                    The UNIVARIATE Procedure
                      Variable:  RESIDUAL

                            Moments

N                            16     Sum Weights                 16
```

[18] If the excess kurtosis is really negative, this would indicate *light* tails relative to the normal distribution. Light tails mean *less* or *smaller* outliers than the normal distribution and are generally not a problem. However, the negative estimate of the excess kurtosis might be due simply to random estimation error since normality tests accept the null hypothesis of normality.

```
Mean                 -0.0194237    Sum Observations    -0.3107791
Std Deviation         0.1338344    Variance             0.01791165
Skewness             0.27064364    Kurtosis            -1.1282057
Uncorrected SS       0.27471116    Corrected SS         0.26867468
Coeff Variation     -689.02644     Std Error Mean        0.0334586
```

Here are the tests of normality from PROC UNIVARIATE. The p-values are all rather large (0.15 or larger), which is more evidence that the residuals are normally distributed.

```
                         Tests for Normality

Test                  --Statistic---     -----p Value------

Shapiro-Wilk          W      0.937735    Pr < W      0.3222
Kolmogorov-Smirnov    D      0.151914    Pr > D     >0.1500
Cramer-von Mises      W-Sq   0.063363    Pr > W-Sq  >0.2500
Anderson-Darling      A-Sq   0.402689    Pr > A-Sq  >0.2500
```

6.13.2 Power transformations

It is commonly the case that the response is right skewed and the conditional response mean is an increasing function of the conditional response variance. In such case, a power transformation with $h(y) = y^\alpha$ with $\alpha < 1$ will remove skewness and stabilize the variance, and the smaller the value of α the greater the effect of the transformation. One can go too far — if the transformed response is *left* skewed or has a conditional variance that is decreasing as a function of the conditional mean, then α has been chosen too small. If α is close to zero, then the power transformation behaves much as the log transformation. For this reason, the power transformation with $\alpha = 0$ is *defined* to be the log transformation.

6.14 The Geometry of Transformations*

Response transformations induce normality of a distribution and stabilize variances because they can stretch apart data in one region and push observations together in other regions. Figure 6.18 illustrates this behavior. On the horizontal axis is a sample of data from a right skewed lognormal distribution. The curve is the log transformation. On the vertical axis the transformed data are plotted. The dashed lines show the transformation of y to $h(y)$. Notice the near symmetry of the transformed data. This symmetry is achieved because the log transformation stretches apart data with small values and shrinks together data with large values. This can be seen by observing the derivative of the log function. The derivative of $\log(y)$ is $1/y$ which is a decreasing function of y. The derivative is of course the slope of the tangent line and the tangent lines at $y = 1$ and $y = 5$ are plotted to show the decrease in the derivative as y increases.

Consider an arbitrary increasing transformation, $h(y)$. If x and x' are two nearby data points that are transformed to $h(x)$ and $h(x')$, respectively, then the distance between transformed values is $|h(x) - h(x')| \approx h^{(1)}(x)|x - x'|$. Therefore, $h(x)$ and $h(x')$ are stretched apart where $h^{(1)}$ is large and pushed together where $h^{(1)}$ is small. A function h is called concave if $h^{(1)}(y)$ is a decreasing function of y. As can be seen in Figure 6.18, concave transformations remove right skewness.

Concave transformations can also stabilize the variance when the untransformed data are such that small observations are less variable than large observations. This is illustrated in Figure 6.19. There are two groups of responses, one with a mean of 1 and a relatively small variance and another with a mean of 5 and a relatively large variance. If the expected value of the response Y_i, conditional on $\boldsymbol{X}_i$, followed a regression model $f(\boldsymbol{X}_i; \boldsymbol{\beta})$, then two groups like these would occur if there were two possible values of $\boldsymbol{X}_i$, one with a small value of $f(\boldsymbol{X}_i; \boldsymbol{\beta})$ and the other with a large value. Because of the concavity of the transformation h, the variance of the group with a mean of 5 is reduced by transformation. After transformation, the groups have nearly the same variance.

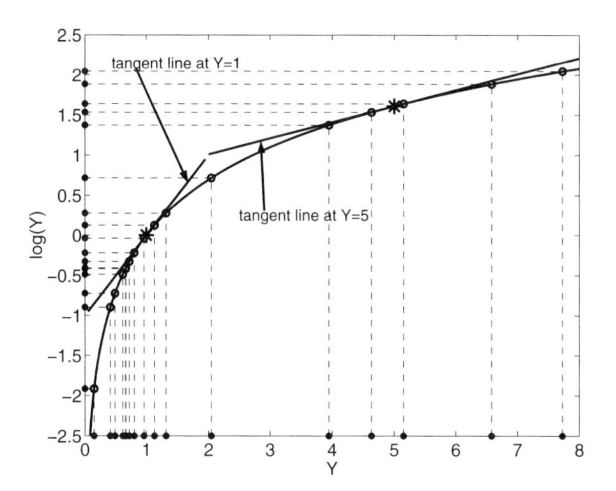

Fig. 6.18. *A symmetrizing transformation. The skewed lognormal data on the horizontal axis are transformed to symmetry by the log transformation.*

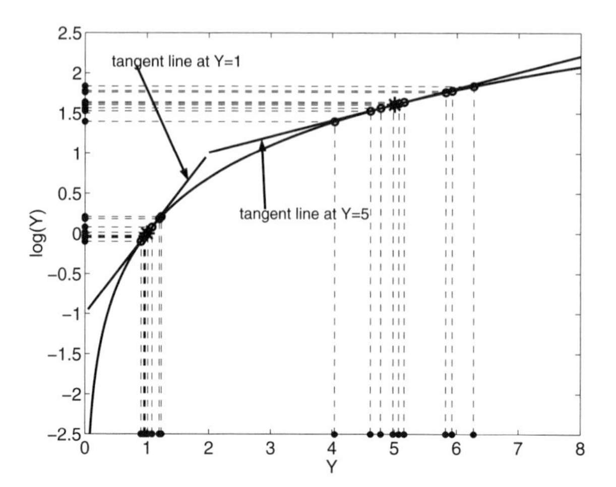

Fig. 6.19. *A variance-stabilizing transformation.*

6.15 Robust Regression*

A robust regression estimator should be relatively immune to two types of outliers. The first are *bad data* meaning *contaminants* that are not part of the population, for example, due to undetected recording errors. The second are outliers due to the noise distribution having heavy tails. There are a large number of robust regression estimators, and their sheer number has been an impediment to their use. Many data analysts are confused as to which robust estimator is best and consequently are reluctant to use any at all. Rather than describe many of these estimators, which might contribute to this problem, I mention just one, the *least trimmed sum of squares estimator*, often called the *LTS*.

Recall the trimmed mean, a robust estimator of "location" for a univariate sample. The trimmed mean is simply the mean of the sample after a certain percentage of the largest observations and the same percentage of the smallest observations have been removed. This trimming removes some nonoutliers which causes some loss of precision, but the loss is not large. The trimming also removes outliers and this causes the estimator to be robust. Trimming is easy for a univariate sample because we know which observations to trim, the very largest and the very smallest. This is not the case in regression. Consider the data in Figure 6.20. There are 21 observations that fall closely along a line plus one *residual outlier* which is far from this line. Notice that the residual outlier has neither an extreme X-value nor an extreme Y-value. It is only outlying relative to the linear regression model. This is always the case in regression. The outliers are only outlying relative to a particular model. We can only identify outliers if we have a model *and* estimates of the parameters in that model. The difficulty is that estimation of the parameters requires the identification of the outliers and vice versa. One can see from the figure that the least squares line is changed by using the residual outlier for estimation. In some cases, though not here, the effect of a residual outlier can be so severe that it totally changes the least squares estimates. This is likely to happen if the residual outlier occurs at a high leverage point.

The LTS estimator simultaneously identifies residual outliers and estimates the parameters of a model. Let $0 < \alpha \leq 1/2$ be the trimming proportion and let k equal $n\alpha$ rounded to an integer, either upward or to the closest integer. The trimmed sum of squares about a set of values of the regression parameters is defined as follows: form the residuals from these parameters, square the residuals, then order the squared residuals and remove the k largest, and finally sum the remaining squared residuals. The LTS estimates are the set of parameter values that minimize the trimmed sum of squares. The trimmed residuals include the outliers, though they also include nonoutliers.

Software to compute the LTS estimate is not as readily available as it should be. IML, the interactive matrix language of SAS, has a routine to compute the LTS estimator, but the use of PROC IML is beyond the scope of this book. The S-PLUS software package also can compute the LTS estimator.

If the noise distribution is heavy-tailed, then an alternative to both robust estimation and least squares is to use a heavy-tailed distribution as a model for the noise and then to estimate the parameters by maximum likelihood. For example, one could assume that the noise has a double exponential or t-distribution. In the latter case, one could either estimate the degrees of freedom or simply fix the degrees of freedom at a low value, which implies heavier tails; see Lange, Little, and Taylor (1989). This strategy is called robust modeling rather than robust estimation. The distinction is that in robust estimation one assumes a fairly restrictive model such as a normal noise distribution but finds a robust alternative to maximum likelihood, while in robust modeling one uses a more flexible model so that maximum likelihood estimation is itself robust. Another possibility is that residual outliers are due to nonconstant standard deviations, with the outliers mainly in the data with a higher noise standard deviation. The remedy to this problem is to model the nonconstant standard deviation, say by one of the GARCH models discussed in Chapter 12. Regression with GARCH noise can be implemented in PROC AUTOREG of SAS, as illustrated in examples in Chapter 12.

6.16 Summary

The available data on each observation are one response variable and p predictor variables. Y_i is the value of the response variable for the ith observation and $X_{i1}, \ldots, X_{ip}$ are the values of predictor variables 1 through p for the ith observation. The straight line regression model is

$$Y_i = \beta_0 + \beta_1 X_i + \epsilon_i.$$

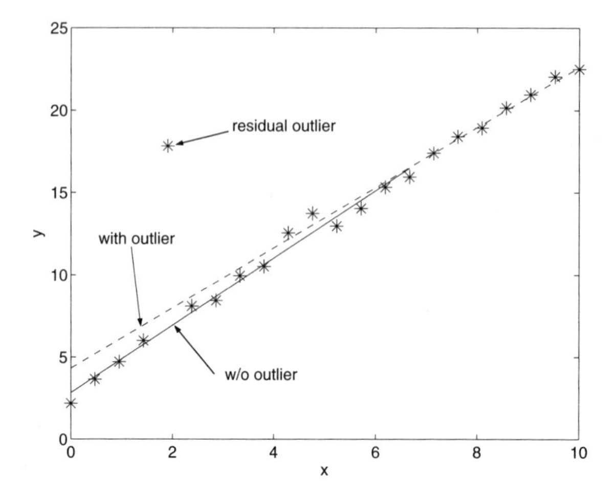

Fig. 6.20. *Straight line regression with one residual outlier.*

The least squares estimate finds $\widehat{\beta}_0, \ldots, \widehat{\beta}_p$ to minimize

$$\sum_{i=1}^{n} \left\{ Y_i - (\widehat{\beta}_0 + \widehat{\beta}_1 X_{i,1} + \cdots + \widehat{\beta}_p X_{ip}) \right\}^2 .$$

The standard error or SE of a coefficient is the estimated standard deviation of its least squares estimator and tells us the precision of that estimator. The t-value and p-value of a coefficient are the t-statistic and p-value for testing the null hypothesis that the coefficient is 0 versus the alternative that it is not 0.

The sums of squares partition is

$$\text{total SS} = \text{regression SS} + \text{residual error SS.}$$

R-squared, denoted by R^2, is

$$R^2 = \frac{\text{regression SS}}{\text{total SS}} = 1 - \frac{\text{residual error SS}}{\text{total SS}}.$$

The regression MS divided by the residual error MS is called the F-statistic, or just F. The F-statistic is used to perform the so-called F-test of the null hypothesis that there is no linear relationship between any of the predictors and Y. The residual error MS is an unbiased estimate of σ_ϵ^2 and is often denoted by $\widehat{\sigma_\epsilon}^2$, s^2, or S^2.

The multiple regression model is

$$Y_i = \beta_0 + \beta_1 X_{i,1} + \cdots + \beta_p X_{i,p} + \epsilon_i .$$

Larger models, that is, models with more predictor variables, will have less bias and they *would* give the best predictions *if* all coefficients could be estimated without error. When unknown coefficients are replaced by estimates, the predictions become less accurate and this effect is worse when there are more coefficients to estimate. Therefore, it is important to choose the right size model. There are a number of tools to help choose a model, but one should not use automatic model selection software blindly. The R^2 statistic can be used to compare models with the same number of parameters, but R^2 is not a useful statistic for comparing models of different sizes because it is biased in favor of large models. The adjusted R^2 statistic does compensate for the biases induced by error in estimation of the regression parameters and the adjusted R^2 *can* be used to select models. C_p is a statistic that estimates how well a model will predict. One should use a model with a large adjusted R^2 or a small C_p. Most regression programs such as SAS's PROC REG will compute *all* possible models and compare models by R^2, C_p, and adjusted R^2.

Residual plotting can detect problems including nonlinearity of the predictor variable effects, correlation of the noise, nonconstant response variance, and nonnormality of the noise. Transform-both-sides regression can induce a normal response distribution and a constant variance the response variance.

6.17 Bibliographic Notes

Neter, Kutner, Nachtsheim, and Wasserman (1996) and Draper and Smith (1998) are good introductions to regression and residual plotting. Atkinson (1985) has nice coverage of transformations and residual plotting and many good examples. For more information on nonlinear regression, see Bates and Watts (1988) and Seber and Wild (1989). Graphical methods for detecting a nonconstant variance, transform-both-sides regression, and weighting are discussed in Carroll and Ruppert (1988). Nonparametric regression is covered in many books, e.g., Bowman and Azzalini (1997) and Ruppert, Wand, and Carroll (2003). This book does not cover multivariate regression where there is more than one response, possibly correlated, but this topic is covered in many textbooks on multivariate analysis such as Mardia, Kent, and Bibby (1979). *Multivariate* regression should not be confused with *multiple* regression which has one response but multiple predictor variables.

Robust regression estimators including the LTS estimator are available in the S-PLUS software package and are discussed in Venables and Ripley (1997). Congdon (2003) has a section on a Bayesian approach to robust estimation.

Comprehensive treatments of regression diagnostics can be found in Belsley, Kuh, and Welsch (1980) and in Cook and Weisberg (1982). Although variance inflation factors detect collinearity, they give little indication of its nature, that is, the correlations that are causing the problem. For this purpose, one should use collinearity diagnostics. These are also discussed in Belsley, Kuh, and Welsch (1980) and can be computed in SAS by PROC REG.

See Tuckman (2002) for more discussion of regression hedging.

6.18 References

Atkinson, A. C. (1985) *Plots, Transformations and Regression*, Clarendon, Oxford.

Bates, D. M. and Watts, D. G. (1988) *Nonlinear Regression Analysis and Its Applications*, Wiley, New York.

Belsley, D. A., Kuh, E., and Welsch, R. E. (1980) *Regression Diagnostics*, Wiley, New York.

Bluhm, C., Overbeck, L., and Wagner, C. (2003) *An Introduction to Credit Risk Modelling*, Chapman & Hall/CRC, Boca Raton, FL.

Bowman, A. W. and Azzalini, A. (1997) *Applied Smoothing Techniques for Data Analysis*, Clarendon, Oxford.

Carroll, R. J. and Ruppert, D. (1988) *Transformation and Weighting in Regression*, Chapman & Hall, New York.

Congdon, P. (2003) *Applied Bayesian Modelling*, Wiley, Chichester.

Cook, R. D. and Weisberg, S. (1982) *Residuals and Influence in Regression*, Chapman & Hall, New York.

Draper, N. R. and Smith, H. (1998) *Applied Regression Analysis, 3rd Edition*, Wiley, New York.

Lange, K. L., Little, R. J. A., and Taylor, J. M. G. (1989) Robust statistical modeling using the t-distribution, *Journal of the American Statistical Association*, **84** , 881–896.

Mardia, K. V., Kent, J. T., and Bibby, J. M. (1979) *Multivariate Analysis*, Academic, London.

Neter, J., Kutner, M. H., Nachtsheim, C. J., and Wasserman, W. (1996) *Applied Linear Statistical Models, 4th ed.*, Irwin, Chicago.

Ruppert, D., Wand, M. P., and Carroll, R. J. (2003) *Semiparametric Regression*, Cambridge University Press, Cambridge, U.K.

Seber, G. A. F. and Wild, C. J. (1989) *Nonlinear Regression*, Wiley, New York.

Tuckman, B. (2002) *Fixed Income Securities, 2nd Ed.*, Wiley, Hoboken, NJ.

Venables, W. N. and Ripley, B. D. (1997) *Modern Applied Statistics with S-PLUS, 2nd ed.*, Springer, New York.

6.19 Problems

1. Suppose that $Y_i = \beta_0 + \beta_1 X_i + \epsilon_i$ where ϵ_i is $N(0, 0.6)$, $\beta_0 = 2$, and $\beta_1 = 1$.
   (a) What are the conditional mean and standard deviation of Y_i given that $X_i = 1$? What is $P(Y_i \le 3 | X_i = 1)$?
   (b) A regression model is a model for the conditional distribution of Y_i given X_i. However, if we also have a model for the marginal distribution of X_i then we can find the marginal distribution of Y_i. Assume that X_i is $N(1, 0.4)$. What is the marginal distribution of Y_i? What is $P(Y_i \le 3)$?

2. Show that if $\epsilon_1, \ldots, \epsilon_n$ are i.i.d. $N(0, \sigma_\epsilon^2)$ then in straight line regression the least squares estimates of β_0 and β_1 are also the maximum likelihood estimates.
   Hint: This problem is similar to the example in Section 2.18.1. The only difference is that in that section, $Y_1, \ldots, Y_n$ are independent $N(\mu, \sigma^2)$, while in this exercise $Y_1, \ldots, Y_n$ are independent $N(\beta_0 + \beta_1 X_i, \sigma_\epsilon^2)$.
3. Use (2.55), (6.1), and (6.2) to show that (6.5) holds.
4. It was stated in Section 6.9 that centering reduces collinearity. As an illustration, consider the example of quadratic polynomial regression where X takes 30 equally spaced values between 1 and 10.
   (a) What is the correlation between X and X^2? What are the VIFs of X and X^2?
   (b) Now suppose that we center X before squaring. What is the correlation between $(X - \overline{X})$ and $(X - \overline{X})^2$? What are the VIFs of $(X - \overline{X})$ and $(X - \overline{X})^2$?
5. A linear regression model with three predictor variables was fit to a data set with 30 observations. The correlation between Y and $\widehat{Y}$ (the predicted values) was 0.5. The total sum of squares was 100.
   (a) What is the value of R^2?
   (b) What is the value of the residual error SS?
   (c) What is the value of the regression SS?
   (d) What is the value of s^2?
6. A data set has 66 observations and five predictor variables. Three models are being considered. One has all five predictors and the others are smaller. Below is residual error SS for all three models. The total SS was 50. Compute C_p and R^2 for all three models. Which model should be used based on this information?

Number of predictors	Residual error SS
3	12.0
4	10.2
5	10.0

7. The quadratic polynomial regression model

$$Y_i = \beta_0 + \beta_1 X_i + \beta_2 X_i^2 + \epsilon_i$$

   was fit to data. The p-value for β_1 was 0.65 and for β_2 was 0.91. Can we accept the hypothesis that β_1 and β_2 are both 0? Discuss.
8. Sometimes it is believed that β_0 is 0 because we think that $E(Y|X = 0) = 0$. Then the appropriate model is

$$y_i = \beta_1 X_i + \epsilon_i.$$

   This model is usually called "regression through the origin" since the regression line is forced through the origin. The least squares estimator of β_1 minimizes

$$\sum_{i=1}^{n} \{Y_i - \beta_1 X_i\}^2.$$

Find a formula that gives $\hat{\beta}_1$ as a function of the Y_is and the X_is.

9. You have just fitted the model $Y_i = \beta_0 + \beta_1 X_i + \epsilon_i$ to your data set, and you observe the following residuals versus fits plot.

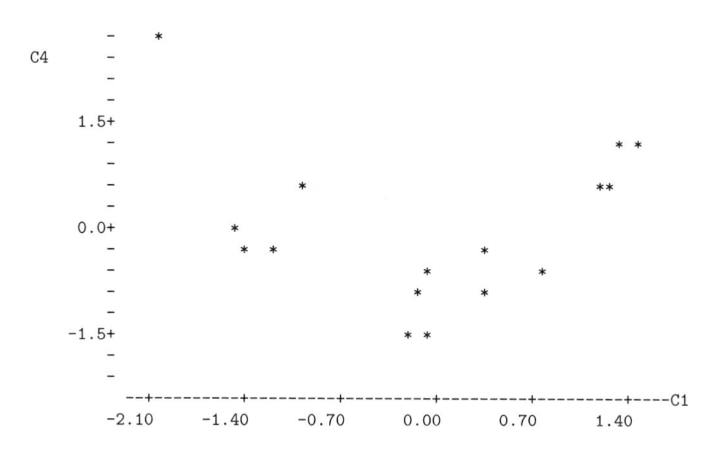

What is this residual plot suggesting to you? What remedial action would you take, if any, and why?

10. Complete the following ANOVA table relative to the model $Y_i = \beta_0 + \beta_1 X_{i,1} + \beta_2 X_{i,2} + \epsilon_i$:

Source	df	SS	MS	F	P
Regression	?	?	?	?	0.05
error	?	3.250	?		
total	14	?			

R-sq = ?

11. Suppose one has a long position of F_{20} face value in 20-year Treasury bonds and wants to hedge this with short positions in both 10- and 30-year Treasury bonds. The prices and durations of 10-, 20-, and 30-year Treasury bonds are P_{10}, DUR_{10}, P_{20}, DUR_{20}, P_{30}, and DUR_{30} and are assumed to be known. A regression of changes in the 20-year yield on changes in the 10- and 30-year yields is $\Delta y_{20} = \widehat{\beta}_0 + \widehat{\beta}_1 \Delta y_{10} + \widehat{\beta}_2 \Delta y_{30}$. The p-value of $\widehat{\beta}_0$ is large and it is assumed that β_0 is close enough to zero to be ignored. What face amounts F_{10} and F_{30} of 10- and 30-year Treasury bonds should be shorted to hedge the long position in 20-year

Treasury bonds? (Express F_{10} and F_{30} in terms of the known quantities P_{10}, P_{20}, P_{30}, DUR_{10}, DUR_{20}, DUR_{30}, $\widehat{\beta}_1$, $\widehat{\beta}_2$, and F_{20}.)

7

The Capital Asset Pricing Model

7.1 Introduction to CAPM

The **CAPM (capital asset pricing model)** has a variety of uses. It provides a theoretical justification for the widespread practice of "passive" investing known as *indexing*. Indexing means holding a diversified portfolio in which securities are held in the same relative proportions as in a broad market index such as the S&P 500. Individual investors can do this easily by holding shares in an *index fund*.[1] CAPM can provide estimates of expected rates of return on individual investments and can establish "fair" rates of return on invested capital in regulated firms or in firms working on a cost-plus basis.[2]

CAPM starts with the question, what would be the risk premiums on securities if the following assumptions were true?

1. The market prices are "in equilibrium." In particular, for each asset, supply equals demand.
2. Everyone has the same forecasts of expected returns and risks.
3. All investors choose portfolios optimally according to the principles of efficient diversification discussed in Chapter 5. This implies that everyone holds a tangency portfolio of risky assets as well as the risk-free asset[3] and only the mix of the tangency portfolio and the risk-free varies between investors.
4. The market rewards people for assuming unavoidable risk, but there is no reward for needless risks due to inefficient portfolio selection. Therefore, the risk premium on a single security is not due to its "stand-alone" risk,

[1] An index fund holds the same portfolio as some index. For example, an S&P 500 index fund holds all 500 stocks on the S&P 500 in the same proportions as in the index. Some funds do not replicate an index exactly, but are designed to have a very high correlation with returns on the index.

[2] Bodie and Merton (2000).

[3] In the CAPM, all risk-free assets are equivalent so we refer to all of them as *the* risk-free asset.

but rather to its contribution to the risk of the tangency portfolio. The various components of risk are discussed in Section 7.4.

The validity of the CAPM can only be guaranteed if all of these assumptions are true, and certainly no one believes that any of them are exactly true. Assumption 3 is at best an idealization. Moreover, some of the conclusions of CAPM are contradicted by the behavior of financial markets; see Section 7.8.2 for an example. Despite its shortcomings, the CAPM is widely used in finance and there is controversy about how serious the evidence against it really is.[4] I believe it is essential for a student of finance to understand the CAPM. Many of its concepts such as the beta of an asset and systematic and diversifiable risks are of great importance and testing CAPM as a null hypothesis has led to new insights about financial markets.

We start with a simple example. Suppose that there are exactly three assets with a total market value of $100 billion:

- Stock A: $60 billion;
- Stock B: $30 billion; and
- risk-free: $10 billion.

The market portfolio of Stock A to Stock B is 2:1. CAPM says that under equilibrium, all investors hold Stock A to Stock B in a 2:1 ratio. Therefore, the tangency portfolio puts weight 2/3 on Stock A and 1/3 on Stock B and *all* investors have two-thirds of their allocation to risky assets in Stock A and one-third in Stock B.

How does it transpire that the market and tangency portfolio are equal? Suppose earlier the tangency portfolio had also been a 2:1 ratio but there had been too little of Stock A and too much of Stock B for everyone to have a 2:1 allocation. For example, suppose that there were one billion shares of each stock and the price per share was $50 for Stock A and $40 for Stock B. Then the market portfolio must have held Stock A to Stock B in a 5:4 ratio, not 2:1. Not everyone could hold the tangency portfolio, though everyone would want to. Thus, prices would be in disequilibrium and would change. The price of Stock A would go up since the supply of Stock A is less than the demand. Similarly the price of Stock B would go down. Therefore, the proportions of the market capitalization in Stock A and Stock B would change, meaning that the market portfolio would change. As the prices change the expected returns might also change and the tangency portfolio might change. Eventually, the market and tangency portfolios would become equal and everyone would hold the tangency portfolio. Or, at least, this is what would happen in the ideal world of the CAPM theory. Of course, the real economic world is more complex than *any* theory, but CAPM still provides a useful way of looking at the real economy. The underlying message from CAPM theory is correct. Prices adjust as all investors look for an efficient portfolio and supply and demand converge.

[4] Campbell, Lo, and MacKinlay (1997).

The market portfolio is 9:1 risky to risk-free. In total, investors must hold risky to risk-free in a 9:1 ratio since they *are* the market. For an individual investor, the risky:risk-free ratio depends on that investor's risk aversion.

- At one extreme, a portfolio of all risk-free has a standard deviation of returns equal to 0.
- At the other extreme, all risky assets, the standard deviation is maximized.[5]

At equilibrium, returns on risky and risk-free assets are such that aggregate demand for risk-free assets equals supply.

7.2 The Capital Market Line (CML)

The **capital market line** (CML) relates the excess expected return on an efficient portfolio to its risk. "Excess expected return" means the amount by which the expected return of the portfolio exceeds the risk-free rate of return and is also called the risk premium. The CML is

$$\mu_R = \mu_f + \frac{\mu_M - \mu_f}{\sigma_M}\sigma_R, \tag{7.1}$$

where R is the return on a given efficient portfolio (mixture of the market portfolio[6] and the risk-free asset), $\mu_R = E(R)$, μ_f is the rate of return on the risk-free asset, R_M is the return on the market portfolio, $\mu_M = E(R_M)$, σ_M is the standard deviation of the return on the market portfolio, and σ_R is the standard deviation of return on the portfolio. The risk premium of R is $\mu_R - \mu_f$ and the risk premium of the market portfolio is $\mu_M - \mu_f$.

In (7.1) μ_f, μ_M, and σ_M are constant. What varies are σ_R and μ_R. These vary as we change the efficient portfolio R. Think of the CML as showing how μ_R depends on σ_R.

The slope of the CML is, of course,

$$\frac{\mu_M - \mu_f}{\sigma_M}$$

which can be interpreted as the ratio of the risk premium to the standard deviation of the market portfolio. This is Sharpe's "reward-to-risk ratio." Equation (7.1) can be rewritten as

[5] This assumes no buying on margin. Recall from Chapter 5 that buying on margin means borrowing at the risk-free rate so one has a negative position in the risk-free asset and using the money to purchase stock. If we allow negative positions in the risk-free asset, then there is no limit to the risk.

[6] Remember that the market portfolio is assumed to be the same as the tangency portfolio.

$$\frac{\mu_R - \mu_f}{\sigma_R} = \frac{\mu_M - \mu_f}{\sigma_M},$$

which says that the reward-to-risk ratio for any efficient portfolio equals that ratio for the market portfolio.

Example 7.1. Suppose that the risk-free rate of interest is $\mu_f = 0.06$, the expected return on the market portfolio is $\mu_M = 0.15$, and the risk of the market portfolio is $\sigma_M = 0.22$. Then the slope of the CML is $(.15 - .06)/.22 = 9/22$. The CML of this example is illustrated in Figure 7.1.

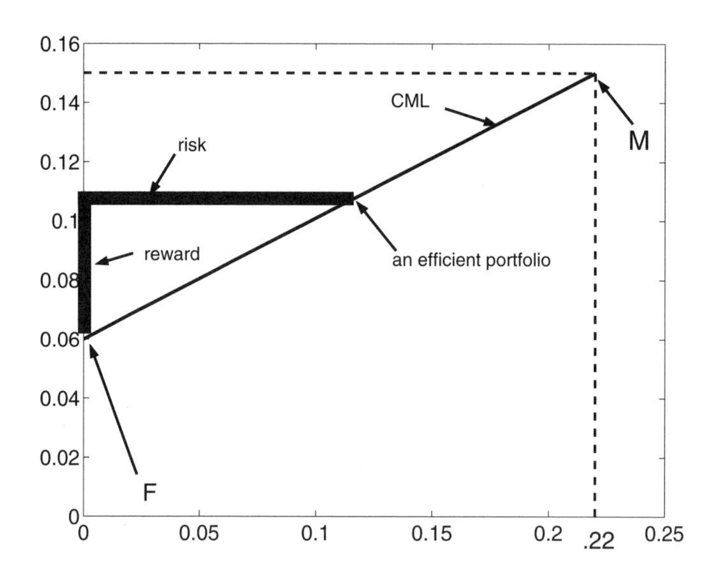

Fig. 7.1. *CML when $\mu_f = 0.06$, $\mu_M = 0.15$, and $\sigma_M = 0.22$. All efficient portfolios are on the line connecting the risk-free asset (F) and the market portfolio (M). Therefore, the reward-to-risk ratio is the same for all efficient portfolios, including the market portfolio. This fact is illustrated by the lengths of the thick gray lines whose lengths are the risk and reward for a typical efficient portfolio.*

The CML is easy to derive. Consider an efficient portfolio that allocates a proportion w of its assets to the market portfolio and $(1 - w)$ to the risk-free asset. Then

$$R = wR_M + (1 - w)\mu_f = \mu_f + w(R_M - \mu_f). \tag{7.2}$$

Therefore, taking expectations in (7.2),

$$\mu_R = \mu_f + w(\mu_M - \mu_f). \tag{7.3}$$

Also from (7.2)

$$\sigma_R = w\sigma_M, \tag{7.4}$$

or

$$w = \frac{\sigma_R}{\sigma_M}. \tag{7.5}$$

Substituting (7.5) into (7.3) gives the CML.

CAPM says that the optimal way to invest is to:

1. Decide on the risk σ_R that you can tolerate, $0 \le \sigma_R \le \sigma_M$;[7]
2. Calculate $w = \sigma_R/\sigma_M$;
3. Invest w proportion of your investment in an index fund, i.e., a fund that tracks the market as a whole;
4. Invest $1 - w$ proportion of your investment in risk-free treasury bills, or a money-market fund.

Alternatively,

1. Choose the reward $\mu_R - \mu_f$ that you want. The only constraint is that $\mu_f \le \mu_R \le \mu_M$ so that $0 \le w \le 1$;[8]
2. Calculate
$$w = \frac{\mu_R - \mu_f}{\mu_M - \mu_f};$$
3. Do Steps 3 and 4 as above.

Instead of specifying the expected return or standard deviation of return, as in Example 5.3 one can find the portfolio with the highest expected return subject to a guarantee that with confidence $1 - \alpha$ the maximum loss is below a prescribed bound M determined, say, by a firm's capital reserves. If the firm invests an amount C, then for the loss to be greater than M the return must be less than $-M/C$. If we assume that the return is normally distributed, then by (2.15), (7.3), and (7.4),

$$P\left(R < -\frac{M}{C}\right) = \Phi\left(\frac{-M/C - \{\mu_f + w(\mu_M - \mu_f)\}}{w\sigma_M}\right). \tag{7.6}$$

Thus, we solve the following equation for w:

$$\Phi^{-1}(\alpha) = \frac{-M/C - \{\mu_f + w(\mu_M - \mu_f)\}}{w\sigma_M}.$$

One can view $w = \sigma_R/\sigma_M$ as an index of the risk aversion of the investor. The smaller the value of w the more risk averse the investor. If an investor has w equal to 0, then that investor is 100% in risk-free assets. Similarly, an investor with $w = 1$ is totally invested in the tangency portfolio of risky assets.[9]

[7] In fact, $\sigma_R > \sigma_M$ is possible by borrowing money to buy risky assets on margin.

[8] This constraint can be relaxed if one is permitted to buy assets on margin.

[9] An investor with $w > 1$ is buying the market portfolio on margin, that is, borrowing money to buy the market portfolio.

7.3 Betas and the Security Market Line

The **Security Market Line** (SML) relates the excess return on an asset to the slope of its regression on the market portfolio. The SML differs from the CML in that the SML applies to all assets while the CML applies only to efficient portfolios.

Suppose that there are many securities indexed by j. Define

$$\sigma_{jM} = \text{covariance between the returns on the } j\text{th security and the market portfolio.}$$

Also, define

$$\beta_j = \frac{\sigma_{jM}}{\sigma_M^2}. \tag{7.7}$$

It follows from the theory of best linear prediction in Section 2.15.1 that β_j is the slope of the best linear predictor of the jth security's returns using returns of the market portfolio as the predictor variable. This fact follows from equation (2.43) for the slope of a best linear prediction equation. In fact, the best linear predictor of R_j based on R_M is

$$\widehat{R}_j = \beta_{0,j} + \beta_j R_M, \tag{7.8}$$

where β_j in (7.8) is the same as in (7.7).

Another way to appreciate the significance of β_j is based on linear regression. As discussed in Section 6.6, linear regression is a method for estimating the coefficients of the best linear predictor based upon data. To apply linear regression, suppose that we have a bivariate time series $(R_{j,t}, R_{M,t})_{t=1}^n$ of returns on the jth asset and the market portfolio. Then, the estimated slope of the linear regression regression of $R_{j,t}$ on $R_{M,t}$ is

$$\hat{\beta}_j = \frac{\sum_{t=1}^n (R_{j,t} - \overline{R}_j)(R_{M,t} - \overline{R}_M)}{\sum_{t=1}^n (R_{M,t} - \overline{R}_M)^2}, \tag{7.9}$$

which, after multiplying numerator and denominator by the same factor n^{-1}, becomes an estimate of σ_{jM} divided by an estimate of σ_M^2 and therefore by (7.7) an estimate of β_j.

Let μ_j be the expected return on the jth security. Then $\mu_j - \mu_f$ is the **risk premium** (or *reward for risk* or **excess expected return**) for that security. Using CAPM, it can be shown that

$$\mu_j - \mu_f = \beta_j(\mu_M - \mu_f). \tag{7.10}$$

This equation, which is called the security market line (SML), is derived in Section 7.5.2. In (7.10) β_j is a variable in the linear equation, not the slope; more precisely, μ_j is a linear function of β_j with slope $\mu_M - \mu_f$. This point is worth remembering. Otherwise, there could be some confusion since β_j was defined earlier as a slope of a regression model. In other words, β_j is a slope in one context but is the independent variable in the SML. One can estimate β_j using (7.9) and then plug this estimate into (7.10).

The SML says that the risk premium of the jth asset is the product of its beta (β_j) and the risk premium of the market portfolio ($\mu_M - \mu_f$). Therefore, β_j measures both the riskiness of the jth asset and the reward for assuming that riskiness. Consequently, β_j is a measure of how "aggressive" the jth asset is. By definition, the beta for the market portfolio is 1; i.e., $\beta_M = 1$. Therefore,

$$\begin{aligned}
\beta_j > 1 &\Rightarrow \text{“aggressive”} \\
\beta_j = 1 &\Rightarrow \text{“average risk”} \\
\beta_j < 1 &\Rightarrow \text{“not aggressive”}.
\end{aligned}$$

Figure 7.2 illustrates the SML and an asset J that is not on the SML. This asset contradicts the CAPM, because according to CAPM all assets are on the SML so no such asset exists.

Consider what would happen if an asset like J did exist. Investors would not want to buy it because, since it is below the SML, its risk premium is too low for the risk given by its beta. They would invest less in J and more in other securities. Therefore the price of J would decline and *after* this decline its expected return would increase. After that increase, the asset J would be on the SML, or so the theory predicts.

7.3.1 Examples of betas

Table 7.1 has some "five-year betas" taken from the Salomon, Smith, Barney website between February 27 and March 5, 2001. The beta for the S&P 500 is given as 1.00; why?

7.3.2 Comparison of the CML with the SML

The CML applies only to the return R of an efficient portfolio. It can be arranged so as to relate the excess expected return of that portfolio to the excess expected return of the market portfolio:

$$\mu_R - \mu_f = \left(\frac{\sigma_R}{\sigma_M}\right)(\mu_M - \mu_f). \tag{7.11}$$

The SML applies to *any* asset and like the CML relates its excess expected return to the excess expected return of the market portfolio:

$$\mu_j - \mu_f = \beta_j(\mu_M - \mu_f). \tag{7.12}$$

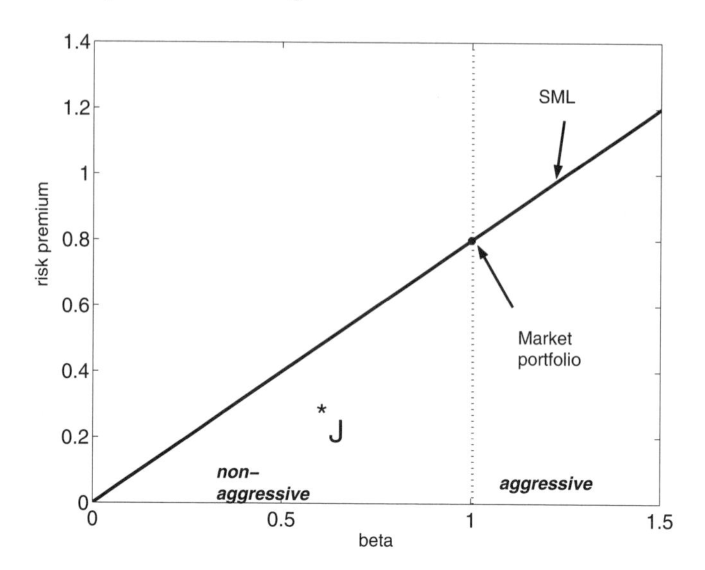

Fig. 7.2. *Security market line (SML) showing that the risk premium of an asset is a linear function of the asset's beta. J is a security not on the line and a contradiction to CAPM. Theory predicts that the price of J decreases until J is on the SML. The vertical dotted line separates the nonaggressive and aggressive regions.*

If we take an efficient portfolio and consider it as an asset, then μ_R and μ_j both denote the expected return on that portfolio/asset. Both (7.11) and (7.12) hold so that

$$\frac{\sigma_R}{\sigma_M} = \beta_R.$$

7.4 The Security Characteristic Line

Let R_{jt} be the return at time t on the jth asset. Similarly, let $R_{M,t}$ and $\mu_{f,t}$ be the return on the market portfolio and the risk-free return at time t. The **security characteristic line** (sometimes shortened to the characteristic line) is a regression model:

$$R_{j,t} = \mu_{f,t} + \beta_j(R_{M,t} - \mu_{f,t}) + \epsilon_{j,t}, \tag{7.13}$$

where $\epsilon_{j,t}$ is $N(0, \sigma^2_{\epsilon,j})$. It is often assumed that the $\epsilon_{j,t}$s are uncorrelated across assets, that is, that $\epsilon_{j,t}$ is uncorrelated with $\epsilon_{j',t}$ for $j \neq j'$. This assumption has important ramifications for risk reduction by diversification; see Section 7.4.1.

Let $\mu_{j,t} = E(R_{j,t})$ and $\mu_{M,t} = E(R_{M,t})$. Taking expectations in (7.13) we get

$$\mu_{j,t} = \mu_{f,t} + \beta_j(\mu_{M,t} - \mu_{f,t}),$$

Table 7.1. Selected stocks and in which industries they are. Betas are given for each stock (Stock's β) and its industry (Ind's β). Betas taken from the Salomon, Smith, Barney website between February 27 and March 5, 2001.

Stock (symbol)	Industry	Stock's β	Ind's β
Celanese (CZ)	Synthetics	0.13	0.86
General Mills (GIS)	Food – major diversif	0.29	0.39
Kellogg (K)	Food – major, diversif	0.30	0.39
Proctor & Gamble (PG)	Cleaning prod	0.35	0.40
Exxon-Mobil (XOM)	Oil/gas	0.39	0.56
7-Eleven (SE)	Grocery stores	0.55	0.38
Merck (Mrk)	Major drug manuf	0.56	0.62
McDonalds (MCD)	Restaurants	0.71	0.63
McGraw-Hill (MHP)	Pub – books	0.87	0.77
Ford (F)	Auto	0.89	1.00
Aetna (AET)	Health care plans	1.11	0.98
General Motors (GM)	Major auto manuf	1.11	1.09
AT&T (T)	Long dist carrier	1.19	1.34
General Electric (GE)	Conglomerates	1.22	0.99
Genentech (DNA)	Biotech	1.43	0.69
Microsoft (MSFT)	Software applic.	1.77	1.72
Cree (Cree)	Semicond equip	2.16	2.30
Amazon (AMZN)	Net soft & serv	2.99	2.46
Doubleclick (Dclk)	Net soft & serv	4.06	2.46

which is equation (7.10), the SML, though in (7.10) it is not shown explicitly that the expected returns can depend on t.[10] The SML gives us information about expected returns, but not about the variance of the returns. For the latter we need the characteristic line. The characteristic line is said to be a "return generating process" since it gives us a probability model of the returns, not just a model of their expected values.

An analogy to the distinction between the SML and characteristic line is this. The regression line $E(Y|X) = \beta_0 + \beta_1 X$ gives the expected value of Y given X but not the conditional probability distribution of Y given X. The regression model

$$Y_t = \beta_0 + \beta_1 X_t + \epsilon_t, \text{ and } \epsilon_t \sim N(0, \sigma^2)$$

does give us this conditional probability distribution.

The characteristic line implies that

$$\sigma_j^2 = \beta_j^2 \sigma_M^2 + \sigma_{\epsilon,j}^2,$$

that

[10] The parameter β_j might also depend on t. This possibility is considered in Section 7.10.

$$\sigma_{jj'} = \beta_j \beta_{j'} \sigma_M^2$$

for $j \neq j'$, and that

$$\sigma_{Mj} = \beta_j \sigma_M^2.$$

The total risk of the jth asset is

$$\sigma_j = \sqrt{\beta_j^2 \sigma_M^2 + \sigma_{\epsilon,j}^2}.$$

The risk has two components: $\beta_j^2 \sigma_M^2$ is called the **market** or **systematic component of risk** and $\sigma_{\epsilon,j}^2$ is called the **unique, nonmarket,** or **unsystematic component of risk**.

7.4.1 Reducing unique risk by diversification

The market component cannot be reduced by diversification, but the unique component can be reduced or even eliminated by sufficient diversification.

Suppose that there are N assets with returns $R_{1,t}, \ldots, R_{N,t}$ for holding period t. If we form a portfolio with weights $w_1, \ldots, w_N$ then the return of the portfolio is

$$R_{P,t} = w_1 R_{1,t} + \cdots + w_N R_{N,t}.$$

Let R_{Mt} be the return on the market portfolio. According to the characteristic line model $R_{j,t} = \mu_{f,t} + \beta_j(R_{M,t} - \mu_{f,t}) + \epsilon_{j,t}$ so that

$$R_{P,t} = \mu_{f,t} + \left(\sum_{j=1}^{N} \beta_j w_j \right) (R_{M,t} - \mu_{f,t}) + \sum_{j=1}^{N} w_j \epsilon_{j,t}.$$

Therefore, the portfolio beta is

$$\beta_P = \sum_{j=1}^{N} w_j \beta_j,$$

and the "epsilon" for the portfolio is

$$\epsilon_{P,t} = \sum_{j=1}^{N} w_j \epsilon_{j,t}.$$

We now assume that $\epsilon_{1t}, \ldots, \epsilon_{Nt}$ are uncorrelated. Therefore, by equation (2.55)

$$\sigma_{\epsilon,P}^2 = \sum_{j=1}^{N} w_j^2 \sigma_{\epsilon,j}^2.$$

Example 7.2. Suppose the assets in the portfolio are equally weighted; that is, $w_j = 1/N$ for all j. Then

$$\beta_P = \frac{\sum_{j=1}^N \beta_j}{N},$$

and

$$\sigma^2_{\epsilon,P} = \frac{N^{-1}\sum_{j=1}^N \sigma^2_{\epsilon,j}}{N} = \frac{\overline{\sigma}^2_\epsilon}{N},$$

where $\overline{\sigma}^2_\epsilon$ is the average of the $\sigma^2_{\epsilon,j}$.

If $\sigma^2_{\epsilon,j}$ is a constant, say σ^2_ϵ, for all j, then

$$\sigma_{\epsilon,P} = \frac{\sigma_\epsilon}{\sqrt{N}}. \tag{7.14}$$

For example, suppose that σ_ϵ is 5%. If $N = 20$, then by (7.14) $\sigma_{\epsilon,P}$ is 1.12%. If $N = 100$, then $\sigma_{\epsilon,P}$ is 0.5%. There are approximately 1600 stocks on the NYSE; if $N = 1600$, then $\sigma_{\epsilon,P} = 0.125\%$.

7.4.2 Can beta be negative?

It is possible for a beta to be negative. Prices of companies mining precious metals tend to move opposite to the market which would cause a negative beta. For example, in March 2003, the betas of the Kinross Gold Corporation, Glamis Gold Ltd., and Vista Gold Corp. were listed as −.29, −.36, and −1.22, respectively, on Yahoo!Financial.

Although we usually think of beta as positive and measuring riskiness, when beta is negative then the interpretation of beta as risk is a bit subtle. The absolute value of beta measures stand-alone riskiness because the stand-alone risk of the jth asset is $\sigma_j = \{\beta_j^2\sigma_M^2 + \sigma^2_{\epsilon,j}\}^{1/2}$. However, the beta for a portfolio is $\beta_P = \sum_{j=1}^N w_j\beta_j$, so a negative beta can reduce the beta, and therefore the risk, of a portfolio. In fact, adding an asset with a negative beta to a portfolio can reduce the portfolio's risk more than adding a risk-free asset. However, according to the SML, the expected return on an asset with a negative beta is *lower* than the risk-free rate. By constructing a well-diversified portfolio of assets with positive and negative beta such that the portfolio beta is 0, one can obtain a portfolio with essentially no risk. Unfortunately, according to the CAPM, this portfolio only earns the risk-free rate. In fact, even without invoking the CAPM theory, we should expect to earn exactly the risk-free rate from any risk-free portfolio. This fact is an example of the "law of one price" which is discussed in Section 8.3.

7.4.3 Are the assumptions sensible?

A key assumption that allows nonmarket risk to be removed by diversification is that $\epsilon_{1,t}, \ldots, \epsilon_{N,t}$ are uncorrelated. This assumption implies that *all* correlation among the cross-section[11] of asset returns is due to a single cause and that cause is measured by the market index. For this reason, the characteristic line is a "single factor" or "single index" model with $R_{M,t}$ being the "factor."

This assumption of uncorrelated ϵ_{jt} would not be valid if, for example, two energy stocks are correlated over and beyond their correlation due to the market index. In this case, unique risk could not be eliminated by holding a large portfolio of all energy stocks. However, if there are many market sectors and the sectors are uncorrelated, then one could eliminate nonmarket risk by diversifying across all sectors. All that is needed is to treat the sectors themselves as the underlying assets and then apply the CAPM theory.

Correlation among the stocks in a market sector can be modeled using a "factor model." Factor models are introduced in Section 7.8.

7.5 Some More Portfolio Theory

In this section we use portfolio theory to show that $\sigma_{j,M}$ quantifies the contribution of the jth asset to the risk of the market portfolio. Also, we derive the SML.

7.5.1 Contributions to the market portfolio's risk

Suppose that the market consists of N risky assets and that $w_{1M}, \ldots, w_{N,M}$ are the weights of these assets in the market portfolio. Then

$$R_{M,t} = \sum_{i=1}^{N} w_{i,M} R_{i,t},$$

which implies that the covariance between the return on the jth asset and the return on the market portfolio is

$$\sigma_{j,M} = \text{Cov}\left(R_{j,t}, \sum_{i=1}^{N} w_{i,M} R_{i,t} \right) = \sum_{i=1}^{N} w_{i,M} \sigma_{i,j}. \tag{7.15}$$

Therefore,

$$\sigma_M^2 = \sum_{j=1}^{N} \sum_{i=1}^{N} w_{j,M} w_{i,M} \sigma_{i,j} = \sum_{j=1}^{N} w_{j,M} \left(\sum_{i=1}^{N} w_{i,M} \sigma_{i,j} \right) = \sum_{j=1}^{N} w_{j,M} \sigma_{j,M}. \tag{7.16}$$

Equation (7.16) shows that the contribution of the jth asset to the risk of the market portfolio is $w_{j,M}\sigma_{j,M}$, where $w_{j,M}$ is the weight of the jth asset in the market portfolio and $\sigma_{j,M}$ is the covariance between the return on the jth asset and the return on the market portfolio.

[11] "Cross-section" of returns means returns across assets within a *single* holding period.

7.5.2 Derivation of the SML

The derivation of the SML is a nice application of portfolio theory, calculus, and geometric reasoning. It is based on a clever idea of putting together a portfolio with two assets, the market portfolio and the ith risky asset, and then looking at the locus in reward-risk space as the portfolio weight assigned to the ith risky asset varies.

Consider a portfolio P with weight w_i given to the ith risky asset and weight $(1 - w_i)$ given to the market portfolio. The return on this portfolio is

$$R_{P,t} = w_i R_{i,t} + (1 - w_i) R_{M,t}.$$

The expected return is

$$\mu_P = w_i \mu_i + (1 - w_i)\mu_M, \tag{7.17}$$

and the risk is

$$\sigma_P = \sqrt{w_i^2 \sigma_i^2 + (1 - w_i)^2 \sigma_M^2 + 2 w_i (1 - w_i) \sigma_{i,M}}. \tag{7.18}$$

As we vary w_i we get the locus of points on (σ, μ) space that is shown as a dashed curve in Figure 7.3.

It is easy to see geometrically that the derivative of this locus of points evaluated at the tangency portfolio (which is the point where $w_i = 0$) is equal to the slope of the CML. We can calculate this derivative and equate it to the slope of the CML to see what we get. The result is the SML.

We have from (7.17)

$$\frac{d\,\mu_P}{d\,w_i} = \mu_i - \mu_M,$$

and from (7.18) that

$$\frac{d\,\sigma_P}{d\,w_i} = \frac{1}{2}\sigma_P^{-1}\left\{2 w_i \sigma_i^2 - 2(1 - w_i)\sigma_M^2 + 2(1 - 2w_i)\sigma_{i,M}\right\}.$$

Therefore,

$$\frac{d\,\mu_P}{d\,\sigma_P} = \frac{d\,\mu_P/d\,w_i}{d\,\sigma_P/d\,w_i} = \frac{(\mu_i - \mu_M)\sigma_P}{w_i \sigma_i^2 - \sigma_M^2 + w_i \sigma_M^2 + \sigma_{i,M} - 2 w_i \sigma_{i,M}}.$$

Next,

$$\left.\frac{d\,\mu_P}{d\,\sigma_P}\right|_{w_i=0} = \frac{(\mu_i - \mu_M)\sigma_M}{\sigma_{i,M} - \sigma_M^2}.$$

Recall that $w_i = 0$ is the tangency portfolio, the point in Figure 7.3 where the dashed locus is tangent to the CML. Therefore,

$$\left.\frac{d\,\mu_P}{d\,\sigma_P}\right|_{w_i=0}$$

must equal the slope of the CML which is $(\mu_M - \mu_f)/\sigma_M$. Therefore,

$$\frac{(\mu_i - \mu_M)\sigma_M}{\sigma_{i,M} - \sigma_M^2} = \frac{\mu_M - \mu_f}{\sigma_M},$$

which, after some algebra, gives us

$$\mu_i - \mu_f = \frac{\sigma_{i,M}}{\sigma_M^2}(\mu_M - \mu_f) = \beta_i(\mu_M - \mu_f),$$

which is the SML given in equation (7.10).

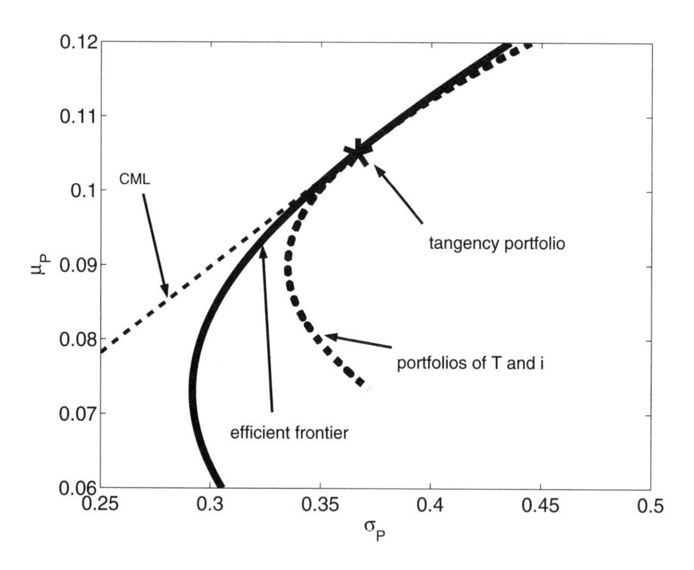

Fig. 7.3. *Derivation of the SML. The market portfolio and the tangency portfolio are equal according to the CAPM. The dashed curve is the locus of portfolios combining asset i and the market portfolio. The dashed curve is to the right of the efficient frontier and intersects the efficient frontier at the tangency portfolio. Therefore, the derivative of the dashed curve at the tangency portfolio is equal to the slope of the CML, since this curve is tangent to the CML at the tangency portfolio.*

7.6 Estimation of Beta and Testing the CAPM

Recall the security characteristic line

$$R_{j,t} = \mu_{f,t} + \beta_j(R_{M,t} - \mu_{f,t}) + \epsilon_{j,t}. \tag{7.19}$$

Let $R^*_{j,t} = R_{j,t} - \mu_{f,t}$ be the excess return on the jth security and let $R^*_{M,t} = R_{M,t} - \mu_{f,t}$, be the excess return on the market portfolio. Then (7.19) can be written as

$$R^*_{j,t} = \beta_j R^*_{M,t} + \epsilon_{j,t}. \tag{7.20}$$

Equation (7.20) is a regression model without an intercept and with β_j as the slope. A more elaborate model is

$$R^*_{j,t} = \alpha_j + \beta_j R^*_{M,t} + \epsilon_{j,t}, \tag{7.21}$$

which includes an intercept. The CAPM says that $\alpha_j = 0$ but by allowing $\alpha_j \neq 0$ we recognize the possibility of mispricing.

Given time series $R_{j,t}$, $R_{M,t}$, and $\mu_{f,t}$ for $t = 1, \ldots, n$, we can calculate $R^*_{j,t}$ and $R^*_{M,t}$ and regress $R^*_{j,t}$ on $R^*_{M,t}$ to estimate α_j, β_j, and $\sigma^2_{\epsilon,j}$. By testing the null hypothesis that $\alpha_j = 0$ we are testing whether the jth asset is mispriced according to the CAPM.

As discussed in Section 6.2.2, when fitting model (7.20) or (7.21) one should use daily data if available, rather than weekly or monthly data. A more difficult question to answer is how long a time series to use. Longer time series give more data, of course, but models (7.20) and (7.21) assume that β is constant and this might not be true over a long time period. Section 7.10 provides a method of testing whether β is constant, and such a test should be done if a very long time series is being used.

Example 7.3. As an example, daily closing prices on Microsoft and the S&P 500 index from November 1, 1993 to April 3, 2003 were used. The S&P 500 was taken as the market price. The log returns were found by differencing the log closing prices. Then three-month T-bill rates were used as the risk-free returns.[12] The excess returns are the log-returns minus the T-bill rates. Here is a SAS program to do the regression. The data already exist as a SAS data set in the library "Sasuser.capm." This data set had been created earlier by reading the data from an Excel file using SAS's Import Wizard. To use the Import Wizard go to the "file" menu within SAS and follow the prompts. The command `set Sasuser.capm` specifies this SAS data set as the input to the data step. In the data step, a new data set is created which is also called "capm" but is only a temporary data set and is not be saved in a library.

```
data capm ;
set Sasuser.capm ;
EXlogR_sp500 = dif(log(close_sp500)) - close_tbill/(100*253) ;
EXlogR_msft = dif(log(close_msft)) - close_tbill/(100*253) ;
run ;
proc reg ;
model EXlogR_msft = EXlogR_sp500 ;
run ;
```

[12] Interest rates are return rates. Thus, we use the T-bill rates themselves as the risk-free returns. One does *not* take logs and difference the T-bill rates as if they were prices. However, the T-bill rates were divided by 100 to convert from a percentage and then by 253 to convert to a daily rate.

Here is the output.

```
                         The REG Procedure
                           Model: MODEL1
                   Dependent Variable: EXlogR_msft

                        Analysis of Variance

                                    Sum of           Mean
Source                   DF        Squares         Square    F Value    Pr > F

Model                     1        0.48843        0.48843    1234.07    <.0001
Error                  2360        0.93407     0.00039579
Corrected Total        2361        1.42250


         Root MSE              0.01989    R-Square     0.3434
         Dependent Mean     0.00081215    Adj R-Sq     0.3431
         Coeff Var          2449.61057


                        Parameter Estimates

                             Parameter       Standard
Variable          DF          Estimate          Error    t Value    Pr > |t|

Intercept          1        0.00069605     0.00040936       1.70      0.0892
EXlogR_sp500       1           1.24449        0.03543      35.13      <.0001
```

For Microsoft, we find that $\hat{\beta} = 1.24451$ and $\hat{\alpha} = 0.0007$. Since the standard error of $\hat{\beta}$ is 0.0354, a 95% confidence interval for β_j is $1.2445 \pm (2)(.0354)$ or $(1.1737, 1.3153)$.[13] The p-value for $\widehat{\beta}$ is less that 0.0001. This p-value is for testing the null hypothesis that $\beta = 0$, so it is not surprising that the null hypothesis is strongly rejected. We do not expect the beta of a stock to be zero, so the outcome of this test should come as no surprise. The estimate of σ_ϵ is the Root MSE which equals 0.01989.

The test that $\alpha = 0$ has a p-value of 0.0892 so we can accept the null hypothesis at 0.05 but not at 0.1. Certainly α is small and for practical purposes equal to 0 even if it is statistically significant at 0.1. This suggests that the data are consistent with the CAPM.

Notice that the R^2 (R-sq) value for the regression is 34.34%. The interpretation of R^2 is the percent of the variance in the excess returns on Microsoft that is due to excess returns on the market. In other words, 34.4% of the risk is due to systematic or market risk $(\beta_j^2 \sigma_M^2)$. The remaining 65.6% is due to unique or nonmarket risk (σ_ϵ^2).

If we assume that $\alpha = 0$, then we can refit the model using a no-intercept model. In PROC REG, the no-intercept model is requested with the "noint" option; the options follow the slash (/) in the model statement. Here is the program.

```
(data step as before)
proc reg ;
model EXlogR_msft = EXlogR_sp500/noint ;
run ;
```

[13] Here "2" is used as an approximate t-value.

The SAS output is as follows.

```
                    Dependent Variable: EXlogR_msft

              NOTE: No intercept in model. R-Square is redefined.

                         Analysis of Variance

                                    Sum of           Mean
Source                    DF       Squares         Square    F Value    Pr > F

Model                      1       0.48885        0.48885    1234.13    <.0001
Error                   2361       0.93521     0.00039611
Uncorrected Total       2362       1.42406


             Root MSE              0.01990    R-Square     0.3433
             Dependent Mean     0.00081215    Adj R-Sq     0.3430
             Coeff Var          2450.59142


                          Parameter Estimates

                            Parameter       Standard
    Variable        DF       Estimate          Error    t Value    Pr > |t|

    EXlogR_sp500     1        1.24498        0.03544      35.13      <.0001
```

Now $\hat{\beta} = 1.24498$ and $\hat{\sigma}_\epsilon = 0.01990$ which are extremely small changes from the intercept model.

Testing that α equals 0 tests only one of the conclusions of the CAPM. Accepting this null hypothesis only means that the CAPM has passed one test, not that we should now accept it as true.[14] To fully test the CAPM, its other conclusions should also be tested. The factor models in Section 7.8 have been used to test the CAPM and fairly strong evidence against CAPM has been found. Fortunately, these factor models do provide a generalization of the CAPM that is likely to be useful for financial decision making.

7.6.1 Regression using returns instead of excess returns

Often, as an alternative to regression using excess returns, the returns on the asset are regressed on the returns on the market. When this is done, an intercept model should be used. In the Microsoft data when using returns instead of excess returns, the estimate of beta changed hardly at all, only from 1.24449 to 1.24568.

7.6.2 Interpretation of alpha

If α is nonzero then the security is mispriced, at least according to CAPM. If $\alpha > 0$ then the security is underpriced; the returns are too large on average.

[14] In fact, acceptance of a null hypothesis should never be interpreted as proof that the null hypothesis is true.

This is an indication of an asset worth purchasing. Of course, one must be careful. If we reject the null hypothesis that $\alpha = 0$, all we have done is to show that the security was mispriced *in the past*. Since for the Microsoft data we accepted the null hypothesis that α is zero, there is no evidence that Microsoft was mispriced.

Warning: If we use returns rather than excess returns, then the intercept of the regression equation does *not* estimate α, so one cannot test whether α is zero by testing the intercept.

7.7 Using CAPM in Portfolio Analysis

Suppose we have estimated beta and σ_ϵ^2 for each asset in a portfolio and also estimated σ_M^2 and μ_M for the market. Then, since μ_f is also known, we can compute the expectations, variances, and covariances of all asset returns by the formulas

$$\mu_j = \mu_f + \beta_j(\mu_M - \mu_f),$$

$$\sigma_j^2 = \beta_j^2\sigma_M^2 + \sigma_{\epsilon j}^2,$$

$$\sigma_{jj'} = \beta_j\beta_{j'}\sigma_M^2 \text{ for } j \neq j'.$$

There is a serious danger here: these estimates depend heavily on the validity of the CAPM assumptions. Any or all of the quantities beta, σ_ϵ^2, σ_M^2, μ_M, and μ_f could depend on time t. However, it is generally assumed that the betas and σ_ϵ^2s of the assets as well as σ_M^2 and μ_M of the market are independent of t so that these parameters can be estimated assuming stationarity of the time series of returns.

7.8 Factor Models

The security characteristic line is a regression model

$$R_{j,t} = \mu_{f,t} + \beta_j(R_{M,t} - \mu_{f,t}) + \epsilon_{j,t}.$$

The variable $(R_{M,t} - \mu_{f,t})$ which is the excess return on the market is sometimes called a **factor** or a **risk factor**. In CAPM, the market risk factor is the only source of risk besides the unique risk of each asset. Because in CAPM the market risk factor is the only risk that any two assets share, it is the sole source of correlation between asset returns. It is important to remember that the lack of other common risk factors is an assumption of CAPM, not an empirically derived fact. Indeed, there is evidence of other common risk factors besides the market risk. Companies within the same country or same industry appear to have common risks beyond the overall market risk. Also, research has suggested that companies with common characteristics such as a high book-to-market value have common risks, though this is controversial.[15]

The **market value** of a stock is the product of the market price per share times the number of shares outstanding. **Book value** is the net worth of the firm according to its accounting balance sheet. There are many reasons why book and market values may differ. One is that book value is determined by accounting methods that do not necessarily reflect market values. For example, if a firm owns raw materials these are valued on the balance sheet at their purchase price minus depreciation, not at the current price at which they could be sold. Also, a stock might have a low book-to-market value because investors expect it to make a high return on equity which increases its market value relative to its book value. Conversely, a high book-to-market value could indicate a firm that is in trouble which decreases its market value. A low market value relative to the book value is an indication of a stock's "cheapness," and stocks with a high market to book value are considered *growth stocks* for which investors are willing to pay a premium because of the promise of higher future earnings. Stocks with a low market to book value are *value stocks* and investing in them is called *value investing*.[16]

Factor models generalize the CAPM by allowing more factors than simply the market risk and the unique risk of each asset. A multifactor model is

$$R_{j,t} - \mu_{f,t} = \beta_{0,j} + \beta_{1,j} F_{1,t} + \cdots + \beta_{p,j} F_{p,t} + \epsilon_{j,t}, \tag{7.22}$$

where $F_{1,t}, \ldots, F_{p,t}$ are the values of p factors at time t. A **factor** can be anything that can be measured and is thought to affect asset returns. Examples of factors include:

- Return on market portfolio (market model, e.g., CAPM);
- Growth rate of GDP;
- Interest rate on short term Treasury bills;
- Inflation rate;
- Interest rate spreads, for example, the difference between long-term Treasury bonds and long-term corporate bonds;
- Return on some portfolio of stocks, for example, all U. S. stocks or all stocks with a high ratio of book equity to market equity — this ratio is called BE/ME in Fama and French (1992, 1995, 1996);
- The difference between the returns on two portfolios, for example, the difference between returns on stocks with high BE/ME values and stocks with low BE/ME values.

With enough factors all commonalities between assets should be accounted for in the models. Then the $\epsilon_{j,t}$ should represent factors truly unique to the individual assets and therefore should be uncorrelated across j (across assets).

[15] See Fama and French (1996) and Shleifer (2000, p. 19–20).
[16] Shleifer (2000, p. 19).

As discussed in Section 7.6, when fitting factor models one should use data at the highest sampling frequency available, which is often daily or weekly.

If a factor model includes the returns on the market portfolio and other factors, then according to the CAPM, the betas on the other factors should be zero, so testing that they are zero also tests the CAPM. In some studies, for example, the Fama-French three-factor model in Section 7.8.2, a number of other factors have been statistically significant, which is evidence against the CAPM.

7.8.1 Estimating expectations and covariances of asset returns

In Section 7.7 we discussed how one could use the CAPM to simplify the estimation of expectations and covariances of asset returns. However, using CAPM for this purpose is dangerous since the estimates depend on the validity of CAPM. Fortunately, it is also possible to estimate return expectations and covariances using a more realistic factor model instead of CAPM.

We start with two factors for simplicity. From (7.22), now with $p = 2$, we have

$$R_{j,t} - \mu_{f,t} = \beta_{0,j} + \beta_{1,j}F_{1,t} + \beta_{2,j}F_{2,t} + \epsilon_{j,t}. \tag{7.23}$$

It follows from (7.23) that

$$E(R_{j,t}) = \mu_f + \beta_{0,j} + \beta_{1,j}E(F_{1,t}) + \beta_{2,j}E(F_{2,t}) \tag{7.24}$$

and

$$\mathrm{Var}(R_{j,t}) = \beta_{1,j}^2 \mathrm{Var}(F_1) + \beta_{2,j}^2 \mathrm{Var}(F_2) + 2\beta_{1,j}\beta_{2,j}\mathrm{Cov}(F_1, F_2) + \sigma_{\epsilon,j}^2.$$

Also, because $R_{j,t}$ and $R_{j',t}$ are two linear combinations of the risk factors, it follows from (2.54) that (see Problem 8)

$$\mathrm{Cov}(R_{j,t}, R_{j',t}) = \beta_{1,j}\beta_{1,j'}\mathrm{Var}(F_1) + \beta_{2,j}\beta_{2,j'}\mathrm{Var}(F_2)$$
$$+(\beta_{1,j}\beta_{2,j'} + \beta_{1,j'}\beta_{2,j})\mathrm{Cov}(F_1, F_2). \tag{7.25}$$

More generally, suppose that $\boldsymbol{\Sigma}_F$ is the $p \times p$ covariance matrix of the p factors. Let

$$\boldsymbol{\beta} = \begin{pmatrix} \beta_{1,1} & \cdots & \beta_{1,j} & \cdots & \beta_{1,n} \\ \vdots & \ddots & \vdots & \ddots & \vdots \\ \beta_{p,1} & \cdots & \beta_{p,j} & \cdots & \beta_{p,n} \end{pmatrix}.$$

Also define $\boldsymbol{\Sigma}_{\boldsymbol{\epsilon}}$ to be the diagonal matrix

$$\boldsymbol{\Sigma}_{\boldsymbol{\epsilon}} = \begin{pmatrix} \sigma_{\epsilon,1}^2 & \cdots & 0 & \cdots & 0 \\ \vdots & \ddots & \vdots & \ddots & \vdots \\ 0 & \cdots & \sigma_{\epsilon,j}^2 & \cdots & 0 \\ \vdots & \ddots & \vdots & \ddots & \vdots \\ 0 & \cdots & 0 & \cdots & \sigma_{\epsilon,n}^2 \end{pmatrix}.$$

If $\boldsymbol{\Sigma}_R$ is the $n \times n$ covariance matrix of the n asset returns, then

$$\text{Var}(R_j) = \boldsymbol{\beta}_j^\mathsf{T} \boldsymbol{\Sigma}_F \boldsymbol{\beta}_j + \sigma_{\epsilon_j}^2$$

and

$$\boldsymbol{\Sigma}_R = \boldsymbol{\beta}^\mathsf{T} \boldsymbol{\Sigma}_F \boldsymbol{\beta} + \boldsymbol{\Sigma}_{\boldsymbol{\epsilon}}.$$

In particular, if $\boldsymbol{\beta}_j = (\, \beta_{1,j} \quad \cdots \quad \beta_{p,j}\,)^\mathsf{T}$ is the jth column of $\boldsymbol{\beta}$, then

$$\text{Cov}(R_j, R_j') = \boldsymbol{\beta}_j^\mathsf{T} \boldsymbol{\Sigma}_F \boldsymbol{\beta}_{j'}.$$

7.8.2 Fama and French three-factor model

Fama and French (1995) have developed a model with three risk factors, the first being the excess return of the market portfolio which we know is the sole factor in the CAPM. The second risk factor, which is called small minus large (SML), is the difference in returns on a portfolio of small stocks and a portfolio of large stock. Here "small" and "large" refer to the size of the market equity which is the price of the stock times the number of shares outstanding. The third factor, HML (high minus low), is the difference in returns on a portfolio of high book-to-market value (BE/ME) stocks and a portfolio of low BE/ME stocks. Fama and French argue that most pricing anomalies that are inconsistent with the CAPM disappear in the three-factor model. Their model of the return on the jth asset for the tth holding period is

$$R_{j,t} - \mu_{f,t} = \beta_{0,j} + \beta_{1,j}(R_{M,t} - \mu_{f,t}) + \beta_{2,j}\text{SML}_t + \beta_{3,j}\text{HML}_t + \epsilon_{j,t},$$

where SML_t and HML_t are the values of SML and HML for the tth holding period.

Notice that this model does *not* use the size or the BE/ME ratio of the jth asset to explain returns. The coefficients $\beta_{2,j}$ and $\beta_{3,j}$ are called the loading of the jth asset on SML and HML. These loadings may, but need not, be related to the size and to the BE/ME ratio of the jth asset. In any event, the loadings are estimated by regression, not by measuring the size or BE/ME of the jth asset. If the loading $\beta_{2,j}$ of the jth asset on SML is high, that may be because the jth asset is small or it may be because that asset is large but, in terms of returns, behaves similarly to small assets. The factors SML_t and HML_t do not depend on j since they are differences between returns on two fixed portfolios, not variables that are measured on the jth asset. This is true in general of the factors and loadings in model (7.22) — only the loadings, that is, the $\beta_{k,j}$ depend on the asset j. The factors are broad macroeconomic variables, linear combinations of returns on portfolios, or other variables that depend only on the market as a whole.

7.8.3 Cross-sectional factor models

Models of form (7.22) are called **time series factor models**. They use time series data on a single asset to estimate the loadings of that asset.

As just discussed, time series factor models cannot make use of variables such as dividend yields, book-to-market value, or other variables specific to the jth firm. An alternative type of model is a **cross-sectional factor model** which is a regression model using data from many assets but from only a single holding period. For example, suppose that R_j, $(B/M)_j$, and D_j are the return, book-to-market value, and dividend yield for the jth asset for some fixed holding period. Then a possible cross-sectional factor model is

$$R_j = \beta_0 + \beta_1 (B/M)_j + \beta_2 D_j + \epsilon_j.$$

The parameters β_1 and β_2 are unknown factors that are estimated by regression and assumed to be a book-to-market value risk factor and a dividend yield risk factor.

So there are two fundamental differences between time series factor models and cross-sectional factor models. The first is that with a times series factor model one estimates parameters one asset at a time using multiple holding periods, while in a cross-sectional model one estimates parameters a single holding period at a time using multiple assets. The other major difference is that in a time series factor model, the factors are directly measured and the loadings are estimated by regression, but in a cross-sectional factor model the opposite is true; that is, the loadings are directly measured and the factors are estimated by regression.

7.9 An Interesting Question*

A student asked me whether it is possible for an asset to have one beta when the market portfolio has a negative return and another for positive returns on the market. The data in Figure 7.4 suggest that this might have occurred with Ford Motor Company, at least between March 1996 and February 2001 when the monthly data in that figure were collected. The curve is a penalized spline fit to the data. Penalized splines, which are discussed in Chapter 13, are similar to lowess in that they provide a "nonparametric" fit to a scatter plot. The noticeable curvature in the spline fit indicates that the slope might be changing around 0 on the horizontal axis. Here is a SAS program that fits a "broken line" model. This model is a straight line from $x = -\infty$ to 0 and a second straight line from $x = 0$ to ∞.[17] The two lines join at $x = 0$ so that the model is continuous. This model can be expressed as a linear regression model with two predictors. The first is the S&P 500 excess return, denoted

[17] Here x is the variable on the horizontal axes, i.e., the excess log return on the S&P 500 index.

by sp in the program. The second, which is denoted by sp2, is equal to sp if sp is positive and is 0 otherwise. Thus, the model

$$Y_i = \beta_0 + \beta_1 \texttt{sp}_i + \beta_2 \texttt{sp2}_i + \epsilon_i$$

is a straight line with slope β_1 to the left of zero and slope $\beta_1 + \beta_2$ to the right of zero. Because β_2 is the change in the slope at zero, testing that β_2 is equal to 0 tests the null hypothesis that there is one beta versus the alternative of two betas, one to the left of zero and the other to the right. Notice that in the data step the variable sp2 is created by first setting it equal to 0 and then changing it to equal sp *if* sp is positive.

```
data capm ;
infile 'C:\book\sas\capm.txt' ;
input SP Microsoft Ford ;
sp2 = 0 ;
if sp > 0 then sp2=sp ;
run ;
```

The remainder of the program fits three models. The first is linear regression with an intercept, the second is the broken line model, and the third is linear regression without an intercept. The latter uses the "noint" option in PROC REG to force SAS to use no intercept. By omitting the intercept we are fitting the security characteristic line, that is, equation (7.13).

```
proc reg ;
model ford = sp ;
model ford = sp sp2 ;
model ford = sp/noint ;
run ;
```

Here is the output, first for linear regression with an intercept.

```
                        Analysis of Variance
                          Sum of           Mean
Source             DF    Squares          Square    F Value   Pr > F

Model               1    0.05866  0.05866             9.14    0.0037
Error              58    0.37213  0.00642
Corrected Total    59    0.43078

         Root MSE                0.08010    R-Square      0.1362
         Dependent Mean          0.01086    Adj R-Sq      0.1213
         Coeff Var             737.68184

                          Parameter Estimates

                     Parameter    Standard
Variable     DF     Estimate       Error   t Value   Pr > |t|
Intercept     1      0.00646     0.01044      0.62    0.5389
SP            1      0.65424     0.21638      3.02    0.00642
```

Next is the output for the broken line model. Notice that the p-value for sp2 is 0.012 which is evidence that there is a change of beta at 0, that is, that the answer to the student's question is "yes."

```
                          Analysis of Variance

                                 Sum of           Mean
Source                   DF     Squares         Square    F Value    Pr > F

Model                     2     0.09804        0.04902       8.40    0.0006
Error                    57     0.33275        0.00584
Corrected Total          59     0.43078


          Root MSE              0.07640    R-Square     0.2276
          Dependent Mean        0.01086    Adj R-Sq     0.2005
          Coeff Var           703.64831


                          Parameter Estimates

                       Parameter    Standard
Variable         DF     Estimate       Error    t Value    Pr > |t|

Intercept         1      0.04003     0.01632       2.45      0.0172
SP                1      1.66452     0.44033       3.78      0.0004
sp2               1     -1.79046     0.68934      -2.60      0.0119
```

Here is the output for the linear regression with no intercept.

```
        NOTE: No intercept in model. R-Square is redefined.

                       Analysis of Variance

                                      Sum of            Mean
Source                       DF      Squares          Square    F Value

Model                         1      0.06328         0.06328       9.97
Error                        59      0.37458         0.00635
Uncorrected Total            60      0.43786


          Root MSE               0.07968    R-Square     0.1445
          Dependent Mean         0.01086    Adj R-Sq     0.1300
          Coeff Var            733.80895

                       Parameter Estimates

                        Parameter         Standard
Variable     DF          Estimate            Error    t Value    Pr > |t|

SP            1           0.67289          0.21314       3.16      0.0025
```

Table 7.2 compares the four models by AIC. The broken line model fits best followed closely by the penalized spline. The fits to the three best fitting models, that is, the spline, the broken line model, and linear regression with an intercept omitted, are shown in Figure 7.5. The broken line and the spline suggest that there is a positive beta for negative excess returns but that beta is essentially 0 for positive returns. Also, the intercept is positive. R^2 is small, e.g., only 22.8% for the broken line model, so much of Ford's risk is due to unique risk.

The S&P 500 and Ford prices are shown in Figure 7.6. The prices have been divided by the price at the end of March 1996, so they are expressed as gross k period returns where k is the number of months since March 1996.

One can see that the S&P 500 rose rather steadily until the summer of 2000, but that most of Ford's rise was concentrated in a shorter period.

This example illustrates the power of the multiple linear regression model. By clever construction of the predictor variables, a wide variety of financial models fall within its domain. This data set also shows that while the security characteristic line might be a useful working assumption, it may be only an approximation to what the financial markets data are doing.

Table 7.2. AIC comparisons between models. "w. int" = linear regression with an intercept, "br. line" = broken line model, "w/o int" is linear regression with no intercept, and "spline" is a penalized spline model. "AIC diff." is the difference between AIC for the given model and AIC for the broken line model.

model	w. int	br. line	w/o int	spline
AIC diff.	4.7111	0	3.1048	0.5189

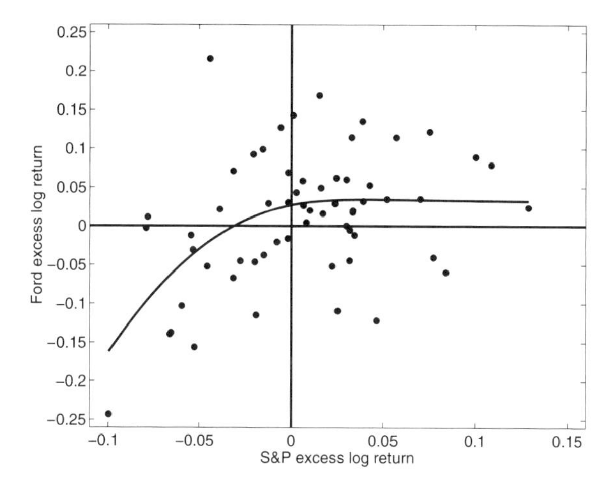

Fig. 7.4. *Monthly excess log returns from March* 1996 *to February* 2001 *for Ford plotted against excess log returns for the S&P* 500 *index. The curve is a penalized spline fit.*

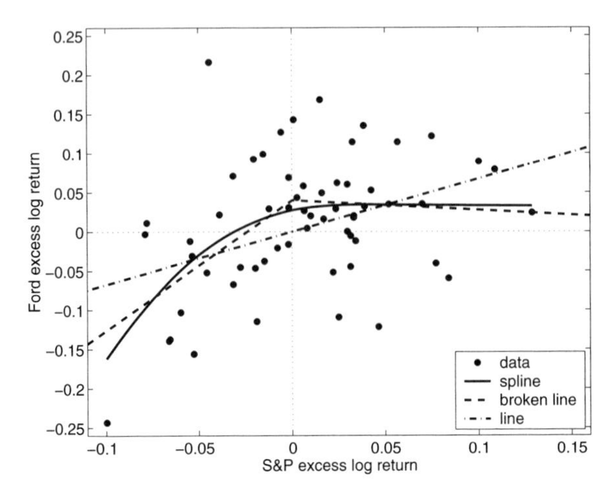

Fig. 7.5. *Excess log returns of Ford plotted against those of the S&P 500 with three fits: a penalized spline, a broken line, and a line with no intercept.*

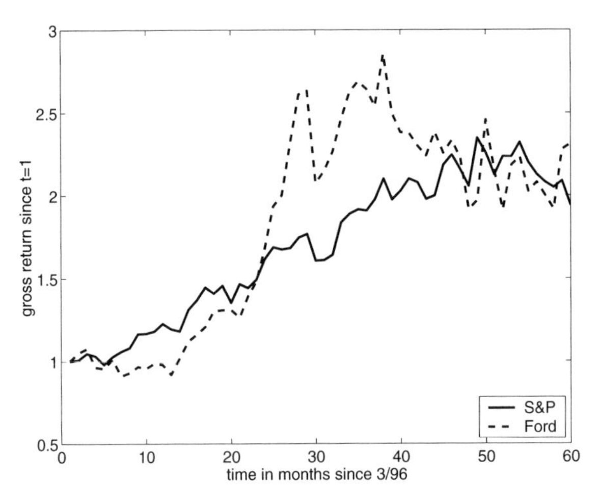

Fig. 7.6. *Gross k period returns of the S&P 500 and Ford since March 1996 plotted against k. Data from March 1996 to February 2001.*

7.10 Is Beta Constant?*

Linear regression is an extremely powerful tool, one that is much more broadly applicable than you might realize at first. By skillful choice of the regression model it is possible to test many interesting hypotheses. This section illustrates one such application of multiple linear regression.

So far, we have been treating an asset's beta as a fixed quantity, but certainly a beta could vary over time. However, it is not immediately obvious how to detect changes over time in beta. Beta is not directly observable;

that is, an asset's beta for a particular holding period cannot be observed. Remember that β_j, the beta of the jth asset, is defined by

$$R^*_{j,t} = \alpha_j + \beta_j R^*_{M,t} + \epsilon_{j,t}.$$

We simplify this equation by assuming that the asset is priced correctly according to the CAPM so that $\alpha_j = 0$. Even with this simplification, both β_j and $\epsilon_{j,t}$ are unknowns in this equation, so we cannot solve for both of them. What we can do is estimate β_j by regression using a time series of data. However, this gives us a β_j that is constant over the time interval during which the data were collected.

Consider a more complex model with a time-varying beta:

$$R^*_{j,t} = \beta_{j,t} R^*_{M,t} + \epsilon_{j,t}, \quad t = 1, \ldots, n, \tag{7.26}$$

where $\beta_{j,t}$ is the asset's beta at time t and n is the length of the time series of data. If we make no further assumptions, then we have n parameters, $\beta_{j,1}, \ldots, \beta_{j,n}$, and we cannot possibly estimate all of them. In fact, we cannot estimate any of them. A reasonable assumption is that $\beta_{j,t}$ varies as a smooth[18] function of time. For example, we could assume that $\beta_{j,t}$ is linear in t so that

$$\beta_{j,t} = b_0 + b_1 t, \tag{7.27}$$

where b_0 is the intercept and b_1 is the slope of the linear relationship between time and beta. Model (7.27) has only two parameters, not n, making estimation feasible. We could use model (7.27) as is, but instead we use a slightly different form of model (7.27) that at first seems more complex but is really simpler then (7.27) in that its parameters are easily interpreted. This model is

$$\beta_{j,t} = \beta_j + \gamma_j \left\{ \frac{t - (n+1)/2}{n-1} \right\}. \tag{7.28}$$

Here $(n+1)/2$ is the midpoint of time variable $t = 1, \ldots, n$ so that β_j has the simple interpretation of being the value of beta when t is halfway between the beginning and end of the data collection period. The denominator $n - 1$ is the length of the interval $[1, n]$. Thus, the predictor $\{t - (n+1)/2\}/(n-1)$ in model (7.28) is $-1/2$ when $t = 1$ and $1/2$ when $t = n$. This means that γ_j has the interpretation of being the change in $\beta_{j,t}$ over the time interval $[1, n]$. Putting models (7.26) and (7.28) together gives model

$$R^*_{j,t} = \beta_j R^*_{M,t} + \gamma_j \left[R^*_{M,t} \left\{ \frac{t - (n+1)/2}{n-1} \right\} \right] + \epsilon_{j,t}. \tag{7.29}$$

Example 7.4. Here is a SAS program to fit (7.29). In the program `sp` is the excess return on the S&P 500 and is used as $R^*_{M,t}$ and `Microsoft` is the excess

[18] A function is called "smooth" if it is at least continuous, preferably having several continuous derivatives.

return on Microsoft and is used as $R^*_{j,t}$. The data set that is used contains monthly returns from March 1996 to February 2001. During a data step, SAS automatically creates a variable _N_ that is the observation number, so _N_ in the program is the variable t in (7.29). The variable `time` in the program is $\{t-(n+1)/2\}/(n-1)$ in (7.29).

```
data capm ;
infile 'book\data\capm.txt' ;
input sp Microsoft ;
time = (_N_  - 30.5)/29.5 ;
sp_time = sp*time ;
run ;
proc reg ;
model Microsoft = sp sp_time/noint ;
run ;
```

The relevant part of the SAS output is below.

Variable	DF	Parameter Estimate	Standard Error	t Value	Pr > |t|
sp	1	1.40641	0.31903	4.41	<.0001
sp_time	1	0.63960	0.58334	1.10	0.2774

The estimate of γ_j is 0.640 and has a standard error of 0.583. The estimate is not much larger than its standard error and its p-value is large, 0.277. Thus, there is little evidence that γ_j is not zero and therefore no reason to believe that $\beta_{j,t}$ is not constant. Of course, we have only tested that $\beta_{j,t}$ is constant versus linear in t. Perhaps $\beta_{j,t}$ is a nonlinear function of t. However, models where $\beta_{j,t}$ is quadratic or cubic in t were also tried and led to the same conclusion that the model where $\beta_{j,t}$ is constant in t cannot be rejected.

In summary, it appears from this analysis that the beta for Microsoft is constant from 1996 to 2001. However, a second analysis was performed using daily Microsoft data from 1993 to 2003. It this case, the p-value for γ_j was 0.0012, which is strong evidence of a nonconstant beta. There are two reasons why beta might appear nonconstant in the second analysis: (1) there are more years of data and beta is more likely to be nonconstant over a longer time period and (2) with daily data we can obtain a much more precise estimate of γ_j so a small deviation of γ_j from 0 is more likely to be detected.

It would be simple to add a nonzero α_j to the model, in which case we could consider both the model with α_j constant and the model where α_j is linear in t. This would allow testing the hypothesis that α_j is constant.

7.11 Summary

The CAPM assumes that prices are in equilibrium, that everyone has the same forecasts of returns, and that everyone uses the principles of portfolio selection introduced in Chapter 5.

The CAPM assumptions imply that everyone holds risk-efficient portfolios that mix the tangency portfolio and risk-free assets. This fact implies that the market portfolio equals the tangency portfolio. A further consequence is that the Sharpe's ratio for any efficient portfolio equals the Sharpe ratio of the market portfolio:

$$\frac{\mu_R - \mu_f}{\sigma_R} = \frac{\mu_M - \mu_f}{\sigma_M}, \tag{7.30}$$

where R is the return on any efficient portfolio. Equation (7.30) can be rearranged to give the capital market line (CML) which is

$$\mu_R = \mu_f + \frac{\mu_M - \mu_f}{\sigma_M}\sigma_R.$$

The CML applies only to efficient portfolios.

Another consequence of CAPM assumptions is the security market line (SML) which applies to any security, for example, the jth, and is

$$\mu_j = \mu_f + (\mu_M - \mu_f)\beta_j.$$

Here β_j is the "independent variable" of this linear relationship and measures the riskiness of the security. The expected return μ_j is the "dependent variable."

The security characteristic line is a model for how actual returns are generated. (The SML only described expected returns.) The security characteristic line is

$$R_{j,t} = \mu_f + \beta_j(R_{M,t} - \mu_f) + \epsilon_{j,t}.$$

The variance of $\epsilon_{j,t}$ is $\sigma^2_{\epsilon,j}$. The security characteristic line implies that the risk of the jth asset can be decomposed into market and nonmarket risks:

$$\sigma_j = \sqrt{\beta_j^2\sigma_M^2 + \sigma^2_{\epsilon,j}},$$

where $\beta_j^2\sigma_M^2$ is the component of the total risk due to market risk and $\sigma^2_{\epsilon,j}$ is the component due to unique risk.

If one assumes that $\epsilon_{j,t}$ is uncorrelated with $\epsilon_{j',t}$ for $j \neq j'$ (i.e., for two different securities), then nonmarket risk can be eliminated by portfolio diversification.

Since the security characteristic line is a regression model, it can be used to estimate β_j and $\sigma^2_{\epsilon,j}$. The R^2 value of the regression estimates the proportion of the total risk (σ_j^2) due to market risk; i.e., it estimates $\beta_j^2\sigma_M^2/\sigma_j^2$. Also, $1 - R^2 = \sigma^2_{\epsilon,j}/\sigma_j^2$ is the proportion of total risk that is nonmarket and can be removed by diversification.

CAPM is a single-factor model with the market index being the factor. In CAPM, all correlations between asset returns are due to this single factor. Multiple-factor models are more realistic and are widely used in practice.

7.12 Bibliographic Notes

The CAPM was developed by Sharpe (1964), Lintner (1965a,b), and Mossin (1966). Introductions to the CAPM can be found in Bodie, Kane, and Marcus (1999), Bodie and Merton (2000), and Sharpe, Alexander, and Bailey (1999). I learned about CAPM from these three textbooks. Campbell, Lo, and MacKinlay (1997) discuss empirical testing of CAPM. This derivation of the SML in Section 7.5.2 was adapted from Sharpe, Alexander, and Bailey (1999). Discussion of factor models can be found in Sharpe, Alexander, and Bailey (1999), Bodie, Kane, and Marcus (1999), and Campbell, Lo, and MacKinlay (1997). The Fama-French three-factor model was introduced by Fama and French (1993) and discussed further in Fama and French (1995, 1996).

7.13 References

Bodie, Z. and Merton, R. C. (2000) *Finance*, Prentice-Hall, Upper Saddle River, NJ.

Bodie, Z., Kane, A., and Marcus, A. (1999) *Investments, 4th Ed.*, Irwin/ McGraw-Hill, Boston.

Campbell, J. Y., Lo, A. W., and MacKinlay, A. C. (1997) *The Econometrics of Financial Markets*, Princeton University Press, Princeton, NJ.

Fama, E. F. and French, K. R. (1992) The cross-section of expected stock returns, *Journal of Finance*, **47**, 427–465.

Fama, E. F. and French, K. R. (1993) Common risk factors in the returns on stocks and bonds, *Journal of Financial Economics*, **33**, 3–56.

Fama, E. F. and French, K. R. (1995) Size and book-to-market factors in earnings and returns, *Journal of Finance*, **50**, 131–155.

Fama, E. F., and French, K. R. (1996) Multifactor explanations of asset pricing anomalies, *Journal of Finance*, **51**, 55–84.

Lintner, J. (1965a) The valuation of risky assets and the selection of risky investments in stock portfolios and capital budgets, *Review of Economics and Statistics*, **47**, 13–37.

Lintner, J. (1965b) Security prices, risk, and maximal gains from diversification, *Journal of Finance*, **20**, 587–615.

Mossin, J. (1966) Equilibrium in capital markets, *Econometrica*, **34**, 768–783.

Sharpe, W. F. (1964) Capital asset prices: A theory of market equilibrium under conditions of risk, *Journal of Finance*, **19**, 425–442.

Sharpe, W. F., Alexander, G. J., and Bailey, J. V. (1999) *Investments, 6th ed.*, Prentice-Hall, Upper Saddle River, NJ.

Shleifer, A. (2000) *Inefficient Markets: An Introduction to Behavioral Finance*, Oxford University Press, Oxford.

7.14 Problems

1. What is the beta of a portfolio if $E(R_P) = 16\%$, $\mu_f = 6\%$, and $E(R_M) = 11\%$?
2. Suppose that the risk-free rate of interest is 0.07 and the expected rate of return on the market portfolio is 0.14. The standard deviation of the market portfolio is 0.12.
   (a) According to the CAPM, what is the efficient way to invest with an expected rate of return of 0.11?
   (b) What is the risk (standard deviation) of the portfolio in part (a)?
3. Suppose that the risk-free interest rate is 0.04, that the expected return on the market portfolio is $\mu_M = 0.10$, and that the volatility of the market portfolio is $\sigma_M = 0.12$.
   (a) What is the expected return on an efficient portfolio with $\sigma_R = 0.05$?
   (b) Stock A returns have a covariance of 0.004 with market returns. What is the beta of Stock A?
   (c) Stock B has beta equal to 1.5 and $\sigma_\epsilon = 0.08$. Stock C has beta equal to 1.8 and $\sigma_\epsilon = 0.10$.
      i. What is the expected return of a portfolio that is one-half Stock B and one-half Stock C?
      ii. What is the volatility of a portfolio that is one-half Stock B and one-half Stock C? Assume that the ϵs of Stocks B and C are independent.
4. Show that equation (7.15) follows from equation (2.54).
5. True or False: The CAPM implies that investors demand a higher return to hold more volatile securities. Explain your answer.
6. Suppose that the riskless rate of return is 5% and the expected market return is 14%. The standard deviation of the market return is 15%. Suppose as well that the covariance of the return on Stock A with the market return is $165\%^2$.[19]
   (a) What is the beta of Stock A?
   (b) What is the expected return on Stock A?
   (c) If the variance of the return on Stock A is $220\%^2$, what percentage of this variance is due to market risk?
7. Suppose there are three risky assets with the following betas and $\sigma^2_{\epsilon_j}$.

j	β_j	$\sigma^2_{\epsilon_j}$
1	0.9	0.010
2	0.7	0.015
3	0.6	0.012

   Suppose also that the variance of $R_{Mt} - \mu_{ft}$ is 0.014.
   (a) What is the beta of an equally weighted portfolio of these three assets?

[19] If returns are expressed in units of percent, then the units of variances and covariances are percent-squared. A variance of $165\%^2$ equals 165/10,000.

   (b) What is the variance of the excess return on the equally weighted portfolio?
   (c) What proportion of the total risk of asset 1 is due to market risk?
8. Verify equation (7.25).

8

Options Pricing

8.1 Introduction

The European call options mentioned in Chapter 1 are one example of the many derivatives now on the market. A **derivative** is a financial instrument whose value is derived from the value of some underlying instrument such as interest rate, foreign exchange rate, or stock price.

A call option gives one the right to buy a specified number[1] of shares of a certain asset such as a stock at the **exercise** or **strike** price, while a **put** option gives one the right to sell the asset at the exercise price. An option has an **exercise date**, which is also called the **strike date, maturity,** or **expiration date**. American options can be exercised at any time up to their exercise date, but European options can be exercised only at their exercise date. European options are easier to price than American options since one does not need to consider the possibility of early exercise. Most options sold on organized exchanges are American.[2]

Many complex types of derivatives cannot be priced by a simple formula in closed form such as that of Black and Scholes for a European call. Rather, these options must be priced numerically, for example, using computer simulation. Numerical pricing is an advanced topic that is not covered in this text.

In this chapter we discuss the main ideas behind the pricing of options. We do not actually prove the Black-Scholes formula for a European call, since that derivation requires advanced mathematics. However, we present a heuristic derivation of that formula to give an intuitive understanding of option pricing. American options and puts are also discussed.

Why do companies purchase options and other derivatives? The answer is simple: to manage risk. In its 2000 Annual Report, the Coca-Cola Company wrote "Our company uses derivative financial instruments primarily to reduce

[1] Options contracts are generally for 100 shares; see Jarrow and Turnbull (2000).

[2] Jarrow and Turnbull (2000).

our exposure to adverse fluctuations in interest rates and foreign exchange rates and, to a lesser extent, adverse fluctuations in commodity prices and other market risks." Derivatives can and have been used to speculate, but that is not their primary purpose. The intent of this quote is clear. The company was assuring its stockholders that it is using derivatives to manage risk, not to gamble.

8.2 Call Options

Suppose that you have purchased a European call option on 100 shares of Stock A with an exercise price of \$70. At the expiration date, suppose that Stock A is selling at \$73. The option allows you to purchase the 100 shares for \$70 and to immediately sell them for \$73, with a gain of \$300 on the 100 shares. Of course, the net profit for purchasing the option isn't \$300 since you had to pay a premium for the option. If the option costs \$2/share, then you paid \$200 for the option. Moreover, you paid the \$200 up front but only got the \$300 at the expiration date. Suppose that the expiration date was 3 months after the purchase date and the continuously compounded risk-free rate is 6% per annum or 1.5% for 3 months. Then the dollar value of your net profit at the time of purchase is

$$\exp(-0.015)300 - 200 = 95.53$$

and is

$$300 - \exp(0.015)200 = 96.98$$

at the exercise date.

A call is never exercised if the exercise price is greater than the price of the stock, since exercising the option would amount to buying the stock for more than it would cost on the market. If a call is not exercised, then one loses the cost of the premium but no more than that.

One can lose money on an option even if it is exercised, because the amount gained by exercising the option might be less than the premium. In the example above, if Stock A were selling for \$71 at the exercise date, then one would exercise the option and gain \$100. This would be less than the \$200 paid for the option. Even though exercising the option results in a loss, the loss is less than it would be if the option were not exercised. An option should always be exercised if $(S_t - K)$ is positive where K is the exercise price.

We use the notation $(x)_+ = x$ if $x > 0$ and $= 0$ if $x \leq 0$. With this notation, the value of a call on its exercise date T is

$$(S_T - K)_+,$$

where S_T is the stock's price on the exercise date.

8.3 The Law of One Price

The **law of one price** states that if two financial instruments have exactly the same payoffs, then they have the same price. This principle is used to price options. To valuate an option, one can find a portfolio or a **self-financing**[3] trading strategy with a *known price* and that has exactly the same payoffs as the option. The price of the option is then known, since it must be the same as the price of the portfolio or self-financing trading strategy.

8.3.1 Arbitrage

Arbitrage means making a guaranteed risk-free profit by trading in the market with no invested capital.[4] In other words, arbitrage is a "free lunch." The **arbitrage price** of a security is the price that guarantees no arbitrage opportunities. The law of one price is equivalent to stating that the market is free of arbitrage opportunities, and arbitrage pricing is another name for pricing by the law of one price.

Example 8.1. Here is a simple example of pricing by the law of one price. Suppose stock in company A sells at \$100/share. The risk-free rate of borrowing is 6% compounded annually. Consider a **forward contract** obliging one party to sell to the other party one share of Company A exactly one year from now at a price P. No money changes hands now. What is the fair market price, i.e., what should P be?

This contract is not an option because the sale *must* take place, so it is easier to price than an option. It would seem that P should depend on the expected price of company A stock one year from now. Perhaps surprisingly, this is not the case. Consider the following strategy. The party that must sell the share one year from now can borrow \$100 and buy one share now; this involves no capital since the purchase is with borrowed money. A year from now that party sells the share for P dollars and pays back \$106 (principal plus interest) to the lender, who could be a third party. The profit is $P - 106$. To avoid arbitrage, the profit must be 0 since no capital was used and there is no risk. Therefore, P should be \$106.

Consider what would happen if P were not \$106. Any other value of P besides \$106 would lead to unlimited risk-free profits on one side of the contract or the other. As investors rushed in to take advantage of this situation, the market would immediately correct the value of P to be \$106.

[3] A trading strategy is "self-financing" if it requires no investment other than the initial investment and allows no withdrawals of cash. After the initial investment, any further purchases of assets are financed by the sale of other assets or by borrowing and the proceeds of any sale are always reinvested.

[4] Investing in risk-free T-bills guarantees a positive net return but is *not* arbitrage since capital is invested.

8.4 Time Value of Money and Present Value

"Time is money" is an old adage that is still true. A dollar a year from now is worth less to us than a dollar now. In finance it is essential that we are able to convert values of future payments to their present values, or vice versa. For example, we saw in Section 8.3 that the arbitrage-enforced future price of a stock is simply the present price converted into a "future value" by multiplying by $1+r$ where r is the risk-free rate. Then the **present value** or **net present value (NPV)** of \$D dollars one year from now is $\$D/(1+r)$ if r is a simple interest rate or $\$D\exp(-r)$ if r is a continuous compounding rate. Another way of stating this is that \$D dollars now is, a year from now, worth $\$(1{+}r)$D dollars without compounding, or $\$\{\exp(r)\}$ D dollars under continuous compounding. When \$D is a future cash flow, then its net present value is also called a **discounted** value and r is the discount rate.

The distinction between simple and compounding is not essential since an interest rate of r without compounding is equivalent to an interest rate of r' with continuous compounding, where

$$1+r=\exp(r'),$$

so that

$$r=\exp(r')-1 \text{ or } r'=\log(1+r).$$

We work with both simple and compound interest, whichever is most convenient.

Example 8.2. If $r = 5\%$, then $r' = \log(1.05) = 0.0488$ or 4.88%. If $r' = 4\%$, then $r = \exp(.04) - 1 = 1.0408 - 1$ or 4.08%. In general, $r > r'$.

Occasionally, wc simplify examples by making the unrealistic assumption that $r = 0$ so that present and future values are equal. This simplifying assumption allows us to focus on concepts other than discounting.

8.5 Pricing Calls — A Simple Binomial Example

We start our study of option pricing with a very simple example illustrated in Figure 8.1. Suppose that a stock is currently selling for \$80 and that at the end of one time period it either has increased to \$100 or decreased to \$60. What is the current value of a call option that allows one to purchase one share of the stock for \$80, the exercise price, after one time period?

At the end of the time period, the call option is worth \$20 $(= (\$100 - \$80)_+)$ if the stock has gone up and worth \$0 $(= (\$60 - \$80)_+)$ if the stock has gone down. See Figure 8.1. However, the question is "What is the option worth *now*?" The answer is, of course, the fair market price for the option at the current time.

One might think that the current value of the option depends on the probability that the stock goes up. However, this is not true. The current value of the option depends only on the rate of risk-free borrowing and the stock price's volatility. For simplicity, we assume that this rate is 0. Later we show how to price options when the rate of interest is positive. It turns out that the value of the option is \$10. Let's see where the \$10 came from.

Consider the following investment strategy. Borrow \$30 and buy one-half of a share of stock. The cost up front is \$40 − \$30 = \$10, so the value now of the portfolio is \$10. If after one time period the stock goes up, then the portfolio is worth $100/2 - 30 = 20$ dollars. If the stock goes down, then the portfolio is worth $60/2 - 30 = 0$ dollars. Thus, after one time period, the portfolio's value is *exactly* the same as the value of the call option, no matter which way the stock moves. By the law of one price, the value of the call option *now* must be the same as the value *now* of the portfolio, which is \$10.

Let's summarize what we have done. We have found a portfolio of the stock and the risk-free asset[5] that *replicates* the call option. The current value of the portfolio is easy to calculate. Since the portfolio replicates the option, the option must have the same value as the portfolio.

Suppose we have just sold a call option. By purchasing this portfolio we have hedged the option. By hedging is meant that we have eliminated all risk, because the net return of selling the option and purchasing the portfolio is exactly 0 no matter what happens to the stock price.

How can one know that the portfolio should be 1/2 share of stock and −\$30 in cash? One does not use trial-and-error, since that would be very tedious, especially in more complicated examples. Rather, one can use the following logic. We call the difference in prices resulting from up and down steps at a node the **volatility** at that node. This definition of volatility is of course a little different from using standard deviation as a measure of volatility as we have done earlier, but the two definitions are similar. The volatility of one share of the stock is (\$100 − \$60) = \$40 while the volatility of the option is (\$20 − \$0) = \$20. The ratio of the volatility of the option to the volatility of the stock is 1/2 and is called the **hedge ratio**. If the portfolio is to exactly replicate the option, then the portfolio must have exactly the same volatility as the option which implies that the portfolio must have a one-half share of stock.

Key point: The number of shares in the portfolio must equal the hedge ratio, where

$$\text{hedge ratio} = \frac{\text{volatility of option}}{\text{volatility of stock}}.$$

So now we know why the portfolio holds a one-half share of stock. How was it determined that \$30 should be borrowed? If the stock goes down, the

[5] Borrowing means a negative position in the risk-free asset.

portfolio is worth \$30 minus the amount borrowed. But we want the portfolio's value to equal that of the option, which is \$0. Thus, the amount borrowed is \$30.

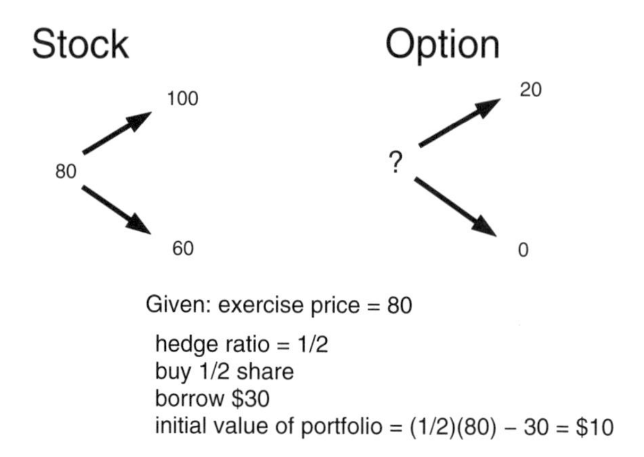

Fig. 8.1. *Example of one-step binomial option pricing. For simplicity, it is assumed that the interest rate is 0. The portfolio of 1/2 share of stock and −\$30 of risk-free assets replicates the call option.*

Key point: We can determine the amount borrowed by equating the value of the portfolio when the stock goes down to the value of the option when the stock goes down. (Alternatively, we could equate the value of the portfolio to the value of the option when the stock goes up. In the example, this would tell us that \$50 minus the amount borrowed equals \$20, or again that \$30 must be borrowed.)

Now suppose that the interest rate is 10%. Then we borrow $\$30/(1.1) = \$27.27$ so that the amount owed after one year is \$30. The cost of the portfolio is $\$40 - (\$30/1.1) = \$12.73$. Thus, the value of the option is \$12.73 if the risk-free rate of interest is 10%. This value is higher than the value of the option when the risk-free rate is 0, because the initial borrowing used by the self-financing strategy is more expensive when the interest rate is higher.

Here's how to price one-step binomial options for other values of the parameters. Suppose the current price is s_1 and after one time period the stock either goes up to s_3 or down to s_2. The exercise price is K. The risk-free rate of interest is r. It is assumed that $s_2 < K < s_3$, so the option is exercised if and only if the stock goes up.[6] Therefore, the hedge ratio is

$$\delta = \frac{s_3 - K}{s_3 - s_2}. \tag{8.1}$$

This is the number of shares of stock that are purchased. The cost is $\delta\, s_1$, the amount borrowed is

$$\frac{\delta s_2}{1+r}, \tag{8.2}$$

and the amount that is paid back to the lender is δs_2. Therefore, the price of the option is

$$\delta \left\{ s_1 - \frac{s_2}{1+r} \right\} = \frac{s_3 - K}{s_3 - s_2} \left\{ s_1 - \frac{s_2}{1+r} \right\}. \tag{8.3}$$

If the stock goes up, then the option is worth $(s_3 - K)$ and the portfolio is also worth $(s_3 - K)$. If the stock goes down, both the option and the portfolio are worth 0. Thus, the portfolio does replicate the option.

Example 8.3. In the example analyzed before, $s_1 = 80$, $s_2 = 60$, $s_3 = 100$, and $K = 80$. Therefore,

$$\delta = \frac{100 - 80}{100 - 60} = \frac{1}{2}.$$

The price of the option is

$$\frac{1}{2} \left\{ 80 - \frac{60}{1+r} \right\},$$

which is \$10 if $r = 0$ and \$12.73 if $r = 0.1$. The amount borrowed is

$$\frac{\delta s_2}{1+r} = \frac{(1/2)60}{1+r} = \frac{30}{1+r},$$

which is \$30 if $r = 0$ and \$27.27 if $r = 0.1$.

8.6 Two-Step Binomial Option Pricing

A one-step binomial model for a stock price may be reasonable for very short maturities. For longer maturities, multiple-step binomial models are needed. A multiple-step model can be analyzed by analyzing the individual steps, going backwards in time.

To illustrate multistep binomial pricing, consider the two-step model of a European call option in Figure 8.2. The option matures after the second step.

[6] This assumption is very reasonable for the following reasons. If $s_2 < s_3 \leq K$, then the option is never exercised under any circumstances. We are certain the option will be worthless, so its price must be 0. If $K \leq s_2 < s_3$, then the option always is exercised so it is not an option but rather a forward contract. We have already seen how to price a forward contract in Section 8.3.

The stock price can either go up \$10 or down \$10 on each step. Initially we assume that $r = 0$ for simplicity.

Using the pricing principles just developed and working backwards, we can fill in the question marks in Figure 8.2. See Figure 8.3 where the nodes have been labeled A, B, and so on. For example, at node B, the hedge ratio is $\delta = 1$ so we need to own one share which at this node is worth \$90. Also, we need to have borrowed $\delta s_2/(1+r) = (1)(80)/(1+0) =$ \$80 so that our portfolio has the same value at nodes E and F as the option; that is, the portfolio should be worth \$0 at node E and \$20 at node D. Since at node B we have stock worth \$90 and the risk-free asset worth $-$\$80, the net value of our portfolio is \$10. By the same reasoning, at node C the hedge ratio is 0 and we should have no stock and no borrowing, so our portfolio is worth \$0.

We can see in Figure 8.3 that at the end of the first step the option is worth \$10 if the stock is up (node B) and \$0 if it is down (node C). Applying one-step pricing at node A, the hedge ratio is 1/2 and we should own 1/2 share of stock (worth \$40) and we should have borrowed \$35. Therefore, the portfolio is worth \$5 at node A, which proves that \$5 is the correct price of the option.

Note the need to work backwards. We could not apply one-step pricing at node A until we had already found the value of the portfolio (and of the option) at nodes B and C.

Let's show that our trading strategy is self-financing. To do this we need to show that we invest no money other than the initial \$5. Suppose that the stock is up on the first step, so we are at node B. Then our portfolio is worth \$90/2 – \$35 or \$10. At this point we borrow \$45 and buy another half-share for \$45; this is self-financing. If the stock is down on the first step, we sell the half share of stock for \$35 and buy off our debt; again the step is self-financing.

8.7 Arbitrage Pricing by Expectation

It was stated earlier that an option is priced by arbitrage; that is, the price is determined by the requirement that the market be arbitrage-free. The expected value of the option is *not* used to price the option. In fact, we do not even need to consider the probabilities that the stock moves up or down. The reason for this is that the option must have the same value as the replicating portfolio regardless of whether the stock moves up or down, so the probabilities of these two events are not needed to determine the arbitrage price of the option.

However, there is a remarkable result showing that arbitrage pricing *can* be done using expectations. More specifically, there exist probabilities of the stock moving up and down such that the arbitrage price of the option *is* equal

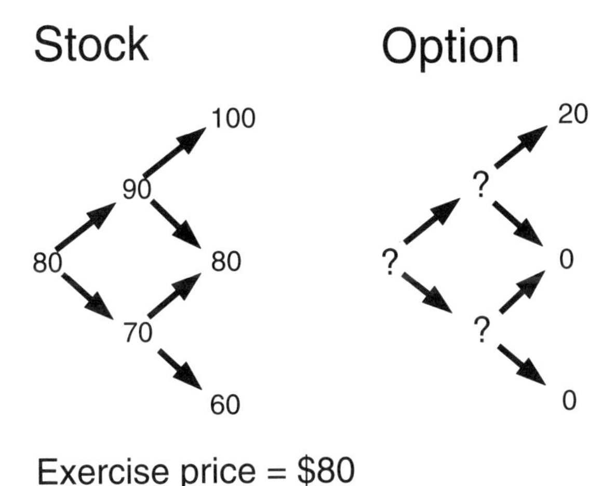

Fig. 8.2. *Two-step binomial model for option pricing.*

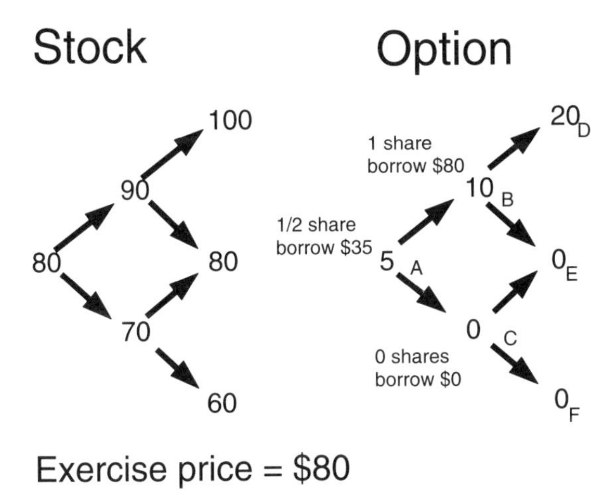

Fig. 8.3. *Pricing the option by backwards induction.*

to the discounted expected value of the option according to these probabilities. Whether these are the "true" probabilities of the stock moving up or down is irrelevant. The fact is that these probabilities give the correct arbitrage price when they are used to calculate expectations.

Let "now" be time 0 and let "one-step ahead" be time 1. Because of the time value of money, the present value of \$D dollars at time 1 is $\$D/(1+r)$, where r is the interest rate.[7] Let $f(2) = 0$ and $f(3) = s_3 - K$ be the values

[7] Assume simple interest here, though a change to compounded interest would not be difficult.

of the option if the stock moves up or down, respectively. We now show that there is a value of q between 0 and 1, such that the present value of the option is

$$\frac{1}{1+r}\{qf(3)+(1-q)f(2)\}. \tag{8.4}$$

The quantity in (8.4) is the present value of the expectation of the option at time 1. To appreciate this, notice that the quantity in curly brackets is the value of the option if the stock goes up times q, which is the arbitrage determined "probability" that the stock goes up, plus the option's value if the stock goes down times $(1-q)$. Thus the quantity in curly brackets is the expectation of the option's value at the end of the holding period. Dividing by $1+r$ converts this to a "present value."

How do we find this wonderful value of q? That's easy. We know that q must satisfy

$$\frac{1}{1+r}\{qf(3)+(1-q)f(2)\} = \frac{s_3-K}{s_3-s_2}\left\{s_1 - \frac{s_2}{1+r}\right\}, \tag{8.5}$$

since the left-hand side of this equation is (8.4) and the right-hand side is the value of the option according to (8.3). Substituting $f(2)=0$ and $f(3)=s_3-K$ into (8.5) we get an equation that can be solved for q to find that

$$q = \frac{(1+r)s_1 - s_2}{s_3 - s_2}. \tag{8.6}$$

We want q to be between 0 and 1 so that it can be interpreted as a probability. From (8.6) one can see that $0 \le q \le 1$ if $s_2 \le (1+r)s_1 \le s_3$.

We show that $s_2 \le (1+r)s_1 \le s_3$ is required in order for the market to be arbitrage-free. If we invest s_1 in a risk-free asset at time 0, then the value of our holdings at time 1 is $(1+r)s_1$. If we invest s_1 in the stock, then the value of our holdings at time 1 is either s_2 or s_3. If $s_2 \le (1+r)s_1 \le s_3$ were not true, then there would be an arbitrage opportunity. For example, if $(1+r)s_1 < s_2 \le s_3$, then we could borrow at the risk-free rate and invest the borrowed money in the stock with a guaranteed profit; at time 1 we would pay back $(1+r)s_1$ and receive *at least* s_2 which is greater than $(1+r)s_1$. To make a guaranteed profit if $s_2 \le s_3 < (1+r)s_1$, sell the stock short and invest the s_1 dollars in the risk-free asset. At the end of the holding period (maturity) receive $(1+r)s_1$ from the risk-free investment and buy the stock for at most $s_3 < (1+r)s_1$. In summary, the requirement that the market be arbitrage-free ensures that $0 \le q \le 1$.

8.8 A General Binomial Tree Model

In this section the one- and two-step binomial tree calculations of the previous sections are extended to an arbitrary number of steps. Consider a possibly nonrecombinant[8] tree as seen in Figure 8.4. Assume that at the jth node the stock is worth s_j and the option is worth $f(j)$, that the jth node leads to either the $2j+1$th node or the $2j$th node after one time "tick," and that the actual time between ticks is δt. Assume as well that interest is compounded continuously at a fixed rate r so that B_0 dollars now are worth $\exp(rn\,\delta t)B_0$ dollars after n time ticks. (Or, B_0 dollars after n ticks are worth $\exp(-rn\delta t)B_0$ dollars now.) Then at node j the value of the option is

$$f(j) = \exp(-r\,\delta t)\Big\{q_j f(2j+1) + (1-q_j)f(2j)\Big\},$$

where the arbitrage-determined q_j is

$$q_j = \frac{e^{r\,\delta t}s_j - s_{2j}}{s_{2j+1} - s_{2j}}. \tag{8.7}$$

The number of shares of stock to be holding is

$$\phi_j = \frac{f(2j+1) - f(2j)}{s_{2j+1} - s_{2j}} = \text{hedge ratio.}$$

Denote the amount of capital to hold in the risk-free asset by ψ_j. Typically ψ_j is negative because money has been borrowed. Since the portfolio replicates the option, at node j the option's value, which is $f(j)$, must equal the portfolio's value, which is $s_j\phi_j + \psi_j$. Therefore,

$$\psi_j = f(j) - \phi_j s_j. \tag{8.8}$$

Also, ψ_j changes in value to $e^{r\,\delta t}\{f(j) - \phi_j s_j\}$ after one more time tick.

Expectations for paths along the tree are computed using the q_js. The probability of any path is just the product of all the probabilities along the path.

Example 8.4. The tree for the example of Section 8.6 is shown in Figure 8.5. This tree is recombinant because nodes 5 and 6 are, in fact, the same. Because $r = 0$ is assumed and because the stock moves either up or down the same amount ($10), the q_j are all equal to 1/2.[9]

The probability of each full path from node 1 to one of nodes 4, 5, 6, or 7 is 1/4. Given the values of the option at nodes 4, 5, 6, and 7, it is easy to compute the expectations of the option's value at other nodes. These expectations are shown in italics in Figure 8.5.

The path probabilities are independent of the exercise price, since they depend only on the prices of the stock at the nodes and on r. Therefore, it is easy to price options with other exercise prices.

---

[8] The tree in Figure 8.4 would be *recombinant* if the stock prices at nodes 5 and 6 were equal so that these two nodes could be combined.

[9] It follows from (8.7) that whenever $r = 0$ and the up moves, which equal $s_{2j+1} - s_j$, and down moves, which equal $s_j - s_{2j}$, are of equal length, then $q_j = 1/2$ for all j.

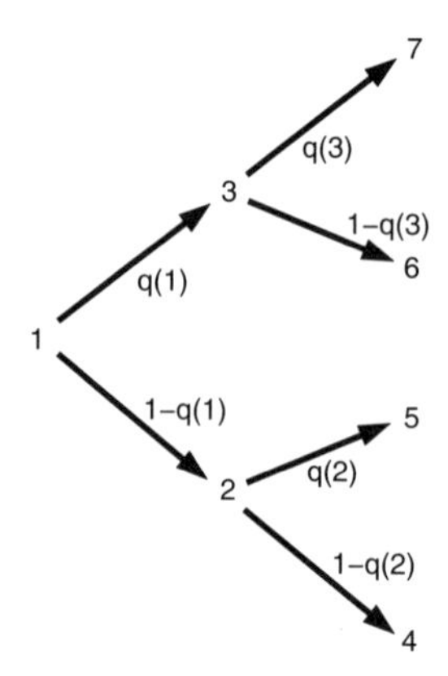

Fig. 8.4. *Nonrecombinant tree. The probability $q(j)$ is the risk-neutral probability at node j of the stock moving upward that is defined in Section 8.4. The tree extends to n steps, though only the first two steps are shown.*

Example 8.5. Assuming the same stock price process as in Figure 8.5, price the call option with an exercise price of \$70.

Answer: Given this exercise price, it is clear that the option is worth \$0, \$10, \$10, and \$30 dollars at nodes 4, 5, 6, and 7, respectively. Then we can use expectations to find that the option is worth \$5 and \$20 at nodes 2 and 3, respectively. Therefore, the option's value at node 1 is \$12.50; this is the price of the option.

8.9 Martingales

A **martingale** is a probability model for a fair game, that is, a game where the expected changes in one's fortune are always zero. More formally, a stochastic process $Y_0, Y_1, Y_2, \ldots$ is a martingale if

$$E(Y_{t+1}|Y_0, \ldots, Y_t) = Y_t$$

for all t.

Example 8.6. Suppose that on each toss of a fair coin we wager half of our fortune that a head appears. If our fortune at time t is Y_t, then we win or lose $Y_t/2$ with probability 1/2 each. Thus, our fortune at time $t+1$ is either $Y_t/2$ or $(3/2)Y_t$, each with probability 1/2. Therefore, $E\{Y_{t+1}|Y_0, \ldots .Y_t\} = E\{Y_{t+1}|Y_t\} = (1/2)(Y_t/2)+(1/2)(3/2)Y_t = Y_t$ so that our sequence of fortunes is a martingale.

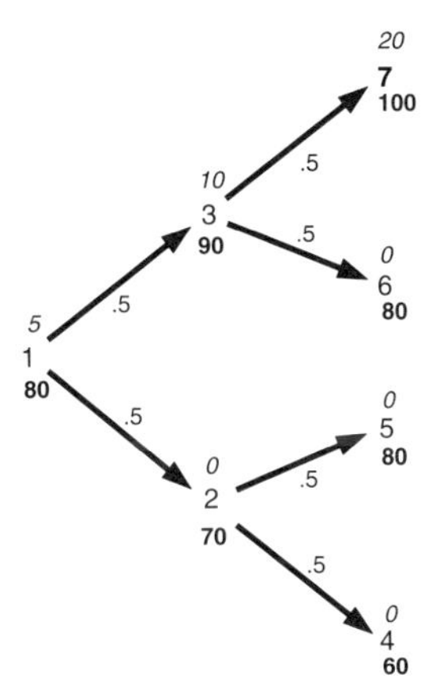

Fig. 8.5. *Two-step example with pricing by probabilities. The number in bold face below a node number is the value of the stock. The number in italics above a node number is the value of the option. Path probabilities are all equal to* 0.5. *The exercise price is* \$80.

Let P_t, $t = 0, 1, \ldots$ be the price of the stock at the end of the tth step in a binomial model. Then $P_t^* = \exp(-rt\,\delta t)P_t$ is called the discounted price process.

Key fact: Under the $\{q_j\}$ probabilities, the discounted price process P_t^* is a martingale.

To see that P_t^* is a martingale, we find using the definition (8.7) of q_j that

$$\begin{aligned} E(P_{t+1}|P_t = s_j) &= q_j s_{2j+1} + (1 - q_j)s_{2j} \\ &= s_{2j} + q_j(s_{2j+1} - s_{2j}) \\ &= s_{2j} + \{\exp(r\,\delta t)s_j - s_{2j}\} = \exp(r\,\delta t)s_j. \end{aligned}$$

This holds for all values of s_j. Therefore, $E(P_{t+1}|P_t) = \exp(r\,\delta t)P_t$, so that

$$E[\exp\{-r(t+1)\,\delta t\}P_{t+1}|P_t] = \exp(-rt\,\delta t)P_t,$$

or $E(P_{t+1}^*|P_t^*) = P_t^*$. This shows that P_t^* is a martingale.

8.9.1 Martingale or risk-neutral measure

Any set of path probabilities $\{p_j\}$ is called a **measure** of the process. The measure $\{q_j\}$ is called the **martingale measure**, the **risk-neutral measure**, or the **pricing measure**. We also call $\{q_j\}$ the risk-neutral path probabilities.

8.9.2 The risk-neutral world

If all investors were risk-neutral, meaning being indifferent to risk, then we might expect that there would be no risk premiums and all expected asset prices would rise at the risk-free rate. Therefore, all discounted asset prices, with discounting at the risk-free rate, would be martingales.

We know that we do not live in such a risk-neutral world, but there is a general principle that expectations taken with respect to a risk-neutral measure give correct, i.e., arbitrage-free, prices of options and other financial instruments.

Example 8.7. In Section 8.3 it was argued that if a stock is selling at \$100/share and the risk-free interest rate is 6%, then the correct future delivery price of a share one year from now is \$106. We can now calculate this value using the risk-neutral measure — in the risk-neutral world, the expected stock price will increase to exactly \$106 one year from now.

8.10 From Trees to Random Walks and Brownian Motion

8.10.1 Getting more realistic

Binomial trees are useful because they illustrate several important concepts, in particular, arbitrage pricing, self-financing trading strategies, hedging, and computation of arbitrage prices by taking expectations with respect to the risk-neutral measure. However, binomial trees are not realistic because stock prices are continuous, or at least approximately continuous. This lack of realism can be alleviated by increasing the number of steps and letting the time between steps shrink. In fact, one can increase the number of steps without limit to derive the Black-Scholes model and formula. That is the aim of Section 8.11. The present section gets closer to that goal.

8.10.2 A three-step binomial tree

Figure 8.6 is a three-step tree where at each step the stock price either goes up \$10 or down \$10. Assume that the risk-free rate is $r = 0$.

Now consider the price of the stock, call it P_t at time t where $t = 0, 1, 2, 3$. Using the risk-neutral path probabilities, which are each 1/2 in this example, P_t is a **stochastic process**, that is, a process that evolves randomly in time. In fact, since P_{t+1} equals $P_t \pm \$10$, this process is a random walk and can be written as

$$P_t = P_0 + (\$10)\{2(W_1 + \cdots + W_t) - t\}, \tag{8.9}$$

where $W_1, \ldots, W_3$ are independent and W_t equals 0 or 1, each with probability 1/2. If W_t is 1, then $2W_t - 1 = 1$ and the price jumps up \$10 on the tth step. If W_t is 0, then $2W_t - 1 = -1$ and the price jumps down \$10. The random sum $W_1 + \cdots + W_t$ is Binomial$(t, 1/2)$ distributed and so has a mean of $t/2$ and variance equal to $t/4$.

The price of the call option at time 0 is

$$E\{(P_3 - K)_+\}, \tag{8.10}$$

where, as before, x_+ equals x if $x > 0$ and equals 0 otherwise. The expectation in (8.10) is with respect to the risk-neutral probabilities. Since $W_1 + W_2 + W_3$ is Binomial$(3, 1/2)$, it equals $0, 1, 2,$ or 3 with probabilities 1/8, 3/8, 3/8, and 1/8, respectively. Therefore,

$$E\{(P_3 - K)_+\} = \frac{1}{8}\Bigg[\{P_0 - 30 - K + (20)(0)\}_+ \\ + 3\{P_0 - 30 - K + (20)(1)\}_+ + 3\{P_0 - 30 - K + (20)(2)\}_+ \\ + \{P_0 - 30 - K + (20)(3)\}_+\Bigg]. \tag{8.11}$$

Example 8.8. If $P_0 = 100$ and $K = 80$, then $P_0 - 30 - K = -10$ and the price of the call option at time 0 is

$$\begin{aligned} E\{(P_3 - K)_+\} &= \frac{1}{8}\Big\{(-10 + 0)_+ + 3(-10 + 20)_+ \\ &\quad + 3(-10 + 40)_+ + (-10 + 60)_+\Big\} \\ &= \frac{1}{8}(0 + 30 + 90 + 50) = \frac{170}{8} = 21.25 \end{aligned}$$

as seen in Figure 8.6. Similarly, if $P_0 = 100$ and $K = 100$, then $P_0 - 30 - K = -30$ and the call's price at time 0 is

$$\begin{aligned} E\{(P_3 - K)_+\} &= \frac{1}{8}\Big\{(-30 + 0)_+ + 3(-30 + 20)_+ \\ &\quad + 3(-30 + 40)_+ + (-30 + 60)_+\Big\} \\ &= \frac{1}{8}(0 + 0 + 30 + 30) = \frac{60}{8} = 7.5. \end{aligned}$$

8.10.3 More time steps

Let's consider a call option with maturity date equal to 1. Take the time interval [0, 1] and divide it into n steps, each of length $1/n$. Suppose that the stock price goes up or down $\sigma/\sqrt{n}$ at each step. The way that the step size $\sigma/\sqrt{n}$ depends on n was carefully chosen, as we soon demonstrate. The parameter σ is the volatility of the stock price, and in applications σ might be estimated from historical data.

The price after m steps $(0 \le m \le n)$ when $t = m/n$ is

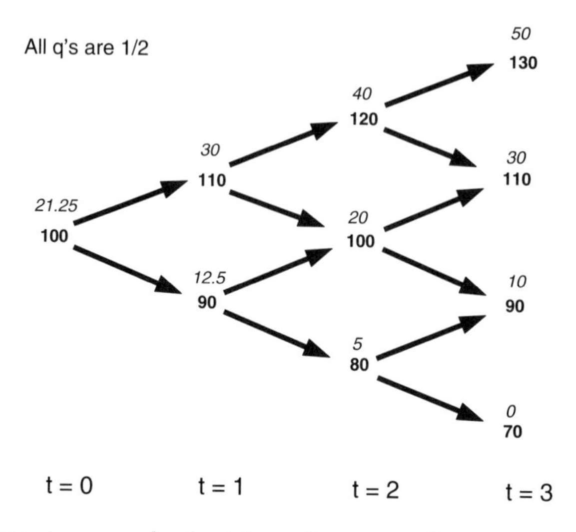

Fig. 8.6. *Three-step example of pricing a European call option by risk-neutral probabilities.* ***Boldface*** *= value of the stock.* ***Italic*** *= value of the option. The exercise price is \$80. The risk-free rate is $r = 0$.*

$$P_t = P_{m/n} = P_0 + \frac{\sigma}{\sqrt{n}}\{2(W_1 + \cdots + W_m) - m\}. \tag{8.12}$$

Since $W_1 + \cdots + W_m$ is Binomial$(m, 1/2)$ distributed, it has expected value $m/2$ and variance $m/4$ and it follows that

$$E(P_t|P_0) = P_0, \tag{8.13}$$

and

$$\text{Var}(P_t|P_0) = \frac{4\sigma^2}{n}\frac{m}{4} = \frac{m}{n}\sigma^2 = t\sigma^2, \tag{8.14}$$

and, in particular,

$$\text{Var}(P_1|P_0) = \sigma^2. \tag{8.15}$$

Moreover, by the central limit theorem, as $n \to \infty$, P_1 converges to a normal random variable with mean P_0 and variance σ^2 which is denoted by $N(P_0, \sigma^2)$ using the notation of Chapter 2.

The step size $\sigma/\sqrt{n}$ was chosen so that $\text{Var}(P_1|P_0) = \sigma^2$ independent of the value of n. As mentioned, σ is just a volatility parameter that could, in practice, be estimated from market data. The factor $\sqrt{n}$ in the denominator was "reverse engineered" to make $\text{Var}(P_t|P_0)$ converge to $t\sigma^2$ as $n \to \infty$.

Let K be the exercise price. Remember that the value of an option is the expectation with respect to the risk-neutral measure of the present value of the option at expiration. Therefore, in the limit, as the number of steps goes to ∞, the price of the option converges to

$$E\{(P_0 + \sigma Z - K)_+\}, \tag{8.16}$$

where Z is $N(0,1)$ so that $P_1 = P_0 + \sigma Z$ is $N(P_0, \sigma^2)$.

For a fixed value of n, P_t is a discrete-time stochastic process since $t = 0, 1/n, 2/n, \ldots, (n-1)/n, 1$. In fact, as we saw before, for any finite value of n, P_t is a random walk. However, in the limit as the number of steps $n \to \infty$, P_t becomes a continuous-time stochastic process. This limit process is called **Brownian motion**.

8.10.4 Properties of Brownian motion

We have seen that Brownian motion is a continuous-time stochastic process that is the limit of discrete-time random walk processes. A Brownian motion process B_t, starting at 0, i.e., with $B_0 = 0$, has the following mathematical properties.

1. $E(B_t) = 0$ for all t.
2. $\text{Var}(B_t) = t\sigma^2$ for all t. Here σ^2 is the volatility of B_t.
3. Changes over nonoverlapping increments are independent. More precisely, if $t_1 < t_2 < t_3 < t_4$ then $B_{t_2} - B_{t_1}$ and $B_{t_4} - B_{t_3}$ are independent.
4. B_t is normally distributed for any t.

If B_0 is not zero, then 1 to 4 hold for the process $B_t - B_0$, which is the change in B_t from times 0 to t. Properties 1 to 3 are shared with random walks with mean-zero steps.

If $\sigma = 1$ then the Brownian motion is said to be standard.

8.11 Geometric Brownian Motion

Random walks are not realistic models for stock prices, since a random walk can go negative. Therefore, (8.16) is close to but not quite the price of the option. To get the price we need to make our model more realistic. We saw in Chapter 3 that geometric random walks are much better than random walks as models for stock prices since geometric random walks are always nonnegative. The limit of a geometric random walk process, as the step size decreases to 0, is called geometric Brownian motion.[10]

We will now introduce a binomial tree model that is a geometric random walk. We do this by making the steps proportional to the current stock price. Thus, if s is the stock price at the current node, then the price at the next node is either s_{up} or s_{down} where

$$s_{\text{up}} = s e^{\mu/n + \sigma/\sqrt{n}}$$

[10] Brownian motion is sometimes called "arithmetic Brownian motion" to distinguish it from geometric Brownian motion.

and

$$s_{\text{down}} = s e^{\mu/n - \sigma/\sqrt{n}}.$$

Notice that the log of the stock price is a random walk since

$$\big(\log(s_{\text{up}}), \log(s_{\text{down}})\big) = \log(s) + \frac{\mu}{n} \pm \frac{\sigma}{\sqrt{n}}.$$

Therefore, the stock price process is a geometric random walk. There is a drift if $\mu \neq 0$, but we show that the amount of drift is irrelevant for pricing. We could have set the drift equal to 0 but we didn't to show later that the drift does *not* affect the option's price.

The risk-neutral probability of an up jump is, by equation (8.7),

$$\begin{aligned} q &= \frac{s\,\exp(r/n) - s_{\text{down}}}{s_{\text{up}} - s_{\text{down}}} \\ &= \frac{\exp(r/n) - \exp(\mu/n - \sigma/\sqrt{n})}{\exp(\mu/n + \sigma/\sqrt{n}) - \exp(\mu/n - \sigma/\sqrt{n})} \\ &\approx \frac{1}{2}\left(1 - \frac{\mu - r + \sigma^2/2}{\sigma\sqrt{n}}\right). \end{aligned} \tag{8.17}$$

The approximation in (8.17) uses the approximations $\exp(x) \approx 1 + x + x^2/2$ and $1/(1+x) \approx 1 - x$. Both these approximation assume that x is close to 0 and are appropriate in (8.17) if the number of steps n is large. Then

$$P_t = P_{m/n} = P_0 \exp\left\{\mu t + \frac{\sigma}{\sqrt{n}} \sum_{i=1}^{m} (2W_i - 1)\right\},$$

where as before W_i is either 0 or 1 (so $2W_i - 1 = \pm 1$).

Using risk-neutral probabilities, we have $E(W_i) = q$ and $\text{Var}(W_i) = q(1 - q)$ for all i. Therefore, using (8.17) and $t = m/n$ we have

$$\begin{aligned} E\left\{\frac{\sigma}{\sqrt{n}} \sum_{i=1}^{m} (2W_i - 1)\right\} &= \frac{\sigma m(2q-1)}{\sqrt{n}} \approx \frac{\sigma}{\sqrt{n}} m \left(\frac{r - \mu - \sigma^2/2}{\sigma\sqrt{n}}\right) \\ &= \left(r - \mu - \frac{\sigma^2}{2}\right)\frac{m}{n} = t(r - \mu - \sigma^2/2) \end{aligned}$$

and

$$\text{Var}\left\{\frac{\sigma}{\sqrt{n}} \sum_{i=1}^{m} (2W_i - 1)\right\} = \frac{4\sigma^2 m q(1-q)}{n} \approx t\sigma^2,$$

since $q \to 1/2$ so that $4q(1-q) \to 1$ as $n \to \infty$. Therefore, in the risk-neutral world

$$P_t \approx P_0 \exp\{(r - \sigma^2/2)t + \sigma B_t\}, \tag{8.18}$$

where B_t is standard Brownian motion and $0 \le t \le 1$. Time could be easily extended beyond 1 by adding more steps. We assume that this has been done.

Notice that (8.18) does not depend on μ, only on σ. The reason is that in the risk-neutral world, the expectation of any asset increases at rate r. The rate of increase in the real world is μ but this is irrelevant for risk-neutral calculations. Remember that risk-neutral expectations do give the correct option price even if they do not correctly describe real-world probability distributions.

If K is the exercise or strike price and T is the expiration date of a European call option, then using (8.18) the value of the option at maturity is

$$(P_T - K)_+ = \left[P_0 \exp\{(r - \sigma^2/2)T + \sigma B_T\} - K\right]_+. \qquad (8.19)$$

Since $B_T \sim N(0,T)$, if we can define $Z = B_T/\sqrt{T}$ then Z is $N(0,1)$. Then the discounted value of (8.19) is

$$\left[P_0 \exp\left\{-\frac{\sigma^2 T}{2} + \sigma\sqrt{T}Z\right\} - \exp(-rT)K\right]_+. \qquad (8.20)$$

We again use the principle that the price of an option is the risk-neutral expectation of the option's discounted value at expiration. By this principle, the call's price at time $t = 0$ is the expectation of (8.20). Therefore,

$$C = \int \left[P_0 \exp\left\{-\frac{\sigma^2 T}{2} + \sigma\sqrt{T}z\right\} - \exp(-rT)K\right]_+ \phi(z)dz, \qquad (8.21)$$

where ϕ is the $N(0,1)$ probability density function.

Computing this integral is a bit complicated, but it can be done. The result is the famous Black-Scholes formula. Let S_0 be the current stock price (we have switched notation from P_0), let K be the exercise price, let r be the continuously compounded interest rate, let σ be the volatility, and let T be the expiration date of a call option. Then by evaluating the integral in (8.21) it can be shown that

$$C = \Phi(d_1)S_0 - \Phi(d_2)K\exp(-rT), \qquad (8.22)$$

where Φ is the standard normal cumulative distribution function,

$$d_1 = \frac{\log(S_0/K) + (r + \sigma^2/2)T}{\sigma\sqrt{T}}, \text{ and } d_2 = d_1 - \sigma\sqrt{T}.$$

Example 8.9. Here is a numerical example. Suppose that $S_0 = 100$, $K = 90$, $\sigma = .4$, $r = .1$, and $T = .25$. Then

$$d_1 = \frac{\log(100/90) + \{0.1 + (0.4)^2/2\}(0.25)}{0.4\sqrt{0.25}} = 0.7518$$

and

$$d_2 = d_1 - 0.4\sqrt{0.25} = 0.5518.$$

Then $\Phi(d_1) = 0.7739$ and $\Phi(d_2) = 0.7095$. Also, $\exp(-rT) = \exp\{(0.1)(0.25)\} = 0.9753$. Therefore,

$$C = (0.7739)(100) - (0.7095)(90)(0.9753) = 15.1.$$

8.12 Using the Black-Scholes Formula

8.12.1 How does the option price depend on the inputs?

Figure 8.7 shows the variation in the price of a call option as the parameters change. The baseline values of the parameters are $S_0 = 100$, $K = 100 \exp(rT)$, $T = 0.25$, $r = 0.06$, and $\sigma = 0.1$. The exercise price K and initial price S_0 have been chosen so that if invested at the risk-free rate, S_0 would increase to K at expiration time. There is nothing special about this choice of S_0 and K and other choices would have been possible and would be interesting to investigate. In each of the subplots in Figure 8.7, one of the parameters is varied while all the others are held at baseline.

One sees that the price of the call increases with σ. This makes sense since in this example, because of the way that K and S_0 have been chosen, $K = S_0 \exp(rT) = E(S_T)$, where $E(S_T)$ is the risk-neutral expectation of S_T. Since $E(S_T)$ equals the strike price, the call is approximately equally likely to be in the money or out of the money.[11] As σ increases, the likelihood that S_T is considerably larger than K also increases. As an extreme case, suppose that $\sigma = 0$. Then in the risk-neutral world $S_T = \exp(rT)S_0 = K$ and the option at expiration is at the money so its value is 0.

The value at maturity is $(S_T - K)_+$ so we expect that the price of the call will increase as S_0 increases and decrease as K increases. This is exactly the behavior seen in Figure 8.7. Also, note that the price of the call increases as either r or T increases. Increasing T has much the same effect as increasing σ since the volatility over the interval $[0, T]$ is $\sigma\sqrt{T}$.

Example 8.10. Table 8.1 gives the exercise price, month of expiration, and the price of call options on GE on February 13, 2001. This information was taken from *The Wall Street Journal*, February 14. Traded options are generally American rather than European and that is true of the options in Table 8.1. However, under the Black-Scholes theory it can be proved that the price of an American call option is identical to the price of an European call option.[12] See Section 8.12.2 for discussion of this point. Since an American call has the same price as a European call, we can use the Black-Scholes formula for European call options to price the options in Table 8.1. We compare the Black-Scholes prices with the actual market prices.

[11] The call is **in the money** if the stock price exceeds the strike price, **out of the money** if the stock price is less than the strike price, and **at the money** if the stock price equals the strike price. The call would be *exactly* equally likely to be in or out of the money if S_T were symmetrically distributed about its expectation, but this is not the case since S_T has a lognormal distribution which is right skewed.

[12] However, as discussed in Section 8.14, American and European put options in general have different prices.

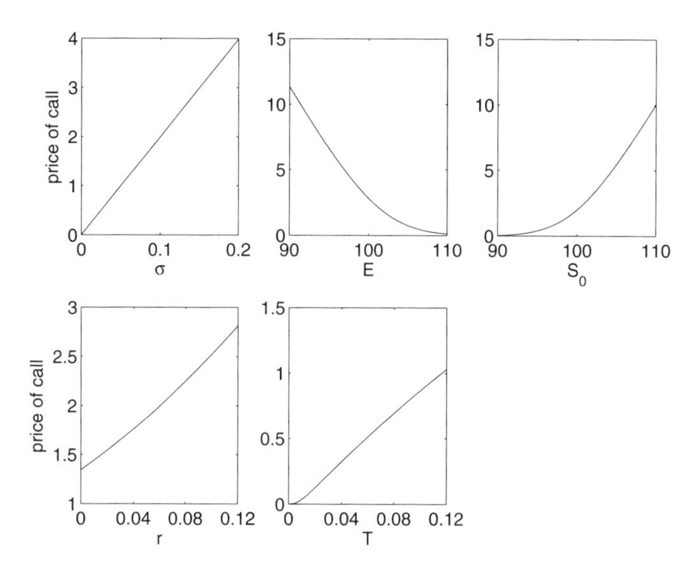

Fig. 8.7. *Price of a call as a function of volatility (σ), exercise price (K), initial price (S_0), risk-free rate (r), and expiration date (T). Baseline values of the parameters are $S_0 = 100$, $K = 100\exp(rT)$, $T = 0.25$, $r = 0.06$, and $\sigma = 0.1$. In each subplot, all parameters except the one on the horizontal axis are fixed at baseline.*

Only the month of maturity is listed in a newspaper. However, maturities (days until expiration) can be determined as follows. An option expires at 10:59 pm Central Time of the Saturday after the third Friday in the month of expiration (Hull, 1995, page 180). February 16, 2001 was the third Friday of its month, so that on February 13, an option with a February expiration date had three trading days (and four calendar days) until expiration. Since there are returns on stocks only on trading days, $T = 3$ for options expiring in February. Similarly, on February 13 an option expiring in March had $T = 23$ trading days until expiration. Since there are 253 trading days/year, there are $253/12 \approx 21$ trading days per month. For June, $T = 23 + (21)(3)$ and for September $T = 23 + (6)(21)$ were used. GE closed at \$47.16 on February 13, so $S_0 = 47.16$. On February 13, the three-month T-bill rate was 4.91%. Thus, the daily rate of return on T-bills was $r = 0.0491/253 = 0.00019470$, assuming that a T-bill has a return only on the 253 trading days per year; see Section 8.12.3. Two values of σ were used. The first, 0.0176, was based on daily GE return from December 1999 to December 2000. The second, 0.025, was chosen to give prices somewhat similar to the actual market prices. The implied volatility values in Table 8.1 are discussed in Section 8.13.

8.12.2 Early exercise of calls is never optimal

It can be proved that early exercise of an American call option is never optimal, at least within the Black-Scholes model where the stock price is geometric

Table 8.1. Actual prices and prices determined by the Black-Scholes formula for options on February 13, 2001. K is the exercise price. T is the maturity.

K	Month of Expiration	T (in days)	Actual Price	B&S calculated price $\sigma = 0.0176$	B&S calculated price $\sigma = 0.025$	Implied Volatility
35	Sep	149	14.90	13.40	14.03	0.0320
40	Sep	149	10.80	9.22	10.37	0.0275
42.50	Mar	23	5.30	5.03	5.38	0.0235
45	Feb	3	2.40	2.22	2.32	0.0290
45	Mar	23	3.40	3.00	3.57	0.0228
50	Feb	3	0.10	0.02	0.09	0.0258
50	Mar	23	0.90	0.64	1.23	0.0209
50	Sep	149	4.70	3.42	5.12	0.0232
55	Mar	23	0.20	0.06	0.28	0.0223
55	Jun	86	1.30	0.92	2.00	0.0204

Brownian motion. The reason is that at any time before the expiration date, the price of the option is higher than the value of the option if exercised immediately. Therefore, it is always better to sell the option rather than to exercise it early.

To see empirical evidence of this principle, consider the first option in Table 8.1. The exercise price is 35 and the closing price of GE was 47.16. Thus, if the option had been exercised at the closing of the market, the option holder would have gained \$(47.16 − 35) = \$12.16. However, the option was selling on the market for \$14.90 that day. Thus, one would gain \$(14.90 − 12.16) = \$2.74 more by selling the option rather than exercising it.

Similarly, the other options in Table 8.1 are worth more if sold than if exercised. The second option is worth \$(47.16 − 40) = \$7.16 if exercised but \$10.80 if sold. The third option is worth \$(47.16 − 42.5) = \$4.66 if exercised but \$5.30 if sold.

Since it is never optimal to exercise an American call option early, the ability to exercise an American call early is not worth anything. This is why American calls are equal in price to European calls with the same exercise price and expiration date.

8.12.3 Are there returns on nontrading days?

We have assumed that there are no returns on nontrading days. For T-bills, this assumption is justified by the way we calculated the daily interest rate. We took the daily rate to be the annual rate divided by 253 on trading days and 0 on nontrading days. If instead we took the daily rate to be the annual rate divided by 365 on every calendar day, then the interest on T-Bills over a year, or a quarter, would be the same.

A stock price is unchanged over a nontrading day. However, the efficient market theory says that stock prices change due to new information. Thus, we might expect that there is a return on a stock over a weekend or holiday but it is realized only when the market reopens. If this were true, then returns from Friday to Monday would be more volatile than returns over a single trading day. Empirical evidence fails to find such an effect. A reason why returns over weekends are not overly volatile might be that there is little business news over a weekend. However, this does not seem to be the explanation. In 1968, the NYSE was closed for a series of Wednesdays. Of course, other businesses were open on these Wednesdays so there was the usual amount of business news during the Wednesdays when the NYSE was closed. For this reason, one would expect increased volatility for Tuesday to Thursday price changes on weeks with a Wednesday closing compared to, say, Tuesday to Wednesday price changes on weeks without a Wednesday market closing. However, no such effect has been detected (French and Roll, 1986). Trading appears to generate volatility by itself. Traders react to each other. Stock prices react to both trading "noise" and to new information, and over short holding periods such as a day the effect of trading noise, which occurs only when the market is open, appears to be larger than the effect of new information; see Chapter 14. Thus, short-term volatility appears to be mostly due to noise trading, which means traders being influenced by each other rather than by information.

8.13 Implied Volatility

Given the exercise price, current price, and maturity of an option and given the risk-free rate, there is some value of σ that makes the price determined by the Black-Scholes formula equal to the current market price. This value of σ is called the **implied volatility**. One might think of implied volatility as the amount of volatility the market believes to exist currently.

How does one determine the implied volatility? The Black-Scholes formula gives price as a function of σ with all other parameters held fixed. What we need is the inverse of this function, i.e., σ as a function of the option price. Unfortunately, there is no formula for the inverse function. The inverse function exists, of course, but there is no explicit formula for it. However, using interpolation one can invert the Black-Scholes formula to get σ as a function of price. Figure 8.8 shows how this could be done for the third option in Table 8.1. The implied volatility in Figure 8.8 is 0.0235 and was determined by MATLAB's interpolation function, interp1.m. The implied volatilities of the other options in Table 8.1 were determined in the same manner. Root-finding algorithms such as Newton-Raphson or bisection could also be used to find the implied volatility.

Notice that the implied volatilities are substantially higher than 0.0176, the average volatility over the previous year. However, there is evidence that the volatility of GE was increasing at the end of last year; see the estimated volatility in Figure 3.4. In that figure, volatility is estimated from December 15, 1999 to December 15, 2000. Volatility is highest at the end of this period and shows some sign of continuing to increase. The estimated volatility on December 15, 2000 was 0.023, which is similar to the implied volatilities in Table 8.1. It would be worthwhile to re-estimate volatility with data from December 15, 2000 to February 13, 2001. It may be that the implied volatilities in Table 8.1 are similar to the observed volatility in early 2001.

The implied volatilities vary among themselves. One potential reason for this variation is that the option prices and the closing price of GE stock are not concurrent. Rather, each price is for the last trade of the day for that option or for the stock. This lack of concurrence introduces some error into pricing by the Black-Scholes formula and therefore into the implied volatilities. However, lack of concurrence should introduce nearly random variation in implied volatility, but in Section 8.13.1 we see that implied volatility varies in a rather systematic way with exercise price and maturity. Another minor problem with these prices is that the Black-Scholes formula assumes that the stock pays no dividends, but GE does pay dividends.[13] A more serious problem is that the Black-Scholes formula assumes that log returns are normally distributed and that the volatility is constant and both these assumptions are at best approximations. In fact, there is good evidence that the problem of systematic variation in implied volatilities is caused by the assumption in the Black-Scholes model of a constant variance. When options are priced by GARCH models that allow a nonconstant variance, this problem seems to disappear; see Section 12.14.

Chance (2003) states that the problem of systematic patterns in implied volatilities is due to other fundamental flaws in the Black-Scholes theory. His argument is that the Black-Scholes theory is "preference free" in that investor needs and preferences for options are ignored. The Black-Scholes model does not explain why anyone would purchase an option in the first place, given that any option can be replicated with a self-financing trading strategy. Chance's argument is interesting but, unfortunately, he does not offer any way to improve upon Black-Scholes pricing whereas the GARCH pricing models do.

8.13.1 Volatility smiles and polynomial regression*

Whatever the reasons why implied volatilities are not constant, it is important to understand how they depend upon exercise price and maturity. This question is addressed in this section.

A plot of implied volatility versus exercise price, with maturity held fixed, is often bowl-shaped with implied volatility highest at the lowest and highest exercise prices. This behavior is called the **volatility smile** and is attributed to nonconstant volatility of the asset returns.

[13] Modifications of the formula to accommodate dividend payments are possible, but we do not pursue that topic here.

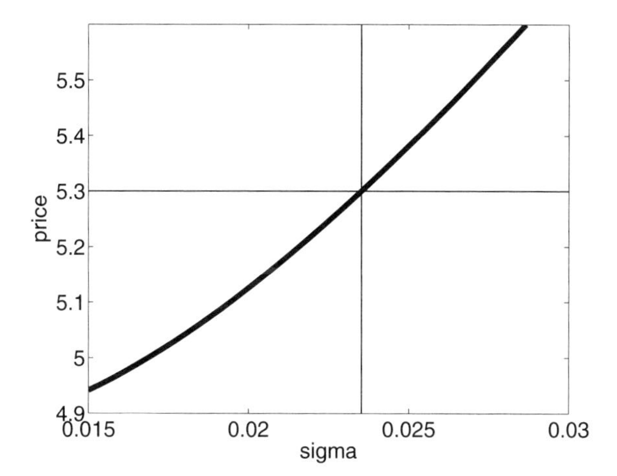

Fig. 8.8. *Calculating the volatility implied by the option with an exercise price of* \$42.50 *expiring in March* 2001. *The price was* \$5.30 *on February* 13, 2001. *The curve is the price given by the Black-Scholes formula as a function of* σ. *The horizontal line is drawn where the price is* \$5.30. *This line intersects the curve at* $\sigma = 0.0242$. *This value of* σ *is the volatility implied by the option's price.*

There are too few options in Table 8.1 to see the volatility smile by plotting implied volatility versus K for a fixed value of T, since there are only three options with $T = 23$, only three with $T = 149$, only two with $T = 3$, and only one with $T = 86$. Instead, we can estimate the volatility smile by modeling the implied volatility as a function of both exercise price and maturity using bivariate polynomial regression, as illustrated in this example. The initial model was

$$V_i = \beta_0 + \beta_1 K_i^c + \beta_2 (K_i^c)^2 + \beta_3 T_i^c + \beta_5 (T_i^c)^2 + \beta_6 K_i^c T_i^c + \epsilon_i,$$

where for the ith option in the table, V_i is the implied volatility, K_i^c is the "centered" exercise price, that is, the exercise price minus its mean $\overline{K}$, and T_i is the centered maturity. The p-value for β_6 is 0.35 so the variable $K_i^c T_i^c$ is dropped. The other four variables provided a reasonable fit with $R^2 = .91$. However, a somewhat better fit was found, with $R^2 = .98$, when the variable $(T_i^c)^3$ was added to the model. The p-value for this variable was small, 0.015, which added further reason to include the variable in the model. The final model to be used for plotting is

$$V_i = \beta_0 + \beta_1 K_i^c + \beta_2 (K_i^c)^2 + \beta_3 T_i^c + \beta_5 (T_i^c)^2 + \beta_6 (T_i^c)^3 + \epsilon_i. \tag{8.23}$$

As a check of model (8.23), C_p was computed for all models using SAS's PROC REG; see the program and output below. Model (8.23) had the second smallest value of C_p. The model with the smallest value of C_p was the same as model (8.23) except the T_i^c was deleted. However, it is unusual to use a model containing power of a predictor variable but without this predictor itself. Therefore, we did include T_i^c in the final model, since $(T_i^c)^2$ is in the model.

Because $T^c = T - \overline{T}$, the estimated value of V when T is fixed at $\overline{T}$ is

$$\widehat{V} = \widehat{\beta}_0 + \widehat{\beta}_1(K - \overline{K}) + \widehat{\beta}_2(K - \overline{K})^2.$$

A plot of $\widehat{V}$ versus K is given in Figure 8.9 where a slight smile can be seen. By the same reasoning, when K is fixed at $\overline{K}$, then

$$\widehat{V} = \widehat{\beta}_0 + \widehat{\beta}_3(T - \overline{T}) + \widehat{\beta}_5(T - \overline{T})^2 + \widehat{\beta}_6(T - \overline{T})^3,$$

which is plotted in Figure 8.10 where again there is a smile.

If T were held fixed at some other value besides $\overline{T}$ then the plot of $\widehat{V}$ versus K would have exactly the same shape as the curve in Figure 8.9 but would be shifted vertically by the amount $\widehat{\beta}_3(T - \overline{T}) + \widehat{\beta}_5(T - \overline{T})^2 + \widehat{\beta}_6(T - \overline{T})^3$. The reason for this is that $\widehat{V}$ is the sum of a function of K and a function of T. Models of this type are called *additive* because the effects of the predictor variables are added together. If we had retained the variable $K_i^c T_i^c$, then the model would not have been additive and the shape of the plot of $\widehat{V}$ versus K would depend on the value at which T were held fixed. The variable $K_i^c T_i^c$ is called an interaction between K and T.

The volatility surface is the plot of implied volatility versus both exercise price and maturity. Using the regression model, the volatility surface can be easily created. Figure 8.11 is the volatility surface for the GE options. This entire surface was created from only 10 option prices and shows the power of regression modeling. However, a word of caution is appropriate. The surface is an extrapolation of implied volatility from 10 values of (K, T) where implied volatility was actually observed to other values of (K, T). Any extrapolation entails some error whose size is difficult to estimate. However, extrapolation is forced upon us by a paucity of data. The extrapolation error has two components. One is modeling error. For example, we did not include interactions in our model, but they might be necessary to have the model fit well in regions where no data were observed, and in this case there is a bias due to not including them. It is difficult to quantify how large this bias might be. The second source of error is due to sampling variation in $\widehat{\boldsymbol{\beta}}$. Fortunately, this error can be estimated by the standard error (SE) of $\widehat{V}$. Figure 8.12 is a plot of this SE. Notice that the SE is largest in the middle of the range of T, say around $T = 90$, because most observed values of T are either small (3 or 23) or large (149). A typical standard error seen in Figure 8.12 is small, around 0.001, only about 1/20th as large as a typical value of the implied volatility.

Here is the SAS program used to compare various possible models. The data step includes statements to create $(K - \overline{K})^2$ and the other polynomials in K and T. Note that $\overline{T} = 63.04$ and $\overline{K} = 46.75$. This is the data step of the program:

```
data ImpVol ;
infile 'C:\book\sas\ImpVolRegData.txt' ;
input K T ImpVol ;
T = T - 63.040 ;
K = K - 46.7500 ;
T2 = T*T ;
K2 = K*K ;
KT = K*T ;
K3 = K*K*K ;
T3 = T*T*T ;
run ;
```

The next part of the program uses PROC REG which is SAS's linear regression procedure. The option "selection = rsquare" chooses the best models of a given size (number of predictors) by R^2 and "best=2" specifies that only the two best models of each size appear in the output. The specifications "adjrsq" and "cp" instruct SAS to include C_p and adjusted R^2 in the output. The proc step of the program is:

```
proc reg;
model ImpVol = K T T2 K2 KT K3 T3/ selection=rsquare adjrsq cp best=2 ;
run ;
```

Here is the SAS output.

```
                        The REG Procedure
                    Dependent Variable: ImpVol
                    R-Square Selection Method
Number in                   Adjusted
  Model      R-Square      R-Square         C(p)    Variables in Model

        1      0.6498        0.6060      62.3013    K3
        1      0.6262        0.5794      66.9068    K
-----------------------------------------------------------------
        2      0.7065        0.6226      53.2406    K T2
        2      0.6970        0.6104      55.0917    T2 K3
-----------------------------------------------------------------
        3      0.8602        0.7903      25.2674    T2 K3 T3
        3      0.8426        0.7639      28.6927    K T2 T3
-----------------------------------------------------------------
        4      0.9760        0.9568       4.6809    K T2 K2 T3
        4      0.9138        0.8449      16.8072    K T T2 K2
-----------------------------------------------------------------
        5      0.9833        0.9624       5.2588    K T T2 K2 T3
        5      0.9827        0.9611       5.3713    K T2 K2 K3 T3
-----------------------------------------------------------------
        6      0.9884        0.9653       6.2568    K T2 K2 KT K3 T3
        6      0.9878        0.9635       6.3729    K T T2 K2 K3 T3
-----------------------------------------------------------------
        7      0.9897        0.9539       8.0000    K T T2 K2 KT K3 T3
```

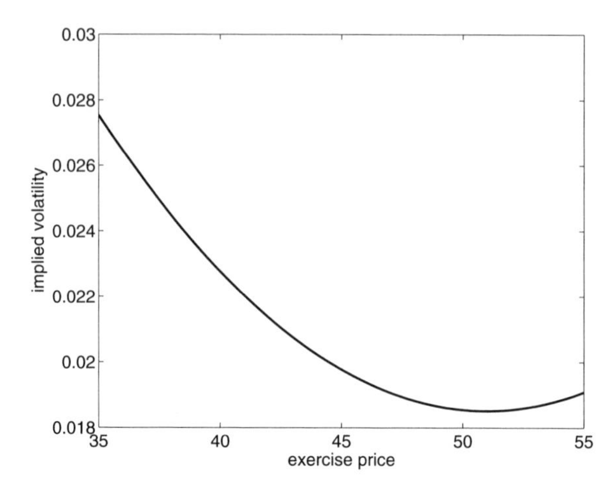

Fig. 8.9. *A plot of* $\widehat{V}$ *versus* K *when* $T = \overline{T}$ *for GE call options.*

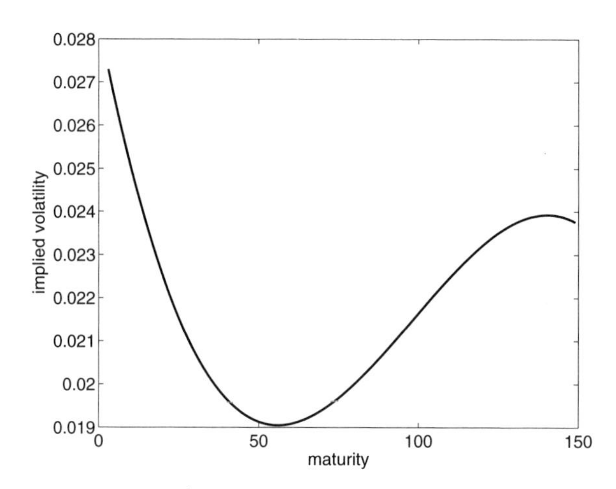

Fig. 8.10. *A plot of* $\widehat{V}$ *versus* T *when* $K = \overline{K}$ *for GE call options.*

8.14 Puts

Recall that a put option gives one the right to sell a share of a stock at the exercise price. The pricing of puts is similar to the pricing of calls, but as we show, there are differences.

8.14.1 Pricing puts by binomial trees

Put options can be priced by binomial trees in the same way as call options. Figure 8.13 shows a two-step binomial tree where the stock price starts at \$100 and increases or decreases by 20% at each step. Assume that the interest rate is 5% compounded continuously and that the exercise price of the put is \$110.

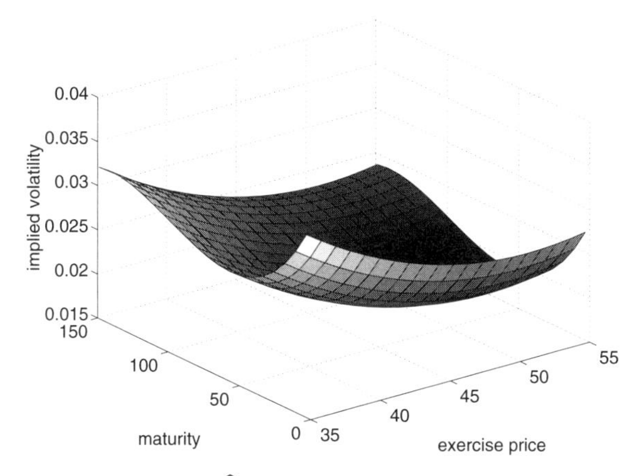

Fig. 8.11. *A plot of $\widehat{V}$ versus T and K for GE call options.*

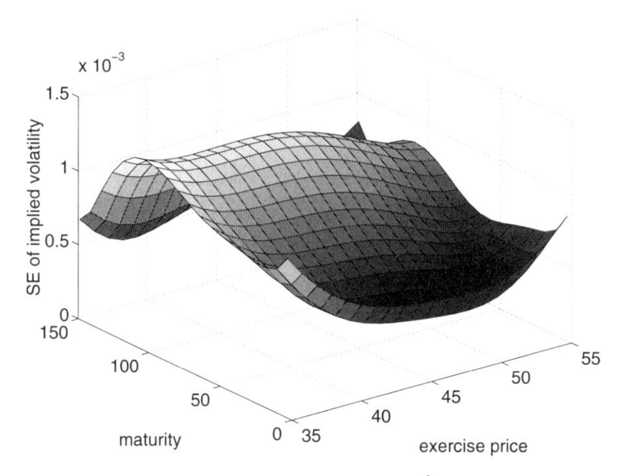

Fig. 8.12. *A plot of the standard error (SE) of $\widehat{V}$ versus T and K for GE call options.*

In this example, European and American puts do not have the same price at all nodes. We start with a European put and then see how an American put differs.

At each step,

$$q = \frac{e^{0.05} - 0.8}{1.2 - 0.8} = .6282.$$

The value of a put after two steps is $(110 - S)_+$ where S is the price of the stock after two steps. Thus the put is worth \$46, \$14, and \$0 at nodes 4, 5, and 6, respectively. The price of the option at node 3 is

$$e^{-.05}\{(q)(0) + (1 - q)(14)\} = e^{-.05}\{(.6282)(0) + (.3718)(14)\} = 4.91,$$

at node 2 is

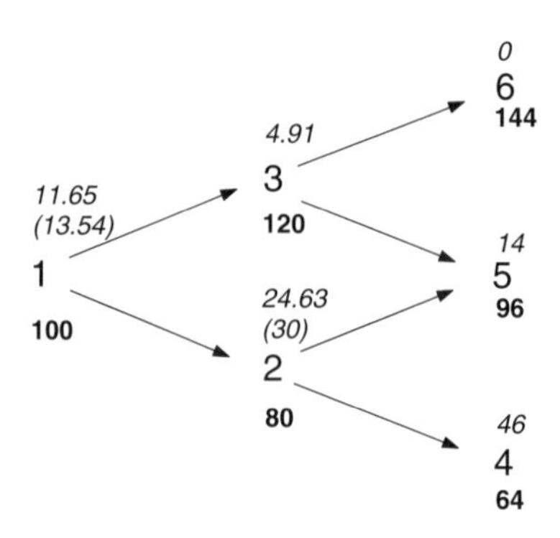

Fig. 8.13. *Pricing a put option. The stock price is in boldface below and the price of a European put is in italics above the node number. The price of an American put is shown in parentheses when it differs from the price of a European put.*

$$e^{-.05}\{(q)(14) + (1-q)(46)\} = 24.63,$$

and at node 1 is

$$e^{-.05}\{(q)(4.91) + (1-q)(24.63)\} = 11.65.$$

Now consider an American option. At nodes 4, 5, and 6 we have reached the expiration time so that the American option has the same value as the European option.

At node 3 the European option is worth \$4.91. At this node, should we exercise the American option early? Clearly not, since the exercise price (\$110) is less than the stock price (\$120). Since early exercise is suboptimal at node 3, the American option is equivalent to the European option at this node and both options are worth \$4.91.

At node 2 the European option is worth \$24.63. The American option can be exercised to earn (\$110 – \$80) = \$30. Therefore, the American option should be exercised early since early exercise earns \$30 while holding the option is worth only \$24.63. Thus, at node 2 the European option is worth \$24.63 but the American option is worth \$30.

At node 1, the American option is worth

$$e^{-.05}\{(q)(4.91) + (1-q)(30)\} = 13.54,$$

which is more than \$11.65, the value of the European option at node 1. The American option should not be exercised early at node 1 since that would earn only \$10. However, the American option is worth more than the European option at node 1 because the American option can be exercised early at node 2 should the stock move down at node 1.

8.14.2 Why are puts different from calls?

We saw that in the Black-Scholes model, it is never optimal to exercise an American call early. Puts are different. In the Black-Scholes model, as in the binomial example just given, early exercise of a put may be optimal.

So why are puts different from calls? The basic idea is this. A put increases in value as the stock price *decreases*. As the stock price decreases, the size of further price changes also decreases, because price changes are proportional to the current price, at least in the Black-Scholes model and in the previous binomial example. At some point we are in the range of diminishing returns where further decreases in the stock price are so small that the put can at best only increase in value at less than the risk-free rate and could decrease in value. At that point it is better to exercise the option and invest the profits in a risk-free asset.

With calls, everything is reversed. A call increases in value as the stock price increases. As the stock price increases, so does the size of future price changes. The expected returns on the call (expectations are with respect to the risk-neutral measure, of course) are always greater than the risk-free rate of return.

8.14.3 Put-call parity

It is possible, of course, to derive the Black-Scholes formula for a European put option by the same reasoning used to price a call. However, this work can be avoided since there is a simple formula relating the price of a European put to that of a call:

$$P = C + e^{-rT}K - S_0, \tag{8.24}$$

where P and C are the prices of a put and of a call with the same expiration date T and exercise price K. Here, the stock price is S_0 and r is the continuously compounded risk-free rate. Thus, the price of a put is simply the price of the call plus $(e^{-rT}K - S_0)$.

Equation (8.24) is derived by a simple arbitrage argument. Consider two portfolios. The first portfolio holds one call and Ke^{-rT} dollars in the risk-free asset. Its payoff at time T is K, the value of the risk-free asset, plus the value of the call, which is $(S_T - K)_+$. Therefore, its payoff is K if $S_T < K$ and S_T if $S_T > K$. In other words, the payoff is either K or S_T, whichever is larger.

The second portfolio holds a put and one share of stock. Its payoff at time T is S_T if $S_T \geq K$ so that the put is not exercised. If $S_T < K$, then the put is exercised and the stock is sold for a payoff of K. Thus, the payoff is K or S_T, whichever is larger, which is the same payoff as for the first portfolio.

Since the two portfolios have the same payoff for all values of S_T, their initial values at time 0 must be equal to avoid arbitrage. Thus,

$$C + e^{-rT}K = P + S_0,$$

which can be rearranged to yield equation (8.24).[14]

Relationship (8.24) holds only for European options. European calls have the same price as American calls so that the right-hand side of (8.24) is the same for European and American options. American puts are worth more than European puts, so the left-hand side of (8.24) is larger for American than for European puts. Thus, (8.24) becomes

$$P > C + e^{-rT}K - S_0, \tag{8.25}$$

for American options, and clearly (8.25) does not tell us the price of an American put but rather only a lower bound on that price.

8.15 The Evolution of Option Prices

As time passes the price of an option changes because of the changing stock price and the decreasing amount of time until expiration. We assume that r and σ are constant, though in the real financial world these could change too and would affect the option price. The Black-Scholes formula remains in effect and can be used to update the price of an option. Suppose that $t = 0$ is when the option was written and $t = T$ is the expiration date. Consider a time point t such that $0 < t < T$. Then the Black-Scholes formula can be used with S_0 in the formula set equal to S $(= S_t)$ and T in the formula set equal to $T - t$ so that

$$C = \Phi(d_1)S - \Phi(d_2)K\exp(-r(T-t)), \tag{8.26}$$

$$d_1 = \frac{\log(S/K) + (r + \sigma^2/2)(T-t)}{\sigma\sqrt{T-t}}, \quad \text{and} \quad d_2 = d_1 - \sigma\sqrt{T-t}.$$

A put option can be priced by put-call parity which now becomes

$$P_t = C_t + e^{-r(T-t)}K - S_t, \tag{8.27}$$

where P_t, C_t, and S_t are the prices at time t of the put, call, and stock.

Figure 8.14 illustrates the evolution of option prices for two simulations of the geometric Brownian motion process of the stock price. Here $T = 1$, $\sigma = .1$, $r = .06$, $S_0 = 100$, and $K = 100$ for both the put and the call. In one case the call was in the money at expiration, while in the second case it was the put that was in the money. For the simulation shown with solid curves the put is out of the money for all time points except around $t = .7$, but the

[14] As usual in these notes, we are assuming that the stock pays no dividend, at least not during the lifetime of the two options. If there are dividends, then a simple adjustment of formula (8.24) is needed. The reason the adjustment is needed is that the two portfolios no longer have exactly the same payoff. One can see that the first portfolio which holds the stock receives a dividend and so receive a higher payoff than the second portfolio which holds no stock and therefore will not receive the dividend.

value of this put only drops to 0 near the expiration date when there is little chance of the put becoming in the money.

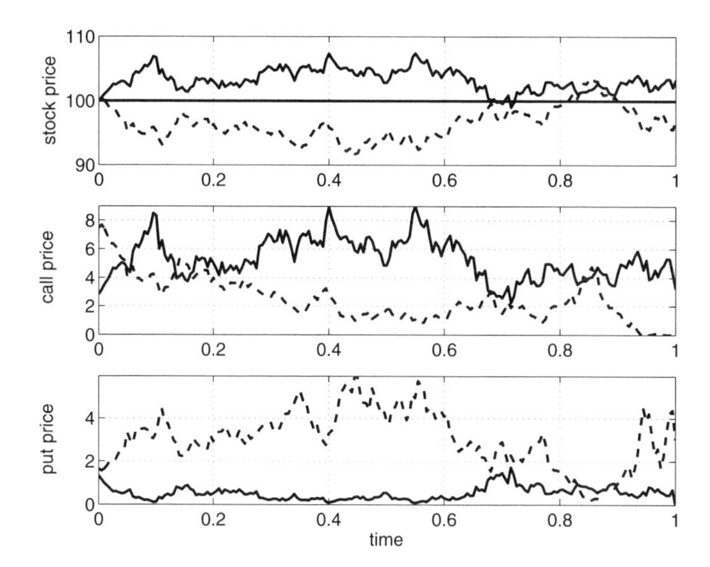

Fig. 8.14. *Evolution of option prices. The stock price is a geometric Brownian motion. Two independent simulations of the stock price are shown with different line types. Here $T = 1$, $\sigma = 0.1$, $r = 0.06$, $S_0 = 100$, and $K = 100$ for both the put and the call. In the simulation shown with a solid (dashed) curve the call (put) is in the money at the expiration date.*

8.16 Leverage of Options and Hedging

Since, as we have just seen, the price of an option changes over time, we can define the return of an option in the same way as returns are defined for any other assets. In particular, the log return of an option is the change in its log price. The return on an option depends on the changes in the volatility σ and price S of the underlying stock, the change in the interest rate r, and the change in the time to maturity $T - t$. Option prices are very sensitive to changes in the price of the stock, a phenomenon called **leverage**.

To appreciate the leverage effect, look at Figure 8.15 which uses the simulated data shown as solid lines in Figure 8.14 and plots the ratio of the call's log return to the stock's log return. There are 200 returns calculated for the periods 0 to 1/200, 1/200 to 2/200, 2/200 to 3/200, and so forth. For time t between 0 and 0.5, this ratio of log returns is approximately 10, meaning that return on the call is roughly 10 times larger than that of the stock. We say that the leverage is 10. As time reaches the expiration date, the leverage increases to over 30. You may be puzzled that the ratio of log returns is occasionally negative, which implies that the stock had a negative return and the call a positive return, or vice versa. This occurs when the return on the stock is very nearly zero but slightly positive and the option has a negative return because of the effect of time to maturity decreasing by 1/200. The plot has been truncated vertically at -30 and 60 to show detail. In fact, the smallest ratio of returns was -242 and the largest was 172, both occurring when the stock's log return was nearly 0.

Figure 8.16 shows the leverage of the put using the same data as in Figure 8.15. In general a put's price increases as the price of the stock decreases, so the leverage is now negative, except occasionally when the stock's log return is nearly zero and the effect of the change in t dominates the effect of the change in S.

Initially, until time t is approximately 0.5, the leverage of the put is about -20. Therefore, a portfolio consisting of the stock and puts on the stock with weights 20/21 and 1/21 will have virtually no risk since any change in the stock price is offset by the resulting change in the put's price. If you owned the stock but wanted to be risk-free over a certain period of time, you could do this by selling the stock at the beginning of the period and then buying it back at the end. An alternative would be to purchase puts at the beginning of the period and then selling them at the end. The alternative would be more attractive if its transaction costs were less than those of selling the stock and then buying it back. Owning the put to offset risk is called **hedging** and the hedge ratio when time is between 0 and 0.5 is approximately 20:1 so the stock in the portfolio should be worth 20 times the value of the puts in the portfolio. One could also hedge by selling calls instead of buying puts, and then the hedge ratio would be approximately 10:1. Of course, one can also use stock to hedge options rather than options to hedge stock. After all, this was the basis for pricing options.

8.17 The Greeks

Let $C(S, T, t, K, \sigma, r)$ be the price of an option as a function of the price of the underlying asset S, the expiration time T, the current time t, the exercise price K, the volatility σ, and the risk-free rate r. The partial derivatives[15] of C with respect to S, t, σ, and r measure the sensitivity of C to changes in

[15] Here "derivative" is being used as in calculus, not as in finance.

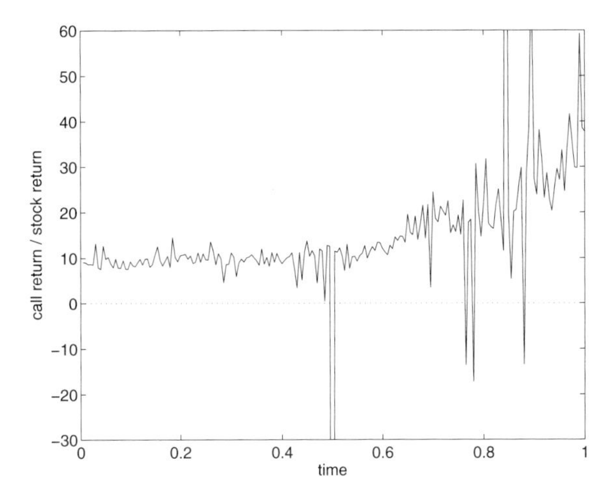

Fig. 8.15. *Ratio of log return on a call to log return on the underlying stock.*

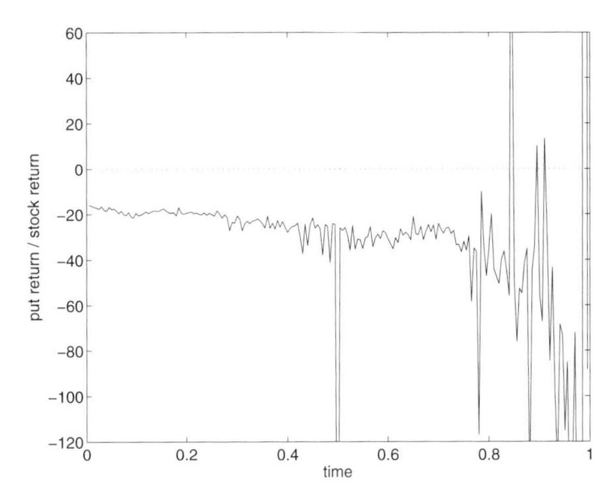

Fig. 8.16. *Ratio of log return on a put to log return on the underlying stock.*

these parameters.[16] These partial derivatives are usually denoted by Greek letters and referred to as "the Greeks." Delta, the sensitivity of the option's price to S, is defined as

$$\Delta = \frac{\partial}{\partial S} C(S, T, t, K, \sigma, r).$$

Similarly, Theta is the sensitivity to t,

$$\Theta = \frac{\partial}{\partial t} C(S, T, t, K, \sigma, r),$$

[16] The exercise price K and expiration time T are fixed, so there is no need to worry about sensitivity to changes in K and T.

Rho is the sensitivity to r,

$$\mathcal{R} = \frac{\partial}{\partial r} C(S, T, t, K, \sigma, r),$$

and Vega[17] is the sensitivity to σ,

$$\mathcal{V} = \frac{\partial}{\partial \sigma} C(S, T, t, K, \sigma, r).$$

In Figures 8.15 and 8.16, σ and r are held fixed so that the returns on the option depend only upon changes in S and t and therefore only upon the sensitivities Δ and Θ. As partial derivatives, the Greeks tells us how an option price will change with *small* changes in S, t, r, and σ. Delta is of particular importance. If the price of the underlying asset changes by a small amount then

$$\text{change in price of option} \approx \Delta \times \text{change in price of asset}. \tag{8.28}$$

Here $\approx$ means that the ratio of the right-hand side to the left-hand side converges to 1 as both price changes converge to zero.

Equation (8.28) implies other approximations that are quite useful. Let P_t^{asset} and P_t^{option} be the price of the asset and the option at time t. Then (8.28) can be written as

$$P_t^{\text{option}} - P_{t-1}^{\text{option}} = \Delta(P_t^{\text{asset}} - P_{t-1}^{\text{asset}})$$

so that

$$\frac{P_t^{\text{option}} - P_{t-1}^{\text{option}}}{P_{t-1}^{\text{option}}} = \left(\Delta \frac{P_{t-1}^{\text{asset}}}{P_{t-1}^{\text{option}}}\right) \frac{P_t^{\text{asset}} - P_{t-1}^{\text{asset}}}{P_{t-1}^{\text{asset}}}. \tag{8.29}$$

Therefore, if we define

$$L = \left(\Delta \frac{P_{t-1}^{\text{asset}}}{P_{t-1}^{\text{option}}}\right), \tag{8.30}$$

then

$$\text{return on option} \approx L \times \text{ return on asset} \tag{8.31}$$

so that L is a measure of the leverage of the asset. By (8.31), the expected value and standard deviation of the return on the option are L times the same

[17] Vega is not a Greek letter and its usual symbol $\mathcal{V}$ is a calligraphic V. I have been unable to determine the origin of this name, though it is thought to have originated in America and been intended to sound Greek. There appears to be no connection with the star Vega, one of the brightest in the night sky, found in the constellation Lyra. Vega was the name of an automobile introduced by General Motors about the time the Black-Scholes formula was developed.

quantities for the asset.[18] More formally, let μ^{asset} and μ^{option} be the expected returns on the asset and option, respectively, and let σ^{asset} and σ^{option} be the standard deviations of these returns. Then

$$\mu^{\text{option}} \approx L \times \mu^{\text{asset}} \text{ and } \sigma^{\text{option}} \approx L \times \sigma^{\text{asset}}. \tag{8.32}$$

Also, suppose that there are two stocks with options on each with leverages L_1 and L_2 and that the covariance between the returns on the stocks is $\sigma_{1,2}$. Then the covariance between the return on the first stock and the return on the option on the second stock is approximately

$$L_2\sigma_{1,2}, \tag{8.33}$$

and the covariance between the two options is approximately

$$L_1 L_2 \sigma_{1,2}. \tag{8.34}$$

Example 8.11. In this example, Δ is computed for a call option whose price is determined by the Black-Scholes formula (8.26) so that

$$C(S,T,t,K,\sigma,r) = \Phi(d_1)S - \Phi(d_2)Ke^{-r(T-t)} \tag{8.35}$$

and consequently

$$\Delta = \Phi(d_1) + \left\{ S\phi(d_1)\left(\frac{d\,d_1}{d\,S}\right) - Ke^{-r(T-t)}\phi(d_2)\left(\frac{d\,d_2}{d\,S}\right)\right\}, \tag{8.36}$$

where $\phi(x) = (1/\sqrt{2\pi})\exp(-x^2/2)$ is the normal PDF. After some calculation it can be shown that the quantity in curly brackets in (8.36) is 0. This calculation is left as an exercise (Problem 8). Thus, $\Delta = \Phi(d_1)$.

Consider the call option in Figures 8.14 and 8.15 where $T = 1$, $\sigma = 0.1$, $r = 0.06$, $S_0 = 100$, and $K = 100$. At $t = 0$ the price of the call is $C(100, 1, 0, 100, 0.1, 0.06) = 7.46$, $d_1 = 0.65$, and $\Delta = \Phi(0.65) = 0.742$. Thus, the revenue from owning a call option on one share is 0.742 times the revenue from owning one share. However, the investment of owning the call on one share is only \$7.46 compared to the investment on one share of \$100. Therefore, the return on the call is $L = (0.742)(100)/7.46 = 9.95$ times the return on the stock; that is, the leverage of the call is 9.95. From Figure 8.15 it can be guessed correctly that the leverage at time $t = 0$ is just under 10.

Example 8.12. For a call option one can show that

$$\Theta = \frac{d\,C(S,T,t,K,\sigma,r)}{dt} = -\frac{S\sigma}{2\sqrt{T-t}}\phi(d_1) - Kre^{-r(T-t)}\Phi(d_2).$$

[18] Although L is a random variable since it depends on S_{t-1} which is random, we treat L as fixed for the purpose of computing the expected return and standard deviation of the return over the period $t-1$ to t conditional on previous returns.

Figure 8.17 shows Delta and Theta for the call option and simulated data shown as solid lines in Figure 8.14. Notice that Theta is always negative so the effect of t getting larger is to decrease the price of the call option. This is why in Figure 8.15 we saw negative returns on the call when the stock had a very slight but positive return.

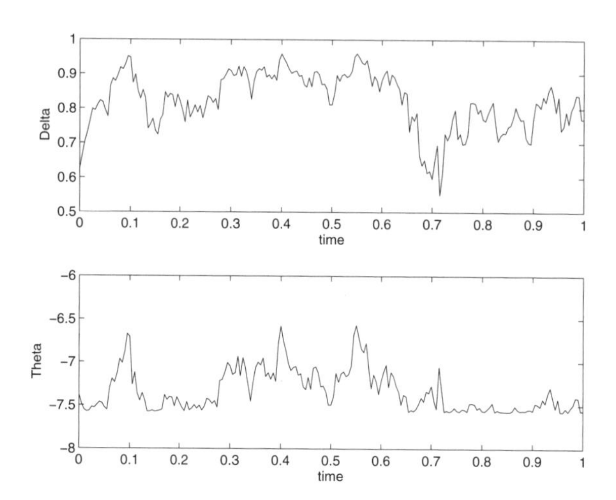

Fig. 8.17. *Delta and Theta for the call option and simulated data shown as solid lines in Figure 8.14.*

8.17.1 Delta and Gamma hedging

Delta-hedging uses Delta to determine the appropriate amount of an option to hold to hedge a stock. We can find a Delta-hedged portfolio by finding the number of shares of stock to hold when the portfolio contains an option on one share; that number turns out to be $-\Delta$. Since the revenue on the option is Δ times the revenue on the stock, a portfolio with Δ shares of stock has the same volatility as a portfolio with an option on one share. Therefore, Δ is the same as the hedge-ratio. Suppose that the option is a call so that $\Delta > 0$. A portfolio that is long Δ shares of stock and short a call on one share, or vice versa, is perfectly hedged against small changes in the share price. For a put, $\Delta < 0$ and the portfolio would be long both $-\Delta$ shares of the stock and one put, or short both $-\Delta$ shares of the stock and one put.

Delta-hedging only hedges against small values of (δS). To guard against larger changes in the stock price, one can also use the second derivative of C with respect to S, which is called Gamma, so that

$$\Gamma = \frac{\partial^2}{\partial S^2} C(S, T, t, K, \sigma, r).$$

Gamma-hedging, which uses both Delta and Gamma to determine the appropriate portfolio weights, is beyond the scope of this text but is covered in most books on derivatives such as Hull (2003).

Regression hedging described in Section 6.5 is an example of Delta-hedging since the hedge is against small changes in yields.

8.18 Intrinsic Value and Time Value*

The intrinsic value of a call is $(S_t - K)_+$, the payoff one would obtain for immediate exercise of the option (which assuming that $t < T$ would be possible only for an American option). When $t < T$, the intrinsic value is less than the price, so early exercise is never optimal. The difference between the intrinsic value and the price is called the *time value of the option.* Time value has two components. The first is a volatility component. The second component is the *time value of money.* If you do exercise the option, it is best to wait until time T so that you delay payment of the exercise price.

The adjusted intrinsic value is $(S_t - e^{-r(T-t)}K)_+$ which is greater than the intrinsic value $(S_t - K)_+$ unless $t = T$. The difference between the price and the adjusted intrinsic value is the volatility component of the time value of the option. As $S_t \to \infty$, the price converges to the adjusted intrinsic value and the volatility component converges to 0. The reason this happens is that as $S_t \to \infty$ you become sure that the option will be in the money at the expiration date, so volatility has no effect.

Figure 8.18 shows the price, intrinsic value, and adjusted intrinsic value of a call option when $t = 0$, $K = 100$, $T = 0.25$, $r = 0.06$, and $\sigma = 0.1$. The intrinsic value of the put is $(K - S_t)_+$, which again is the payoff one would obtain for immediate exercise of the option, if that is possible (American option). The intrinsic value is sometimes greater than the price, in which case early exercise is optimal. The adjusted intrinsic value is $(e^{-r(T-t)}K - S_t)_+$. As $S_t \to 0$, the likelihood that the option will be in the money at the expiration date increases to 1 and the price converges to the adjusted intrinsic value.

Figure 8.19 shows the price, intrinsic value, and adjusted intrinsic value of a put option when $t = 0$, $K = 100$, $T = .25$, $r = 0.06$, and $\sigma = 0.1$.

8.19 Summary

An option gives the holder the right but not the obligation to do something, for example, to purchase a certain amount of a stock at a specified price within a fixed time frame. A call option gives one the right to purchase (call

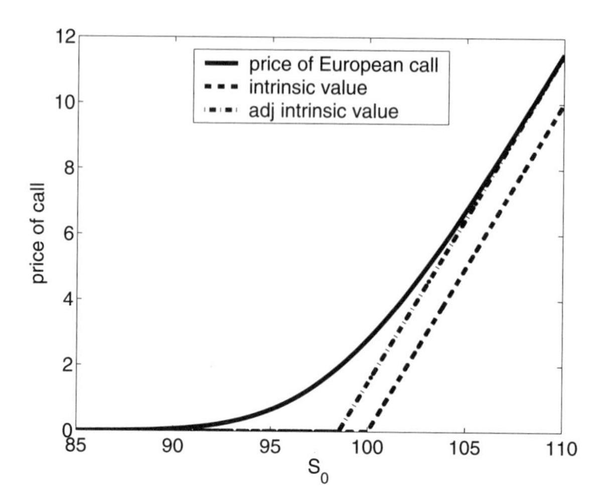

Fig. 8.18. *Price (for European or American option), intrinsic value, and adjusted intrinsic value of a call option. The intrinsic value is the payoff if one exercises early. Here* $K = 100$, $T = .25$, $t = 0$, $r = 0.06$, *and* $\sigma = 0.1$.

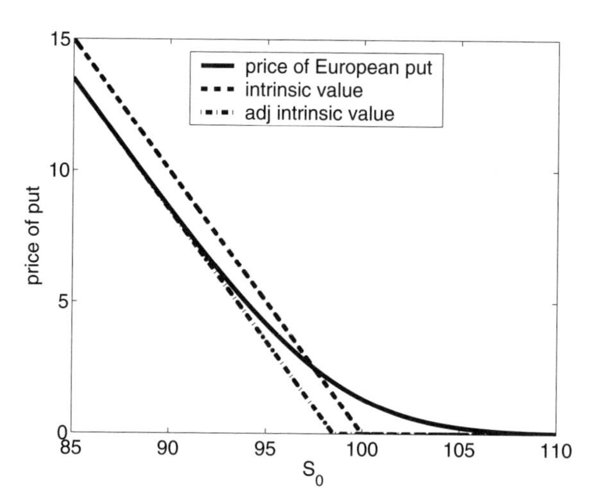

Fig. 8.19. *Price (for European option), intrinsic value, and adjusted intrinsic value of a put option. The intrinsic value is the payoff if one exercises early. The price of an American put would be either the price of the European put or the intrinsic value, whichever is larger. Here* $S_0 = 100$, $K = 100$, $T = .25$, $r = 0.06$, *and* $\sigma = 0.1$.

in) an asset. A put gives one the right to sell (put away) an asset. European options can be exercised only at their expiration date. American options can be exercised on or before their expiration date.

Arbitrage means making a guaranteed profit without investing capital. Arbitrage pricing determines the price of a financial instrument which guarantees that the market is free of arbitrage opportunities. Options can be priced by arbitrage using binomial trees.

The "measure" of a binomial tree model or other stochastic process model gives the set of path probabilities of that model. There exists a risk-neutral measure such that expected prices calculated with respect to this measure are equal to the prices that make the market arbitrage-free.

In a binomial tree model with price changes proportional to the current price, as the number of steps increases the limit process is a geometric Brownian motion and the price of the option in the limit is given by the Black-Scholes formula.

To price an option by the Black-Scholes formula, one needs an estimate of the stock price's volatility. This can be obtained from historical data. Conversely, the implied volatility of a stock is the volatility that makes the actual market price equal to the price given by the Black-Scholes formula. Implied volatilities depend on the exercise price and expiration date, which is an indication that the assumptions behind the Black-Scholes formula are not completely true. When implied volatility is plotted against the exercise price or expiration date, usually a pattern appears that is called the volatility smile.

Within the Black-Scholes model, the early exercise of calls is never optimal but the early exercise of puts is sometimes optimal. Therefore, European and American calls have equal prices, but American puts are generally worth more than European puts.

Put-call parity is the relationship

$$P = C + e^{-r(T-t)}K - S_t$$

between P, the price of a European put, and C, the price of a European call, all at time t. It is assumed that the put and the call have the same exercise price K and expiration date T, and S_0 is the price of the stock.

The "Greeks" are the sensitivities of a derivative to its parameters and are used to develop hedging strategies. Delta is the sensitivity to changes in the price of the underlying asset.

8.20 Bibliographic Notes

Baxter and Rennie (1998) discuss the mathematics of many types of derivatives, and I learned much of the material in this chapter from their book. Bernstein (1992) tells the fascinating story behind the Black-Scholes formula. Jarrow and Turnbull (2000) and Hull (2003) offer comprehensive and excellent treatments of derivatives covering binomial pricing, the Black-Scholes formula, and hedging, and describes a wide variety of derivatives. Hull (1995) is a nice overview of the options markets and the uses of options.

8.21 References

Baxter, M. and Rennie, A. (1998) *Financial Calculus: An Introduction to Derivative Pricing, Corrected Reprinting*, Cambridge University Press, Cambridge.

Bernstein, P. (1992) *Capital Ideas: The Improbable Origins of Modern Wall Street*, Free Press, New York.

Chance, D. (2003) Rethinking implied volatility, *Financial Engineering News*, January/February 2003, 7, 20.

French, K. R. and Roll, R. (1986) Stock return variances; The arrival of information and the reaction of traders, *Journal of Financial Economics*, **17**, 5–26.

Hull, J. C. (1995) *Introduction to Futures and Options Markets, 2nd Ed.*, Prentice-Hall, Englewood Cliffs, NJ.

Hull, J. C. (2003) *Options, Futures, and Other Derivatives, 5th Ed.*, Prentice Hall, Upper Saddle River, NJ.

Jarrow, R. and Turnbull, S. (2000) *Derivative Securities, 2nd Ed.*, South-Western College Publishing, Cincinnati, OH.

Merton, R. C. (1992) *Continuous-Time Finance, Revised Ed.*, Blackwell, Cambridge, MA and Oxford, UK.

8.22 Problems

1. Consider a three-step binomial tree model. The stock price starts at \$50. At each step the price increases 10% or decreases 10%; i.e., the return is $\pm 10\%$. The exercise price of a European call is \$55. Assume that the continuously compounded risk-free rate is 5% and that the time between steps is $\delta t = 1$, so that one dollar increases in value to $e^{.05}$ dollars after one time step.
   (a) What is the initial price of the option?
   (b) If the stock goes up on the first step, what is the option worth?
   (c) How much stock and risk-free asset would the replicating portfolio hold initially?
   (d) How would the portfolio be adjusted if the stock went up on the first step?
   (e) What are the risk-neutral probabilities? Are they the same at all nodes?

2. Now consider a 20-step binomial tree model. The stock price starts at \$50. At each step the price increases 1% or decreases 1%; i.e., the return is $\pm 1\%$. The exercise price is \$53. Assume that the continuously compounded risk-free rate is 5% per unit of time and that the time between steps is $\delta t = 1/20$, so that one dollar increases in value to $e^{.05/20}$ dollars after one time step. What is the initial price of the option? (You should consider using MATLAB, Excel, or other software to do this problem.)
3. Suppose that a stock is currently selling at 92 and that the risk-free continuously compounded annual rate is 0.018.
   (a) What is the price of a call option on this stock with an expiration date $T = 0.5$ (time in years) and with an exercise price of 98? Assume that the volatility of annual log-returns is $\sigma = 0.2$.
   (b) What are the intrinsic value and adjusted intrinsic values of the call?
   (c) What is the price of a put option on the same stock with the same exercise price and expiration date as the call?
   (d) If, one quarter from now, the stock price is 95 and the interest rate and volatility are unchanged, what is the price of the call?
4. Suppose a stock is selling at 40 and the risk-free rate is 0.02. A call option with an expiration date of 0.5 (in years) and exercise price of 43 is selling at 4. What is the implied volatility of this stock? (I suggest you do this problem in MATLAB or Excel. You will need to use interpolation.)
5. What would happen if the exercise price of a call option were raised by \$1 with everything else unchanged? Would the option price go up or down? Would the change be \$1, less than \$1, or more than \$1? Explain your answer.
6. The price of a stock at time t is S_t. Suppose

$$S_t = 100 + X_1 + X_2 + \cdots + X_t, \ t = 0, 1, 2, \ldots,$$

   where $\mathrm{P}(X = 1) = .6$ and $\mathrm{P}(X_t = -1) = 0.4$.
   (a) Let $E(S_4|S_0, S_1, S_2)$ be the conditional expectation of S_4 given S_0, S_1, S_2. Find $E(S_4|S_0, S_1, S_2)$.
   (b) What is $\mathrm{Var}\{E(S_4|S_0, S_1, S_2)\}$?
   (c) What is $E\{S_4 - E(S_4|S_0, S_1, S_2)\}^2$?
   (d) What is the value at $t = 0$ of a call option on S_2 with a exercise price of 101? Your answer should be a function of r.
7. Consider a two-step binomial tree for a stock price. The stock price starts at 100 and goes up or down by 20% on each step. The risk-free continuously compounded interest rate is 3% per time step. The stock price at time t is denoted by S_t.
   (a) What is the value at time 0 of a European put option maturing at time 2 with an exercise price of 110?
   (b) Would early exercise ever be optimal for an American put option maturing at time 2 with an exercise price of 110? If so, why?
   (c) What is the value at time 0 of an American put option maturing at time 2 with an exercise price of 110?

(d) What is the value at time 0 of a derivative that pays at time 2 the amount $\{(S_1 - 110)_+ + (S_2 - 110)_+\}$?

8. Show that the quantity in curly brackets in (8.36) is 0.
9. Show that Gamma of a call option priced by the Black-Scholes formula (8.35) is

$$\Gamma = \frac{\phi(d_1)}{S\sigma\sqrt{T-t}}.$$

9

Fixed Income Securities

9.1 Introduction

Corporations finance their operations by selling stock and bonds. Owning a share of stock means partial ownership of the company. You share in both the profits and losses of the company, so nothing is guaranteed.

Owning a bond is different. When you buy a bond you are loaning money to the corporation. The corporation is obligated to pay back the principal and to pay interest as stipulated by the bond. You receive a fixed stream of income, unless the corporation defaults on the bond. For this reason, bonds are called "fixed-income" securities.

It might appear that bonds are risk-free, almost stodgy. This is not the case. Many bonds are long-term, e.g., 20 or 30 years. Even if the corporation stays solvent or if you buy a U.S. Treasury bond where default is for all intents and purposes impossible, your income from the bond is guaranteed only if you keep the bond to maturity. If you sell the bond before maturity, your return will depend on changes in the price of the bond. Bond prices move in opposite direction to interest rates, so a decrease in interest rates will cause a bond "rally" where bond prices increase. Long-term bonds are more sensitive to interest rate changes than short-term bonds. The interest rate on your bond is fixed, but in the market interest rates fluctuate. Therefore, the market value of your bond fluctuates too. For example, if you buy a bond paying 5% and the rate of interest increases to 6% then your bond is inferior to new bonds offering 6%. Consequently, the price of your bond will decrease. If you sell the bond you would lose money. So much for a "fixed income" stream!

If you ever bought a CD,[1] which really is a bond that you buy from a bank or credit union, you will have noticed that the interest rate depends on the

[1] A CD or certificates of deposit is a type of deposit available from banks in the United States. Generally, the bank pays principal and interest only at the time of maturity. Small denomination certifcates of deposit are popular with retail investors, partly because they are guaranteed up to $100,000 by the Federal Deposit

maturity of the CD. This is a general phenomenon. For example, on March 28, 2001, the interest rate of Treasury bills[2] was 4.23% for three-month bills. The yields on Treasury notes and bonds were 4.41%, 5.01%, and 5.46% for 2-, 10-, and 30-year maturities, respectively. The **term structure** of interest rates describes how rates of interest change with the maturity of bonds.

In this chapter we will study how bond prices fluctuate due to interest rate changes and how the term structure of interest rates can be determined.

9.2 Zero-Coupon Bonds

Zero-coupon bonds, also called pure discount bonds and sometimes known as "zeros," pay no principal or interest until maturity. A "zero" has a **par value** or **face value** which is the payment made to the bond holder at maturity. The zero sells for less than the par value, which is the reason it is a "discount bond."

For example, consider a 20-year zero with a par value of $1000 and 6% interest compounded annually. The market price is the present value of $1000 with 6% annual discounting. That is, the market price is

$$\frac{\$1000}{(1.06)^{20}} = \$311.80.$$

If the interest is 6% but compounded every six months, then the price is

$$\frac{\$1000}{(1.03)^{40}} = \$306.56,$$

and if the interest is 6% compounded continuously then the price is

$$\frac{\$1000}{\exp\{(.06)(20)\}} = \$301.19.$$

9.2.1 Price and returns fluctuate with the interest rate

For concreteness, assume semi-annual compounding. Suppose you bought the zero for $306.56 and then six months later the interest rate increased to 7%. The market price would now be

$$\frac{\$1000}{(1.035)^{39}} = \$261.41,$$

Insurance Corporation, and CD rates are often posted in prominent locations within banks.

[2] Treasury bills have maturities of one year or less, Treasury notes have maturities from 1 to 10 years, and Treasury bonds have maturities from 10 to 30 years.

so the value of your investment would drop by ($306.56 − $261.41) = $45.15. You will still get your $1000 if you keep the bond for 20 years, but if you sold it now you would lose $45.15. This is a return of

$$\frac{-45.15}{306.56} = -14.73\%$$

for a half-year or −29.46% per year. And the interest rate only changed from 6% to 7%! Notice that the interest rate went up and the bond price went down. This is a general phenomenon. Bond prices always move in the opposite direction of interest rates.

If the interest rate dropped to 5% after six months, then your bond would be worth

$$\frac{\$1000}{(1.025)^{39}} = \$381.74.$$

This would be an annual rate of return of

$$2\left(\frac{381.74 - 306.56}{306.56}\right) = 49.05\%.$$

If the interest rate remained unchanged at 6%, then the price of the bond would be

$$\frac{\$1000}{(1.03)^{39}} = \$315.75.$$

The annual rate of return would be

$$2\left(\frac{315.75 - 306.56}{306.56}\right) = 6\%.$$

Thus, if the interest rate does not change, you can earn a 6% annual rate of return, the same return rate as the interest rate, by selling the bond before maturity. If the interest rate does change, however, the 6% annual rate of return is guaranteed only if you keep the bond until maturity.

General formula

The price of a zero-coupon bond is given by

$$\text{PRICE} = \text{PAR}(1 + r)^{-T}$$

if T is the time to maturity in years and the annual rate of interest is r with annual compounding. If we assume semi-annual compounding, then the price is

$$\text{PRICE} = \text{PAR}(1 + r/2)^{-2T}. \tag{9.1}$$

9.3 Coupon Bonds

Coupon bonds make regular interest payments.[3] Coupon bonds generally sell at par value when issued. At maturity, one receives the principal and the final interest payment.

As an example, consider a 20-year coupon bond with a par value of $1000 and 6% annual interest with semi-annual coupon payments, so that the 6% is compounded semi-annually. Each coupon payment will be $30. Thus, the bond holder receives 40 payments of $30, one every six months plus a principal payment of $1000 after 20 years. One can check that the present value of all payments, with discounting at the 6% annual rate (3% semi-annual), equals $1000:

$$\sum_{t=1}^{40} \frac{30}{(1.03)^t} + \frac{1000}{(1.03)^{40}} = 1000.$$

After six months if the interest rate is unchanged, then the bond (including the first coupon payment which is now due) is worth

$$\sum_{t=0}^{39} \frac{30}{(1.03)^t} + \frac{1000}{(1.03)^{39}} = (1.03)\left(\sum_{t=1}^{40} \frac{30}{(1.03)^t} + \frac{1000}{(1.03)^{40}}\right) = 1030,$$

which is a 6% annual return as expected. If the interest rate increases to 7%, then after six months the bond (plus the interest due) is only worth

$$\sum_{t=0}^{39} \frac{30}{(1.035)^t} + \frac{1000}{(1.035)^{39}} = (1.035)\left(\sum_{t=1}^{40} \frac{30}{(1.035)^t} + \frac{1000}{(1.035)^{40}}\right) = 924.49.$$

This is an annual return of

$$2\left(\frac{924.49 - 1000}{1000}\right) = -15.1\%.$$

If the interest rate drops to 5% after six months then the investment is worth

$$\sum_{t=0}^{39} \frac{30}{(1.025)^t} + \frac{1000}{(1.025)^{39}} = (1.025)\left(\sum_{t=1}^{40} \frac{30}{(1.025)^t} + \frac{1000}{(1.025)^{40}}\right) = 1,153.70, \tag{9.2}$$

and the annual return is

$$2\left(\frac{1153.6 - 1000}{1000}\right) = 30.72\%.$$

[3] At one time actual coupons were attached to the bond, one coupon for each interest payment. When a payment was due, its coupon could be clipped off and sent to the issuing company for payment.

9.3.1 A general formula

Let's derive some useful formulas. If a bond with a par value of PAR matures in T years and makes semi-annual coupon payments of C and the discount rate (rate of interest) is r per half-year, then the value of the bond when it is issued is

$$\sum_{t=1}^{2T} \frac{C}{(1+r)^t} + \frac{\text{PAR}}{(1+r)^{2T}} = \frac{C}{r}\{1-(1+r)^{-2T}\} + \frac{\text{PAR}}{(1+r)^{2T}}$$
$$= \frac{C}{r} + \left\{\text{PAR} - \frac{C}{r}\right\}(1+r)^{-2T}. \tag{9.3}$$

Derivation of (9.3)

The summation formula for a finite geometric series is

$$\sum_{i=0}^{T} r^i = \frac{1-r^{T+1}}{1-r}, \tag{9.4}$$

provided that $r \neq 1$. Therefore,

$$\sum_{t=1}^{2T} \frac{C}{(1+r)^t} = \frac{C}{1+r}\sum_{t=0}^{2T-1}\left(\frac{1}{1+r}\right)^t = \frac{C\{1-(1+r)^{-2T}\}}{(1+r)(1-(1+r)^{-1})}$$
$$= \frac{C}{r}\{1-(1+r)^{-2T}\}. \tag{9.5}$$

9.4 Yield to Maturity

Suppose a bond with $T = 30$ and $C = 40$ is selling for \$1200, \$200 above par value. If the bond were selling at par value, then the interest rate would be 0.04/half-year (= 0.08/year). The 4%/half-year rate is called the **coupon rate**.

But the bond is *not* selling at par value. If you purchase the bond at \$1200 you will make *less* than 8% per year interest. There are two reasons that the rate of interest is less than 8%. First, the coupon payments are \$40 or 40/1200 = 3.333%/half-year (or 6.67%/year) for the \$1200 investment; 6.67%/year is called the **current yield**. Second, at maturity you only get back \$1000, not the entire \$1200 investment. The current yield of 6.67%/year, though less than the coupon rate of 8%/year, overestimates the return since it does not account for this loss of capital.

The **yield to maturity**, often shortened to simply **yield,** is a measure of the average rate of return, including the loss (or gain) of capital because the bond was purchased above (or below) par. For this bond, the yield to maturity is the value of r that solves

$$1200 = \frac{40}{r} + \left\{1000 - \frac{40}{r}\right\}(1+r)^{-60}. \qquad (9.6)$$

The right-hand side of (9.6) is (9.3) with $C = 40$, $T = 30$, and $\text{PAR} = 1000$. It is easy to solve equation (9.6) numerically. The MATLAB program "yield.m" in Section 9.4.2 does the following:

- Computes the bond price for each r value on a grid;
- Graphs bond price versus r (this is not necessary but it is fun to see the graph); and
- Interpolates to find the value of r such that bond value equals 1200.

One finds that the yield to maturity is 0.0324, that is, 3.24%/half-year. Figure 9.1 shows the graph of bond price versus r and shows that $r = .0324$ maps to a bond price of $1200.

The yield to maturity of 0.0324 is less than the current yield of 0.0333 which is less than the coupon rate of $40/1000 = 0.04$. (All three rates are rates per half-year.) Whenever, as in this example, the bond is selling above par value, then the coupon rate is greater than the current yield because the bond sells above par value, and the current yield is greater than the yield to maturity because the yield to maturity accounts for the loss of capital when at the maturity date you only get back the par value, not the entire investment. In summary,

$$\text{price} > \text{par} \Rightarrow \text{coupon rate} > \text{current yield} > \text{yield to maturity}.$$

Everything is reversed if the bond is selling below par value. For example, if the price of the bond were only $900, then the yield to maturity would be 0.0448 (as before, this value can be determined by "yield.m" using interpolation), the current yield would be $40/900 = 0.0444$, and the coupon rate would still be $40/1000 = 0.04$. In general,

$$\text{price} < \text{par} \Rightarrow \text{coupon rate} < \text{current yield} < \text{yield to maturity}.$$

9.4.1 General method for yield to maturity

The yield to maturity (on a semi-annual basis) of a coupon bond is the value of r that solves

$$\text{PRICE} = \frac{C}{r} + \left\{\text{PAR} - \frac{C}{r}\right\}(1+r)^{-2T}. \qquad (9.7)$$

Here PRICE is the market price of the bond, PAR is the par value, C is the semi-annual coupon payment, and T is the time to maturity in years.

For a zero-coupon bond, $C = 0$ and (9.7) becomes

$$\text{PRICE} = \text{PAR}(1+r)^{-2T}. \qquad (9.8)$$

Comparing (9.8) and (9.1) we see that for a zero-coupon bond, the yield to maturity is the interest rate.

9.4.2 MATLAB functions

The MATLAB function "bondvalue.m" computes (9.3). The call to this function is `bondvalue(c,T,r,par)`. The program is listed below. As you can see, all but the last line is a comment explaining the program.

```
function bv = bondvalue(c,T,r,par) ;
%
%   Computes bv = bond values (current prices) corresponding
%       to all values of yield to maturity in the
%       input vector r
%
%       INPUT
%        c = coupon payment (semi-annual)
%        T = time to maturity (in years)
%        r = vector of yields to maturity (semi-annual rates)
%        par = par value
%
bv = c./r + (par - c./r) .* (1+r).^(-2*T) ;
```

Here is the program "yield.m" that uses "bondvalue.m" and interpolation to find a yield to maturity. The first part of the program specifies the values of price, C, T, and par value. These could be changed to find the yield to maturity of other bonds.

```
%   Computes the yield to maturity of a bond paying semi-annual
%   coupon payments
%
%   price, coupon payment, and time to maturity (in years)
%   are set below
%
%   Uses the function "bondvalue"
%
price = 1200    %   current price of the bond
C = 40          %   coupon payment
T= 30           %   time to maturity
par = 1000      %   par value of the bond
```

Next `r`, which is a grid of 300 values of the interest rate between 2% and 5%, is created and the value of the bond at each of these rates is computed by a call to "bondvalue.m". Finally, the yield to maturity is found by interpolation.

```
r = linspace(.02,.05,300)' ;
value = bondvalue(C,T,r,par) ;  %
yield2M = interp1(value,r,price,'cubic') ;
```

The remainder of the program produces Figure 9.1.

```
p=plot(r,value) ;
xlabel('yield to maturity') ;
ylabel('price of bond') ;
title(['par='  num2str(par)  ', coupon payment='  num2str(C) ...
   ', T=' num2str(T)]) ;
line( [.02 .05], [price price]) ;
line( [yield2M yield2M] , [200 1800]) ;
```

Since the grid of values of `r` ranges from 0.02 to 0.05, this grid would need to be expanded if interest rates were over 5% or under 2%.

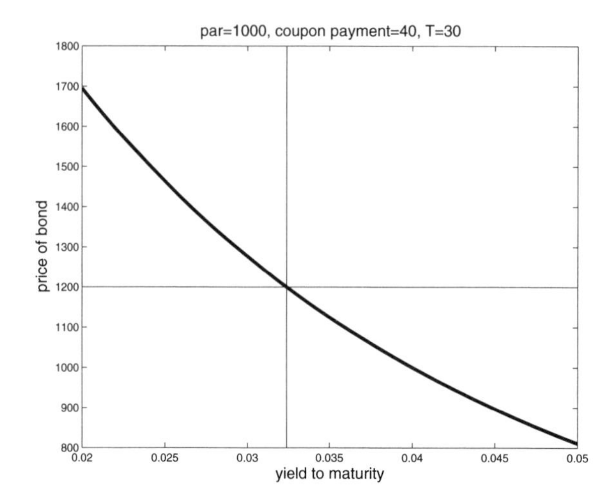

Fig. 9.1. *Bond price versus the interest rate r and determining by interpolation the yield to maturity when the price equals \$1200.*

9.4.3 Spot rates

The yield to maturity of a zero-coupon bond of maturity n years is called the n-year **spot rate** and is denoted by y_n. In Section 8.4 when we discussed net present values, also known as discounted values, we assumed that there was a constant discount rate r. This was an oversimplification. One should use the n-year spot rate to discount a payment n years from now, so a payment of \$1 to be made n years from now has a net present value of $\$1/(1+y_n)^n$ if y_n is the spot rate per annum or $\$1/(1+y_n)^{2n}$ if y_n is a semi-annual rate.

A coupon bond is a bundle of zero-coupon bonds, one for each coupon payment and a final one for the principal payment. The component zeros have different maturity dates and therefore different spot rates. The yield to maturity of the coupon bond is, thus, a complex "average" of the spot rates of the zeros in this bundle.

Example 9.1. Consider the simple example of a one-year coupon bond with semi-annual coupon payments of \$40 and a par value of \$1000. Suppose that the one-half year spot rate is 5%/year and the one-year spot rate is 6%/year. Think of the coupon bond as being composed of two zero-coupon bonds, one with $T = 1/2$ and a par value of 40 and the second with $T = 1$ and a par value of 1040. The price of the bond is the sum of the prices of these two zeros. Applying (9.8) twice to obtain the prices of these zeros and summing, we obtain the price of the zero-coupon bond

$$\frac{40}{1.025} + \frac{1040}{(1.03)^2} = 1019.32.$$

The yield to maturity on the coupon bond is the value of y that solves

$$\frac{40}{1+y} + \frac{1040}{(1+y)^2} = 1019.32.$$

The solution is $y = 0.0299$. This is the rate per half-year. Thus, the annual yield to maturity is twice 0.0299 or 5.98%/year.

General formula

Here is a formula that generalizes the example above. Suppose that a coupon bond pays semi-annual coupon payments of C, has a par value of PAR, and has T years until maturity. Let $r_1, r_2, \ldots, r_{2T}$ be the half-year spot rates for zero-coupon bonds of maturities $1/2, 1, 3/2, \ldots, T$ years. Then the yield to maturity (on a half-year basis) of the coupon bond is the value of y that solves:

$$\frac{C}{1+r_1} + \frac{C}{(1+r_2)^2} + \cdots + \frac{C}{(1+r_{2T-1})^{2T-1}} + \frac{\mathrm{PAR}+C}{(1+r_n)^{2T}}$$
$$= \frac{C}{1+y} + \frac{C}{(1+y)^2} + \cdots + \frac{C}{(1+y)^{2T-1}} + \frac{\mathrm{PAR}+C}{(1+y)^{2T}}. \tag{9.9}$$

The left-hand side of equation (9.9) is the price of the coupon bond, and the yield to maturity is the value of y that makes the right-hand side of (9.9) equal to the price.

9.5 Term Structure

9.5.1 Introduction: Interest rates depend upon maturity

On January 26, 2001, the 1-year T-bill rate was 4.83% and the 30-year Treasury bond rate was 6.11%. This is typical. Short- and long-term rates usually do differ. Often short-term rates are lower than long-term rates. This makes sense since long-term bonds are riskier. Long-term bond prices fluctuate more with interest rate changes and these bonds are often sold before maturity. However, during periods of very high short-term rates, the short-term rates may be higher than the long-term rates. The reason is that the market believes that rates will return to historic levels and no one will commit to the high interest rate for, say, 20 or 30 years. Figure 9.2 shows weekly values of the 90-day, 10-year, and 30-year Treasury rates from 1970 to 1993, inclusive. Notice that the 90-day rate is more volatile than the longer term rates and is usually less than them. However, in the early 1980s when interest rates were very high, the short-term rates were higher than the long-term rates. These data were taken from the Federal Reserve Bank of Chicago's website.

The **term structure** of interest rates is a description of how, *at a given time*, yield to maturity depends on maturity.

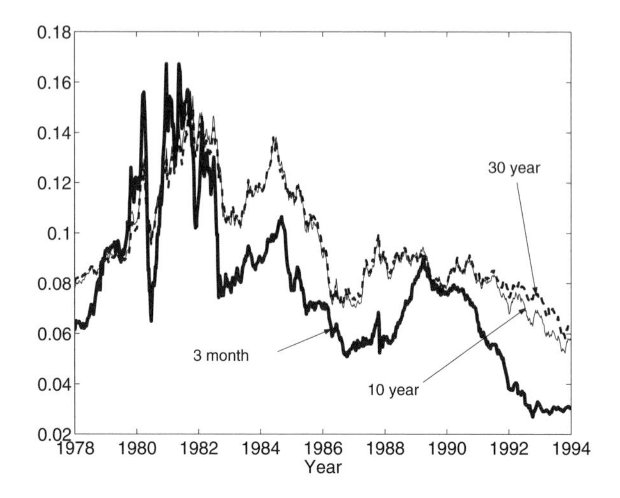

Fig. 9.2. *Treasury rates of three maturities. Weekly time series. The data were taken from the website of the Federal Reserve Bank of Chicago.*

9.5.2 Describing the term structure

Term structure for all maturities up to n years can be described by any one of the following:

- Prices of zero-coupon bonds of maturities 1-year, 2-years, ..., n-years denoted here by $P(1), P(2), \ldots, P(n)$;
- Spot rates (yields of maturity of zero-coupon bonds) of maturities 1-year, 2-years, ..., n-years denoted by $y_1, \ldots, y_n$;
- Forward rates $r_1, \ldots, r_n$ where r_i is the forward rate paid in the ith future year ($i = 1$ for next year, and so on).

As discussed each of the sets $\{P(1), \ldots, P(n)\}$, $\{y_1, \ldots, y_n\}$, and $\{r_1, \ldots, r_n\}$ can be computed from either of the other sets. For example, equation (9.11) below gives $\{P(1), \ldots, P(n)\}$ in terms of $\{r_1, \ldots, r_n\}$, and equations (9.12) and (9.13) below give $\{y_1, \ldots, y_n\}$ in terms of $\{P(1), \ldots, P(n)\}$ or $\{r_1, \ldots, r_n\}$, respectively.

Term structure can be described by breaking down the time interval between the present time and the maturity time of a bond into short time segments with a constant interest rate within each segment, but with interest rates varying between segments. For example, a three-year loan can be considered as three consecutive one-year loans.

Example 9.2. As an illustration, suppose that loans have the forward interest rates listed in Table 9.1. Using the forward rates in the table, we see that a par \$1000 1-year zero would sell for

$$\frac{1000}{1 + r_1} = \frac{1000}{1.06} = \$943.40 = P(1).$$

A par $1000 2-year zero would sell for

$$\frac{1000}{(1+r_1)(1+r_2)} = \frac{1000}{(1.06)(1.07)} = \$881.68 = P(2),$$

since the rate r_1 is paid the first year and r_2 the following year. Similarly, a par $1000 3-year zero would sell for

$$\frac{1000}{(1+r_1)(1+r_2)(1+r_3)} = \frac{1000}{(1.06)(1.07)(1.08)} = 816.37 = P(3).$$

Table 9.1. Forward interest rate example.

Year (i)	**Interest rate** (r_i)(%)
1	6
2	7
3	8

The general formula for the present value of $1 paid n periods from now is

$$\frac{1}{(1+r_1)(1+r_2)\cdots(1+r_n)}. \tag{9.10}$$

Here r_i is the **forward interest rate** during the ith period. The price of an n-year par $1000 zero-coupon bond $P(n)$ is $1000 times the discount factor in (9.10); that is,

$$P(n) = \frac{1000}{(1+r_1)\cdots(1+r_n)}. \tag{9.11}$$

Example 9.3. (Back to the Example 9.2) Now we show how to find the yields to maturity. For a 1-year zero, the yield to maturity y_1 solves

$$\frac{1000}{(1+y_1)} = 993.40,$$

which implies that $y_1 = 0.06$. For a 2-year zero, the yield to maturity is y_2 that solves

$$\frac{1000}{(1+y_2)^2} = 881.68.$$

Thus,

$$y_2 = \sqrt{\frac{1000}{881.68}} - 1 = 0.0650.$$

It is easy to show that y_2 is also given by

$$y_2 = \sqrt{(1+r_1)(1+r_2)} - 1 = \sqrt{(1.06)(1.07)} - 1 = 0.0650.$$

For a 3-year zero, the yield to maturity y_3 solves

$$\frac{1000}{(1+y_3)^3} = \frac{1000}{881.68}.$$

Also,

$$y_3 = \{(1+r_1)(1+r_2)(1+r_3)\}^{1/3} - 1$$
$$= \{(1.06)(1.07)(1.08)\}^{1/3} - 1 = 0.0700,$$

or, more precisely 0.069969. Thus, $(1+y_3)$ is the geometric average of 1.06, 1.07, and 1.08 and approximately equal to their arithmetic average of 1.07.

Recall that $P(n)$ is the price of a par \$1000 n-year zero-coupon bond. The general formulas for the yield to maturity y_n of an n-year zero are

$$y_n = \left\{\frac{1000}{P(n)}\right\}^{1/n} - 1, \tag{9.12}$$

and

$$y_n = \{(1+r_1)\cdots(1+r_n)\}^{1/n} - 1. \tag{9.13}$$

Equations (9.12) and (9.13) give the yields to maturity in terms of the bond prices and forward rates, respectively. Also,

$$P(n) = \frac{1000}{(1+y_n)^n}, \tag{9.14}$$

which gives $P(n)$ in terms of the yield to maturity.

As mentioned before, interest rates for future years are called **forward rates**. A forward contract is an agreement to buy or sell an asset at some fixed future date at a fixed price. Since $r_2, r_3, \ldots$ are rates that can be locked in now for future borrowing, they are forward rates.

The general formula for determining forward rates from yields to maturity is

$$r_1 = y_1, \tag{9.15}$$

and

$$r_n = \frac{(1+y_n)^n}{(1+y_{n-1})^{n-1}} - 1. \tag{9.16}$$

Now suppose that we only observed bond prices. Then we can calculate yields to maturity and forward rates using (9.12) and then (9.16).

Example 9.4. Suppose that one-, two-, and three-year par 1000 zeros are priced as given in Table 9.2. Then using (9.12), the yields to maturity are:

$$y_1 = \frac{1000}{920} - 1 = 0.087,$$

Table 9.2. Bond price example.

maturity	price
1 year	\$920
2 year	\$830
3 year	\$760

$$y_2 = \left\{ \frac{1000}{830} \right\}^{1/2} - 1 = 0.0976,$$

$$y_3 = \left\{ \frac{1000}{760} \right\}^{1/3} - 1 = 0.096.$$

Then, using (9.15) and (9.16)

$$r_1 = y_1 = 0.087,$$

$$r_2 = \frac{(1+y_2)^2}{(1+y_1)} - 1 = \frac{(1.0976)^2}{1.0876} - 1 = 0.108,$$

and

$$r_3 = \frac{(1+y_3)^3}{(1+y_2)^2} - 1 = \frac{(1.096)^3}{(1.0976)^2} - 1 = 0.092.$$

The formula for finding r_n from the prices of zero-coupon bonds is

$$r_n = \frac{P(n-1)}{P(n)} - 1, \tag{9.17}$$

which can be derived from

$$P(n) = \frac{1000}{(1+r_1)(1+r_2)\cdots(1+r_n)},$$

and

$$P(n-1) = \frac{1000}{(1+r_1)(1+r_2)\cdots(1+r_{n-1})}.$$

To calculate r_1 using (9.17), we need $P(0)$, the price of a 0-year bond, but $P(0)$ is simply the par value.[4]

Example 9.5. Thus, using (9.17),

$$r_1 = \frac{1000}{920} - 1 = 0.087,$$

$$r_2 = \frac{920}{830} - 1 = 0.108,$$

and

$$r_3 = \frac{830}{760} - 1 = 0.092.$$

[4] Trivially, a bond that must be paid back immediately is worth exactly its par value.

9.6 Continuous Compounding

Now we assume continuous compounding with forward rates $r_1, \ldots, r_n$. We show that the use of continuously compounded rates simplifies the relationships between the forward rates, the yields to maturity, and the prices of zero-coupon bonds.

If $P(n)$ is the price of a \$1000 par value n-year zero-coupon bond, then

$$P(n) = \frac{1000}{\exp(r_1 + r_2 + \cdots + r_n)}. \tag{9.18}$$

Therefore,

$$\frac{P(n-1)}{P(n)} = \frac{\exp(r_1 + \cdots + r_n)}{\exp(r_1 + \cdots + r_{n-1})} = \exp(r_n), \tag{9.19}$$

and

$$\log\left\{\frac{P(n-1)}{P(n)}\right\} = r_n. \tag{9.20}$$

The yield to maturity of an n-year zero-coupon bond solves the equation

$$P(n) = \frac{1000}{\exp(n y_n)},$$

and is easily seen to be

$$y_n = (r_1 + \cdots + r_n)/n. \tag{9.21}$$

Therefore, $\{r_1, \ldots, r_n\}$ is easily found from $\{y_1, \ldots, y_n\}$ by the relationship

$$r_1 = y_n,$$

and

$$r_n = n y_n - (n-1) y_{n-1} \text{ for } n > 1.$$

Example 9.6. Using the prices in Table 9.2 we have $P(1) = 920$, $P(2) = 830$, and $P(3) = 760$. Therefore, using (9.20),

$$r_1 = \log\left\{\frac{1000}{920}\right\} = 0.083,$$

$$r_2 = \log\left\{\frac{920}{830}\right\} = 0.103,$$

and

$$r_3 = \log\left\{\frac{830}{760}\right\} = 0.088.$$

Also, $y_1 = r_1 = 0.083$, $y_2 = (r_1 + r_2)/2 = 0.093$, and $y_3 = (r_1 + r_2 + r_3)/3 = 0.091$.

9.7 Continuous Forward Rates

So far, we have assumed that forward interest rates vary from year to year but are constant within each year. This assumption is, of course, unrealistic and was made only to simplify the introduction of forward rates. Forward rates should be modeled as a function varying continuously in time.

To specify the term structure in a realistic way, we assume that there is a function $r(t)$ called the **forward rate function** such that the current price of a zero-coupon bond of maturity T and with par value equal to 1 is given by

$$D(T) = \exp\left\{ - \int_0^T r(t)dt \right\}. \tag{9.22}$$

$D(T)$ is called the discount function and the price of any zero-coupon bond is given by discounting its par value by multiplication with the discount function; i.e.,

$$P(T) = \text{PAR} \times D(T), \tag{9.23}$$

where $P(T)$ is the price of a zero-coupon bond of maturity T with par value equal to PAR. Formula (9.22) is a generalization of formula (9.18). To appreciate this, suppose that $r(t)$ is the piecewise constant function

$$r(t) = r_k \text{ for } k - 1 < t \leq k.$$

With this piecewise constant r, for any integer T, we have

$$\int_0^T r(t)dr = r_1 + r_2 + \cdots + r_T,$$

so that

$$\exp\left\{ - \int_0^T r(t)dt \right\} = \exp\{-(r_1 + \cdots + r_T)\}$$

and therefore (9.18) agrees with (9.22) in this special situation. However, (9.22) is a more general formula since it applies to noninteger T and to arbitrary $r(t)$, not only piecewise constant functions.

The yield to maturity of a zero-coupon bond with maturity date T is defined to be

$$y_T = \frac{1}{T} \int_0^T r(t)\, dt. \tag{9.24}$$

Thinking of the right-hand side of (9.24) as the average of $r(t)$ over the interval $0 \leq t \leq T$, we see that (9.24) is the analogue of (9.21). From (9.22) and (9.24) it follows that the discount function can be obtained from the yield to maturity by the formula

$$D(T) = \exp\{-T y_T\}, \tag{9.25}$$

so that the price of a zero-coupon bond maturing at time T is the same as it would be if there were a constant forward interest rate equal to y_T.

Example 9.7. Suppose the forward rate is the linear function $r(t) = 0.03 + 0.0005\,t$. Find $r(15)$, y_{15}, and $D(15)$.

Answer: $r(15) = 0.03 + (0.0005)(15) = 0.0375$,

$$y_{15} = (15)^{-1}\int_0^{15}(0.03{+}0.0005\,t)dt = (15)^{-1}(0.03\,t{+}0.0005\,t^2/2)\Big|_0^{15} = 0.03375,$$

and $D(15) = \exp(-15y_{15}) = \exp\{-(15)(0.03375)\} = \exp(0.5055) = 0.6028$.

The discount function $D(T)$ and forward rate function $r(t)$ in formula (9.22) depend on the current time, which is taken to be zero in that formula. However, we could be interested in how the discount function and forward rate function change over time. In that case we define the discount function $D(s,T)$ to be the price at time s of a zero-coupon bond, with a par value of \$1, maturing at time T. Then

$$D(s,T) = \exp\left\{-\int_s^T r(s,t)dt\right\}. \tag{9.26}$$

Since $r(t)$ and $D(t)$ in (9.22) are $r(0,t)$ and $D(0,t)$ in our new notation, (9.22) is the special case of (9.26) with $s = 0$. However, for the remainder of this chapter we assume that $s = 0$ and return to the simpler notation of $r(t)$ and $D(t)$.

9.8 Sensitivity of Price to Yield

As we have seen, bonds are risky because bond prices are sensitive to interest rates. This problem is called interest-rate risk. This section describes a traditional method of quantifying interest-rate risk.

Using equation (9.25), we can approximate how the price of a zero-coupon bond changes if there is a small change in yield. Suppose that y_T changes to $y_T + \delta$ where the change in yield δ is small. Then the change in $D(T)$ is approximately δ times

$$\frac{d}{dy_T}\exp\{-Ty_T\} \approx -T\exp\{-Ty_T\} = -TD(T). \tag{9.27}$$

Therefore, by equation (9.23), for a zero-coupon bond of maturity T,

$$\frac{\text{change bond price}}{\text{bond price}} \approx -T \times \text{change in yield}. \tag{9.28}$$

In this equation "$\approx$" means that the ratio of the right- to left-hand sides converges to 1 as $\delta \to 0$.

Equation (9.28) is worth examining. The minus sign on the right-hand side shows us something we already knew, that bond prices move in the opposite direction to interest rates.[5] Also, the relative change in the bond price, which

[5] Remember that yield is an average interest rate.

is the left-hand side of the equation, is proportional to T which quantifies the principle that longer-term bonds have higher interest-rate risks than short-term bonds.

9.8.1 Duration of a coupon bond

Remember that a coupon bond can be considered as a bundle of zero-coupon bonds of various maturities. The **duration** of a coupon bond, which we denote by DUR, is the weighted average of these maturities with weights in proportion to the net present value of the cash flows (coupon payments and par value at maturity). Now assume that all yields change by a constant amount δ, that is, y_T changes to $y_T + \delta$ for all T. Then equation (9.28) applies to each of these cash flows and averaging them with these weights gives us that for a coupon bond

$$\frac{\text{change bond price}}{\text{bond price}} \approx -\text{DUR} \times \text{change in yield}. \tag{9.29}$$

The details of the derivation of (9.29) are left as an exercise (Problem 10). **Duration analysis** uses (9.29) to approximate the effect of a change in yield on bond prices.

We can rewrite (9.29) as

$$\text{DUR} \approx \frac{-1}{\text{price}} \times \frac{\text{change in price}}{\text{change in yield}} \tag{9.30}$$

and use (9.30) as a *definition* of duration. Notice that "bond price" has been replaced by "price." The reason for this is that (9.30) can define the durations of not only bonds but also of derivative securities whose prices depend on yield, for example, call options on bonds. When this definition is extended to derivatives, duration has nothing to do with maturities of the underlying securities. Instead, duration is solely a measure of sensitivity of price to yield. Tuckman (2002) gives an example of a 10-year coupon bond with a duration of 7.79 years and a call option on this bond with a duration of 120.82. These durations show that the call is much riskier than the bond since it is 15.5 (= 129.82/7.79) times more sensitive to changes in yield.

Unfortunately, the underlying assumption behind (9.29) that all yields change by the same amount is not realistic, so duration analysis is falling into disfavor and Value-at-Risk is replacing duration analysis as a method for evaluating interest-rate risk.[6]

9.9 Estimation of a Continuous Forward Rate*

In practice, the forward rate function $r(t)$ is unknown. Only bond prices are known. However, we can estimate $r(t)$ from the bond prices using nonlinear

[6] See Dowd (1998).

regression. An example of estimating $r(t)$ was given in Section 6.10 assuming that $r(t)$ was constant and using as data the prices of zero-coupon bonds of different maturities. In this section, we estimate $r(t)$ without assuming it is constant.

Suppose that $r(t)$ is linear so that

$$r(t) = \beta_0 + \beta_1 t,$$

for some unknown parameters β_0 and β_1. Then

$$\int_0^T r(t)dt = \beta_0 T + \beta_1 T^2/2.$$

The nonlinear regression model states that the price of the ith bond in the sample with maturity T_i expressed as a fraction of par value is

$$P_i = D(T_i) + \epsilon_i = \exp\left\{ - \left(\beta_0 T_i + \frac{\beta_1 T_i^2}{2} \right) \right\} + \epsilon_i, \tag{9.31}$$

where D is the discount function and ϵ_i is an "error" due to problems such as prices being somewhat stale[7] and the bid-ask spread.[8]

We now look at an example using data on U.S. STRIPS, a type of zero-coupon bond. STRIPS is an acronym for "Separate Trading of Registered Interest and Principal of Securities." The interest and principal on Treasury bills, notes, and bonds are traded separately through the Federal Reserve's book-entry system, in effect creating zero-coupon bonds by repackaging coupon bonds. [9]

The data are from December 31, 1995. The prices are given as a percentage of par value. For example, a price of 0.95 means that the bond is selling at 95% of its par value, e.g., for $950 if the par value is $1000. Table 9.3 contains the first five records of data. The variables are current date, maturity date, price, and record number (order of the observation in this data set). You can see that the current data are always Dec, 31, 1995. The first bond matures on Feb 15, 1996 and is selling at 99.393% of par value.

Price is plotted against maturity in years in Figure 9.3. Maturities are nearly equally spaced from 0 to 30 years. We can see that the price drops smoothly with maturity and that there is not much noise in the data.

Here is a SAS program to fit this model to the U.S. STRIPS data. In the search for good starting values for the iterative search, the parameter β_0 varies on a grid from 0.01 to 0.09. The idea here is that β_0 is the "constant" in the

---

[7] Prices are "stale" because they can be determined only at the time of the last sale.

[8] A bond dealer buys bonds at the bid price and sell them at the ask price which is slightly higher than the bid price. The difference is called the bid-ask spread and covers the trader's administrative costs and profit.

[9] Jarrow (2002, p. 15).

interest rate model. The interest rates should be close to β_0 and 0.01 to 0.09 is the typical range of interest rates. The product of 30 and β_1 determines how much the interest rates deviate from a constant value over the 30 years. We expect this deviation to be small and β_1 to be even smaller. Also, the deviation could be either positive or negative. Therefore, the initial grid of values of β_1 was chosen to be -0.005 to 0.005 by 0.001. The reason for the "100" in the model for price is that in the data set the prices were a percentage, rather than a fraction, of par value.

```
data USStrips ;
infile 'C:\book_finance\strips_dec95.txt' ;
input time2mat price;
run ;
title 'NL regression using US Strips data, linear';
proc nlin ;
parm  beta0= .01 to .09 by .01 beta1 = -.005 to .005 by .001;
model price = 100*exp(-beta0*time2mat - (beta1*time2mat**2)/2) ;
output out=outdata p=pred r=resid ;
run ;
```

Here is the relevant portion of the SAS output.

```
                                Sum of        Mean              Approx
Source                    DF   Squares      Square  F Value  Pr > F

Regression                 2    334862      167431   398547  <.0001
Residual                 115   48.3119      0.4201
Uncorrected Total        117    334910
Corrected Total          116   69756.0

                               Approx        Approximate 95%
Parameter      Estimate     Std Error      Confidence Limits

beta0            0.0532      0.000311      0.0526      0.0538
beta1          0.000746      0.000035    0.000677    0.000815
```

Note that the confidence interval for β_1 does not contain 0. Therefore, we can conclude with 95% confidence that β_1 is not 0 and therefore the forward rate function is not constant. If the forward curve is linear, then it appears to be increasing since $\beta_1 > 0$ (with 95% confidence). However, there is no compelling reason to assume that the forward function is linear. There may be curvature to this function. To check for curvature, we can fit a quadratic model for the forward rate function. That is, we fit the model

$$r(t) = \beta_0 + \beta_1 t + \beta_2 t^2,$$

which gives us the new regression model

$$P_i = \exp\left\{ -\left(\beta_0 T_i + \frac{\beta_1 T_i^2}{2} + \frac{\beta_2 T_i^3}{3} \right) \right\} + \epsilon_i.$$

Here is the SAS program to fit a quadratic forward rate function. The data step is deleted since it is the same as in the previous SAS program.

```
proc nlin ;
parm beta0 = .02 to .09 by .005 beta1 = -.005 to .005 by .001
     beta2 = -.001 to .001 by .0001 ;
model price = 100*exp(-beta0*time2mat - (beta1*(time2mat)**2)/2
     - (beta2*(time2mat)**3)/3) ;
output out=outdata p=pred r=resid ;
run ;
```

Here are selected parts of the output.

```
                                    Sum of        Mean               Approx
Source                      DF     Squares      Square    F Value    Pr > F

Regression                   3      334905      111635    2725982    <.0001
Residual                   114      4.6686      0.0410
Uncorrected Total          117      334910

Corrected Total            116     69756.0


                                     Approx          Approximate 95%
  Parameter      Estimate         Std Error         Confidence Limits

  beta0            0.0475          0.000199        0.0471        0.0479
  beta1           0.00240          0.000052       0.00230       0.00251
  beta2          -0.00008          2.285E-6      -0.00008      -0.00007
```

The confidence interval for β_2 contains only negative numbers indicating that with 95% confidence we can conclude that β_2 is negative. This is strong evidence that the forward rate curve is not linear, but rather is concave.[10]

Figure 9.4 shows the fitted constant, linear, and quadratic forward curves. There is also a fitted curve using a spline that is discussed later. AIC was found to be 89, −98, and −368 for constant, linear, and quadratic rate curves, respectively. Since small values of AIC indicate a good model, AIC chooses the quadratic model over the constant and linear models.

Of course, there is no guarantee that the forward curve is really quadratic, only that a quadratic model fits better than a constant or linear model. Perhaps a higher degree polynomial or a nonpolynomial function would be an even better model. A flexible way of modeling an arbitrary smooth curve is with a spline. A spline is a function that is created by piecing together polynomials. The spline in Figure 9.4 is composed of two quadratic functions, one for $T < 15$ and the other for $T > 15$. The AIC for the spline model is −625, much smaller than for the quadratic model, indicating that the spline model fits better than any of the polynomial models that were tried before.

The two quadratic functions that make up the spline function are pieced together so that at $T = 15$ the spline function is continuous with a continuous first derivative. This means that the two functions join together without a "gap" that would occur if the spline jumped in value at $T = 15$. Also, they join together without a "kink" that would occur if the first derivative jumped in value at $T = 15$. The second derivative is discontinuous and in fact jumps in

[10] A function is concave if its second derivative is always nonpositive. This means that its first derivative is always constant or decreasing. For example, a parabola that opens downward is concave.

value at $T = 15$ causing a change in curvature at $T = 15$. Quadratic functions have a constant second derivative[11] so the second derivative of the spline has a constant value for $T < 15$ and then jumps to a new constant value for $T > 15$. Second derivatives measure curvature, and you can see that the fitted spline in Figure 9.4 has one constant curvature to the left of 15 and a greater constant curvature to the right of 15.

The location, $T = 15$, where the spline changes from one quadratic function to another is called a **knot** of the spline. Putting the knot at $T = 15$ was somewhat arbitrary, though 15 is the midpoint of the range of T.

Ideally we would like the shape of the fitted spline to depend only on the data, not on the number of knots and the locations of the knots. This can be accomplished by using a relatively large number of knots, say 10 to 20, and placing then evenly over the range of T. Figure 9.5 shows spline fits to the STRIPS data with 1, 10, and 20 knots. A 5-knot spline was also fit but not included in the figure since it was similar to the 1-knot spline and the two splines were difficult to distinguish visually. However, the extra flexibility of the 5-knot spline allows it to fit the data somewhat better than the 1-knot spline. A potential problem with a large number of knots is that the fitted forward curve can oscillate too much. This can be seen in Figure 9.5 when there are 10 or 20 knots, especially the latter case. The problem with 10 or more knots is that the data are **overfit**, meaning that there are so many parameters than the random noise in the data is being fit by the spline.

The overfitting problem can be solved by penalizing large jumps in the second derivative; see Section 13.8. This penalization of overfitting prevents the fitted forward curve from being too wiggly. The overfitting penalty is analogous to the penalties on the number of parameters in a model that are incorporated into the AIC and C_p criteria as discussed in Sections 4.8.1 and 6.7.

Table 9.3. Prices of U.S. STRIPS. Only the first five records are shown here.

Current Data	Maturity Date	Price	Record Number
1995-12-31	1996-02-15	0.99393	1
1995-12-31	1996-08-15	0.96924	2
1995-12-31	1997-02-15	0.94511	3
1995-12-31	1997-08-15	0.92070	4
1995-12-31	1998-02-15	0.89644	5

[11] Note that $a + bx + cx^2$ has a second derivative equal to $2c$.

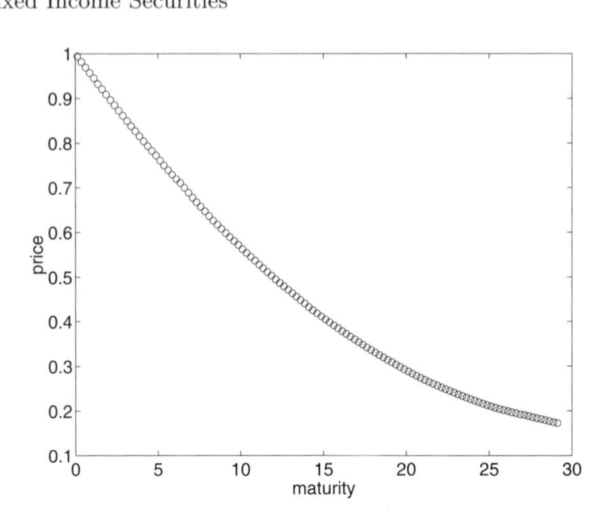

Fig. 9.3. *Prices of the U.S. STRIPS plotted as a function of maturity.*

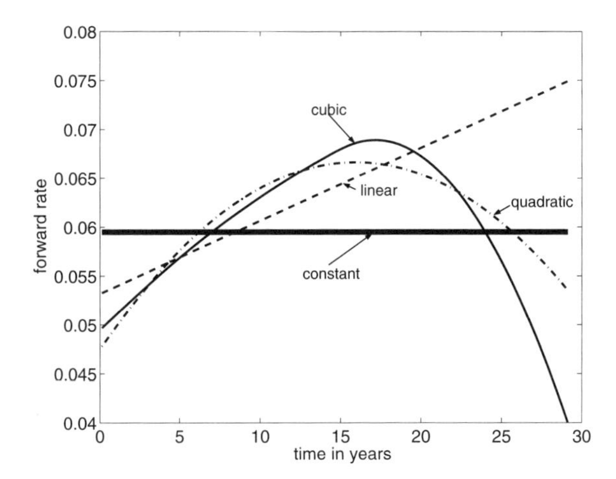

Fig. 9.4. *Polynomial and spline estimates of forward rates of U.S. Treasury bonds.*

9.10 Summary

Buying a bond means making a loan to a corporation, government, or other issuing agency that is obligated to pay back the principal and interest (unless it defaults). You receive a fixed stream of income and bonds are called "fixed-income" securities. Bond prices fluctuate with the interest rates and for long-term bonds your income is guaranteed only if you keep the bond to maturity.

Zero-coupon bonds pay no principal or interest until maturity. Par value is the payment made to the bond holder at maturity and a zero sells for less than par value. Coupon bonds make regular interest payments and generally sell around par value.

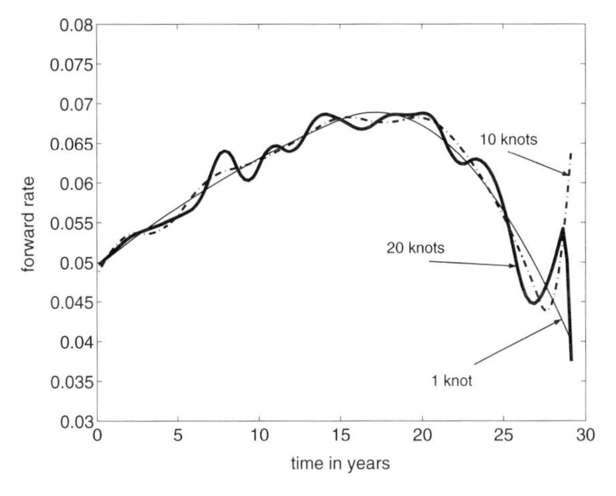

Fig. 9.5. *Splines with* 1, 10, *or* 20 *knots fit to U.S. STRIPS data.*

The yield to maturity is the average rate of return. The yield to maturity of a zero-coupon bond of maturity n years is called the n-year **spot rate**. A coupon bond is a bundle of zeros, each with a different maturity and therefore a different spot rate. The yield to maturity of a coupon bond is a complex "average" of these different spot rates. Term structure is a description of how, *at a given time*, yield to maturity depends on maturity.

Continuous compounding simplifies the relationships between forward rates, spot rates, and prices of zeros. Realistic models of term structure allow the forward rate, yield to maturity, and bond prices to depend continuously upon maturity. Continuous forward rates can be estimated by nonlinear regression.

9.11 Bibliographic Notes

Tuckman (2002) is an excellent comprehensive treatment of fixed-income securities, which is written at an elementary mathematical level and is highly recommended for readers wishing to learn more about this topic. Bodie, Kane, and Marcus (1999), Sharpe, Alexander, and Bailey (1999), and Campbell, Lo, and MacKinlay (1997) provide good introductions to fixed-income securities, with the last named being at a more advanced level. James and Webber (2000) is an advanced book on interest rate modeling. Jarrow (2002) covers many advanced topics that are not included in this book, including modeling the evolution of term structure, bond trading strategies, options and futures on bonds, and interest-rate derivatives.

9.12 References

Bodie, Z., Kane, A., and Marcus, A. (1999) *Investments, 4th Ed.*, Irwin/McGraw-Hill, Boston.

Campbell, J. Y., Lo, A. W., and MacKinlay, A. C. (1997) *Econometrics of Financial Markets.* Princeton University Press, Princeton, NJ.

Dowd, K. (1998) *Beyond Value At Risk*, Wiley, Chichester.

James, J. and Webber, N. (2000) *Interest Rate Modeling*, Wiley, Chichester.

Jarrow, R. (2002) *Modeling Fixed-Income Securities and Interest Rate Options, 2nd Ed.*, Stanford University Press, Stanford, CA.

Sharpe, W., Alexander, G., and Bailey, J. (1999) *Investments, 6th Ed.*. Prentice-Hall, Englewood Cliffs, NJ.

Tuckman, B. (2002) *Fixed Income Securities, 2nd Ed.*, Wiley, Hoboken, NJ.

9.13 Problems

1. Suppose that the forward rate is $r(t) = 0.022 + 0.0003t$.
   (a) What is the yield to maturity of a bond maturing in 20 years?
   (b) What is the price of a par $1000 zero-coupon bond maturing in 25 years?
2. Verify the following equality

$$\sum_{t=1}^{2T} \frac{C}{(1+r)^t} + \frac{\text{PAR}}{(1+r)^{2T}}$$

$$= \frac{C}{r} + \left\{\text{PAR} - \frac{C}{r}\right\}(1+r)^{-2T}. \tag{9.32}$$

3. One year ago a par $1000 20-year coupon bond with semi-annual coupon payments was issued. The annual interest rate at that time was 10%. Now, a year later, the annual interest rate is 8%.
   (a) What are the coupon payments?
   (b) What is the bond worth now? Assume that the second coupon payment was just received, so the bond holder receives an additional 38 coupon payments, the next one in six months.
   (c) What would the bond be worth if instead the second payment were just about to be received?
4. A par $1000 zero-coupon bond that matures in five years sells for $848. Assume that there is a constant continuously compounded forward rate r.
   (a) What is r?
   (b) Suppose that one year later the forward rate r is still constant but has changed to be 0.04. Now what is the price of the bond?
   (c) If you bought the bond for the original price of $848 and sold it one year later for the price computed in part (b), then what is the net return?

5. A coupon bond with a par value of \$1000 and a 10-year maturity pays semi-annual coupons of \$18.
   (a) Suppose the current interest rate for this bond is 4% per year compounded semi-annually. What is the price of the bond?
   (b) Is the bond selling above or below par value? Why?
6. Suppose that a coupon bond with a par value of \$1000 and a maturity of 5 years is selling for \$1100. The semi-annual coupon payments are \$25.
   (a) Find the yield to maturity of this bond. (You will not be able to give a numerical answer. Rather, you should give an equation which, when solved by interpolation or some other method, gives the yield to maturity.)
   (b) What is the current yield on this bond?
   (c) Is the yield-to-maturity less or greater than the current yield? Why?
7. Suppose that the continuous forward rate is $r(t) = 0.03 + 0.001t$. What is the current value of a par \$100 zero-coupon bond with a maturity of 15 years?
8. Suppose that the continuous forward rate is $r(t) = .02 + .001t - .0005(t - 10)_+$. What is the yield-to-maturity on a 20-year zero-coupon bond?
9. An investor is considering the purchase of zero-coupon bonds with maturities of 1, 3, or 5 years. Currently the spot rates for 1-, 2-, 3-, 4-, and 5-year zero-coupon bonds are, respectively, 0.02, 0.03, 0.04, 0.0425, and 0.045 per year with semi-annual compounding. A financial analyst has advised this investor that interest rates will increase during the next year and the analyst expects all spot rates to increase by the amount 0.005, so that the one-year spot rate will become 0.025 and so forth. The investor plans to sell the bond at the end of one year and wants the greatest return for the year. This problem does the bond math to see which maturity, 1, 3, or 5 years, will give the best return under two scenarios, interest rates are unchanged and interest rates increase as forecast by the analyst.
   (a) What are the current prices of 1-, 3-, and 5-year zero-coupon bonds with par values of 1000?
   (b) What will be the prices of these bonds one year from now if spot rates remain unchanged?
   (c) What will be the prices of these bonds one year from now if spot rates each increase by 0.005?
   (d) If the analyst is correct that spot rates will increase by 0.005 in one year, which maturity, 1, 3, or 5 years, will give the investor the greatest return when the bond is sold after one year? Justify your answer.
   (e) If instead the analyst is incorrect and spot rates remain unchanged then which maturity, 1, 3, or 5 years, earn the highest return when the bond is sold after one year? Justify your answer.
   (f) The analyst also said that if the spot rates remain unchanged then the bond with the highest spot rate will earn the greatest one-year return. Is this correct? Why?

(*Hint:* Be aware that a bond will not have the same maturity in one year as it has now, so the spot rate that applies to that bond will change.)

10. Suppose that a bond pays a cash flow C_i at time T_i for $i = 1, \ldots, N$. Then the net present value (NPV) of cash flow C_i is

$$\mathrm{NPV}_i = C_i \exp(-T_i \, y_{T_i}).$$

Define the weights

$$\omega_i = \frac{\mathrm{NPV}_i}{\sum_{j=1}^{N} \mathrm{NPV}_j}$$

and define the duration of the bond to be

$$\mathrm{DUR} = \sum_{i=1}^{N} \omega_i T_i,$$

which is the weighted average of the times of the cash flows. Show that

$$\frac{d}{d\delta} \sum_{i=1}^{N} C_i \exp\{-T_i(y_{T_i} + \delta)\} \bigg|_{\delta=0} = -\mathrm{DUR} \sum_{i=1}^{N} C_i \exp\{-T_i \, y_{T_i}\}$$

and use this result to verify equation (9.29).

10

Resampling

10.1 Introduction

Computer simulation is widely used in all areas of operations research and, in particular, applications of simulation to statistics have become very widespread. In this chapter we apply a simulation technique called the "bootstrap" or "resampling" to study the effects of estimation error on portfolio selection. The term "bootstrap" was coined by Bradley Efron and comes from the phrase "pulling oneself up by one's bootstraps" that apparently originated in the eighteenth century story "Adventures of Baron Munchausen" by Rudolph Erich Raspe.[1] In this chapter, "bootstrap" and "resampling" are treated as synonymous.

When statistics are computed from a randomly chosen sample, then these statistics are random variables. Students often do not appreciate this fact. After all, what could be random about $\overline{X}$? We just averaged the data, so what is random? The point is that the sample is only one of many possible samples. Each possible sample gives a possible value of $\overline{X}$. Thus, although we only see one value of $\overline{X}$, it was selected at random from the many possible values and therefore $\overline{X}$ is a random variable.

Methods of statistical inference such as confidence intervals and hypothesis tests are predicated on the randomness of statistics. For example, the confidence coefficient of a confidence interval tells us the probability that an interval constructed from a random sample contains the parameter. Confidence intervals are usually derived using probability theory. Often, however, the necessary probability calculations are intractable, and in that case we can replace theoretical calculations by Monte Carlo simulation.

But how do we simulate sampling from an *unknown* population? The answer, of course, is that we cannot do this exactly. However, a sample is a good representative of the population, and we can simulate sampling from the pop-

[1] Efron and Tibshirani (1993).

ulation by sampling from the sample, which is usually called "resampling," and is illustrated in Figure 10.1.

Each resample has the same sample size n as the original sample. The reason for this is that we are trying to simulate the original sampling, so we want the resampling to be as similar as possible to the original sampling. Moreover, the resamples are drawn *with replacement.* Why? The reason is that only sampling with replacement gives independent observations, and we want the resamples to be i.i.d. just as the original sample. In fact, if the resamples were drawn without replacement then every resample would be exactly the same as the original sample, so the resamples would show no random variation. This would not be very satisfactory, of course.

The number of resamples taken should, in general, be large. Just how large depends on context and is discussed more fully later. Sometimes thousands or even tens of thousands of resamples are used. We let B denote the number of resamples.

There is some good news and some bad news about the bootstrap. The good news is that computer simulation replaces difficult mathematics. The bad news is that resampling is a new concept with which it takes some time to become comfortable. The problem is not that resampling is all that conceptually complex. Rather, the difficulty is that students do not have a great deal of experience with drawing even a single random sample from a population. Resampling is even more complex than that, with two layers of sampling and multiple resamples.

10.2 Confidence Intervals for the Mean

Before applying resampling to portfolio analysis, we look at a simpler problem. Suppose we wish to construct a confidence interval for the population mean based on a random sample. One starts with the so-called "t-statistic"[2] which is

$$t = \frac{\mu - \overline{X}}{\frac{s}{\sqrt{n}}}. \tag{10.1}$$

The denominator of t, $s/\sqrt{n}$, is just the standard error of the mean so that the denominator estimates the standard deviation of the numerator.

If we are sampling from a normally distributed population, then the probability distribution of t is known to be the t-distribution with $n-1$ degrees of freedom. Using the notation of Section 2.9.2, we denote by $t_{\alpha/2}$ the $\alpha/2$ upper t-value, that is, the $1-\alpha/2$ quantile of this distribution. Thus t in

[2] Actually, t is not quite a statistic since it depends on the unknown μ, whereas a statistic, by definition, is something that depends only on the sample, not on unknown parameters.

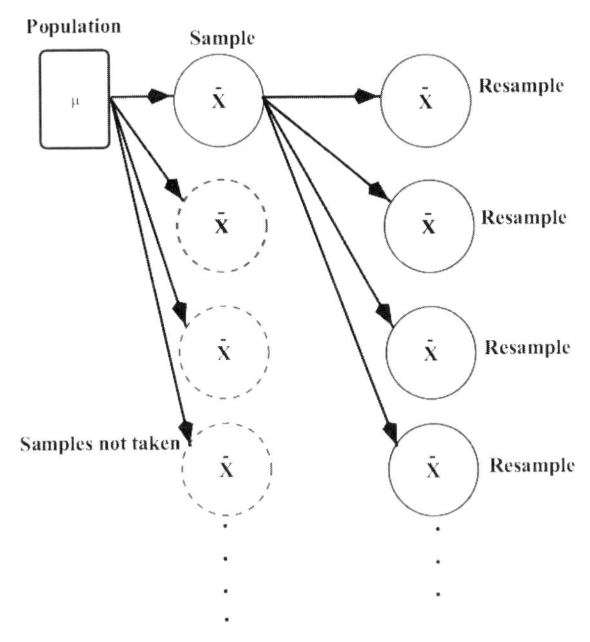

Fig. 10.1. *Resampling. There is a single population with mean μ. A single sample is taken and has mean $\overline{X}$. Other samples could have been taken but were not. Thus, $\overline{X}$ is the observed value of a random variable. We want to know the probability distribution of $\overline{X}$ or, perhaps, the t-statistic $= (\overline{X} - \mu)/s_{\overline{X}}$. To approximate this distribution, we resample from the sample. Each resample has a mean, $\overline{X}_{boot}$.*

(10.1) has probability $\alpha/2$ of exceeding $t_{\alpha/2}$. Because of the symmetry of the t-distribution, the probability is also $\alpha/2$ that t is less than $-t_{\alpha/2}$.

Therefore, for normally distributed data, the probability is $1 - \alpha$ that

$$-t_{\alpha/2} \le t \le t_{\alpha/2}. \tag{10.2}$$

Substituting (10.1) into (10.2), after a bit of algebra we find that the probability is $1 - \alpha$ that

$$\overline{X} - t_{\alpha/2}\frac{s}{\sqrt{n}} \le \mu \le \overline{X} + t_{\alpha/2}\frac{s}{\sqrt{n}}, \tag{10.3}$$

which shows that

$$\overline{X} \pm \frac{s}{\sqrt{n}} t_{\alpha/2}$$

is a $1 - \alpha$ confidence interval for μ, assuming normally distributed data. This is the confidence interval given by equation (2.72).

What if we are not sampling from a normal distribution? In that case, the distribution of t (10.1) *not* is the t-distribution, but rather some other distribution that is not known to us. There are two problems. First, we do not know the distribution of the population. Second, even if the population distribution were known it is a difficult, usually intractable, probability calculation to get the distribution of the t-statistic from the distribution of the population. This calculation has only been done for normal populations. Considering the difficulty of these two problems, can we still get a confidence interval? The answer is "yes, by resampling." We can take a large number, say B, of resamples from the original sample.

Let $\overline{X}_{\text{boot},b}$ and $s_{\text{boot},b}$ be the sample mean and standard deviation of the bth resample, $b = 1, \ldots, B$. Define

$$t_{\text{boot},b} = \frac{\overline{X} - \overline{X}_{\text{boot},b}}{\frac{s_{\text{boot},b}}{\sqrt{n}}}. \tag{10.4}$$

Notice that $t_{\text{boot},b}$ is defined in the same way as t except for two changes. First, $\overline{X}$ and s in t are replaced by $\overline{X}_{\text{boot},b}$ and $s_{\text{boot},b}$ in $t_{\text{boot},b}$. Second, μ in t is replaced by $\overline{X}$ in $t_{\text{boot},b}$. The last point is a bit subtle, and you should stop to think about it. A resample is taken using the original sample as the population. Thus, for the resample, the population mean is $\overline{X}$!

Because the resamples are independent of each other, the collection $t_{\text{boot},1}$, $t_{\text{boot},2}, \ldots$ can be treated as a random sample from the distribution of the t-statistic. After B values of $t_{\text{boot},b}$ have been calculated, one from each resample, we find the $100\alpha\%$ and $100(1-\alpha)\%$ percentiles of this collection of $t_{\text{boot},b}$ values. Call these percentiles t_L and t_U. More specifically, we find t_L and t_U as follows. The B values of $t_{\text{boot},b}$ are sorted from smallest to largest. Then we calculate $B\alpha/2$ and round to the nearest integer. Suppose the result is K_L. Then the K_Lth sorted value of $t_{\text{boot},b}$ is t_L. Similarly, let K_U be $B(1-\alpha/2)$ rounded to the nearest integer and then t_U is the K_Uth sorted value of $t_{\text{boot},b}$.

If the original population is skewed, then there is no reason to suspect that the $100\alpha\%$ percentile is minus the $100(1-\alpha)\%$ percentile, as happens for symmetric populations such as the t-distribution. In other words, we do not necessarily expect that $t_L = -t_U$. However, this fact causes us no problem since the bootstrap allows us to estimate t_L and t_U without assuming any relationship between them. Now we replace $-t_{\alpha/2}$ and $t_{\alpha/2}$ in the confidence interval (10.3) by t_L and t_U, respectively. Finally, the bootstrap confidence interval for μ is

$$\left(\overline{X} + t_L \frac{s}{\sqrt{n}},\ \overline{X} + t_U \frac{s}{\sqrt{n}}\right).$$

The bootstrap has solved both problems mentioned above. We do not need to know the population distribution since we can estimate it by the sample.[3] Moreover, we don't need to calculate the distribution of the t-statistic using probability theory. Instead we can simulate from this distribution.

[3] There is something a bit clever going on here. A sample isn't a probability distribution. What we are doing is creating a probability distribution, called the **empirical distribution**, from the sample by giving each observation in the sample probability $1/n$ where n is the sample size.

We use the notation

$$\mathrm{SE} = \frac{s}{\sqrt{n}}$$

and

$$\mathrm{SE}_{\text{boot}} = \frac{s_{\text{boot}}}{\sqrt{n}}.$$

Example 10.1. We start with a very small sample of size six to illustrate how the bootstrap works. The sample is 82, 93, 99, 103, 104, 110, $\overline{X} = 98.50$, and the SE is 4.03.

The first bootstrap sample is 82, 82, 93, 93, 103, 110. In this bootstrap sample, 82 and 93 were sampled twice, 103 and 110 were sampled once, and the other elements of the original sample were not sampled. This happened by chance, of course, since the bootstrap sample was taken at random, with replacement, from the original sample. For this bootstrap sample $\overline{X}_{\text{boot}} = 93.83$, $\mathrm{SE}_{\text{boot}} = 4.57$, and $t_{\text{boot}} = (98.5 - 93.83)/4.57 = 1.02$.

The second bootstrap sample is 82, 103, 110, 110, 110, 110. In this bootstrap sample, 82 and 103 were sampled twice, 110 was sampled four times, and the other elements of the original sample were not sampled. It may seem strange at first that 110 was resampled four times. However, this event happened by chance and, in fact, is not highly unlikely and can be expected to happen occasionally.[4] For this bootstrap sample $\overline{X}_{\text{boot}} = 104.17$, $\mathrm{SE}_{\text{boot}} = 4.58$, and $t_{\text{boot}} = (98.5 - 104.17)/4.58 = -1.24$.

The third bootstrap sample is 82, 82, 93, 99, 104, 110. For this bootstrap sample $\overline{X}_{\text{boot}} = 95.00$, $\mathrm{SE}_{\text{boot}} = 4.70$, and $t_{\text{boot}} = (98.5-95.00)/4.570 = 1.02$.

If this example were to continue, more bootstrap samples would be drawn and all bootstrap t-values would be saved in order to compute quantiles of the bootstrap t-values. However, since the sample size is so small, this example is not very realistic and we do not continue it.

Example 10.2. Suppose that we have a random sample with a more realistic size of 40 from some population and $\overline{X} = 107$ and $s = 12.6$. Let's find the "normal theory" 95% confidence interval for the population mean μ. With 39 degrees of freedom, $t_{.025} = 2.02$. Therefore, the confidence interval for μ is

$$107 \pm 2.02 \frac{12.6}{\sqrt{40}} = (102.97, 111.03).$$

[4] The number of times that 110 appears in any resample is a binomial random variable with parameters $p = 1/6$ and $n = 6$. Therefore, the probability that 110 occurs exactly four times in the sample is $\{6!/(4!\,2!)\}(1/6)^4(5/6)^2 = 0.00804$. Of course, there is nothing special about 110 and we might have been surprised if any element of the sample appeared four times in a resample. In fact, however, we should not be surprised. The probability that one of the six elements of the original sample occurs exactly four times in a resample is $(6)(.00804) = 0.0482$. Thus, we see some element of the sample resampled exactly four times in about one bootstrap resample out of 21, so this is not a particularly rare event.

Suppose that we use resampling instead of normal theory and that we use 1000 resamples. This gives us 1000 values of $t_{\text{boot},b}$. We rank them from smallest to largest. The 2.5% percentile is the one with rank $25 = (1000)(.025)$. Suppose the 25th smallest value of $t_{\text{boot},b}$ is -1.98. The 97.5% percentile is the value of $t_{\text{boot},b}$ with rank 975. Suppose that its value is 2.25. Then $t_L = -1.98$, $t_U = 2.25$, and the 95% confidence interval for μ is

$$\left(107 - 1.98\frac{12.6}{\sqrt{40}},\ 107 + 2.25\frac{12.6}{\sqrt{40}}\right) = (103.06,\ 111.48).$$

Example 10.3. (Log-returns for MSCI-Argentina) MSCI-Argentina is the Morgan Stanley Capital Index for Argentina and is roughly comparable to the S&P 500 for the U.S. The log-returns for this index from January 1988 to January 2002, inclusive, are used in this example. Normal probability plots of these returns and of other returns used in the next section are found in Figure 10.4. There is evidence of nonnormality, in particular, that the log-returns are heavy-tailed, especially the left tail.

The bootstrap was implemented with $B = 10,000$. The t-values were $t_L = -1.93$ and $t_U = 1.98$. To assess the Monte Carlo variability, the bootstrap was repeated two more times with the results first that $t_L = -1.98$ and $t_U = 1.96$ and then that $t_L = -1.94$ and $t_U = 1.94$. We see that $B = 10,000$ gives reasonable accuracy but that the third significant digit is still uncertain. Also, using normal theory, $t_L = -t_{.025} = -1.974$ and $t_U = t_{.025} = 1.974$, which are similar to the bootstrap values that do not assume normality. Therefore, the use of the bootstrap in this example confirms that the normal theory t-confidence interval is satisfactory. This is not too surprising since the validity of the normal theory confidence interval depends on normality of $\overline{X}$, not on normality of the individual data. With a sample size as large as here ($n = 169$), the central limit theorem suggests that $\overline{X}$ is nearly normally distributed. In other examples, particularly with more strongly skewed data and small sample sizes, the normal theory t-confidence interval is less satisfactory.

10.3 Resampling and Efficient Portfolios

10.3.1 The global asset allocation problem

One application of optimal portfolio selection is allocation of capital to different market segments. For example, Michaud (1998) discusses a global asset allocation problem where capital must be allocated to "U.S. stocks and government/corporate bonds, Euros, and the Canadian, French, German, Japanese, and U.K. equity markets" and Britten-Jones (1999) analyzes a similar global allocation problem. Here we look at a similar example where we allocate capital to the equity markets of 10 different countries. Monthly log-returns for these markets were calculated from: MSCI Hong Kong, MSCI Singapore, MSCI Brazil, MSCI Argentina, MSCI UK, MSCI Germany, MSCI Canada,

MSCI France, MSCI Japan, and the S&P 500. "MSCI" means "Morgan Stanley Capital Index." The data came from the online financial database "Datastream" and are from January 1988 to January 2002, inclusive, so there are 169 months of data.

Figure 10.2 shows time series plots of the returns for each of these countries. The plots look stationary though there may be GARCH effects. Figure 10.3 shows autocorrelation plots of the returns. There is no evidence of nonstationarity though there is some short-term autocorrelation. Figure 10.4 shows normal probability plots of the returns. There is evidence of heavy tails, especially for the left tails.

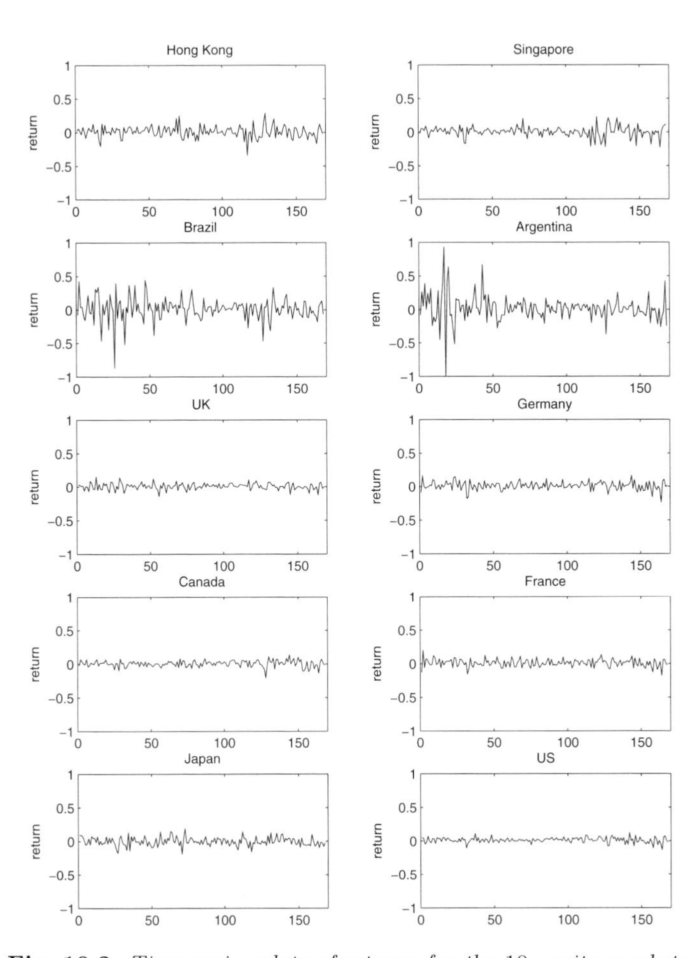

Fig. 10.2. *Time series plots of returns for the* 10 *equity markets.*

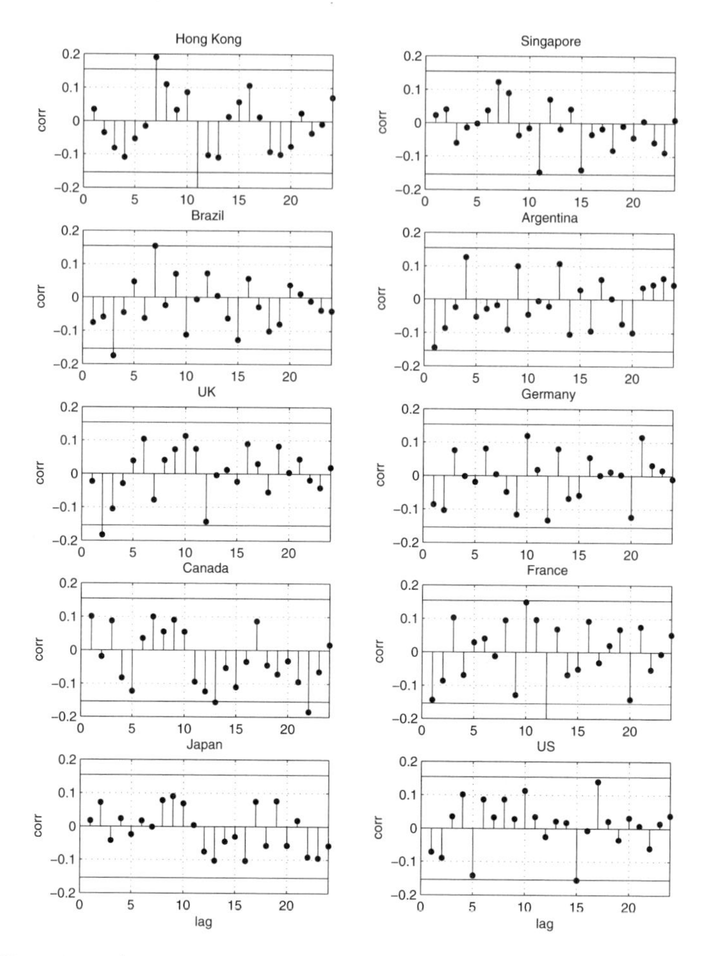

Fig. 10.3. *Autocorrelation plots of returns for the* 10 *equity markets.*

10.3.2 Uncertainty about mean-variance efficient portfolios

If we are planning to invest in a number of international capital markets, then to allocate our investment among them efficiently we want to know the efficient frontier. Of course, we can never know it exactly. At best, we can estimate the efficient frontier using estimated expected returns[5] and the estimated covariance matrix of returns. But how close is the estimated efficient frontier

[5] In this example, we use log-returns exclusively, so "return" always means "log-return."

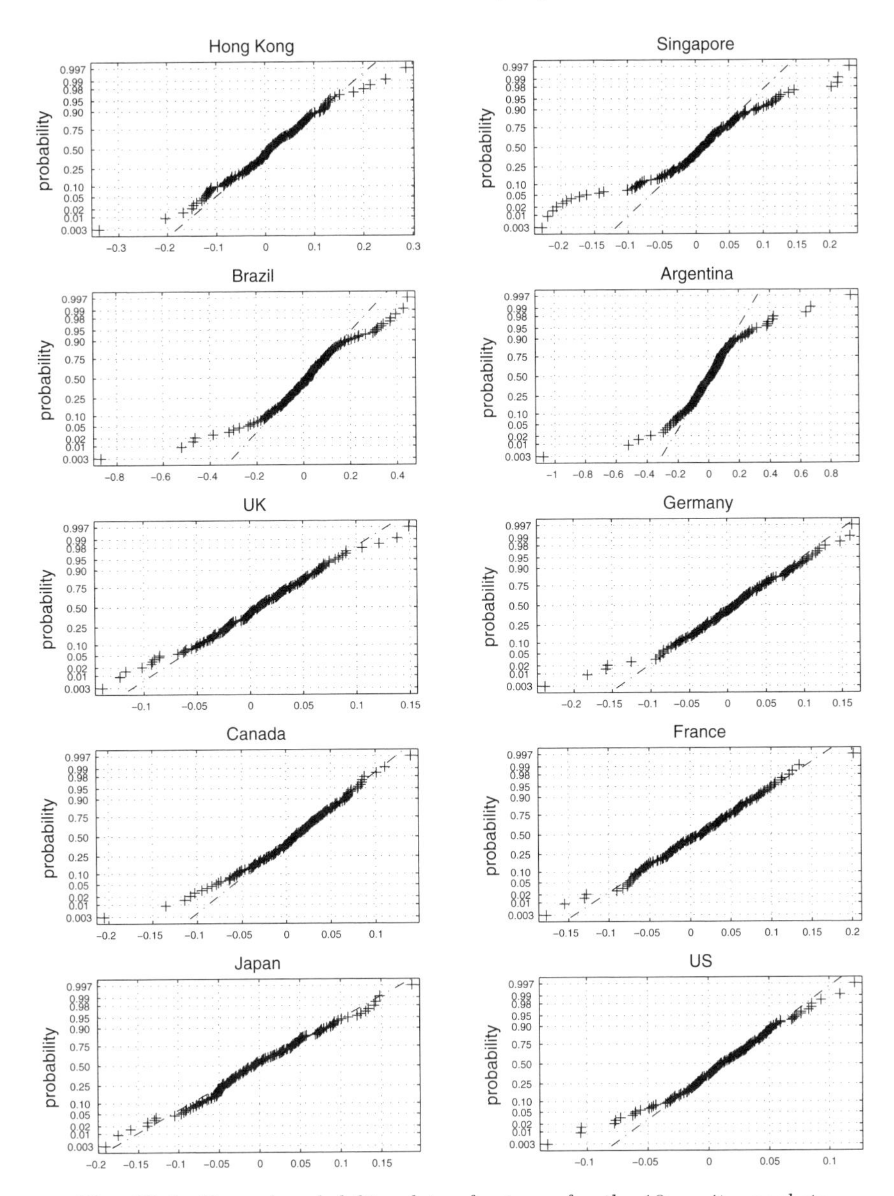

Fig. 10.4. *Normal probability plots of returns for the* 10 *equity markets.*

to the unknown true efficient frontier? This question can be addressed by resampling.

Each resample consists of 168 returns drawn with replacement from the 168 returns of the original sample.[6] From the resampling perspective, the original sample is treated as the population and the efficient frontier calculated from the original sample is viewed as the true efficient frontier. We can repeatedly recalculate the efficient frontier using each of the resamples and compare these re-estimated efficient frontiers to the "true efficient frontier." This is done for six resamples in Figure 10.5. In each subplot, the "true efficient frontier" is shown as a solid curve and is labeled "optimal." These curves do not vary from subplot to subplot since they are all calculated from the original sample. The dashed curves labeled "achieved" are the efficient frontiers calculated from the resamples. "Achieved" means that we calculated the true mean and standard deviation actually achieved by efficient frontiers calculated from the resamples. "True mean and standard deviation" means that we use the original sample (the "population" from the resampling perspective). We can see that there is a fair amount of variability in the resampled efficient frontiers.

To be more precise about how the "optimal" and "achieved" curves were created, let $\widehat{\boldsymbol{\mu}}$ and $\widehat{\boldsymbol{\Omega}}$ be the mean vector and covariance matrix estimated from the original sample. For a given target for the expected portfolio return μ_P, let $\widehat{\boldsymbol{\omega}}_{\mu_P}$ be the efficient portfolio weight vector given by equation (5.23) with $\boldsymbol{g}$ and $\boldsymbol{h}$ estimated from the original sample. Let $\widehat{\boldsymbol{\omega}}_{\mu_P,b}$ be the efficient portfolio weight vector with $\boldsymbol{g}$ and $\boldsymbol{h}$ estimated from the bth resample. Then the solid "optimal" curves in Figure 10.5 are $\widehat{\boldsymbol{\omega}}_{\mu_P}^{\mathsf{T}}\widehat{\boldsymbol{\mu}} = \mu_P$ plotted against

$$\sqrt{\widehat{\boldsymbol{\omega}}_{\mu_p}^{\mathsf{T}}\widehat{\boldsymbol{\Omega}}\widehat{\boldsymbol{\omega}}_{\mu_P}}$$

for a grid of μ_P values. The dashed curves are expected portfolio returns $\widehat{\boldsymbol{\omega}}_{\mu_P,b}^{\mathsf{T}}\widehat{\boldsymbol{\mu}}$ plotted against portfolio risks

$$\sqrt{\widehat{\boldsymbol{\omega}}_{\mu_P,b}^{\mathsf{T}}\widehat{\boldsymbol{\Omega}}\widehat{\boldsymbol{\omega}}_{\mu_P,b}}\,.$$

The dashed resampled efficient frontier curve lies below and to the right of the solid true efficient frontier curve, because estimation error makes the estimated efficient frontiers suboptimal. Also, $\widehat{\boldsymbol{\omega}}_{\mu_P,b}^{\mathsf{T}}\widehat{\boldsymbol{\mu}}$ does not equal μ_P; that is, we do not achieve the expected return μ_P that we have targeted because of error when estimating $\boldsymbol{\mu}$. To see this, look at Figure 10.6. In that figure we concentrate on estimating only a single point on the efficient frontier, the point where the expected portfolio return is 0.012.[7] This point is shown in the upper subplot as a large dot and is the point

[6] There are 169 months of data but we lose one month's data when differencing log prices.

[7] There is nothing special about 0.012, but some value of the expected return was needed.

$$\left(\sqrt{\widehat{\omega}^{\mathsf{T}}_{.012}\widehat{\Omega}\widehat{\omega}_{0.012}},\ 0.012\right). \tag{10.5}$$

Each small dot in the upper subplot is the estimate of point (10.5) from a resample and is

$$\left(\sqrt{\widehat{\omega}^{\mathsf{T}}_{.012,b}\widehat{\Omega}\widehat{\omega}_{.012,b}},\ \widehat{\omega}^{\mathsf{T}}_{.012,b}\widehat{\mu}\right).$$

The middle subplot of Figure 10.6 is a histogram of the values of

$$\frac{\sqrt{\widehat{\omega}^{\mathsf{T}}_{.012,b}\widehat{\Omega}\widehat{\omega}_{.012,b}}}{\sqrt{\widehat{\omega}^{\mathsf{T}}_{.012}\widehat{\Omega}\widehat{\omega}_{.012}}},$$

the ratio of achieved to optimal risk. The lower subplot of Figure 10.6 is a histogram of the values of

$$\frac{\widehat{\omega}^{\mathsf{T}}_{.012,b}\widehat{\mu}}{.012},$$

the ratio of achieved to targeted expected return. Notice that this ratio is often less than one; that is, the expected return is less than what was sought.

10.3.3 What if we knew the expected returns?

An interesting question is, "What is the main problem here, mis-estimation of the expected returns, mis-estimation of the covariance matrix of the returns, or both?" One of the interesting things we can do with resampling is to play the game of "what if?" In particular, we can ask, "What would happen if we knew the true expected returns and only had to estimate the covariance matrix?" We can also ask the opposite question, "What would happen if we knew the true covariance matrix and only had to estimate the expected returns?" By playing these "what if" games, we can address our question about the relative effects of mis-estimation of μ and mis-estimation of the covariance matrix.

Figure 10.7 addresses the first of these questions. In that figure, when we estimate the efficient frontier for each resample, we use the mean returns from the original sample, which from the resampling perspective are the population values. Only the covariance matrix is estimated from the resample. Because the true expected returns are known, the estimated efficient portfolio from a resample always does achieve an expected return of 0.012. The standard deviations of the estimated efficient portfolios are larger than optimal because of estimation error when the covariance matrix is estimated. This can be seen in both the upper subplot and also in the lower subplot which is a histogram of the values of the ratios

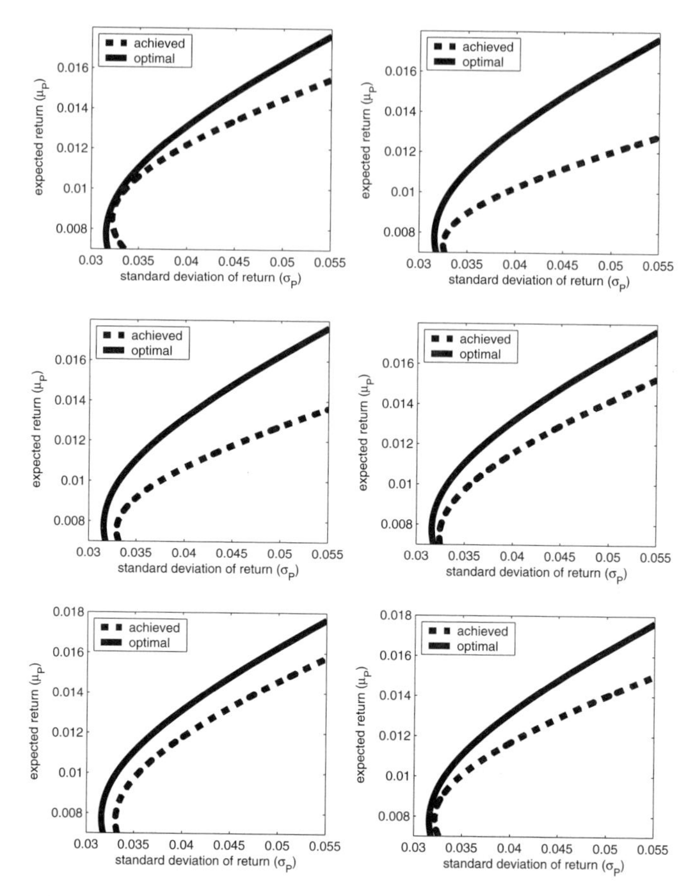

Fig. 10.5. *Actual efficient frontier for the sample (optimal) and bootstrap efficient frontier (achieved) for each of six bootstrap resamples.*

$$\frac{\sqrt{\widehat{\omega}^{\mathsf{T}}_{.012,b}\widehat{\Omega}\widehat{\omega}_{.012,b}}}{\sqrt{\widehat{\omega}^{\mathsf{T}}_{.012}\widehat{\Omega}\widehat{\omega}_{.012}}},$$

of achieved to optimal risk.

10.3.4 What if we knew the covariance matrix?

Figure 10.8 addresses the second of these questions, what if the covariance matrix were known and only the expected returns needed to be estimated? In this figure, when we estimate the efficient frontier for each resample, we use the covariance matrix from the original sample. Only the expected returns are estimated from the resample. Thus, we see what happens when the "true" covariance matrix is known. Figure 10.8 is surprisingly similar to Figure 10.6.

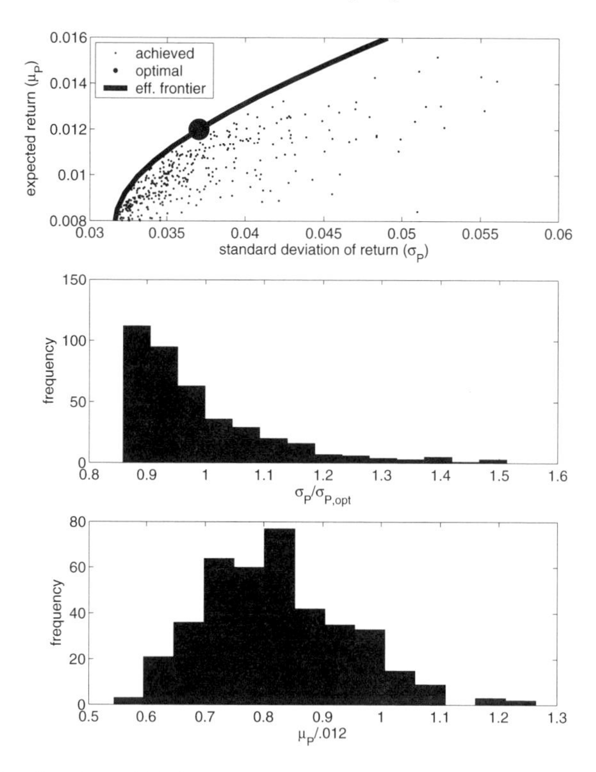

Fig. 10.6. *Results from 400 bootstrap resamples. For each resample, the efficient portfolio with a mean return of 0.012 is estimated. In the upper subplot, the actual mean return and standard deviation of the return are plotted as a small dot. The large dot is the point on the efficient frontier with a mean return of 0.012. Histograms of the ratios of σ_P to optimal (middle subplot) and of μ_P to 0.012 (lower subplot) are also shown.*

This is an indication that knowing the true covariance matrix does not help us much.

The conclusion from our "what if" games is that the real problem when we estimate the efficient frontier is estimation error in the expected returns. This is clearly seen by comparing Figures 10.6 and 10.7. We get very close to the efficient portfolio when the true expected returns are known and only the covariance matrix is estimated.

10.3.5 What if we had more data?

Here is another "what if" game. This time the question is, "What would happen if we had more data?" It is disappointing that the efficient frontier is

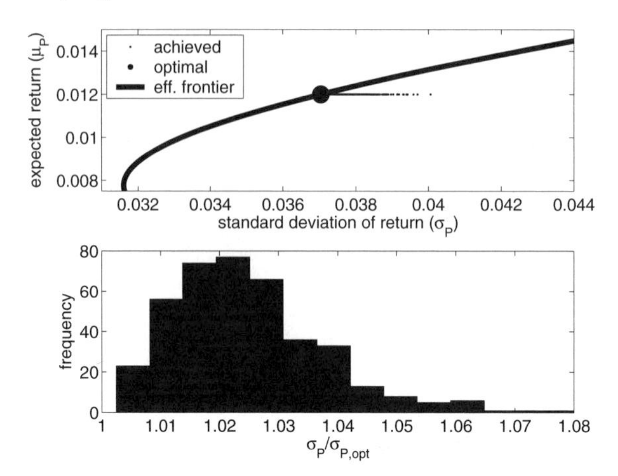

Fig. 10.7. *Results from* 400 *bootstrap resamples assuming that the vector of mean returns is known. For each resample the efficient frontier with a mean return of* 0.012 *is estimated using the mean returns from the* sample *and the covariance matrix of returns from the* resample. *In the upper subplot, the actual mean return, which is always* 0.012, *and standard deviations of the return are plotted as a small dot. The large dot is the point on the efficient frontier with a mean return of* 0.012. *A histogram of the ratios of achieved to optimal* σ_P *is also shown (lower subplot).*

not estimated more accurately, so it is natural to wonder whether it would be worthwhile to use a longer time series. Without even bothering to collect more data we can assess the effect of more data on accuracy. This assessment is done by using resamples that are *larger* than the sample size n.[8] In Figure 10.9 we repeat the simulation in Figure 10.6 but with the resample size equal to $3n$, that is, three times as large as the actual sample. Comparing Figures 10.6 and 10.9 we see that more data do help, though perhaps not as much as we would like. Of course, using a longer time series carries the danger that the data may not be stationary. Another problem is that the MSCI indices may not be available for earlier time periods.

10.4 Bagging*

So far, we have investigated the bootstrap as a means to assess estimation accuracy. An interesting question is whether the bootstrap can *improve* accuracy. *Bagging* is an attempt to do this. Bagging is a acronym meaning

[8] It was stated earlier that the resample size should equal the sample size. That is true, in general, but here we see that using a resample size different from n is useful if we want to know what would have happened if the sample size had been different.

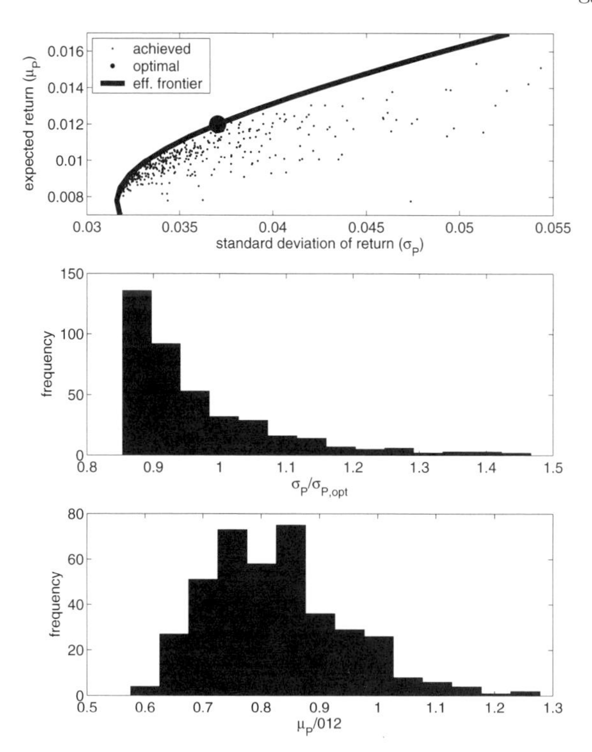

Fig. 10.8. *Results from* 400 *bootstrap resamples assuming that the covariance matrix of the returns is known. For each resample the efficient frontier with a mean return of* 0.012 *is estimated using the mean returns from the resample and the covariance matrix of returns from the sample. The actual mean return and standard deviation of the return are plotted as a small dot. The large dot is the point on the efficient frontier with a mean return of* 0.012.

"bootstrap aggregation." The idea is simple. Suppose we are estimating some quantity, e.g., the optimal portfolio weights to achieve an expected return of 0.012. We have one estimate from the original sample, and this estimate is the one often used. However, we also have B additional estimates, one from each of the bootstrap samples. The bagging estimate is the average of all of these bootstrap estimates. Hastie, Tibshirani, and Friedman (2001) state that "bagging can dramatically reduce the variance of unstable procedures like trees, leading to improved prediction." Richard Michaud has patented methods of portfolio rebalancing that essentially are an application of bagging to the efficient frontier, though Michaud's work appears to have been done earlier than, or at least independently of, Breiman's (1996) paper on bagging. See Michaud (1998). I do not know of any published studies as to whether bagging does improve the estimation of portfolio weights, but these weights are so sensitive to the expected returns and the expected returns are so difficult to estimate accurately that it seems prudent to remain skeptical

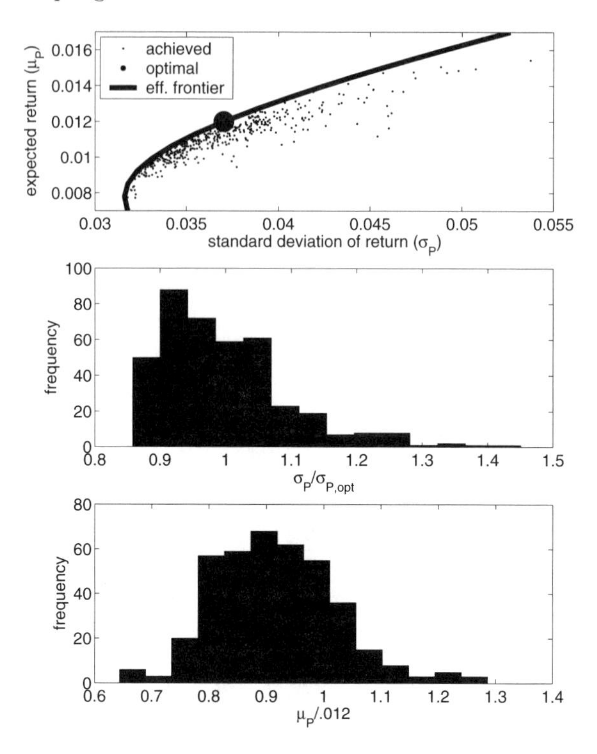

Fig. 10.9. *Results from* 400 *bootstrap resamples with the resample size equal to* $3n$.

about the extent to which bagging or any other estimation method leads to satisfactory results for portfolio optimization.

10.5 Summary

Resampling, which is also called the bootstrap, means taking samples with replacement from the original sample. Resampling the t-statistic allows us to construct a coefficient interval for the population mean without assuming that the population has a normal distribution.

Resampling the efficient frontier shows that the efficient frontier is poorly estimated. By experimenting with certain hypothetical scenarios, e.g., that the expected returns are known, we can pinpoint the cause of the error in estimation of the efficient frontier. This problem is mainly due to estimation error in the expected returns.

10.6 Bibliographic Notes

Efron and Tibshirani (1993) and Davison and Hinkley (1997) are good introductions to the bootstrap. Efron (1979) introduced the name "bootstrap" and did much to popularize resampling methods. Jobson and Korkie (1980) and Britten-Jones (1999) discuss the statistical issue of estimating the efficient frontier; see the latter for additional recent references. Britten-Jones (1999) shows that the tangency portfolio can be estimated by regression analysis and hypotheses about the tangency portfolio can be tested by regression F-tests. Michaud (1998) applies the bootstrap to portfolio management. His work has been awarded several patents and he has founded a company to market this technology.

Bagging is an acronym was introduced by Breiman (1996). Hastie, Tibshirani, and Friedman (2001) have a section on bagging including an example of bagging a decision tree.

10.7 References

Breiman, L. (1996) Bagging predictors, *Machine Learning*, **26**, 123–140.

Britten-Jones, M. (1999) The sampling error in estimates of mean-variance efficient portfolio weights, *The Journal of Finance*, **54**, 655–671.

Davison, A. C. and Hinkley, D. V. (1997) *Bootstrap Methods and Their Applications*, Cambridge University Press, Cambridge, UK.

Efron, B. (1979) Bootstrap methods: Another look at the jackknife, *The Annals of Statistics*, **7**, 1–26.

Efron, B. and Tibshirani, R. (1993) *An Introduction to the Bootstrap*, Chapman & Hall, New York.

Hastie, T., Tibshirani, R., and Friedman, J. (2001) *The Elements of Statistical Learning*, Springer, New York.

Jobson, J. D. and Korkie, B. (1980) Estimation for Markowitz efficient portfolios, *Journal of the American Statistical Association*, **75**, 544–554.

Michaud, R. O. (1998) *Efficient Asset Management*, Harvard Business School Press, Boston.

10.8 Problems

1. In Figure 10.7 the ratio $\sigma_{\mathrm{P}}/\sigma_{\mathrm{P,opt}}$ of achieved risk to optimal risk always exceeds 1, so there is extra risk due to estimation error. However, in Figures 10.6 and 10.8 $\sigma_{\mathrm{P}}/\sigma_{\mathrm{P,opt}}$ is sometimes less than 1. How does it happen that $\sigma_{\mathrm{P}}/\sigma_{\mathrm{P,opt}}$ can be less than 1, i.e., that estimation error results in *less* risk than "optimal"?

2. To estimate the risk of a stock, a sample of 50 log-returns was taken and s was 0.28. To get a confidence interval for σ, 10,000 resamples were taken. Let $s_{b,\text{boot}}$ be the sample standard deviation of the bth resample. The 10,000 values of $s_{b,\text{boot}}/s$ were sorted and the table below contains selected values of $s_{b,\text{boot}}/s$ ranked from smallest to largest (so rank 1 is the smallest and so forth).

Rank	Value of $s_{b,\text{boot}}/s$
250	0.5
500	0.7
1000	0.8
9000	1.3
9500	1.6
9750	2.1

Find a 90% confidence interval for σ.

3. Suppose one has a sample of monthly log-returns on two stocks with sample means of 0.003 and 0.007, sample variances of 0.01 and 0.02, and a sample covariance of 0.005. For purposes of resampling, consider these to be the "true population values."
A bootstrap resample has sample means of 0.004 and 0.0065, sample variances of 0.012 and 0.023, and a sample covariance of 0.0051.
(a) Using the resample, estimate the efficient portfolio of these two stocks that has an expected return of 0.005; that is, give the two portfolio weights.
(b) What is the estimated variance of the return of the portfolio in part (a) using the resample variances and covariances?
(c) What are the actual expected return and variance of return for the portfolio in (a) when calculated with the true population values (e.g., with using the original sample means, variances, and covariance)?

11

Value-At-Risk

11.1 The Need for Risk Management

The financial world has always been risky, but for a variety of reasons the risks have increased over the last few decades. One reason is an increase in volatility. Equity returns are more volatile, as can be seen in Figure 11.1 where the average absolute value of daily log returns of the S&P 500 has approximately doubled over the period from 1993 to 2003. Foreign exchange rates are more volatile now than before the breakdown in the 1970s of the Bretton Woods agreement of fixed exchange rates.[1] Interest rates rose to new levels in the late 1970s and early 1980s, have risen and fallen several times since then, and are now (in 2003) extremely low. Figure 4.7 shows that interest rate volatility has itself varied over time but has certainly been higher since 1975 than before.

Another reason that risks have increased is the explosion in the size of the derivatives market, which in notational amounts was $50 trillion in 1995. This large-scale trading of derivatives is a rather recent phenomenon. In fact, the Chicago Board of Options Exchange only began to sell call options in 1973. As we saw in Section 8.16, options' returns can be much more volatile than the returns of the underlying stock, so investing in derivatives is potentially much riskier than investing in stocks. Moreover, the risks of derivatives are not easily understood, a problem called **opacity**.

There are a number of different types of risk, but only three are mentioned here. **Market risk** is risk due to changes in prices. **Credit risk** is the risk that a counterparty does not meet contractual obligations, for example, that interest or principal on a bond is not paid. **Liquidity risk** is the risk due to the extra cost of liquidating a position because buyers are difficult to locate.

It is essential that senior management understand the risks that junior managers and traders are assuming for the firm. Rogue traders sometimes

[1] This information and several other facts mentioned in this chapter are from Dowd (1998).

gamble large amounts of the firm's assets. For example, a trader at the Daiwa Bank in Japan lost over \$1.1 billion during an 11-year period and was only discovered when he confessed. Similarly, Nick Leeson caused the Barings Bank to go under due to huge losses from unauthorized trading in derivatives. Orange County, CA went bankrupt in 1994 because of adverse interest-rate changes against a highly leveraged derivatives position assumed by the County's treasurer.

Early attempts to measure risk such as duration analysis, which is discussed in Section 9.8.1 and is used for estimating the market risk of fixed-income securities, were somewhat primitive and of only limited applicability. In contrast, **value-at-risk,** which is usually abbreviated to **VaR**, has become very widely used because it can be applied to all types of risks and all types of securities including complex portfolios.

Riskmetrics, a VaR system, started at J.P. Morgan as an internal system of risk disclosure when the Chairman asked for a daily estimate of potential losses over the next 24-hour period on the bank's entire portfolio. The details of RiskMetrics were made public in 1994 and this led to a huge increase in interest in VaR.

VaR uses two parameters, the horizon and the confidence level, which are denoted by T and $1-\alpha$, respectively. Given these, the VaR is a bound such that the loss over the horizon is less than this bound with probability equal to the confidence coefficient. For example, if the horizon is one week, the confidence coefficient is 99% (so $\alpha = 0.01$), and the VaR is \$5 million, then there is only a 1% chance of a loss exceeding \$5 million over the next week. We sometimes use the notation VaR(α) or Var(α, T) to indicate the dependence of VaR on α or on both α and the horizon T. VaR can be best understood through examples, so we turn to these now.

11.2 VaR with One Asset

To illustrate the techniques for estimating VaR, we begin with the simple case of a single asset. In this section, VaR is estimated using historic data to estimate the distribution of returns, so we make the assumption that returns are stationary, at least over the historic period we use. Two cases are considered, first without and then with the assumption that the returns are normally distributed.

11.2.1 Nonparametric estimation of VaR

We start in Example 11.1 with a *nonparametric* estimate of VaR, meaning that the loss distribution is not assumed to be in a parametric family of distributions such as the normal distributions.

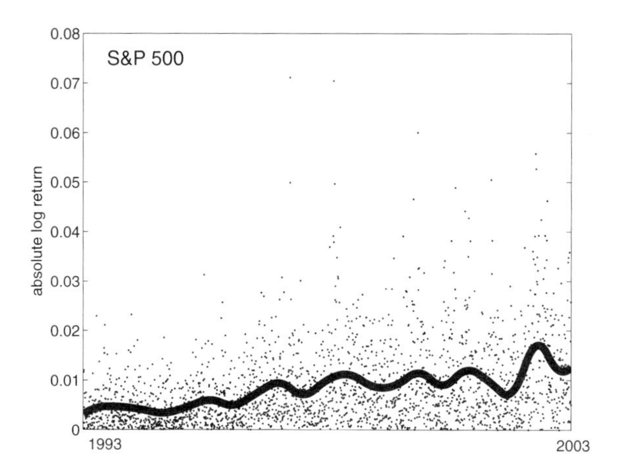

Fig. 11.1. *Absolute daily log returns of S&P 500 from Nov 1, 1993 until Mar 4, 2003 with a scatterplot smooth.*

Example 11.1. As a simple example, suppose that you hold a \$20,000 position in an S&P 500 index fund, so your returns are those of this index, and that you want a 24-hour VaR. We estimate this VaR using the 1000 daily returns on the S&P 500 for the period ending on Mar 4, 2003. These are approximately the last four years of daily returns in Figure 11.1.

Suppose you want 95% confidence. Since 5% of 1000 is 50, the estimate of VaR(0.05) can be the 50th smallest daily return which is -0.0227. In other words, a daily return of -0.0227 or less occurred only 5% of the time in the historic data, so we estimate that there is a 5% chance of a return of that size occurring during the next 24 hours. A return of -0.0227 on a \$20,000 investment yields a revenue of $-\$454.00$, and therefore the estimated VaR, or more precisely VaR(0.05, 24 hour), is \$454.00.[2]

Let's generalize what was done in this example. We want a confidence coefficient of $1 - \alpha$, where $\alpha = 0.05$ in the example. Therefore, we estimate the α-quantile of the return distribution. In the nonparametric method, this quantile is estimated as the α-quantile of a sample of historic returns. The revenue on the investment is the initial investment times the return, so we estimate VaR by the product of the initial investment (\$20,000 in the example) and this sample quantile. More precisely, suppose there are n returns $R_1, \ldots, R_n$ in the historic sample and let K be $(n\alpha)$ rounded to the nearest integer. The α-quantile of the sample is the Kth smallest return, that is, the Kth order statistic $R_{(K)}$ of the sample of returns. If S is the size of the initial investment then

$$\text{VaR}(\alpha) = -S \times R_{(K)},$$

[2] Since VaR is a *loss*, it is minus the revenue.

with the minus sign converting revenue (return times initial investment) to a loss.

Estimation using sample quantiles is only feasible if the sample size is large. If we were interested in a quarterly rather than a daily holding period, then there would have only been 16 observations using the last four years of returns. We could, of course, use more years but this would not increase the sample size substantially and Figure 11.1 shows that volatility has been increasing, so we would bias our estimate if we used more than five or six years of return data.

11.2.2 Parametric estimation of VaR

VaR can best be estimated from a small sample by parametric techniques such as assuming a normal distribution. Moreover, parametric estimation is much easier than nonparametric when there are a large number of assets in the portfolio rather than just one as assumed in this section. Parametric estimation with one asset is illustrated in the next example.

Example 11.2. Although parametric estimation of VaR is best suited for small sample sizes, in this example we use the same large dataset as in Example 11.1 so that parametric and nonparametric estimates can be compared. As mentioned in Section 2.8.2, the α-quantile of an $N(\mu, \sigma^2)$ distribution is $\mu + \Phi^{-1}(\alpha)\sigma$, so $\mathrm{VaR}(\alpha)$ is $-S \times \{\mu + \Phi^{-1}(\alpha)\sigma\}$ and can be estimated by

$$\mathrm{VaR}(\alpha) = -S \times \{\overline{X} + \Phi^{-1}(\alpha)s\}, \tag{11.1}$$

where $\overline{X}$ and s are the mean and standard deviation of a sample of returns and S is the size of the initial investment. For the 1000 daily S&P 500 returns, $\overline{X} = -3.107 \times 10^{-4}$ and $s = 0.0141$. Also, $\Phi^{-1}(0.05) = -1.645$. Therefore, the 0.05-quantile of the probability distribution of returns is estimated as $-3.107 \times 10^{-4} - (1.645)(0.0141) = -0.0236$ and the parametric estimate of $\mathrm{VaR}(0.05)$ is $\$471 = (0.0236) \times (\$20{,}000)$.

Notice that the nonparametric estimate of the VaR(0.05), \$454, is smaller than the parametric estimate, \$471. The reason for the disagreement between the two estimates is that the daily returns appear nonnormal. This can be seen in the normal plot, Figure 11.2. The daily returns are heavier-tailed than a normal distribution, and the outlying returns inflate the estimate of σ making the parametric estimate of VaR(0.05) larger. The nonparametric estimate of VaR is the 50th order statistic, which is not one of the outliers, so the nonparametric estimate is not inflated by the outliers. However, if we make the confidence coefficient smaller which would make K smaller, then the $X_{(K)}$ order statistic might be among the outliers and then the nonparametric estimate of VaR could be larger than the parametric estimate. This happens if $\alpha = 0.01$, in which case the nonparametric estimate, \$679, is slightly larger than the parametric estimate, \$664. If $\alpha = 0.005$, then the nonparametric estimate of VaR(0.005), \$767, is noticeably larger than the parametric estimate, \$734.

In this example, $\overline{X}$, the estimate of the expected return, is negative because the particular years used include a prolonged bear market. However, we certainly don't expect future returns to be, on average, negative, or if we did then we should not have \$20,000 invested in the market. One could consider replacing $\overline{X}$ by some other estimate such as the average daily return over a much longer time period, say the last 20 years or an analyst's forecast of future returns. However, the expected return for a single day is certainly small and the VaR depends mostly on the estimate of σ, so changing the estimate of μ has relatively little effect; see Problem 2.

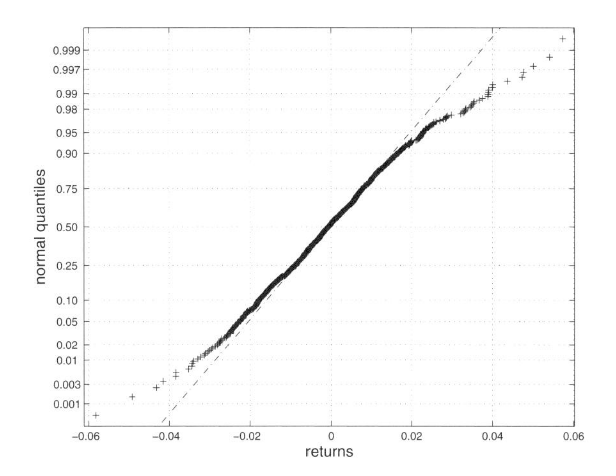

Fig. 11.2. *Normal plot of daily returns of S&P* 500 *for* 1000 *trading days ending on Mar* 4, 2003.

11.2.3 Estimation of VaR assuming Pareto tails*

There is an interesting compromise between using a totally nonparametric estimator of VaR as in Section 11.2.1 and a parametric estimator as in Section 11.2.2. Recall that the nonparametric estimator is feasible for large α, but not for small α. The idea is to assume that the return distribution has a Pareto left tail, or equivalently that the loss distribution has a Pareto right tail. Then it is possible to use a nonparametric estimate of $\text{VaR}(\alpha_0)$ for a *large* value of α_0 to obtain estimates of $\text{VaR}(\alpha_1)$ for *small* values of α_1. It is assumed here that $\text{VaR}(\alpha_1)$ and $\text{VaR}(\alpha_0)$ have the same horizon.

Because the return distribution is assumed to have a Pareto left tail, for $x > 0$,

$$P(R \leq -x) = L(x)x^{-a}, \tag{11.2}$$

where $L(x)$ is slowly varying at infinity and a is the tail index. Therefore, if $x_1 > 0$ and $x_2 > 0$ then

$$\frac{P(R < -x_1)}{P(R < -x_2)} = \frac{L(x_1)}{L(x_2)} \left(\frac{x_1}{x_2}\right)^{-a}. \tag{11.3}$$

Now suppose that $x_1 = \text{VaR}(\alpha_1)$ and $x_1 = \text{VaR}(\alpha_1)$, where $0 < \alpha_1 < \alpha_0$. Then (11.3) becomes

$$\frac{\alpha_1}{\alpha_0} = \frac{P\{R < -\text{VaR}(\alpha_1)\}}{P\{R < -\text{VaR}(\alpha_0)\}} = \frac{L\{\text{VaR}(\alpha_1)\}}{L\{\text{VaR}(\alpha_0)\}} \left(\frac{\text{VaR}(\alpha_1)}{\text{VaR}(\alpha_0)}\right)^{-a}. \tag{11.4}$$

Because L is slowly varying at infinity and $\text{VaR}(\alpha_1)$ and $\text{VaR}(\alpha_0)$ are assumed to be reasonably large, we make the approximation that

$$\frac{L\{\text{VaR}(\alpha_1)\}}{L\{\text{VaR}(\alpha_0)\}} = 1$$

so (11.4) simplifies to

$$\frac{\text{VaR}(\alpha_1)}{\text{VaR}(\alpha_0)} = \left(\frac{\alpha_0}{\alpha_1}\right)^{1/a}$$

so, now dropping the subscript "1" of α_1, we have

$$\text{VaR}(\alpha) = \text{VaR}(\alpha_0) \left(\frac{\alpha_0}{\alpha}\right)^{1/a}. \tag{11.5}$$

Equation (11.5) becomes an estimate of $\text{VaR}(\alpha)$ when $\text{VaR}(\alpha_0)$ is replaced by a nonparametric estimate and the tail index a is replaced by an estimate such as the Hill estimator or an alternative to the Hill estimator described in Example 11.3 below. Notice another advantage of (11.5), that it provides an estimate of $\text{VaR}(\alpha)$ not just for a single value of α but for all values. This is useful if one wants to compute and compare $\text{VaR}(\alpha)$ for a variety of values of α, as is illustrated in Example 11.3 below.

Estimator (11.5) is called a *semiparametric estimator* because the tail index is specified by a parameter but the slowly varying function $L(x)$ is not assumed to be in a parametric family and so is modeled nonparametrically. A model combining parametric and nonparametric components is called semiparametric.

11.2.4 Estimating the tail index*

It follows from (11.2) that

$$\log\{P(R \leq -x)\} = \log\{L(x)\} - a\log(x). \tag{11.6}$$

Since $L(x)$ is assumed to be slowly varying at infinity, we make the assumption that for large values of x, $L(x)$ is a constant L. If $R_{(1)}, \ldots, R_{(n)}$ are the order statistics of the returns, then the number of observed returns less than or equal to $R_{(k)}$ is k, so we estimate $\log\{P(R \leq R_{(k)})\}$ to be $\log(k/n)$. Then from (11.6) we have

$$\log(k/n) \approx \log(L) - a\log(-R_{(k)}). \tag{11.7}$$

The approximation (11.7) depends on the accuracy of the approximation that $L(x)$ is constant and so (11.7) is expected to be accurate only if $-R_{(k)}$ is large, which means k is small, perhaps only 5%, 10%, or perhaps 20% of the sample size n. If we plot the points $\left[\{\log(k/n), \log(-R_{(k)})\}\right]_{k=1}^{m}$ for m a small percentage of n, say 10%, then we should see these points fall on roughly a straight line. Moreover, if we fit a straight line to these points by least squares then minus the slope estimates the tail index a. We call this estimator the *regression estimator of the tail index*. The intercept estimates $\log(L)$, though this parameter is not often of interest.

Example 11.3. This example uses the 1000 daily S&P 500 returns in Figure 11.2. These returns were sorted and the $m = 100$ smallest returns (100 largest losses) were used for plotting and estimation of the tail index. This value of m was selected by plotting $\left[\{\log(k/n), \log(-R_{(k)})\}\right]_{k=1}^{m}$ for various values of m and choosing the largest value of m giving a roughly linear plot. A least squares line was fit to these 100 points and R^2 was 0.99, indicating a good fit to a straight line. The slope of the line was -3.82 so a was estimated to be 3.82. The plotted points and the least squares line can be seen in Figure 11.3.

Suppose we have invested \$10,000 in an S&P 500 index fund. Then VaR(0.1, 24 hours) is estimated to be \$10,000 times minus the 100th of the 1000 returns when the returns are sorted from smallest to largest. This return was -0.0178 so VaR(0.1, 24 hours) is estimated to be \$178. Using (11.5) and $a = 3.82$ or $1/a = 0.262$ we have

$$\mathrm{VaR}(\alpha) = 178\left(\frac{0.1}{\alpha}\right)^{0.262}. \tag{11.8}$$

The solid curve in Figure 11.4 is a plot of $\mathrm{VaR}(\alpha)$ for $0.005 \leq \alpha \leq 0.1$ using (11.8). The $\mathrm{VaR}(\alpha)$ values for α equal to 0.005, 0.01, 0.025, and 0.05 are shown on the graph.

As mentioned, the value of m was chosen to be 100 because the plot looked roughly linear for this value. The plots for larger m, for example, $m = 150$, were rather nonlinear. However, the plot for $m = 50$ was even closer to linear with $R^2 = 0.997$. As a sensitivity analysis, the estimation of the tail index and of $\mathrm{VaR}(\alpha)$ was repeated using $m = 50$. In this case $1/\widehat{a} = 0.238$ rather than 0.262 as before. The estimate of VaR(0.05) changed from 325.7 when $m = 100$ to 300.9 when $m = 50$. In a more extreme case, VaR(0.005) changed from 390.2 to 352.4. Unfortunately, this amount of uncertainty in the estimate of VaR seems inevitable. I prefer using $m = 100$ because the plot in Figure 11.3 is rather linear and using more data increases the precision of the estimate.

The parametric estimates of VaR(α), which are also shown in Figure 11.4, are similar to the estimates assuming a Pareto tail when α is between 0.005 and 0.01, but as can be seen on the extreme left of the figure, the estimates assuming a Pareto tail become much larger than the parametric estimates as α gets very small. This is to be expected because the Pareto tail is heavier than that of the normal distribution which is used for the parametric estimates.

The Hill estimate of a was also implemented. Figure 11.5 is a Hill plot, that is, a plot of the Hill estimate $\widehat{a}$ versus $n(c)$, the number of extreme negative returns used to estimate a. In this context, the Hill estimator given by (2.63) is

$$\widehat{a}_{\text{Hill}}(c) = \frac{n(c)}{\sum_{-R_i \geq c} \log(-R_i/c)}, \tag{11.9}$$

where $n(c)$ is the number of $-R_i$ that exceed c. In the Hill plot, 50 equally spaced values of c between 0.01 and 0.03 were used and $n(c)$ ranged from 226 for $c = 0.01$ to 16 for $c = 226$. The Hill plot is shown in Figure 11.5 and, unfortunately, does *not* show the hoped-for region of stability where $\widehat{a}$ is nearly constant as a function of $n(c)$. The Hill estimates vary from 5.08 to 1.81 as c varies over its range. Thus, the Hill estimates are similar to the regression estimates of the tail index. The advantage of the regression estimate is that one can use the linearity of the plots of $\{(\log(k/n), -R_{(k)})\}_{k=1}^{m}$ for different m to guide the choice of m.

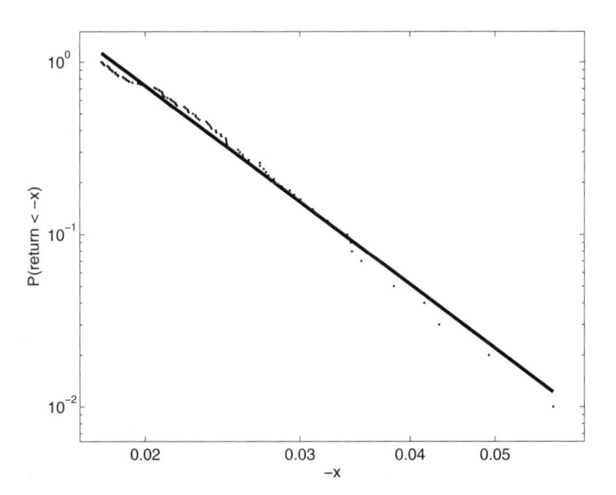

Fig. 11.3. *Estimation of tail index using* 100 *smallest returns* (100 *largest losses) among the daily returns of S&P* 500 *for* 1000 *consecutive trading days ending on Mar* 4, 2003. *Dots:* $-R_{(k)}$ *is plotted on the x-axis, and* k/n *on the y-axis, both with log scales. Solid line: least-squares fit of* $\log(k/n)$ *(dependent of Y-variable) regressed on* $\log(-R_{(k)})$ *(independent or X-variable).*

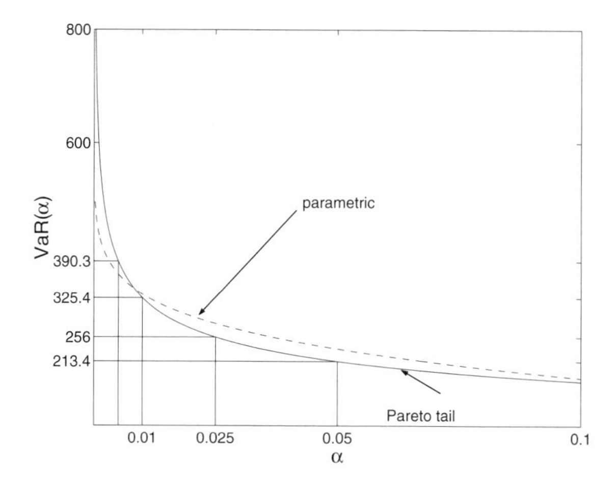

Fig. 11.4. *Estimation of VaR(α) using formula* (11.8) *(solid) and using the parametric estimator (dashed).*

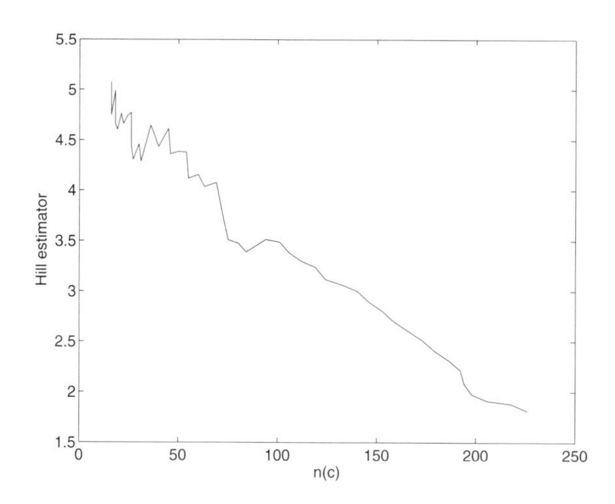

Fig. 11.5. *Estimation of tail index by applying a Hill plot to the daily returns of S&P* 500 *for* 1000 *consecutive trading days ending on Mar* 4, 2003.

11.2.5 Confidence intervals for VaR using resampling

The estimate of VaR is precisely that, just an estimate. If we had used a different sample of historic data, then we would have gotten a different estimate of VaR. We might want to know the precision of this estimate. Fortunately, a confidence interval for VaR is rather easily obtained by resampling. The idea is to take a large number, B, of resamples of the returns data. Then a VaR estimate is computed from each resample and for the original sample. The confidence interval can be based upon either a parametric or nonparametric estimator of VaR to give either a parametric or nonparametric confidence interval. The B bootstrap estimates of VaR are ordered from smallest to largest. Suppose that we want the confidence coefficient of the interval to be $1 - \gamma$. This confidence coefficient should not be confused with the confidence coefficient of VaR, which we denote by $1 - \alpha$. Let J equal $B\gamma/2$ rounded to the nearest integer. Then the Jth smallest and Jth largest bootstrap estimates of VaR are the limits of the confidence interval.

It is worthwhile to restate the meanings of α and γ, since it is easy to confuse these two confidence coefficients, but they must be distinguished since they have rather different interpretations. VaR is defined so that the probability of a loss being greater than VaR is α. On the other hand, γ is the confidence coefficient for the confidence interval for VaR. If many confidence intervals are constructed, then approximately γ of them do not contain the VaR. Thus, α is about the loss from the investment while γ is about the confidence interval being correct.

Example 11.4. In this example, we continue Example 11.1 and find a confidence interval for VaR. We use $\alpha = 0.05$ as before. Suppose we want $\gamma = 0.02$. $B = 50,000$ resamples were taken. Therefore, $J = (50,000)(.02)/2 = 500$. The 500th smallest bootstrap nonparametric VaR estimate was \$412 and the 500th largest was \$495 so that the nonparametric confidence interval for VaR is \$412 to \$495. The parametric confidence interval is \$433 to \$510. Notice that the two confidence intervals have sizeable overlap, so the difference between the parametric and nonparametric estimates might be due to sampling variation.

A rather large value of B, 50,000, was used in this example. In general, B should be large when the purpose of resampling is to create a confidence interval. However, $B = 10,000$ is generally large enough for setting confidence intervals unless γ is very small. One can determine whether the value of B in use is large enough by repeating the bootstrap several times. This gives several confidence intervals. The same original sample is used in all of the bootstraps, so only resampling error can account for variation among the confidence intervals. If the variation in the endpoints of the intervals is small relative to the width of the intervals, then B is large enough. In our example, this is true for $B = 10,000$ and $B = 50,000$ gives us even more precision at the cost of more computation time. However, using $B = 50,000$ takes only 33 seconds to compute both the parametric and nonparametric confidence intervals on a 2.2 GHz PC using MATLAB, so excess computation time is not a problem.

11.2.6 VaR for a derivative

Suppose that instead of a stock, one owns a derivative whose value depends on the stock. One can estimate a VaR for this derivative by first estimating a VaR for the asset and then using (8.31). The result is that

$$\text{VaR for derivative} = L \times \text{VaR for asset}, \tag{11.10}$$

where L is defined by equation (8.30). In (11.10), the holding period, S, and confidence coefficient for the derivative are the same as for the asset. The beauty of equation (11.10) is that it allows one to estimate VaR for a derivative using only historic data for the underlying asset without needing data on the derivative itself. That latter would be difficult if not impossible to find since one would need past data on derivatives with exactly the same maturity, exercise price, and other parameters as the derivative for which an estimate of VaR is sought.

11.3 VaR for a Portfolio of Assets

When VaR is estimated for a portfolio of assets rather than a single asset, parametric estimation based on the assumption of normally distributed returns is very convenient and this is the approach taken here. VaR has much in common with the portfolio theory developed in Chapters 5 and 7, but one major difference is that portfolio theory uses the standard deviation of the return distribution to measure risk while VaR uses a quantile of that distribution. Also, VaR is the product of that quantile and the initial investment so VaR expresses absolute monetary loss rather than the relative loss expressed by a return.

Estimating VaR becomes complex when the portfolio contains stocks, bonds, options, foreign exchange positions, and other assets. However, when a portfolio contains only stocks, then VaR is relatively straightforward to estimate, so we turn our attention to that case first. Then we look at two simple cases of a portfolio with stocks and other types of assets, (1) a portfolio of a stock plus an option on that stock and (2) a portfolio of a stock plus an option on a different stock. Although these two cases are too simple to be realistic, they illustrate general principles that can be readily applied to more complex portfolios.

11.3.1 Portfolios of stocks only

With a portfolio of only stocks, means, variances, and covariances of returns could be estimated directly from a sample of returns as discussed in Section 5.3.2, or using CAPM or another factor model as discussed in Sections 7.7 and 7.8.1. Once these estimates are available, they can be plugged into equations

(5.5) and (5.6) to obtain estimates of the expected value and variance of the return on the portfolio, which are denoted by $\widehat{\mu}_P$ and $\widehat{\sigma}_P^2$. Then, analogous to (11.1), VaR can be estimated assuming normally distributed returns on the portfolio by

$$\mathrm{VaR}_P = -\{\widehat{\mu}_P + \Phi^{-1}(\alpha)\widehat{\sigma}_P\}S, \tag{11.11}$$

where S is the size of the initial investment, i.e., the initial value of the entire portfolio.

11.3.2 Portfolios of one stock and an option on that stock

Suppose a portfolio contains a stock and an option on that stock with weights w and $1-w$. Let R^{stock} be the return on the stock. By (8.31), the return on the portfolio is approximately

$$wR + (1-w)LR = \{w + (1-w)L\}R,$$

where L is the leverage of the option, so that the expected return on the portfolio is $\{w+(1-w)L\}\mu^{\text{stock}}$ and the standard deviation of the return on the portfolio is $\{w+(1-w)L\}\sigma^{\text{stock}}$, where μ^{stock} and σ^{stock} are the expected value and standard deviation of the return on the stock. Therefore, it is easy to check that the VaR of the portfolio is $\{w+(1-w)L\}$ times the VaR of the same initial investment made in the stock only; i.e.,

$$\mathrm{VaR}^{\text{portfolio}} = \{w + (1-w)L\} \times \mathrm{VaR}^{\text{stock}}.$$

A special case is a portfolio of the option only, so that $w = 0$ and then

$$\mathrm{VaR}^{\text{option}} = L \times \mathrm{VaR}^{\text{stock}}.$$

11.3.3 Portfolios of one stock and an option on another stock

Let μ_1^{stock} and μ_2^{stock} be the expected returns on two stocks and let σ_1^{stock} and σ_2^{stock} be the standard deviations of the returns. Finally, let $\sigma_{1,2}^{\text{stock}}$ be the covariance of these returns. Suppose a portfolio contains the first stock plus an option on the second stock with weights w and $1-w$. Let L_2 be the leverage defined by (8.30) for this option. Then the expected return on the portfolio is

$$\mu_P = w\mu_1^{\text{stock}} + (1-w)L_2\mu_2^{\text{stock}}$$

and using (5.3) and (8.33) the standard deviation of the return is

$$\sigma_P = \left\{w^2(\sigma_1^{\text{stock}})^2 + (1-w)^2L_2^2(\sigma_2^{\text{stock}})^2 + 2w(1-w)L_2\sigma_{1,2}^{\text{stock}}\right\}^{1/2}. \tag{11.12}$$

Let $\widehat{\mu}_P$ and $\widehat{\sigma}_P$ be estimates of the portfolio expected return and standard deviation of returns obtained by plugging in estimates of μ_1^{stock}, μ_2^{stock}, σ_1^{stock}, σ_2^{stock}, and $\sigma_{1,2}^{\text{stock}}$ into these two formulas. Notice that one only needs returns on the underlying stocks, not on the option. Then, as before, the VaR of the portfolio is

$$\mathrm{VaR}_P = -\{\widehat{\mu}_P + \Phi^{-1}(\alpha)\widehat{\sigma}_P\}S.$$

11.4 Choosing the Holding Period and Confidence

The choice of holding period and confidence coefficient are somewhat interdependent and depend on the eventual use of the VaR estimate. For shorter holding periods such as one day, a large α (small confidence coefficient) would result in frequent losses exceeding VaR. For example, $\alpha = 0.05$ would result in a loss exceeding VaR approximately once per month since there are approximately 20 trading days in a month.

There is, of course, no need to restrict attention to only one holding period or confidence coefficient. When VaR is estimated parametrically by (11.11) it is easy to re-estimate VaR with a different holding period or confidence coefficient. The latter only requires a change of α. Changing the holding period is nearly as easy. Suppose that $\widehat{\mu}_P^{\text{1day}}$ and $\widehat{\sigma}_P^{\text{1day}}$ are the estimated mean and standard deviation of the return for one day. Assuming the random walk hypothesis, returns are independent from day to day so that the mean and standard deviation for M days are

$$\widehat{\mu}_P^{\text{M days}} = M\widehat{\mu}_P^{\text{1 day}} \tag{11.13}$$

and

$$\widehat{\sigma}_P^{\text{M days}} = \sqrt{M}\widehat{\sigma}_P^{\text{1 day}}. \tag{11.14}$$

Therefore, the VaR for M days is

$$\text{VaR}_P^{\text{M days}} = -\left\{M\widehat{\mu}_P^{\text{1 day}} + \sqrt{M}\Phi^{-1}(\alpha)\widehat{\sigma}_P^{\text{1 day}}\right\} S, \tag{11.15}$$

where S is the size of the initial investment. The power of equation (11.15) is, for example, that it allows one to change from a daily to a weekly holding period without re-estimating the mean and standard deviation with weekly instead of daily returns. Instead, one simply uses (11.15) with $M = 5$. The danger in using (11.15) is that it assumes no autocorrelation of the daily returns. If there is positive autocorrelation, then (11.15) underestimates the M-day VaR.

11.5 VaR and Risk Management

We have discussed the measurement of risk but have said little about its management. In fact, risk management is too large a topic to cover in this book, but a discussion about the appropriateness of VaR for risk management may be in order here. Any single measure of risk has limitations and risk management should never be based solely on VaR. To indicate one serious problem with VaR, we start with some definitions. A *tail event* occurs if the loss exceeds the VaR. Another measure of risk is the conditional expectation of the loss given that a tail event occurs, which is called the *expected loss given a tail event*, **expected shortfall**, or simply the *tail loss* or *shortfall*. The abbreviation *ES* (expected shortfall) is sometimes used.

Dowd (1998) cautions against focusing on VaR and calls tail loss a "useful supplementary statistic." Basak and Shapiro (2001) go much farther than this. They prove that within their mathematical model risk management focusing solely on VaR can lead to a much higher expected loss than is necessary, because VaR ignores the magnitude of extreme losses provided these happen with low enough probabilities. Rather than risk management based on VaR, Basak and Shapiro recommend "limited-expected-losses-based risk management" (LEL-RM) which is a strategy to constrain both the probability of a tail event and the expected loss given a tail event.

A serious problem with VaR is that it may *discourage* diversification. Artzner, Delbaen, Eber, and Heath (1997, 1999) ask the question, what properties can be reasonably required of a risk measure? They list four properties that any risk measure should have, and they call a risk measure *coherent* if it has all four. One property that is very desirable is *subadditivity*. Let $R(P)$ be a measurable of risk of a portfolio P, for example, VaR or ES. Then R is said to be subadditive if for any two portfolios P_1 and P_2, $R(P_1+P_2) \leq R(P_1)+R(P_2)$. Subadditivity says that the risk for the combination of two portfolios is at most the sum of their individual risks, which means that diversification reduces risk or at least does not increase risk. For example, if a bank has two traders, then the risk of them combined is less than or equal to the sum of their individual risks, at least if risk is subadditive. Subadditivity also implies that for m portfolios, $P_1, \ldots, P_m$,

$$R(P_1 + \cdots + P_m) \leq R(P_1) + \cdots + R(P_m).$$

Suppose a firm has 100 traders and monitors the risk of each trader's portfolio. If the firm uses a subadditive risk measure, then it can be sure that the total risk of the 100 traders is at most the sum of the 100 individual risks. If the risk measure uses by the firm is not subadditive, then there is no such guarantee.

Unfortunately, VaR is *not* subadditive and therefore is incoherent. To appreciate this fact, consider the following example. Two corporations each have a 4% chance of going bankrupt and whether or not one corporation goes bankrupt is independent of whether the other goes bankrupt. Each corporation has outstanding bonds. If one purchases a bond from one of them, then the return is 0 if the corporation does not go bankrupt and -1 if it does go bankrupt.[3] Suppose one buys \$1000 worth of bonds of corporation #1, which is called portfolio P_1. Let R be VaR(0.05). Then $R(P_1) = 0$ since there is a 0.96 chance of no loss. If P_2 is \$1000 worth of bonds of corporation #2, then $R(P_2) = 0$ for the same reason. However, $P_1 + P_2$ is a portfolio of \$1000 worth of bonds in each of the two corporations. Using the binomial distribution with $n = 2$ and $p = 0.96$, the probability that neither goes

[3] This example is oversimplified to emphasize a point. Risks due to changes in interest rates are being ignored as is the positive return due to interest.

bankrupt is $(0.96)^2 = 0.9216$, the probability that exactly one goes bankrupt is $(2)(0.96)(0.04) = 0.0768$, and the probability that both go bankrupt is $(0.04)^2 = 0.0016$. In this example, the loss is \$2000 if there are two bankruptcies, \$1000 if there is exactly one bankruptcy, and \$0 if there are no bankruptcies among the two corporations. It follows that $R(P_1+P_2) = 1000$ since there is more than 5% change of one bankruptcy but less than 5% chance of two. Thus, $R(P_1 + P_2) = 1000 > R(P_1) + R(P_2) = 0 + 0 = 0$, which shows that VaR is not subadditive. VaR encourages putting one's entire portfolio into the bonds of a single corporation because the probability of a single corporation going bankrupt is small (0.04) but the probability of at least one of two corporations going bankrupt is nearly twice as large ($0.0768 + 0.0016 = 0.0784$). Although VaR is often considered to be the industry standard for risk management, Artzner, Delbaen, Eber, and Heath (1997) make an interesting observation. They note that when setting margin requirements, an exchange should use a subadditive risk measure so that the aggregate risk due to all customers is guaranteed to be smaller than the sum of the individual risks. Apparently no organized exchanges use quantiles of loss distributions to set margin requirements. Thus, exchanges do seem aware of the shortcomings of VaR, and VaR is not the standard for measuring risk within exchanges.

Tail loss is subadditive, which is a strong reason for preferring tail loss to VaR. In the example, if R is tail loss rather than VaR and if we still use $\alpha = 0.05$, then $R(P_1) = 1000(0.04) = 40$ and $R(P_2) = 40$ while $R(P_1 + P_2) = 1000(0.0768) + 2000(0.0016) = 76.8 + 3.2 = 80$ so that $R(P_1 + P_2) = R(P_1) + R(P_2)$.

Unfortunately, estimation of tail loss is a problem that has received relatively little attention, at least compared to estimation of VaR, though research in this area is beginning.[4] Tail loss is very sensitive to the probabilities of extreme losses and these probabilities are notoriously difficult to estimate, more difficult certainly than VaR. Whether a strategy like Basak and Shapiro's LEL-RM is something that could be implemented in practice is an open question. The current standard practice in industry is to focus on VaR. An alternative to looking at ES would be to examine VaR(α) for several value of α including some very small values so that the probabilities of extreme losses are taken into account.

11.6 Summary

Investing has always been risky but is perhaps riskier now than in the past. Senior management needs a method to measure the risks that traders in their company are assuming. VaR has become a standard tool for risk measurement and management. A VaR depends upon two parameters that one is free to choose, the holding period and the confidence coefficient. The VaR is such

[4] See Scaillet (2003).

that the loss during the holding period is equal to or less than the VaR with probability equal to the confidence coefficient. One should not focus exclusively on a single risk measure and VaR should be supplemented by other risk measures such as the expected loss given a tail event.

11.7 Bibliographic Notes

Value-at-risk is an enormous subject and we have only touched upon a few aspects. We have not considered portfolios with bonds, foreign exchange positions, interest rate derivatives, or credit derivatives. We also have not considered risks other than market risk or how VaR can be used for risk management. To cover VaR thoroughly requires at least a book-length treatment of that subject. Fortunately, excellent books about VaR exist, for example, Dowd (1998) and Jorion (2001) which are standard references on VaR. I first learned the theory of VaR from reading Dowd (1998). Alexander (2001), Hull (2003), and Gourieroux and Jasiak (2001) have chapters on VaR and risk management. The semiparametric method of estimation based on the assumption of a Pareto tail and equation (11.5) are from Gourieroux and Jasiak (2001).

11.8 References

Alexander, C. (2001) *Market Models: A Guide to Financial Data Analysis*, Wiley, Chichester.

Artzner, P., Delbaen, F., Eber, J-M., and Heath, D. (1997) Thinking coherently, *RISK*, 10, 68–71.

Artzner, P., Delbaen, F., Eber, J-M., and Heath, D. (1999) Coherent measures of risk, *Mathematical Finance*, 9, 203–238.

Basak, S. and Shapiro, A. (2001) Value-at-risk-based management: Optimal policies and asset prices, *The Review of Financial Studies*, **14**, 371–405.

Dowd, K. (1998) *Beyond Value At Risk*, Wiley, Chichester.

Gourieroux, C. and Jasiak, J. (2001) *Financial Econometrics*, Princeton University Press, Princeton, NJ.

Hull, J. C. (2003) *Options, Futures, and Other Derivatives, 5th Ed.*, Prentice-Hall, Upper Saddle River, NJ.

Jorion, P. (2001) *Value At Risk*, McGraw-Hill, New York.

Scaillet, O. (2003) Nonparametric estimation of conditional expected shortfall, manuscript.

11.9 Problems

1. Verify equation (11.12).

2. Repeat the VaR estimation in Example 11.2 but with a positive estimate of μ. Choose an estimate of μ that you feel is reasonable and explain your choice. Compare your estimate with the one in the problem. Do you think there is a significant difference between the two estimates?
3. Assume that the loss distribution has a Pareto tail and an estimate of a is 4.11. If $\text{VaR}(0.05) = \$211$, what is $\text{VaR}(0.005)$?
4. Find a source of stock price data on the Internet and obtain daily prices for a stock of your choice over the last 1000 days.
   (a) Assuming that the loss distribution is normal, find the parametric estimate of $\text{VaR}(0.025, 24 \text{ hours})$.
   (b) Find the nonparametric estimate of $\text{VaR}(0.025, 24 \text{ hours})$.
   (c) Use a normal plot to decide if the normality assumption is reasonable.
   (d) Estimate the tail index assuming a Pareto tail and then use the estimate of $\text{VaR}(0.025, 24 \text{ hours})$ from part (a) to estimate $\text{VaR}(0.0025, 24 \text{ hours})$.
5. Suppose Stock A has a daily return distribution that is normal with mean 0.0001 and standard deviation 0.017. Currently the price of Stock A is \$115 and the annual risk-free rate is 0.02.
   (a) What is the price of a call option on Stock A with an expiration of 90 days and a strike price of \$113?
   (b) What is Δ for this option?
   (c) What is L for this option?
   (d) What is $\text{VaR}(0.05, 24 \text{ hours})$ for a portfolio holding \$1000 worth of this option and nothing else?
   (e) What is $\text{VaR}(0.05, 24 \text{ hours})$ for a portfolio holding \$1000 worth of this option and \$1000 worth of Stock A?
   (f) Daily returns of Stock B have a mean of 0.0002, a standard deviation of 0.019, and a correlation of 0.3 with daily returns on Stock A. What is $\text{VaR}(0.05, 24 \text{ hours})$ for a portfolio holding \$1000 worth of the call option on Stock A and \$1000 worth of Stock B?
   (g) What is $\text{VaR}(0.05, 24 \text{ hours})$ for a portfolio holding \$1000 worth of the call option on Stock A, \$1000 worth of Stock A, and \$1000 worth of Stock B?
6. Suppose the risk measure R is $\text{VaR}(\alpha)$ for some α. Let P_1 and P_2 be two portfolios whose returns have a joint normal distribution with means μ_1 and μ_2, standard deviations σ_1 and σ_2, and correlation ρ. Suppose the initial investments are S_1 and S_2. Show that $R(P_1+P_2) \leq R(P_1)+R(P_2)$.[5]
7. Suppose that you have sold a call for 100 shares of company A with a strike price of 108 and bought a call for 100 shares of the same company with a strike price of 112. Both calls have the same maturity T.

[5] This result shows that VaR is subadditive on a set of portfolios that have a joint normal distribution, as might be true for portfolios containing only stocks. However, portfolios containing derivatives or bonds with non-zero probabilities of default generally do not have normally distributed returns.

(a) What is the loss on this portfolio as a function of the stock's price S_T at time T?
(b) Suppose that you know the distribution of S_T and have simulated a sample of size 20,000 from this distribution. How would you convert this sample into estimates of VaR(0.05) and ES(0.05)?[6]
(c) Suppose that the current price of the stock is $S_0 = 105$ and $\log(S_T/S_0)$ is $N(0.1, 5^2)$. Simulate 20,000 values from the distribution of S_T and estimate VaR(0.05) and ES(0.05).

[6] ES(0.05) is the expected shortfall using $\alpha = 0.05$.

12

GARCH Models

12.1 Introduction

Figure 12.1 illustrates how volatility can vary dramatically over time in financial markets. This figure is a semilog plot of the absolute values of weekly changes in AAA bond interest rates. Larger absolute changes occur in periods of higher volatility. In fact, the expected absolute change is proportional to the standard deviation. Because many changes were zero, 0.005% was added so that all data could plot on the log scale. A spline was added to show changes in volatility more clearly. The volatility varies by an order of magnitude over time; e.g., the spline (without the 0.005% added) varies between 0.017% and 0.20%. Accurate modeling of time-varying volatility is of utmost importance in financial engineering. The ARMA time series models studied in Chapter 4 are unsatisfactory for modeling volatility changes and other models are needed when volatility is not constant.

Despite the popularity of ARMA models, they have a significant limitation — they assume a constant volatility. In finance, this can be a severe handicap. In this chapter we look at GARCH time series models that are becoming widely used in econometrics and finance because they have randomly varying volatility.

ARMA models are used to model the conditional expectation of the current observation Y_t, given the past observations. ARMA models do this by writing Y_t as a linear function of the past plus a white noise term. ARMA models also allow us to predict future observations given the past and present. The prediction of Y_{t+1} given $Y_t, Y_{t-1} \ldots$ is simply the conditional expectation of Y_{t+1} given $Y_t, Y_{t-1} \ldots$.

However, ARMA models have rather boring conditional variances — the conditional variance of Y_t given the past is always a constant. What does this mean for, say, modeling stock returns? Suppose we have noticed that recent daily returns have been unusually volatile. We might suppose that tomorrow's return is also more variable than usual. However, if we are modeling returns as an ARMA process, we cannot capture this type of behavior because the conditional variance is constant. So we need better time series models if we want to model the nonconstant volatility often seen in financial time series.

ARCH is an acronym meaning AutoRegressive Conditional Heteroscedasticity.[1] In ARCH models the conditional variance has a structure very similar to the structure of the conditional expectation in an AR model. We first study the ARCH(1) model, which is the simplest GARCH model and similar to an AR(1) model. Then we look at ARCH(p) models that are analogous to AR(p) models. Finally, we look at GARCH (Generalized ARCH) models that model conditional variances much like the conditional expectation of an ARMA model.

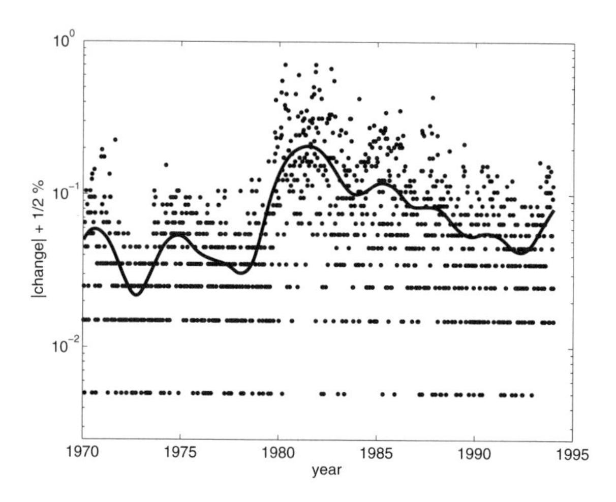

Fig. 12.1. *Absolute weekly changes in AAA bond rates with* 0.005% *added so that a log scale could be used on the vertical axis.*

12.2 Modeling Conditional Means and Variances

Before looking at GARCH models, we study some general principles about modeling nonconstant conditional variance.

Consider modeling with a *constant* conditional variance, $\mathrm{Var}(Y_t|X_{1,t}, \ldots, X_{p,t}) = \sigma^2$. Then the general form for the regression of Y_t on $X_{1.t}, \ldots, X_{p,t}$ is

$$Y_t = f(X_{1,t}, \ldots, X_{p,t}) + \epsilon_t, \tag{12.1}$$

where ϵ_t has expectation equal to 0 and a constant variance σ^2. The function f is the conditional expectation of Y_t given $X_{1,t}, \ldots, X_{p,t}$. To appreciate this

[1] Recall that heteroscedasticity is just a fancy way of saying nonconstant variance.

fact, notice that if we take the conditional (given the $X_{i,t}$ values) expectation of (12.1), $f(X_{1,t}, \ldots, X_{p,t})$ is treated as a constant and the conditional expectation of ϵ_t is 0. Moreover, the conditional variance is simply the variance of ϵ_t, that is, σ^2. Frequently, f is linear so that

$$f(X_{1,t}, \ldots, X_{p,t}) = \beta_0 + \beta_1 X_{1,t} + \cdots + \beta_p X_{p,t}.$$

Principle: To model the conditional mean of Y_t given $X_{1.t}, \ldots, X_{p,t}$, write Y_t as the conditional mean *plus* white noise.

Equation (12.1) can be modified to allow a nonconstant conditional variance, that is, conditional heteroscedasticity. Let $\sigma^2(X_{1,t}, \ldots, X_{p,t})$ be the conditional variance of Y_t given $X_{1,t}, \ldots, X_{p,t}$. Then the model

$$Y_t = f(X_{1,t}, \ldots, X_{p,t}) + \sigma(X_{1,t}, \ldots, X_{p,t})\, \epsilon_t \tag{12.2}$$

gives the correct conditional mean and variance.

Principle: To allow a nonconstant conditional variance in the model, *multiply* the white noise term by the conditional standard deviation. This product is added to the conditional mean as in the previous principle.

The function $\sigma(X_{1,t}, \ldots, X_{p,t})$ should be nonnegative since it is intended to be a standard deviation.[2] If the function $\sigma(\cdot)$ is linear, then its coefficients must be constrained to ensure nonnegativity, so nonlinear functions are usually used. Modeling nonconstant conditional variances in regression is treated in depth in the book by Carroll and Ruppert (1988). Models for conditional variances are often called "variance function models." The GARCH models of this chapter are a special class of variance function models.

12.3 ARCH(1) Processes

Suppose $\epsilon_1, \epsilon_2, \ldots$ is Gaussian white noise with unit variance; that is, let this process be independent N(0,1). Then

$$E(\epsilon_t | \epsilon_{t-1}, \ldots) = 0,$$

and

$$\text{Var}(\epsilon_t | \epsilon_{t-1}, \ldots) = 1. \tag{12.3}$$

Property (12.3) is called **conditional homoscedasticity**.[3]

The process a_t is an ARCH(1) process if

[2] If $\sigma(X_{1,t}, \ldots, X_{p,t})$ were negative then $|\sigma(X_{1,t}, \ldots, X_{p,t})|$ would be the conditional standard deviation of Y given $X_{1,t}, \ldots, X_{p,t}$.

[3] Recall that homoscedasticity means constant variance.

$$a_t = \epsilon_t \sqrt{\alpha_0 + \alpha_1 a_{t-1}^2}. \qquad (12.4)$$

We require that $\alpha_0 \geq 0$ and $\alpha_1 \geq 0$ because a standard deviation cannot be negative. It is also required that $\alpha_1 < 1$ in order for a_t to be stationary with a finite variance. If $\alpha_1 = 1$ then a_t is stationary but its variance is ∞; see Section 12.10 below. Equation (12.4) can be written as

$$a_t^2 = (\alpha_0 + \alpha_1 a_{t-1}^2)\,\epsilon_t^2,$$

which is very much like an AR(1) but in a_t^2, not a_t and with multiplicative noise with a mean of 1 rather than additive noise with a mean of 0. In fact, the ARCH(1) model induces an ACF for a_t^2 that is the same as an AR(1)'s ACF.

Define

$$\sigma_t^2 = \text{Var}(a_t | a_{t-1}, \ldots)$$

to be the conditional variance of a_t given past values. Since ϵ_t is independent of a_{t-1} and $E(\epsilon_t^2) = \text{Var}(\epsilon_t) = 1$,

$$E(a_t | a_{t-1}, \ldots) = 0, \qquad (12.5)$$

and

$$\begin{aligned} \sigma_t^2 &= E\{(\alpha_0 + \alpha_1 a_{t-1}^2)\,\epsilon_t^2 | a_{t-1}, a_{t-2}, \ldots\} \\ &= (\alpha_0 + \alpha_1 a_{t-1}^2) E\{\epsilon_t^2 | a_{t-1}, a_{t-2}, \ldots\} \\ &= \alpha_0 + \alpha_1 a_{t-1}^2. \end{aligned} \qquad (12.6)$$

Understanding equation (12.6) is crucial to understanding how GARCH processes work. This equation shows that if a_{t-1} has an unusually large deviation from its expectation of 0 so that a_{t-1}^2 is large, then the conditional variance of a_t is larger than usual. Therefore, a_t is also expected to have an unusually large deviation from its mean of 0. This volatility propagates since a_t having a large deviation makes σ_{t+1}^2 large so that a_{t+1} tends to be large and so on. Similarly, if a_{t-1}^2 is unusually small, then σ_t^2 is small, and a_t^2 is also expected to be small, and so forth. Because of this behavior, unusual volatility in a_t tends to persist, though not forever. The conditional variance tends to revert to the unconditional variance provided that $\alpha_1 < 1$ so that the process is stationary with a finite variance.

The unconditional, i.e., marginal, variance of a_t denoted by $\gamma_a(0)$ is gotten by taking expectations in (12.6) which give us

$$\gamma_a(0) = \alpha_0 + \alpha_1 \gamma_a(0).$$

This equation has a positive solution if $\alpha_1 < 1$:

$$\gamma_a(0) = \alpha_0/(1 - \alpha_1).$$

If $\alpha_1 \geq 1$ then $\gamma_a(0)$ is infinite. It turns out that a_t is stationary nonetheless. The integrated GARCH model (I-GARCH) has $\alpha_1 = 1$ and is discussed in Section 12.10.

Straightforward calculations using (12.6) show that the ACF of a_t is

$$\rho_a(h) = 0 \text{ if } h \neq 0.$$

In fact, any process such that the conditional expectation of the present observation given the past is constant is an uncorrelated process. In introductory statistics courses, it is often mentioned that independence implies zero correlation but not vice versa. A process, such as the GARCH processes, where the conditional mean is constant but the conditional variance is nonconstant is a good example of a process that is uncorrelated but not independent. The dependence of the conditional variance on the past is the reason the process is not independent. The independence of the conditional mean on the past is the reason that the process is uncorrelated.

Although a_t is uncorrelated just like the white noise process ϵ_t, the process a_t^2 has a more interesting ACF: if $\alpha_1 < 1$ then

$$\rho_{a^2}(h) = \alpha_1^{|h|}, \quad \forall\, h.$$

If $\alpha_1 \geq 1$, then a_t^2 is nonstationary, so of course it does not have an ACF.

Example 12.1. A simulated ARCH(1) process is shown in Figure 12.2. The top left panel shows the independent white noise process, ϵ_t. The top right panel shows $\sigma_t = \sqrt{1 + 0.95 a_{t-1}^2}$, the conditional standard deviation process. The bottom left panel shows $a_t = \sigma_t \epsilon_t$, the ARCH(1) process. As discussed in the next section, an ARCH(1) process can be used as the noise term of an AR(1) process. This is shown in the bottom right panel. The AR(1) parameters are $\mu = .1$ and $\phi = .8$. The variance of a_t is $\gamma_a(0) = 1/(1 - .95) = 20$ so the standard deviation is $\sqrt{20} = 4.47$.

The processes were all started at 0 and simulated for 70 observations. The first 10 observations were treated as a burn-in period where the process was converging to its stationary distribution. In the figure, only the last 60 observations are plotted.

The white noise process in the top left panel is normally distributed and has a standard deviation of 1, so it is less than 2 in absolute value about 95% of the time. Notice that just before $t = 10$, the process is a little less than -2 which is a somewhat large deviation from the mean of 0. This deviation causes the conditional standard deviation (σ_t) shown in the top right panel to increase and this increase persists for about 10 observations though it slowly decays. The result is that the ARCH(1) process exhibits more volatility than usual when t is between 10 and 15.

Figure 12.3 shows a simulation of 600 observations from the same processes as in Figure 12.2. A normal probability plot of a_t is also included. Notice that this ARCH(1) process exhibits extreme nonnormality. This is typical of

ARCH processes. Conditionally they are normal with a nonconstant variance, but their marginal distribution is nonnormal with a constant variance.

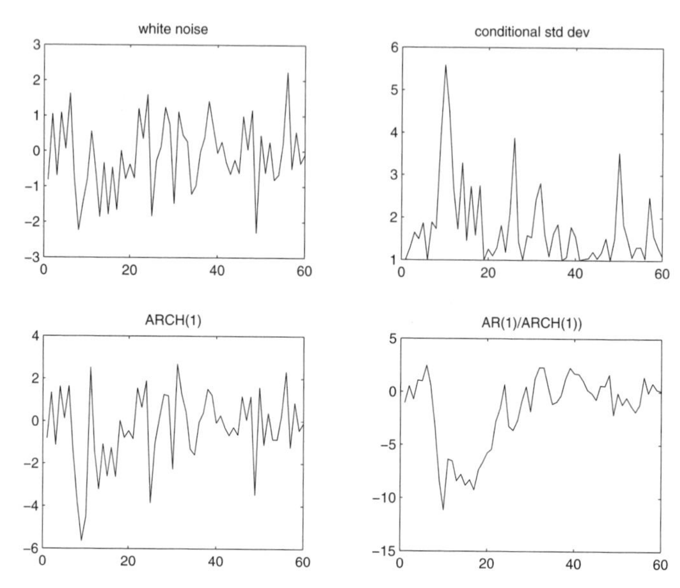

Fig. 12.2. *Simulation of* 60 *observations from an ARCH(*1*) process and an AR(*1*)/ARCH(*1*) process. The parameters are* $\alpha_0 = 1$, $\alpha_1 = 0.95$, $\mu = 0.1$, *and* $\phi = 0.8$.

12.4 The AR(1)/ARCH(1) Model

As we have seen, an AR(1) has a nonconstant conditional mean but a constant conditional variance, while an ARCH(1) process is just the opposite. If we think that both the conditional mean and variance of a process depend on the past then we need the features of both the AR and ARCH models. Fortunately, we can combine the two models. In fact, we can combine any ARMA model with any GARCH model. In this section we start simple and combine an AR(1) model with an ARCH(1) model.

Let a_t be an ARCH(1) process so that $a_t = \epsilon_t \sqrt{\alpha_0 + \alpha_1 a_{t-1}^2}$ where ϵ_t is WhiteNoise(0, 1), and suppose that

$$u_t - \mu = \phi(u_{t-1} - \mu) + a_t.$$

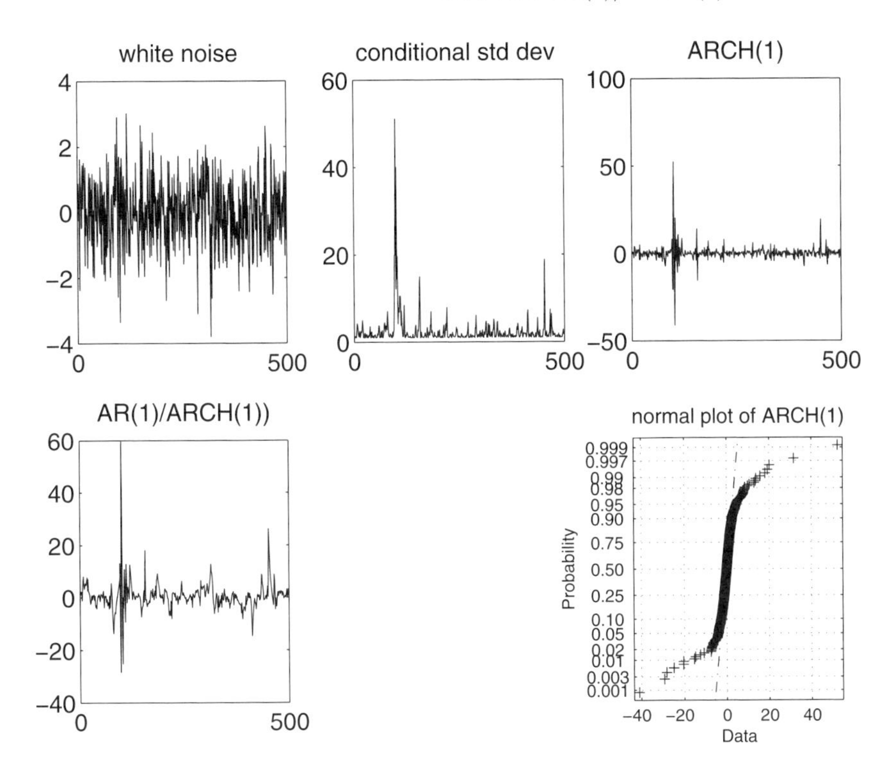

Fig. 12.3. *Simulation of 600 observations from an ARCH(1) process and an AR(1)/ARCH(1) process. The parameters are* $\alpha_0 = 1$, $\alpha_1 = 0.95$, $\mu = 0.1$, *and* $\phi = 0.8$.

The process u_t looks like an AR(1) process, except that the noise term is not independent white noise but rather an ARCH(1) process.

Because a_t is an uncorrelated process, a_t has the same ACF as independent white noise and therefore u_t has the same ACF as an AR(1) process:

$$\rho_u(h) = \phi^{|h|} \quad \forall\, h.$$

Moreover, a_t^2 has the ARCH(1) ACF:

$$\rho_{a^2}(h) = \alpha_1^{|h|} \quad \forall\, h.$$

We need to assume that both $|\phi| < 1$ and $\alpha_1 < 1$ in order for u to be stationary with a finite variance. Of course, $\alpha_0 \geq 0$ and $\alpha_1 \geq 0$ are also assumed.

The process u_t is such that its conditional mean and variance, given the past, are both nonconstant so a wide variety of real time series can be modeled.

Example 12.2. A simulation of an AR(1)/ARCH(1) process is shown in the bottom right panel of Figure 12.2. Notice that when the ARCH(1) noise term in the bottom left panel is more volatile, then the AR(1)/ARCH(1) process moves more rapidly.

12.5 ARCH(q) Models

As before, let ϵ_t be Gaussian white noise with unit variance. Then a_t is an ARCH(q) process if

$$a_t = \sigma_t \epsilon_t,$$

where

$$\sigma_t = \sqrt{\alpha_0 + \sum_{i=1}^{q} \alpha_i a_{t-i}^2}$$

is the conditional standard deviation of a_t given the past values $a_{t-1}, a_{t-2}, \ldots$ of this process. Like an ARCH(1) process, an ARCH(q) process is uncorrelated and has a constant mean (both conditional and unconditional) and a constant unconditional variance, but its conditional variance is nonconstant. In fact, the ACF of a_t^2 is the same as the ACF of an AR(q) process.

12.6 GARCH(p, q) Models

A deficiency of ARCH(q) models is that the conditional standard deviation process has high frequency oscillations with high volatility coming in short bursts. This behavior can be seen in the top right plot in Figure 12.3. GARCH models permit a wider range of behavior, in particular, more persistent volatility. The GARCH(p, q) model is

$$a_t = \epsilon_t \sigma_t,$$

where

$$\sigma_t = \sqrt{\alpha_0 + \sum_{i=1}^{q} \alpha_i a_{t-i}^2 + \sum_{i=1}^{p} \beta_i \sigma_{t-i}^2}. \tag{12.7}$$

Because past values of the σ_t process are fed back into the present value, the conditional standard deviation can exhibit more persistent periods of high or low volatility than seen in an ARCH process. The process a_t is uncorrelated with a stationary mean and variance and a_t^2 has an ACF like an ARMA process. GARCH models include ARCH models as a special case, and we use the term "GARCH" to refer to both ARCH and GARCH models. A very general time series model lets a_t be GARCH(p_G, q_G) and uses a_t as the noise term in an ARIMA(p_A, d, q_A) model.[4]

Figure 12.4 is a simulation of 600 observations from a GARCH(1,1) process and from a AR(1)/ GARCH(1,1) process. The GARCH parameters are $\alpha_0 =$ 1, $\alpha_1 = 0.08$, and $\beta_1 = 0.9$. The large value of β_1 causes σ_t to be highly correlated with σ_{t-1} and gives the conditional standard deviation process a

[4] We use subscripts on p and q to distinguish between the GARCH (G) and ARIMA (A) parameters.

relatively long-term persistence, at least compared to its behavior under an ARCH model. In particular, notice that the conditional standard deviation is less "bursty" than for the ARCH(1) process in Figure 12.3.

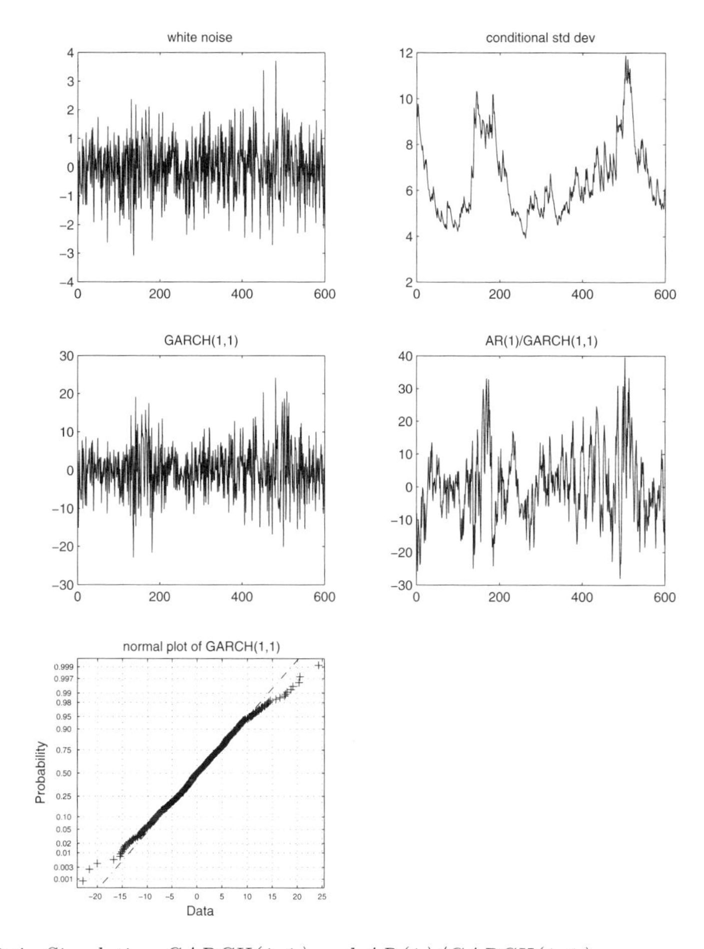

Fig. 12.4. *Simulation GARCH(1,1) and AR(1)/GARCH(1,1) processes. The parameters are* $\alpha_0 = 1$, $\alpha_1 = 0.08$, $\beta_1 = 0.9$, *and* $\phi = 0.8$.

12.7 GARCH Processes Have Heavy Tails

Researchers have long noticed that stock returns have "heavy-tailed" or "outlier-prone" probability distributions. As discussed in Section 2.12, this means that they have more extreme outliers than expected from a normal distribution. One reason for outliers may be that the conditional variance is not constant, and the outliers occur when the variance is large as in the normal mixture example of Section 2.12. In fact, GARCH processes exhibit heavy tails for this reason. Therefore, when we use GARCH models we can model both the conditional heteroscedasticity and the heavy-tailed distributions of financial market data. This makes GARCH models especially useful in financial applications.

12.8 Comparison of ARMA and GARCH Processes

Table 12.1 compares Gaussian white noise, ARMA, GARCH, and ARMA/GARCH processes according to various properties: conditional means, conditional variances, conditional distributions, marginal means, marginal variances, and marginal distributions.

Table 12.1. Comparison of ARMA and GARCH models.

Property	Gaussian Wh. Noise	ARMA	GARCH	ARMA/ GARCH
Cond. mean	constant	nonconst	0	nonconst
Cond. var	constant	constant	nonconst	nonconst
Cond. dist'n	normal	normal	normal	normal
Mar. mean & var.	constant	constant	constant	constant
Mar. dist'n	normal	normal	heavy-tail	heavy-tail

All the processes are stationary so that their marginal means and variances are constant. Gaussian white noise is the "baseline" process. Because it is an independent process the conditional distributions are the same as the marginal distribution. Thus, its conditional means and variances are constant and both its conditional and marginal distributions are normal. Gaussian white noise is the "driver" or "source of randomness" behind all the other processes. Therefore, they all have normal conditional distributions just like Gaussian white noise.

12.9 Fitting GARCH Models

A time series was simulated using the same program that generated the data in Figure 12.2, the only difference being that 300 observations were generated rather than only 60 as in the figure. The data were analyzed with SAS using the following program.

```
data arch ;
infile 'c:\book\sas\garch02.txt' ;
input y ;
run ;
title 'Simulated ARCH(1)/AR(1) data' ;
proc autoreg ;
model y =/nlag = 1  archtest garch=(q=1);
run ;
```

This program uses the PROC AUTOREG command that fits AR models. Since nlag = 1, an AR(1) model is being fit. However, the noise is not modeled as independent white noise. Rather an ARCH(1) model is used because of the specification "garch=(q=1)" in the "model" statement below the PROC AUTOREG command. More complex GARCH models can be fit using, for example, "garch=(p=2,q=1)." The parameters p and q have the same definition in PROC AUTOREG as in equation (12.7). The specification "archtest" requests tests of ARCH effects, that is, tests of the null hypothesis of conditional homoscedasticity versus the alternative of conditional heteroscedasticity.

Selected parts of the output from this SAS program are listed below.

```
          The AUTOREG Procedure
         Dependent Variable    y
      Q and LM Tests for ARCH Disturbances
Order             Q     Pr > Q           LM     Pr > LM

1         119.7578    <.0001      118.6797     <.0001
2         137.9967    <.0001      129.8491     <.0001
3         140.5454    <.0001      131.4911     <.0001
4         140.6837    <.0001      132.1098     <.0001
5         140.6925    <.0001      132.3810     <.0001
6         140.7476    <.0001      132.7534     <.0001
7         141.0173    <.0001      132.7543     <.0001
8         141.5401    <.0001      132.8874     <.0001
9         142.1243    <.0001      132.8879     <.0001
10         142.6266    <.0001      132.9226     <.0001
11         142.7506    <.0001      133.0153     <.0001
12         142.7508    <.0001      133.0155     <.0001
```

There are two types of ARCH tests, Q and LM (Lagrange multiplier) tests, and each of these has an order parameter, which varies from 1 to 12 in this output. All tests of conditional homoscedasticity reject with p-values of 0.0001 or smaller.

Next, we have the results of fitting the AR(1)/ARCH(1) model. The AIC and SBC statistics are useful for comparisons between this model and other models that might be contemplated. The normality test has the null hypothesis that the conditional distribution is normal and this hypothesis is accepted with a p-value of 0.4692.

```
          The AUTOREG Procedure
           GARCH Estimates
SSE               1056.42037   Observations               300
MSE                  3.52140   Uncond Var          3.72785257
Log Likelihood    -549.43844   Total R-Square          0.6077
SBC               1121.69201   AIC                 1106.87688
Normality Test        1.5134   Pr > ChiSq              0.4692

                                 Standard               Approx
Variable       DF    Estimate       Error    t Value   Pr > |t|
```

```
Intercept        1      0.4810     0.3910      1.23     0.2187
AR1              1     -0.8226     0.0266    -30.92     <.0001
ARCH0            1      1.1241     0.1729      6.50     <.0001
ARCH1            1      0.6985     0.1167      5.98     <.0001
```

The estimate of the AR parameter is -0.8226 (0.8226 in our notation) and has a small standard error of only 0.0266. The estimates of the ARCH parameters are $\widehat{\alpha}_0 = 1.12$ and $\widehat{\alpha}_1 = 0.70$. Since the data were simulated, the true parameter values are known and they are $\alpha_0 = 1$ and $\alpha_1 = 0.95$. The standard errors of the ARCH parameters are rather large. This is a general phenomenon because time series data usually have less information about variance parameters than about the parameters specifying the conditional expectation. An approximate 95% confidence interval for α_1 is

$$0.70 \pm (2)(0.117) = (0.446, 0.934),$$

which does not quite include the true parameter, 0.95. This could have just been bad luck, though it may indicate that $\widehat{\alpha}_1$ is downward biased. The confidence interval is based on the assumption of unbiasedness and is not valid if there is a sizeable bias.

Example 12.3. (S&P 500 *returns)* This is an example of regression where the "errors" in the model are not assumed to be white noise but rather are assumed to be an AR/GARCH process. Example 10.5 in Pindyck and Rubinfeld (1998) contains monthly observations from 1960 to 1996 on these variables:

- the S&P 500 index (FSPCOM),
- the return on the S&P 500 (RETURNSP),
- the dividend yield on the S&P 500 index (FSDXP),
- the three-month T-bill rate (R3),
- the change in the three-month T-bill rate (DR3),
- the wholesale price index (PW), and
- the rate of wholesale price inflation (GPW).

In this analysis, only RETURNSP, DR3, and GPW are used.

It is expected that variations in stock returns are in part caused by changes in interest rates and changes in the rate of inflation. Therefore, a regression model where RETURNSP is regressed on DR3 and GPW is used. Regression models that regress returns on macroeconomic variables in this way are called "factor models" — see Section 7.8. Figure 12.5 shows the residuals from this regression. The residuals represent the part of the S&P 500 returns that cannot be explained by changes in interest rates and the inflation rate. In the figure, there is some sign of nonconstant volatility, because we see periods of relatively small fluctuations, for example, after 1990, and periods of higher fluctuations, for example, around 1975. Also, there is no reason to assume that the residuals are uncorrelated as is assumed in a standard regression model. To allow for possible nonconstant volatility and correlated errors, we assume that

$$\text{RETURNSP} = \gamma_0 + \gamma_1 \text{DR3} + \gamma_2 \text{GPW} + u_t, \tag{12.8}$$

where u_t is an AR(1)/GARCH(1,1) process.[5] Therefore,

$$u_t = \phi_1 u_{t-1} + a_t,$$

where a_t is a GARCH(1,1) process so that

$$a_t = \epsilon_t \sigma_t,$$

where

$$\sigma_t = \sqrt{\alpha_0 + \alpha_1 a_{t-1}^2 + \beta_1 \sigma_{t-1}^2}$$

and ϵ_t is white noise.

Below is a listing of the SAS program used to fit the regression model with AR(1)/GARCH(1,1) errors. The first part of the program creates the SAS data set. Notice that the variable DR3 which is the change in the variable R3 is created by the differencing function "dif" in SAS.

```
data arch ;
infile 'c:\book\sas\pindyck105.dat' ;
input month year RETURNSP FSPCOM FSDXP R3 PW GPW;
DR3 = dif(R3) ;
run ;
title 'S&P 500 monthly data from Pindyck & Rubinfeld, Ex 10.5' ;
```

The regression model with AR(1)/GARCH(1,1) errors is specified by the commands:

```
title2 'Regression model with AR(1)/GARCH(1,1)' ;
proc autoreg ;
model returnsp = DR3 gpw/nlag = 1  archtest garch=(p=1,q=1);
run ;
```

In these commands, the statement "returnsp = DR3 gpw" specifies the regression model, that is, that "returnsp" is the dependent variable and "DR3" and "gpw" are the independent variables, "nlag = 1" specifies the AR(1) structure, "garch=(p=1,q=1)" specifies the GARCH(1,1) structure, and "archtest" specifies that tests of conditional heteroscedasticity be performed. We could have deleted "archtest" if we were not interested in testing for GARCH effects, for example, because we were already certain that they existed.

The first part of the output consists of the Q and LM tests for GARCH effects.

```
                 Q and LM Tests for ARCH Disturbances

      Order           Q     Pr > Q           LM     Pr > LM

        1       26.8804     <.0001      26.5159      <.0001
        2       27.1508     <.0001      27.1519      <.0001
        3       28.2188     <.0001      28.4391      <.0001
```

[5] We denote the regression coefficients by gamma rather than beta, as is somewhat standard, because beta is used for parameters in the GARCH model for a_t.

4	28.6957	<.0001	28.4660	<.0001
5	33.4112	<.0001	32.6168	<.0001
6	34.0892	<.0001	32.6962	<.0001
7	34.4187	<.0001	32.9617	<.0001
8	34.6542	<.0001	32.9636	<.0001
9	35.2228	<.0001	33.3330	0.0001
10	35.3047	0.0001	33.4174	0.0002
11	35.8274	0.0002	33.9440	0.0004
12	36.0142	0.0003	33.9507	0.0007

The p-values of the Q and LM tests are all very small, less than 0.0001. Therefore, the errors in the regression model appear conditionally heteroscedastic. SAS first fits an ordinary regression model (assuming white noise errors)[6] with the following results.

Ordinary Least Squares Estimates

SSE	0.46677572	DFE	430
MSE	0.00109	Root MSE	0.03295
SBC	-1711.5219	AIC	-1723.7341
Regress R-Square	0.0551	Total R-Square	0.0551
Durbin-Watson	1.5203		

The R-square value is only 0.0551, so the regression has little predictive value.

For the regression with AR(1)/GARCH(1,1) errors we get the following results.

GARCH Estimates

SSE	0.44176656	Observations	433
MSE	0.00102	Uncond Var	0.00104656
Log Likelihood	889.071523	Total R-Square	0.1058
SBC	-1735.6479	AIC	-1764.143
Normality Test	43.0751	Pr > ChiSq	<.0001

Notice that AIC is much smaller than for ordinary regression which indicates that the AR(1)/GARCH(1,1) model fits the errors much better than the white noise model. Thus, AIC is in agreement with the Q and LM test results.

The estimated parameters of the regression with AR(1)/GARCH(1,1) errors are:

Variable	DF	Estimate	Standard Error	t Value	Approx Pr > |t|
Intercept	1	0.0125	0.001875	6.66	<.0001
DR3	1	-1.0665	0.3282	-3.25	0.0012
GPW	1	-0.7239	0.1992	-3.63	0.0003

Since all p-values are small, both independent variables are significant.

The estimated AR and GARCH parameters are:

AR1	1	-0.2016	0.0603	-3.34	0.0008
ARCH0	1	0.000147	0.0000688	2.14	0.0320
ARCH1	1	0.1337	0.0404	3.31	0.0009
GARCH1	1	0.7254	0.0918	7.91	<.0001

[6] One can tell that a standard regression model with white noise errors is being used because the output is labeled "Ordinary Least Squares".

The estimate of ϕ is $-.2016$ in SAS's notation but $+.2016$ in our notation. Thus, there is a *positive* association between returns and lagged returns. The GARCH(1) estimate (0.7254) is larger than the ARCH(1) (0.1337) estimate. This implies that the conditional variance exhibits reasonably long persistence of volatility; see the discussion in the next section of the persistence of volatility and how it depends upon the parameters of the GARCH model. Since all p-values are small, all AR and GARCH parameters are significant.

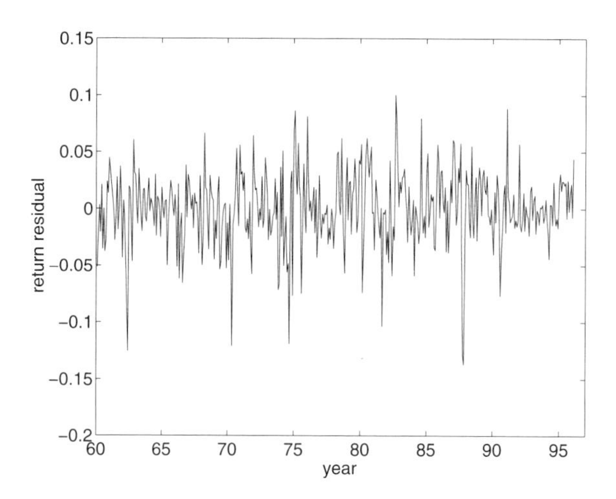

Fig. 12.5. *Residuals when the S&P 500 returns are regressed against the change in the three-month T-bill rates and the rate of inflation.*

12.10 I-GARCH Models

Integrated GARCH (I-GARCH) processes were designed to model data that have persistent changes in volatility. A GARCH(p, q) process is stationary with a finite variance if

$$\sum_{i=1}^{q} \alpha_i + \sum_{i=1}^{p} \beta_i < 1.$$

A GARCH(p, q) process is called an I-GARCH process if

$$\sum_{i=1}^{q} \alpha_i + \sum_{i=1}^{p} \beta_i = 1.$$

I-GARCH processes are either nonstationary or if they are stationary they have an infinite variance.

Infinite variance implies heavy tails, though a distribution can be heavy-tailed with a finite variance. To appreciate what infinite variance processes can look like, we do some simulation. Figure 12.6 shows 40,000 observations of ARCH(1) processes with $\alpha_1 = 0.9$, 1, and 1.8. The same white noise process is used in each of the ARCH(1) processes. All three ARCH(1) processes are stationary but only the one with $\alpha_1 = 0.9$ has a finite variance. The second process is an I-GARCH process.[7] The third process has $\alpha_1 > 1$ and so has more extreme outliers than an I-GARCH process. Notice how all three processes revert to their conditional mean of 0 as expected of stationary processes. The larger the value of α_1 the more the volatility comes in sharp bursts. The processes with $\alpha_1 = 0.9$ and $\alpha_1 = 1$ look similar; there is no sudden change in behavior when the variance becomes infinite. The process with $\alpha_1 = 0.9$ already has a heavy tail despite having a finite variance. Increasing α_1 from 0.9 to 1 does not increase the tail-weight dramatically.

Normal plots of the simulated data in Figure 12.6 are shown in Figure 12.7. Clearly, the larger the value of α_1, the heavier the tails of the marginal distribution.

None of the processes in Figure 12.6 shows much persistence of higher volatility. To model persistence of higher volatility, one needs a GARCH (p, q) or I-GARCH(p, q) process with $q \geq 1$. Figure 12.8 shows simulations from I-GARCH(1, 1) processes. Since $\alpha_1 + \beta_1 = 1$ for these processes, $\beta_1 = 1 - \alpha_1$, and the process is completely specified by α_0 and α_1. In this figure, α_0 is fixed at 1 and α_1 is varied. Notice that the conditional variance is very bursty when $\alpha_1 = .95$. When $\alpha_1 = .02$, the conditional standard deviation looks much smoother and high or low volatility is more persistent. The point here is that with both the ARCH parameter α_1 and the GARCH parameter β_1 available, the model has the flexibility to model whatever behavior is seen in the data.

I-GARCH processes can be fit by SAS by adding the specification "type = integrated" into the program, e.g., for the previous example with S&P 500 returns:

```
proc autoreg ;
model returnsp =/nlag = 1 garch=(p=1,q=1,type=integrated);
run ;
```

For this example, the I-GARCH(1,1) model seems to fit somewhat worse than a GARCH (1,1) model according to AIC and so a GARCH rather than I-GARCH model would be recommended for this data set; see Section 12.4.

[7] Actually, it is an I-ARCH process since $q = 0$.

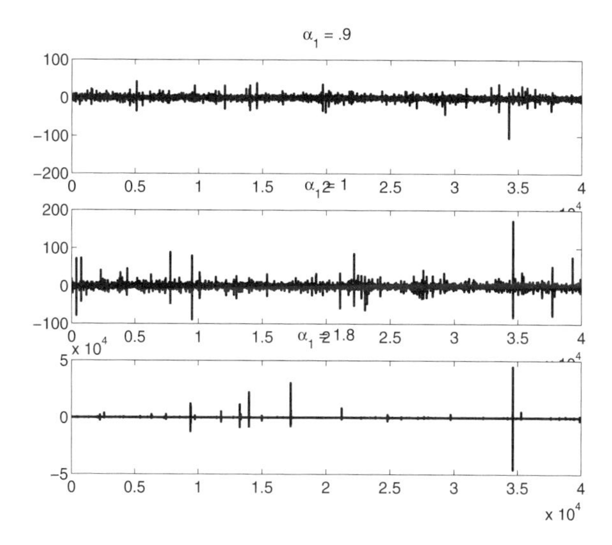

Fig. 12.6. *Simulated ARCH(1) processes with $\alpha_1 = 0.9$, 1, and 1.8. Notice that the vertical scale depends on α_1 because larger values of α_1 result in more extreme outliers.*

12.10.1 What does it mean to have an infinite variance?

As discussed in Section 2.3.6, a random variable need not have a well-defined and finite expected value or variance. GARCH models are an example where these problems can occur. In this section some of the effects of the expected value or variance not being well-defined or being infinite are described.

One consequence of the expectation not existing is this. Suppose we have a sample of i.i.d. random variables with density f_X. The law of large numbers says that the sample mean converges to $E(X)$ as the sample size goes to infinity. However, the law of large numbers holds only if $E(X)$ is defined. Otherwise, there is no point to which the sample mean can converge and it just wanders without converging.

Figure 12.9 shows the sample mean of the first t observations plotted against t for the data in Figure 12.6. The sample mean appears to converge to 0 when $\alpha_1 = 0.9$ or 1, but when $\alpha_1 = 1.8$ it is unclear what the sample mean is doing. The sample mean decays towards 0 when the process is not in a high volatility period, but can shoot up or down during a burst of volatility.

Now suppose that the expectation of X exists, is finite, and equals μ_X. Then the variance of X is defined as

$$\int_{-\infty}^{\infty} (x - \mu_X)^2 f_X(x) dx. \tag{12.9}$$

If this integral is $+\infty$, then the variance is infinite. If the expectation of X is not defined or is infinite, then the variance is also not defined since either

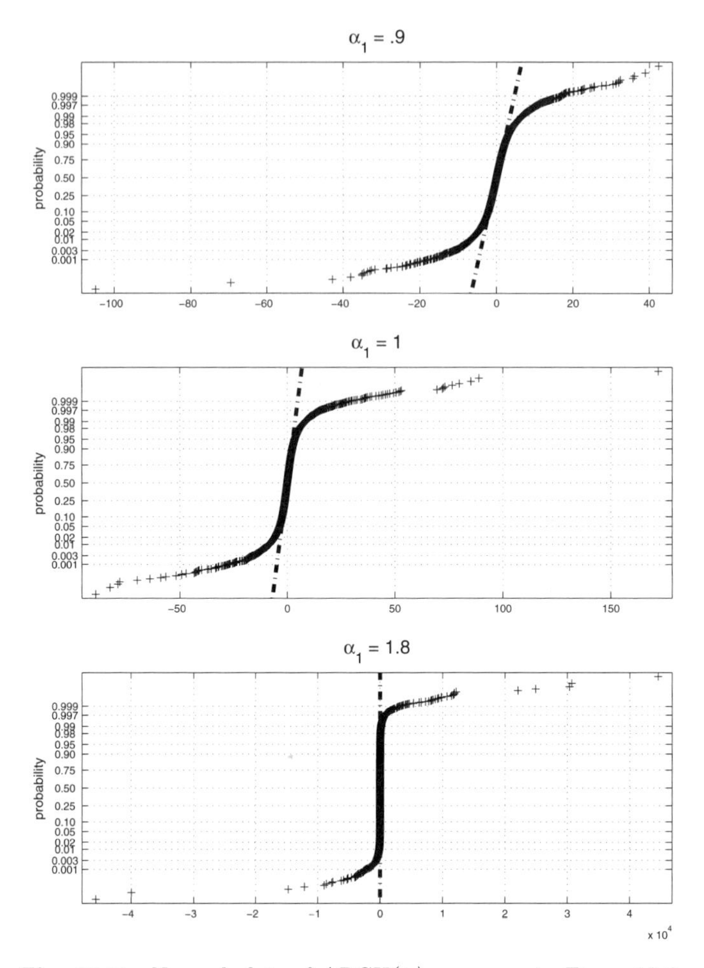

Fig. 12.7. *Normal plots of ARCH(1) processes in Figure* 12.6.

there is no μ_X to use in (12.9) or μ_X is either $+\infty$ or $-\infty$ and then (12.9) is not defined.

The law of large numbers also implies that the sample variance converges to the variance of X as the sample size increases. If the variance of X is infinity, then the sample variance converges to infinity.

Figure 12.10 shows the sample variance of the first t observations plotted against t for the data in Figure 12.6. In the top panel, the sample variance converges to $10 = (1 - \alpha_1)^{-1}$, but the convergence may be very slow and it is hard to see even with 40,000 observations. In the middle and bottom panels the variance is infinity so the sample variance converges to infinity. This convergence does appear to be happening in the bottom panel, but it is hard to see in the middle panel. Of course, in the middle panel the value of

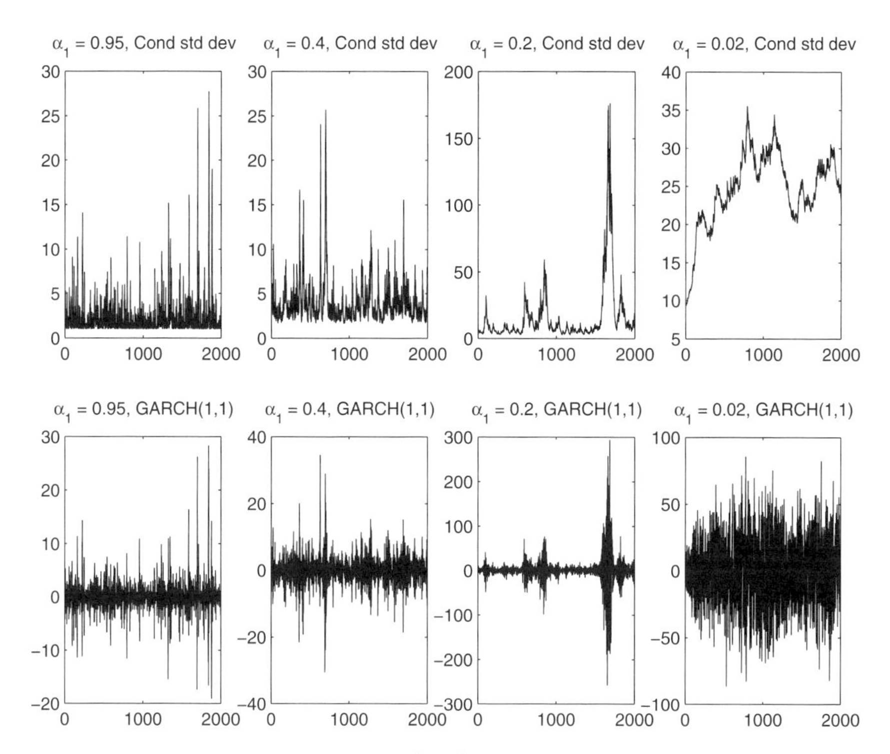

Fig. 12.8. *Simulations of I-GARCH*$(1,1)$ *processes. The conditional standard deviation processes are on top and the I-GARCH*$(1,1)$ *processes are in the bottom row. The parameters* (α_1, β_1) *vary from* $(0.95, 0.05)$ *on the left to* $(0.02, 0.98)$ *on the right. In each case* $\beta_1 = 1 - \alpha_1$.

α_1 is on the borderline between finite and infinite variance, and the infinite variance may take a very long time to have its effect.

12.11 GARCH-M Processes

We have seen that one can fit regression models with AR/GARCH errors. In fact, we have done that with the S&P 500 data. In some examples, it makes sense to use the conditional standard deviation as one of the regression variables. For example, when the dependent variable is a return we might expect that higher conditional variability causes higher returns, because the market demands a higher risk premium for higher risk.

Models where the conditional standard deviation is a regression variable are called GARCH-in-mean, or GARCH-M, models. They have the form

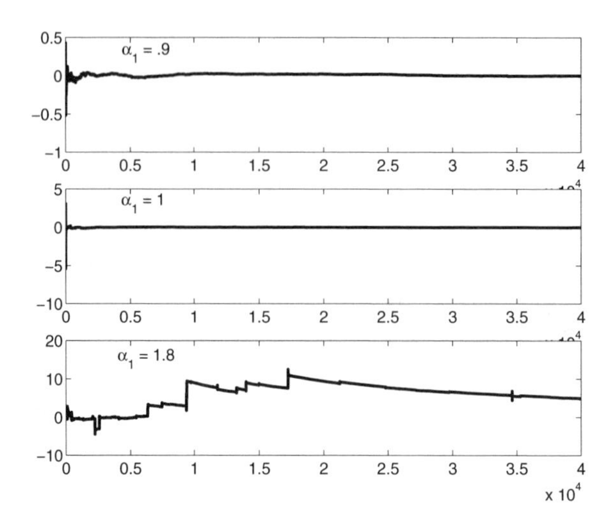

Fig. 12.9. *Sample means of simulated ARCH(1) processes with* $\alpha_1 = 0.9$, 1, *and* 1.8.

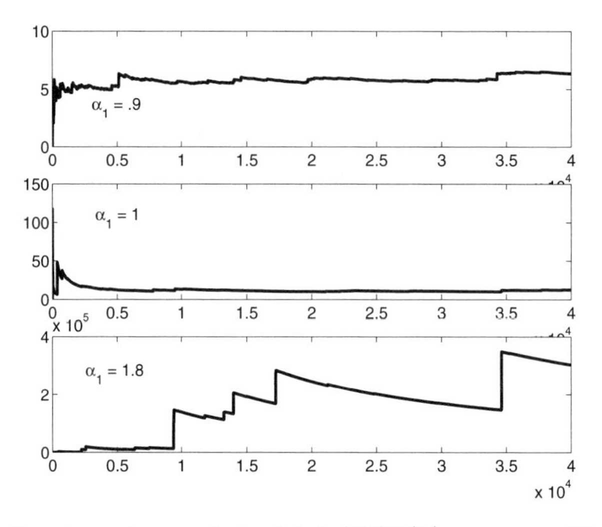

Fig. 12.10. *Sample variances of simulated ARCH(1) processes with* $\alpha_1 = 0.9$, 1, *and* 1.8.

$$Y_t = \boldsymbol{X}_t^{\mathsf{T}}\boldsymbol{\gamma} + \delta\sigma_t + a_t,$$

where a_t is a GARCH process with conditional standard deviation process σ_t. In this model, σ_t and the components of $\boldsymbol{X}_t$ are the predictor variables and δ and the components of $\boldsymbol{\gamma}$ are the regression coefficients. However, the predictor σ_t is unlike the predictors in $\boldsymbol{X}_t$ in that $\boldsymbol{X}_t$ contains observable variables whereas σ_t must be estimated.

GARCH-M models can be fit in SAS by adding the keyword "mean" to the GARCH specification, e.g.,

```
proc autoreg ;
model returnsp =/nlag = 1 garch=(p=1,q=1,mean);
run ;
```

or for I-GARCH-M

```
proc autoreg ;
model returnsp =/nlag = 1 garch=(p=1,q=1,mean,type=integrated);
run ;
```

For the S&P 500 returns data, a GARCH(1,1)-M was fit in SAS. The estimate of $\widehat{\delta}$ was 0.5150 with a standard error of 0.3695. This gives a t-value of 1.39 and a p-value of 0.1633. Since the p-value is reasonably large we could accept the null hypothesis that $\delta = 0$. Therefore, we see no strong evidence that there are higher returns during times of higher volatility. The volatility of the S&P 500 is *market risk* so this finding is a bit surprising. It may be that the effect is small ($\widehat{\delta}$ is positive, after all) and cannot be detected with certainty. The AIC criterion *does* select the GARCH-M model; see Section 12.4.

12.12 E-GARCH

In finance, the **leverage effect** predicts that an asset's returns may become more volatile when its price decreases. E-GARCH processes were designed to model the leverage effect. The exponential GARCH, or E-GARCH, model is

$$\log(\sigma_t) = \alpha_0 + \sum_{i=1}^{q} \alpha_1 g(\epsilon_{t-i}) + \sum_{i=1}^{p} \beta_i \log(\sigma_{t-i}), \tag{12.10}$$

where

$$g(\epsilon_t) = \theta \epsilon_t + \gamma\{|\epsilon_t| - E(|\epsilon_t|)\}$$

and $\epsilon_t = a_t/\sigma_t$. Since $\log(\sigma_t)$ can be negative, there is no need to constrain the parameters to keep the right-hand side of (12.10) nonnegative as is the case for ordinary GARCH models. Computations are easier since constrained optimization is not needed when computing the maximum likelihood estimate.

Notice that

$$g(\epsilon_t) = -\gamma E(|\epsilon_t|) + (\gamma + \theta)|\epsilon_t| \text{ if } \epsilon_t > 0,$$

and

$$g(\epsilon_t) = -\gamma E(|\epsilon_t|) + (\gamma - \theta)|\epsilon_t| \text{ if } \epsilon_t < 0.$$

In Problem 1 it is shown that $E(|\epsilon_t|) = \sqrt{2/\pi} = 0.7979$.

Typically, $-1 < \widehat{\theta} < 0$ so that $0 < \gamma + \theta < \gamma - \theta$. For example, $\widehat{\theta} = -0.7$ in the S&P 500 example; see below. The function g with $\theta = -0.7$ is plotted in the top left panel of Figure 12.11. Notice that $g(\epsilon_t)$ is negative if $|\epsilon_t|$ is close to zero so that small values of noise decrease σ_t. If $|\epsilon_t|$ is large, then σ_t increases. With a negative value of θ, σ_t increases more rapidly as a function of $|\epsilon_t|$ when ϵ_t is negative than when ϵ_t is positive,

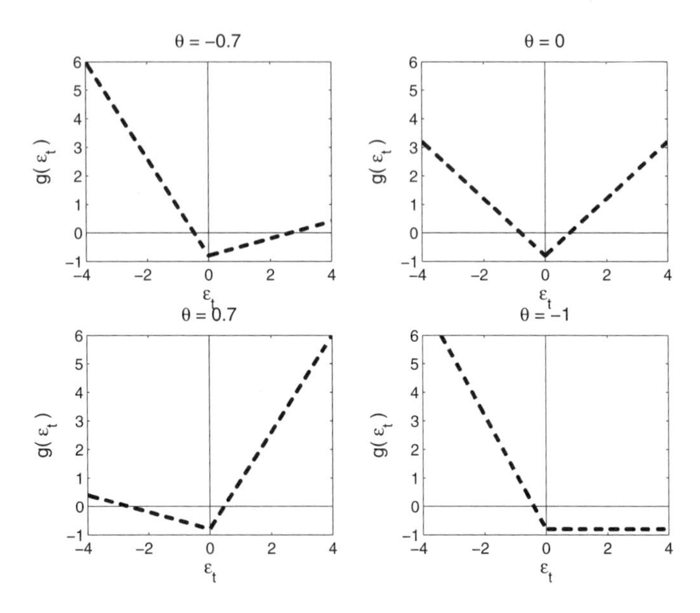

Fig. 12.11. *The g function for the S&P* 500 *data (top left panel) and several other values of* θ.

As mentioned previously, the leverage effect occurs when an asset's returns become more volatile as its price decreases. This is the type of behavior seen in an E-GARCH process when $\theta < 0$. The ability to accommodate leverage effects was the reason that the E-GARCH model was introduced by Daniel Nelson.

The function g for several other values of θ are also shown in Figure 12.11. When $\theta = 0$ (top right) the function is symmetric about 0. The bottom right panel where $\theta = -1$ shows an extreme case where $g(\epsilon_t)$ is negative for all positive ϵ_t.

SAS fits the E-GARCH model with γ fixed at 1 and θ estimated.[8] The E-GARCH model is specified by using "type=exp" as in

```
proc autoreg ;
model returnsp =/nlag = 1 garch=(p=1,q=1,mean,type=exp);
run ;
```

[8] The data can only determine the ratio γ/θ, so one of γ or θ must be fixed at an arbitrary value. SAS uses $\gamma = 1$.

This command specifies both a GARCH-in-mean effect and the E-GARCH model. Omitting "mean" removes the GARCH-in-mean effect.

Example 12.4. (Back to the S&P 500 Example) SAS can fit six different AR(1)/GARCH(1,1) models since SAS allows "type" to be "integrated," "exp," or "nonneg." The last is the default and specifies a GARCH model with nonnegativity constraints. Moreover, for each of these three types we can specify that a GARCH-in-mean effect be included or not. Table 12.2 contains the AIC statistics for the six models. The models are ordered from best fitting to worse fitting according to AIC — remember that a smaller AIC is better.

Table 12.2. AIC statistics for six AR(1)/GARCH(1,1) models fit to the S&P 500 returns data. ΔAIC is the change in AIC between a given model and the E-GARCH-M model which has the smallest AIC value among the six models.

Model	AIC	ΔAIC
E-GARCH-M	−1783.9	0.0
E-GARCH	−1783.1	0.8
GARCH-M	−1764.6	19.3
GARCH	−1764.1	19.8
I-GARCH-M	−1758.0	25.9
I-GARCH	−1756.4	27.5

By AIC the E-GARCH-M model is best, though the E-GARCH model fits nearly as well. The E-GARCH-M model is used in the remaining discussion. To see if more AR or GARCH parameters would improve the fit, an AR(2) model instead of AR(1) and E-GARCH(1,2)-M, E-GARCH(2,1)-M, and E-GARCH(2,2)-M models instead of E-GARCH(1,1)-M were tried, but none of these lowered AIC or had all parameters significant at $p = .1$. Thus, AR(1)/E-GARCH(1,1) appears to be a good fit to the noise and adding a GARCH-in-mean term to the regression model seems reasonable although it does not improve the fit very much.

The fit to this model is in the SAS output listed below.

```
          S&P 500 monthly data from Pindyck & Rubinfeld, Ex 10.5
               Regression model with AR(1)/E-GARCH(1,1)-M
                         The AUTOREG Procedure
                    Exponential GARCH Estimates

    SSE                 0.44211939   Observations              433
    MSE                    0.00102   Uncond Var                  .
    Log Likelihood      900.962569   Total R-Square         0.1050
    SBC                 -1747.2885   AIC                -1783.9251
    Normality Test         24.9607   Pr > ChiSq             <.0001

                                        Standard               Approx
   Variable        DF     Estimate        Error    t Value    Pr > |t|

   Intercept        1    -0.003791       0.0102      -0.37      0.7095
```

```
DR3        1   -1.2062   0.3044   -3.96   <.0001
GPW        1   -0.6456   0.2153   -3.00   0.0027
AR1        1   -0.2376   0.0592   -4.01   <.0001
EARCH0     1   -1.2400   0.4251   -2.92   0.0035
EARCH1     1    0.2520   0.0691    3.65   0.0003
EGARCH1    1    0.8220   0.0606   13.55   <.0001
THETA      1   -0.6940   0.2646   -2.62   0.0087
DELTA      1    0.5067   0.3511    1.44   0.1490
```

12.13 The GARCH Zoo*

There are many more types of GARCH models than the few we have mentioned. Only the most widely used models that can be fit in SAS have been discussed so far. The number of models seems limited only by the number of letters in the alphabet, not the imagination of econometricians. Here's a sample of other GARCH models mentioned in Bollerslev, Engle, and Nelson (1994) and elsewhere:

- A-GARCH = asymmetric GARCH,
- M-GARCH = multivariate GARCH,
- NGARCH = nonlinear asymmetric GARCH,[9]
- QARCH = quadratic ARCH,
- TARCH = threshold ARCH,
- STARCH = structural ARCH,
- SWARCH = switching ARCH,
- QTARCH = quantitative threshold ARCH,
- vector ARCH,
- diagonal ARCH, and
- factor ARCH,

12.14 Applications of GARCH in Finance

GARCH models were developed by econometricians working with business and finance data, and the applications to finance have been extensive. The review paper by Bollerslev, Engle, and Nelson (1994) lists hundreds of references. Finance models such as the CAPM and the Black-Scholes model for option pricing assume a constant conditional variance. When this assumption is false, use of these models can lead to serious errors. Therefore, generalization of finance models to include GARCH errors has been a very active research topic.

The use of GARCH models for pricing options is a particularly promising area. Ritchken and Trevor (1999) use a multinomial tree method to price European and American options assuming that the log price is a GARCH process. This model is a generalization of the usual geometric random walk

[9] Engle and Ng (1993).

for prices. In their multinomial tree method each node leads not to two new nodes as in a binomial tree but rather to $2n + 1$ new nodes for some integer $n \geq 1$. Thus, each node leads to at least three new nodes (the case where $n = 1$). Ritchken and Trevor's algorithm, which is described in more detail in Section 12.15, is only one example of many methods that have been developed for pricing options when the price of the underlying asset follows a GARCH model. See Heston and Nandi (2000) for another example.

Once a method has been developed for pricing an option under a GARCH model, one can then find the *implied GARCH parameters* by fitting the model to options data by nonlinear regression. This is similar to finding the implied volatility of the Black-Scholes model, though there are some differences. One difference is that in the Black-Scholes model, there is only one parameter, the constant volatility, whereas in a GARCH model there are several parameters that determine how the conditional volatility evolves. Another difference is that in the Black-Scholes model, each option has its own implied volatility which is the value of σ that makes the Black-Scholes price exactly equal to the market price. When fitting GARCH models, one does not look for implied GARCH parameters corresponding to each option. Instead, implied GARCH parameters are obtained for a large set of options by minimizing the sum of squared residuals, which are the differences between the prices given by the GARCH pricing model and the observed market prices. One could, of course, find a single implied volatility for a set of options by minimizing the sum of squared residuals, the differences between the Black-Scholes prices and the observed prices. However, this leads to large pricing biases because of the volatility smile.

One of the real success stories for GARCH pricing is that it "explains" the volatility smile. When options are priced by GARCH pricing models, there is little evidence of bias. This is an good indication that the volatility smile is caused by applying the geometric random walk model with a constant variance to asset prices that do not follow this model.[10]

12.15 Pricing Options Under Generalized GARCH Processes*

In Chapter 8, binomial trees were used to price options when the log price of the underlying asset followed a random walk with the steps being ordinary white noise with a constant conditional variance. Ritchken and Trevor (1999) have generalized the binomial tree algorithm to price options when the asset's log price follows a random walk with GARCH noise. The risk-neutral measure of the GARCH process is another GARCH process. Their algorithm is interesting in several respects. Since the conditional variance of a GARCH process is not constant, the algorithm must keep track of both the log price of

[10] See Duan (1995), Heston and Nandi (2000), and Hsieh and Ritchken (2000).

the asset and its conditional variance. At each node of the tree, there is only one possible value of the log price, but the conditional variance takes many possible values. Also, the size of the jumps must adjust so that larger jumps are allowed when the conditional variance is large. To describe these interesting features, the algorithm is discussed rather fully here, but not all details of the algorithm are mentioned and a reader interested in implementing the algorithm is advised to consult the original paper.

Ritchken and Trevor assume that S_t, the price of an asset on "day"[11] t, evolves according to a random walk whose noise is a GARCH process:

$$\log(S_{t+1}/S_t) = r + \lambda\sqrt{h_t} - h_t/2 + \sqrt{h_t}\nu_{t+1}, \tag{12.11}$$

$$h_{t+1} = \beta_0 + \beta_1 h_t + \beta_2 h_t(\nu_{t+1} - c)^2. \tag{12.12}$$

In these equations r is the risk-free interest rate per day, λ is the risk premium per unit of standard deviation, ν_t is Gaussian WhiteNoise(0, 1), h_t is the conditional variance of the noise $\sqrt{h_t}\nu_{t+1}$ given the past, and c is a nonnegative parameter that models possible asymmetries such as the leverage effect. This model is called a *nonlinear asymmetric GARCH* (NGARCH) model. Under the risk-neutral measure, S_t evolves according to a slightly different random walk with generalized GARCH noise:

$$\log(S_{t+1}/S_t) = (r - h_t/2) + \sqrt{h_t}\epsilon_{t+1}, \tag{12.13}$$

$$h_{t+1} = \beta_0 + \beta_1 h_t + \beta_2 h_t(\epsilon_{t+1} - c^*)^2, \tag{12.14}$$

where $c^* = c + \lambda$ and ϵ_t is WhiteNoise(0, 1). Notice the similarity between the term $r - h_t/2$ in (12.13) and $r - \sigma^2/2$ in equation (8.18). If c^* equals zero, then (12.14) is a standard GARCH(1,1) process but with slightly different notation than has been used in this chapter. In particular, $\sqrt{h_t}$, $\sqrt{h_t}\epsilon_{t+1}$, β_0, and β_2 of Ritchken and Trevor are our σ_{t+1}, a_{t+1}, α_0, and α_1, respectively.[12] This model has five parameters, β_0, β_1, β_2, c^*, and the initial value of the variance h_0. One can estimate these parameters by fitting equations (12.11) and (12.12) to a time series of asset prices and then determine the price of an option with the pricing algorithm discussed below. Alternatively, one can find values of these parameters so that the prices determined by the algorithm match as closely as possible the prices of a set of options on the asset, e.g., calls of various maturities and strike prices. These *implied GARCH parameters* are a generalization of implied volatility. Using the implied GARCH parameters, one can then price other options.

Ritchken and Trevor approximate the GARCH process S_t with another process S_t^a, where a superscript a denotes "approximate." The process S_y^a

[11] "Day" is in quotes because in actuality it could be any unit of time.

[12] It is rare to find standardized notation in statistics, economics, finance, or any other field of study that uses mathematics. When writing this section, I decided to use the notation of Ritchken and Trevor so that students could easily consult that paper for more information.

evolves according to a multinomial tree where instead of only two possible jumps at each node (as in a binomial tree) there are $2n + 1$ jumps for some positive integer n. For simplicity, only the case $n = 1$, which gives *trinomial trees*, is discussed here. Ritchken and Trevor found that using just three jumps ($n = 1$) works satisfactorily for longer maturities such as 100 or 200 days. For shorter maturities, more possible jumps are needed. Using 11 possible jumps ($n = 5$) produces accurate results across all maturities.

Under the approximate GARCH process, the log price $y_t^a = \log(S_t^a)$ takes values on a lattice $\{y_0^a + k\gamma : k = \ldots, -2, -1, 0, 1, 2, \ldots\}$. Here γ is a parameter determining the fineness of the lattice and y_0^a is the known initial price of the asset. Ritchken and Trevor suggest using a value of γ equal to $\sqrt{h_0}$, the initial standard deviation of the log prices. On the tth step, the log price either moves up or down by η steps on the lattice (i.e., by the amount $\gamma\eta$) or does not move. The parameter η is needed so that the standard deviation of the trinomial distribution can equal the conditional standard deviation $\sqrt{h_t}$; larger values of the standard deviation require larger values of η. Thus, there are three parameters, the probability p_u of an up step, the probability p_d of a down step, and η, that can be determined so that the mean of the trinomial distribution at the tth step equals

$$E_t(y_{t+1}^a) = y_t^a + r - h_t^a/2, \tag{12.15}$$

and the variance of the trinomial distribution equals

$$\mathrm{Var}_t(y_{t+1}^a) = h_t^a. \tag{12.16}$$

Here "E_t" and "Var_t" denote conditional expectation and variance given the information set at day t. Since there are three parameters but only two equations to satisfy, there are many possible solutions. Ritchken and Trevor choose η to be the smallest integer such that there are values of p_u and p_d between 0 and 1 that solve (12.15) and (12.16). These values are

$$p_u = \frac{h_t^a}{2\eta^2\gamma^2} + \frac{(r - h_t^a/2)}{2\eta\gamma} \text{ and } p_d = \frac{h_t^a}{2\eta^2\gamma^2} - \frac{(r - h_t^a/2)}{2\eta\gamma}. \tag{12.17}$$

The approximate GARCH process y_t^a evolves according to

$$y_{t+1}^a = y_t^a + j\eta\gamma, \tag{12.18}$$

where j is -1, 0, or 1, and

$$h_{t+1}^a = \beta_0 + \beta_1 h_t^a + \beta_2 h_t^a(\epsilon_{t+1}^a - c^*)^2, \tag{12.19}$$

where

$$\epsilon_{t+1}^a = \frac{j\eta\gamma - (y_t + r - h_t/2)}{\sqrt{h_t^a}} \tag{12.20}$$

is the move from y_t^a to y_{t+1}^astandardized by its mean and standard deviation, $E_t(y_{t+1}^a)$ and $\sqrt{\mathrm{Var}_t(y_{t+1}^a)}$.

The trinomial tree has been constructed so that it is recombinant. A node of the tree is a pair (t, i), where t is the day and i is the net number[13] of up-steps since day 0. The log price of the asset at node (t, i) is $y_0^a + \gamma i$. There are many paths to any single node, but at the node there is a single value of the log price. However, the value of h_t^a at node (t, i) depends upon that path to that node. The algorithm would become extremely complex if it kept track of all paths to a node, since the number of paths from the initial node $(0, 0)$ to a node (t, i) grows geometrically with t and gets quite large as the algorithm proceeds. Ritchken and Trevor have a clever solution to this problem. They keep track of the minimum and maximum conditional variance h_t^a that can occur at each node and compute option prices for K values of the variance that are equally spaced between the minimum and maximum. Here K is a parameter of the algorithm that is selected as a trade-off between speed and parsimony (K small) and accuracy (K large). Ritchken and Trevor found that a value of K equal to 20 appears satisfactory and increasing K from 20 to 40 produces almost no change in computed option prices. What makes this component of the algorithm computationally efficient is that the maximum and minimum conditional variances at a node (t, i) are easily determined from the maximum and minimum of the conditional variances at nodes on day $t-1$.

Option prices are computed on the tree using the standard method of starting at terminal nodes and working backwards. At each node there are K option prices, one for each value of the conditional variance at that node. At the terminal nodes, all K prices are the same since the value of the option at maturity depends only upon the asset's price, not its conditional variance. However, at nodes before the terminal node, the option's price depends on both the asset's price and its conditional variance.

As just mentioned, a node is denoted by (t, i), where t is day and i is the net number of up-steps since day 0. At node (t, i), let $C_t^a(i, k)$ be the option price corresponding to the kth possible variances $h_t^a(i, k)$, $k = 1, \ldots, K$. For a call option with maturity T and strike price[14] X one has

$$C_T^a(i, k) = (S_T^a(i) - X)_+, \text{ for } k = 1, \ldots, K. \tag{12.21}$$

At day t, $t < T$, each node (t, i) has three successor nodes. For each successor node there are K prices depending on variance. The jump from a node at time t to a node at time $t + 1$ typically ends up with a variance at time $t + 1$ that is not one of the variances for which an option price has been determined. In this case, the option price is found by linear interpolation. Then the expected price of the unexercised option can be found as a weighted average of the option prices at the three successor nodes using p_u, p_d, and $(1 - p_u - p_d)$ as the probabilities of up, down, and no moves. This quantity

[13] The net number of up-steps is the number of up-steps minus the number of down-steps.

[14] The usual notation K for strike price has another meaning in the model of Ritchken and Trevor.

is discounted by e^{-r}. For a European option, this is the expected discounted option price. For an American option the expected discounted option price is the maximum of this value and the value of the option if exercised, which is $(S_t^a(i) - X)_+$ for a call. The expected discounted option price at time t is computed K times, once for each possible variance at the node at time t and thus gives $C_t^a(i,k)$, $k = 1, \ldots, K$.

The algorithm was described by Ritchken and Trevor for the NGARCH model (12.11) and (12.12), but as they mention it can be used with other GARCH models as well.

12.16 Summary

The marginal, or unconditional, distribution of a stationary process is the distribution of an observation from the process given no information about the previous or future observations. Under the assumption of stationarity the marginal distribution must, by definition of stationarity, be constant. In particular, the marginal mean and variance are constant.

Besides the marginal distribution, we are interested in the conditional distribution of the next observation given the current information set of present and past values of the process, and perhaps of other processes. For ARMA processes the conditional mean is nonconstant but the conditional variance is constant. The constant conditional variance of ARMA processes makes them unsuitable for modeling the volatility of financial markets. GARCH processes have nonconstant conditional variance and were developed to model changing volatility. GARCH processes can be used as the "noise" term of an ARMA process. ARMA/GARCH processes have both a nonconstant conditional mean and a nonconstant conditional variance.

GARCH and ARMA/GARCH processes can be estimated by maximum likelihood. PROC AUTOREG in SAS fits AR/GARCH models. The simple ARCH(q) models have bursts of volatility but cannot model persistent volatility. The generalized ARCH (GARCH) models can model persistent volatility. The marginal distribution of a GARCH process has heavier tails than the normal distribution. In fact, for certain parameter values a GARCH process has an infinite variance, which is an extreme case of heavy tails. I-GARCH (integrated GARCH) models are examples of GARCH models with infinite variance.

If the marginal variance is infinite, then the sample variance converges to infinity as the sample size increases. For extremely heavy tails, the marginal expectation may not exist. Then there exists no point to which the sample mean can converge and the sample mean wanders aimlessly.

ARMA/GARCH processes can be used as the noise term in regression models. SAS's PROC AUTOREG can use an AR/GARCH noise term in a regression model. The GARCH-M models use the conditional standard deviation as an independent variable in the regression. The "leverage effect" occurs when a negative return (drop in price) increases the volatility of future returns because the denominator of those returns is smaller. E-GARCH models were designed to capture the leverage effect. In an E-GARCH model, the log of the conditional standard deviation is modeled as an ARMA process but with the white noise process ϵ_t replaced by another white noise process $g(\epsilon_t) = \theta\epsilon_t + \gamma\{|\epsilon_t| - E(|\epsilon_t|)\}$. There is no need for nonnegativity constraints on the parameters, such as those in an ordinary GARCH model, since the log standard deviation can be negative. The parameter θ in an E-GARCH model determines the leverage effects.

There are many of other GARCH models in the literature, but the ones discussed here, ARCH(q), GARCH(p, q), E-GARCH, GARCH-M, and I-GARCH, can model a wide variety of data types and can be fit by SAS. There is a large and growing literature on financial models with returns following GARCH processes.

12.17 Bibliographic Notes

There is a vast literature on GARCH processes beginning with Engle (1982) where ARCH models were introduced. Enders (1995), Pindyck and Rubinfeld (1998), Gourieroux and Jasiak (2001), Alexander (2001), and Tsay (2002) have chapters on GARCH models. There are many review articles including Bollerslev (1986), Bera and Higgins (1993), Bollerslev, Engle, and Nelson (1994), and Bollerslev, Chou, and Kroner (1992). Jarrow (1998) and Rossi (1996) contain a number of papers on volatility in financial markets. SAS Institute (1993) describes the use of PROC AUTOREG for GARCH modeling. Duan (1995), Ritchken and Trevor (1999), Heston and Nandi (2000), Hsieh and Ritchken (2000), Duan and Simonato (2001), and many other authors study the effects of GARCH errors on options pricing and Bollerslev, Engle, and Wooldridge (1988) use GARCH models in the CAPM.

12.18 References

Alexander, C. (2001) *Market Models: A Guide to Financial Data Analysis*, Wiley, Chichester.

Bera, A. K., and Higgins, M. L. (1993) A survey of Arch models, *Journal of Economic Surveys*, **7**, 305–366. (Reprinted in Jarrow (1998).)

Bollerslev, T. (1986) Generalized autoregressive conditional heteroskedasticity, *Journal of Econometrics*, **31**, 307–327.

Bollerslev, T. and Engle, R. F. (1993) Common persistence in conditional variances, *Econometrica*, **61**, 167–186.

Bollerslev, T., Chou, R. Y., and Kroner, K. F. (1992) ARCH modelling in finance, *Journal of Econometrics*, **52**, 5–59. (Reprinted in Jarrow (1998))

Bollerslev, T., Engle, R. F., and Nelson, D. B. (1994) ARCH models, In *Handbook of Econometrics, Vol IV*, Engle, R.F., and McFadden, D.L., Elsevier, Amsterdam.

Bollerslev, T., Engle, R. F., and Wooldridge, J. M. (1988) A capital asset pricing model with time-varying covariances, *Journal of Political Economy*, **96**, 116–131.

Carroll, R. J. and Ruppert, D. (1988) *Transformation and Weighting in Regression*, Chapman & Hall, New York.

Duan, J.-C., (1995) The GARCH option pricing model, *Mathematical Finance*, **5**, 13–32. (Reprinted in Jarrow (1998).)

Duan, J-C. and Simonato, J. G. (2001) American option pricing under GARCH by a Markov chain approximation, *Journal of Economic Dynamics and Control*, **25**, 1689–1718.

Enders, W. (1995) *Applied Econometric Time Series*, Wiley, New York.

Engle, R. F. (1982) Autoregressive conditional heteroskedasticity with estimates of variance of U.K. inflation, *Econometrica*, **50**, 987–1008.

Engle, R. F. and Ng, V. (1993) Measuring and testing the impact of news on volatility, *Journal of Finance*, **4**, 47–59.

Gourieroux, C. and Jasiak, J. (2001) *Financial Econometrics*, Princeton University Press, Princeton, NJ.

Heston, S. and Nandi, S. (2000) A closed form GARCH option pricing model, *The Review of Financial Studies*, **13**, 585–625.

Hsieh, K. C. and Ritchken, P. (2000) An empirical comparison of GARCH option pricing models, working paper.

Jarrow, R. (1998) *Volatility: New Estimation Techniques for Pricing Derivatives*, Risk Books, London. (This is a collection of articles, many on GARCH models or on stochastic volatility models, which are related to GARCH models.)

Pindyck, R. S. and Rubinfeld, D.L. (1998) *Econometric Models and Economic Forecasts*, Irwin/McGraw Hill, Boston.

Ritchken, P. and Trevor, R. (1999) Pricing options under generalized GARCH and stochastic volatility processes, *Journal of Finance*, **54**, 377–402.

Rossi, P. E. (1996) *Modelling Stock Market Volatility*, Academic, San Diego.

SAS Institute (1993) *SAS/ETS User's Guide, Version 6, 2nd Ed.*, SAS Institute, Cary, NC.

Tsay, R. S. (2002) *Analysis of Financial Time Series*, Wiley, New York.

12.19 Problems

1. Let Z have a N(0,1) distribution. Show that

$$E(|Z|) = \int_{-\infty}^{\infty} \frac{1}{\sqrt{2\pi}} |z| e^{-z^2/2} dz = 2 \int_0^{\infty} \frac{1}{\sqrt{2\pi}} z e^{-z^2/2} dz = \sqrt{\frac{2}{\pi}}.$$

Hint: $\frac{d}{dz}\, e^{-z^2/2} = -z e^{-z^2/2}$.

2. Suppose that $f_X(x) = 1/4$ if $|x| < 1$ and $f_X(x) = 1/(4x^2)$ if $|x| \geq 1$. Show that
$$\int_{-\infty}^{\infty} f_X(x)dx = 1$$
so that f_X really is a density, but that
$$\int_{-\infty}^{0} x f_X(x)dx = -\infty$$
and
$$\int_{0}^{\infty} x f_X(x)dx = \infty,$$
so that a random variable with this density does not have an expected value.
3. Suppose that ϵ_t is a WhiteNoise(0, 1) process, that
$$a_t = \epsilon_t \sqrt{1 + 0.5 a_{t-1}^2},$$
and that
$$u_t = 3 + 0.7 u_{t-1} + a_t.$$
(a) Find the mean of u_t.
(b) Find the variance of u_t.
(c) Find the autocorrelation function of u_t.
(d) Find the autocorrelation function of a_t^2.
4. Let u_t be the AR(1)/ARCH(1) model
$$a_t = \epsilon_t \sqrt{\alpha_0 + \alpha_1 \, a_{t-1}^2},$$
$$(u_t - \mu) = \phi(u_{t-1} - \mu) + a_t,$$
where ϵ_t is WhiteNoise(0,1). Suppose that $\mu = 0.7$, $\phi = 0.5$, $\alpha_0 = 1$, and $\alpha_1 = 0.3$.
(a) Find $E(u_2 | u_1 = 1, u_0 = 0.2)$.
(b) Find $\text{Var}(u_2 | u_1 = 1, u_0 = 0.2)$.
5. Suppose that ϵ_t is white noise with mean 0 and variance 1, that $a_t = \epsilon_t \sqrt{7 + a_{t-1}^2/2}$, and that $Y_t = 3 + 0.6 Y_{t-1} + a_t$.
(a) What is the mean of Y_t?
(b) What is the ACF of Y_t?
(c) What is the ACF of a_t?
(d) What is the ACF of a_t^2?
6. Let Y_t be a stock's return in time period t and let X_t be the inflation rate during this time period. Assume the GARCH-M model
$$Y_t = \beta_0 + \beta_1 X_t + \delta \sigma_t + a_t,$$

where

$$a_t = \epsilon_t \sqrt{1 + 0.5 a_{t-1}^2}.$$

Here the ϵ_t are independent $N(0,1)$ random variables. Assume that $\beta_0 = .05$, $\beta_1 = .3$, and $\delta = 0.2$.

(a) What is $E(Y_t | X_t = .1 \text{ and } a_{t-1} = 0.6)$?
(b) What is $\text{Var}(Y_t | X_t = .1 \text{ and } a_{t-1} = 0.6)$?
(c) Is the conditional distribution of Y_t given X_t and a_{t-1} normal? Why or why not?
(d) Is the marginal distribution of Y_t normal? Why or why not?

13

Nonparametric Regression and Splines

13.1 Introduction

As we have seen in Chapter 6, regression is about modeling the conditional expectation of a response given predictor variables. The conditional expectation is called the regression function and is the best possible predictor of the response based upon the predictor variables. *Linear regression* assumes that the regression function is a linear function and estimates the intercept and slope, or slopes if there are multiple predictors. *Nonlinear parametric regression*[1] does not assume linearity but does assume that the regression function is of a *known* parameter form, for example, an exponential function. In this chapter, we study **nonparametric regression** where the form of the regression function is also nonlinear but, unlike nonlinear regression, not specified by a model but rather estimated from data. Nonparametric regression is used when we know or suspect that the regression function is curved, but we do not have a model for the curve.

There are many techniques for nonparametric regression, but in my experience splines are easy to use as well as to understand because they are a natural extension of linear regression. As mentioned in Section 9.9, a spline is a function constructed by piecing together polynomial functions. The spline modeling techniques studied in this chapter can be used in a wide variety of practical problems including modeling financial markets data. Another simple and effective method of nonparametric regression is local polynomial regression. Estimates using local polynomial regression are usually similar to spline estimates, and I believe it is best to understand one of the two methods well rather than to have some familiarity with both. For this reason, only splines are discussed here. Good introductions to local polynomial regression can be found in Wand and Jones (1995) and Fan and Gijbels (1996).

[1] Nonlinear parametric regression is generally called nonlinear regression, but the adjective "parametric" has been added to emphasize the difference with nonparametric regression which is also nonlinear.

Models for evolution of short-term interest rates are important in finance, for example, because they are needed for the pricing of interest rate derivatives. In this chapter, we use an example of short-term Euro rates introduced by Yau and Kohn (2001), two statisticians at the Australian Graduate School of Management in Sydney. Figure 13.1 shows data on the interest rate of Eurodollar[2] deposits of one-month maturity. The top plot is the weekly time series of the Euro interest rate. The bottom plot is the time series of weekly changes in the Euro rate which were found by differencing the original series. What interests us here is how the volatility of the changes in the interest rate depends on the current value of the rate. To begin addressing this question, Figure 13.2 shows the differences and squared differences of the interest rate plotted again the rate itself, that is, $(r_{t+1} - r_t)$ and $(r_{t+1} - r_t)^2$ plotted against r_t where r_t is the interest rate at time t.

A common model for changes in short-term interest rates is

$$\Delta r_t = \mu(r_{t-1}) + \sigma(r_{t-1})\epsilon_t, \tag{13.1}$$

where $\Delta r_t = r_t - r_{t-1}$, $\mu(\cdot)$ is the drift function, $\sigma(\cdot)$ is the volatility function, also called the diffusion function, and ϵ_t is $N(0,1)$ noise. As discussed in Yau and Kohn (2001), many different parametric models have been proposed for $\mu(\cdot)$ and $\sigma(\cdot)$, for example, by Merton (1973), Vasicek (1977), Cox, Ingersoll, and Ross (1985), and Chan et al. (1992). The simplest model, the one due to Merton (1973), is that $\mu(\cdot)$ and $\sigma(\cdot)$ are constant. Chan et al. (1992) assume that $\mu(r) = \beta(r - \alpha)$ and $\sigma(r) = \theta r^\gamma$ where α, β, θ, and γ are unknown parameters. The approach of Yau and Kohn (2001) that is followed here is to model both $\mu(\cdot)$ and $\sigma(\cdot)$ nonparametrically. Doing this allows one to check which parametric models fit the data, if any, and to have a nonparametric alternative if none of the parametric models fits well.

The solid curves in Figure 13.2 are estimates of $\mu(\cdot)$ and $\sigma^2(\cdot)$ based on a nonparametric regression method called *penalized splines*.[3] The estimate of $\mu(\cdot)$ seems to be zero or nearly so. It is assumed that $\mu(\cdot)$ is 0, in which case

$$E\{(\Delta r_t)^2 | r_{t-1}\} = \sigma^2(r_{t-1}).$$

Therefore, $\sigma^2(\cdot)$ is estimated by regressing $(\Delta r_t)^2$ on r_{t-1}. If μ appeared to be nonzero, then $\sigma^2(\cdot)$ would be estimated by regressing $\{\Delta r_t - \widehat{\mu}(r_t)\}^2$ on r_{t-1}. The goal of this chapter is to explain how curves such as the ones in Figure 13.2 can be obtained.

[2] A Eurodollar is a dollar held on deposit at a bank outside the United States (Marshall, 2000).

[3] Yau and Kohn (2001) also use spline-based estimates though not penalized splines.

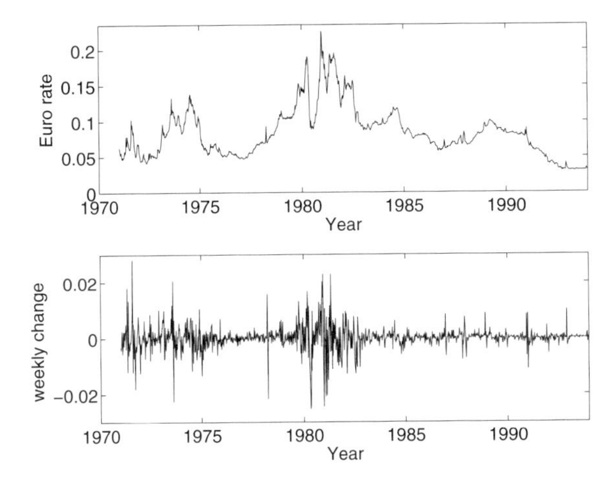

Fig. 13.1. *Study of volatility in weekly changes of the one-month Euro bond rate. The top plot is the time series of the weekly values of the interest rate. The bottom plot is the time series of changes in the interest rate.*

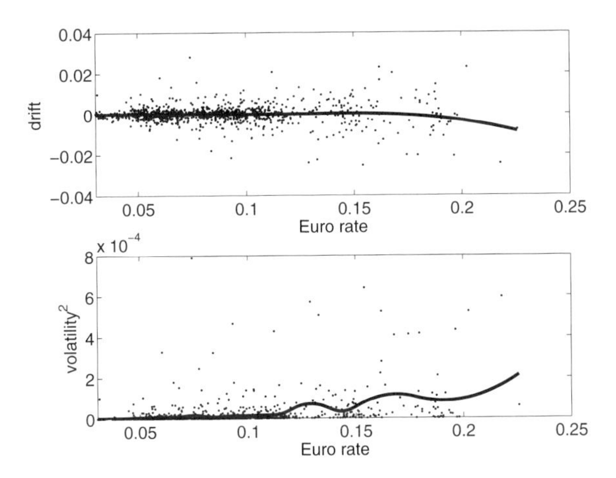

Fig. 13.2. *Study of volatility and drift in weekly changes of the Euro bond rate. The top plot is a plot of weekly rate changes against the rate itself. The solid curve is a 20-knot P-spline estimate of the drift function $\mu(\cdot)$. The bottom plot is a plot of squared weekly rate changes against the rate. The solid curve is a 20-knot P-spline estimate of the squared volatility function $\sigma^2(\cdot)$. The penalty parameters of both P-splines were selected by GCV.*

13.2 Choosing a Regression Method

Here are some guidelines to help decide when to use linear, nonlinear parametric, or nonparametric regression.

13.2.1 Nonparametric regression

Nonparametric regression can be used when there is no theory to suggest the functional form of the regression function.

Nonparametric regression is most useful when there is enough information in the data that the form of the regression function can be recovered without use of prior knowledge or theory. In particular, it is desirable that there are enough data points, that the noise in the data is sufficiently small, or that the predictor variable(s) vary over a wide enough range that nonlinearities in the regression function can be detected. When none of these three conditions are met, then nonparametric is not necessarily inappropriate but may not be much different from using linear regression.

When one has little idea about the shape of the regression function, then nonparametric regression is a good place to start. If the regression function really is linear, then the linearity should be apparent in the nonparametric fit. It is common to compare linear and nonparametric regression estimates before deciding which to use.

13.2.2 Linear

Linear regression is often used when there is no theory to suggest the shape of the regression function and the data show little or no evidence of nonlinearity. If a scatterplot of the response versus the predictor variable appears linear, then either the regression function is linear or it is nonlinear but the data are not informative enough to reveal the nonlinearity. The data can be noninformative about the shape of the regression function if the data are very noisy, there are too few data, and the predictor variable(s) varies too little. In this case, there is no guarantee that the regression function is linear, but if the data do not reveal nonlinearities then there is little point in not using linear regression.

More rarely, linear regression is used because there is a theoretical model that suggests that the regression function is, in fact, linear.

13.2.3 Nonlinear parametric regression

As discussed in Section 6.10, nonlinear parametric regression is used when there is a theory that suggests a specific nonlinear model, for example, equation (9.31) for the price of a zero-coupon bond. Sometimes a nonlinear model is used, not because it was derived from theory, but because experience has shown that the model generally fits data of a particular type.

Nonlinear parametric regression is also used when the range of the response is restricted, for example, when the response is a proportion so that it lies between 0 and 1. In such cases, one can use a nonlinear regression function whose range is the same as the range of the response.

13.2.4 Comparison of linear and nonparametric regression

Figures 13.3 to 13.7 use simulated data to illustrate when linear regression is appropriate and when it is preferable to use nonparametric regression. In all five figures the regression function is nonlinear, $2x^2 \sin(4x)$, but it is assumed that this functional form of the curve is unknown to the data analyst. The figures differ in their values of σ, n, and the range of x. In each figure, a fit by linear regression is shown along with a nonparametric fit using a quadratic P-spline (penalized spline). P-splines are introduced in Section 13.8.

In Figure 13.3, $\sigma = .4$, $n = 15$, and the range of x is $[0, 0.7]$. With this relatively large value of σ, small sample size, and short range of x, the nonlinearity of the regression function is difficult to detect. The linear fit and the nonparametric fit are similar and neither shows any of the curvature seen in the true regression function. Although in this example nonparametric regression cannot successfully recover the curvature in the regression function, using nonparametric regression is not inappropriate here. Nonparametric regression does as well as possible with this data set and gives essentially the same answer as linear regression so one cannot say that using linear regression would be a much better alternative.

In each of the next three figures, we vary one of the parameters in Figure 13.3 so that

1. The noise σ is smaller,
2. The range of x is larger, or
3. The sample size n is larger.

Any one of these changes allows us to detect the curvature in the regression function.

Figure 13.4 is similar to Figure 13.3 except that σ is much smaller, only 0.05, in Figure 13.4 compared to 0.4 in Figure 13.3. Because of the low noise, the nonlinearity in the regression can be detected in Figure 13.4, and compared to the linear fit the nonparametric fit is much closer to the true regression function. Notice that the nonparametric fit comes quite close to the true regression function without having any prior knowledge of the shape of that curve. Figure 13.5 is similar to Figure 13.3 except that the range of x is larger in Figure 13.5 so that the nonlinearity in the regression function can be detected. In Figure 13.6, the range of x is short and σ is large, but the nonlinearity can be detected because the sample size is large, $n = 300$.

Bear in mind that all differentiable functions look like straight lines over very short ranges of x. For example, Figure 13.7 is similar to Figure 13.4 except the range of x is very short, $[0, 0.2]$. Within this short range, the regression function is relatively linear. Even though σ here is the same small value as in Figure 13.4, the deviation from linearity is no longer obvious. Therefore, in Figure 13.7 the linear and nonparametric fits are similar.

Although the nonparametric fit in Figure 13.7 shows some curvature, the curvature is due to the nonparametric fit following the noise in the data, not the signal. For example, by chance the three leftmost data points in Figure 13.7 all lie below the true curve. The nonparametric fit dips down to follow these data points and misses the true curve by a substantial amount. The linear fit cannot dip down like this and on the left side of the plot, compared to the nonparametric fit, the linear fit is somewhat closer to the true curve.

It never hurts to look at a nonparametric regression estimate and to compare this estimate with a parametric estimate.

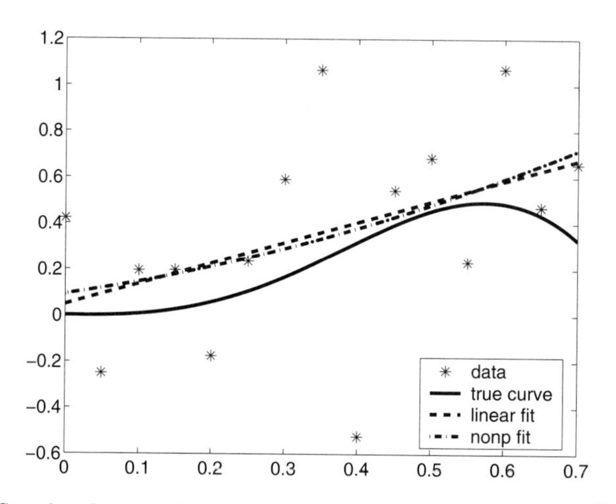

Fig. 13.3. *Simulated regression example with regression function* $2x^2 \sin(4x)$. *Here* $\sigma = 0.4$, $n = 15$, *and the range of* x *is* $[0, 0.7]$. *In this example,* σ *is relatively large,* n *is small, and the range of* x *is relatively small. Therefore, the nonlinearity in the regression function cannot be detected from the data and the nonparametric and linear regression fits are similar.*

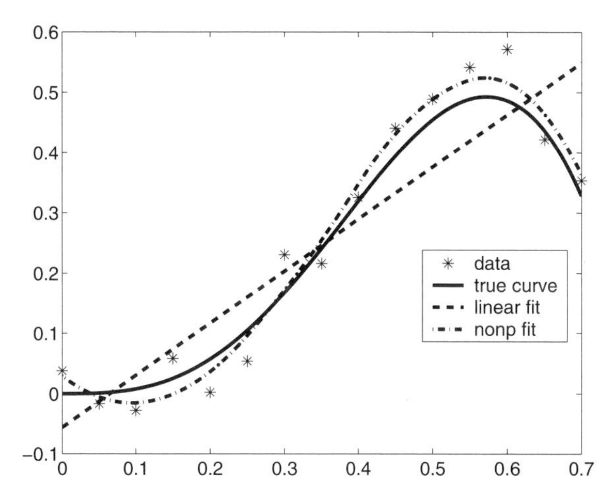

Fig. 13.4. *Simulated regression example with regression function* $2x^2\sin(4x)$*. Here* $\sigma = 0.05$*,* $n = 15$*, and the range of* x *is* $[0, 0.7]$*. Because of the low noise, the nonlinearity of the regression function is evident and the nonparametric fit is close to the true curve.*

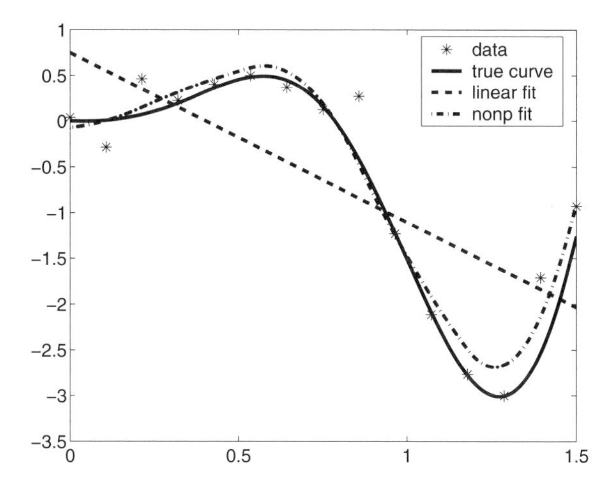

Fig. 13.5. *Simulated regression example with regression function* $2x^2\sin(4x)$*. Here* $\sigma = 0.4$*,* $n = 15$*, and the range of* x *is* $[0, 1.5]$*. Because of the wide range of* x*, the nonlinearity is seen in the data and captured by the nonparametric fit.*

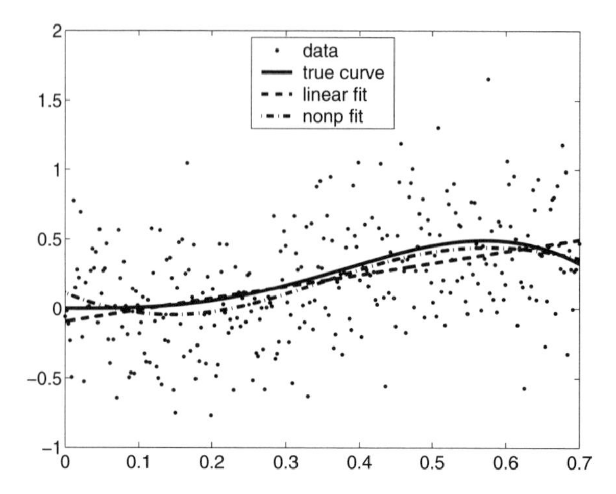

Fig. 13.6. *Simulated regression example with regression function* $2x^2\sin(4x)$. *Here* $\sigma = 0.4$, $n = 300$, *and the range of* x *is* $[0, 0.7]$. *Because of the large sample size, the nonlinearity in the regression function can be detected.*

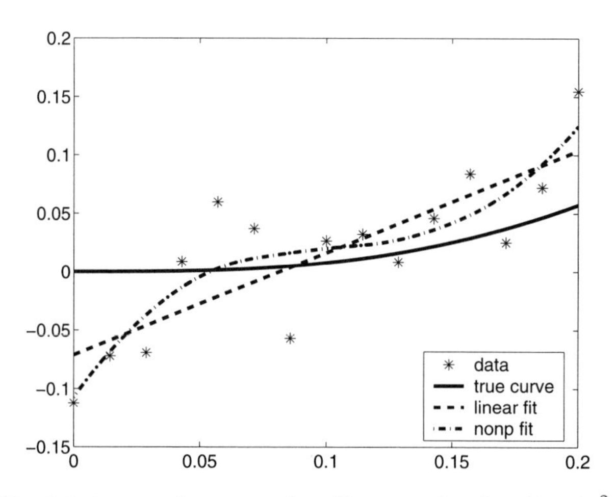

Fig. 13.7. *Simulated regression example with regression function* $2x^2\sin(4x)$. *Here* $\sigma = 0.05$, $n = 15$, *and the range of* x *is* $[0, 0.2]$. *Because of the short range of* x, *the nonlinearity of the regression function is difficult to detect. By chance, the three leftmost Ys are all less than the regression function which causes the nonparametric fit to be worse than the linear fit.*

13.3 Linear Splines

A linear spline is constructed by piecing together linear functions so that they join together at specified locations called "knots." Many of the concepts needed to understand splines can be introduced in the simple case of a linear spline with only a single knot.

13.3.1 Linear splines with one knot

We start simple, a linear spline with one knot. Figure 13.8 illustrates such a spline. This spline is defined as

$$f(x) = \begin{cases} 0.5 + 0.2x, & x < 2, \\ -0.5 + 0.7x, & x \geq 2. \end{cases}$$

Because $0.5 + 0.2x = 0.9 = -0.5 + 0.7x$ when $x = 2$, the two linear components are equal at the point $x = 2$ so that they join together there.

The point $x = 2$ where the spline switches from one linear function to the other is called a **knot**. A linear spline with a knot at the point t can be constructed as follows. The spline is defined to be $s(x) = a + bx$ for $x < t$ and $s(x) = c + dx$ for $x > t$. The parameters a, b, c, and d can be chosen arbitrarily except that they must satisfy the equality constraint

$$a + bt = c + dt, \tag{13.2}$$

which assures us that the two lines join together at t. Solving for c in (13.2), we get $c = a + (b - d)t$. Substituting this expression for c into the definition of $s(x)$ and doing some rearranging, we have

$$s(x) = \begin{cases} a + bx, & x < t, \\ a + bx + (d - b)(x - t), & x \geq t. \end{cases} \tag{13.3}$$

Recall the definition that for any number y

$$(y)_+ = \begin{cases} 0, & y < 0, \\ y, & y \geq 0. \end{cases}$$

By this definition

$$(x - t)_+ = \begin{cases} 0, & x < t, \\ x - t, & x \geq t. \end{cases}$$

We call $(x - t)_+$ a linear *plus function* with a knot at t. The spline $s(x)$ in (13.3) can be written using this plus function:

$$s(x) = a + bx + (d - b)(x - t)_+.$$

In summary, if we want a linear spline that equals $a + bx$ for $x < t$ and then has its slope jump from b to d at $x = t$, then the spline is $a + bx$ plus the jump $(d - b)$ times the plus function $(x - t)_+$.

Figure 13.9 illustrates a linear plus function with a knot at 1 and its first derivative. Notice that

$$\frac{d}{dx}(x-t)_+ = \begin{cases} 0, & x < t, \\ 1, & x \geq t. \end{cases}$$

A linear spline with a single knot was used in Section 7.9 and can be seen in Figure 7.5. The plus function was created as a variable named sp2 in the data step of the SAS program beginning on page 247.

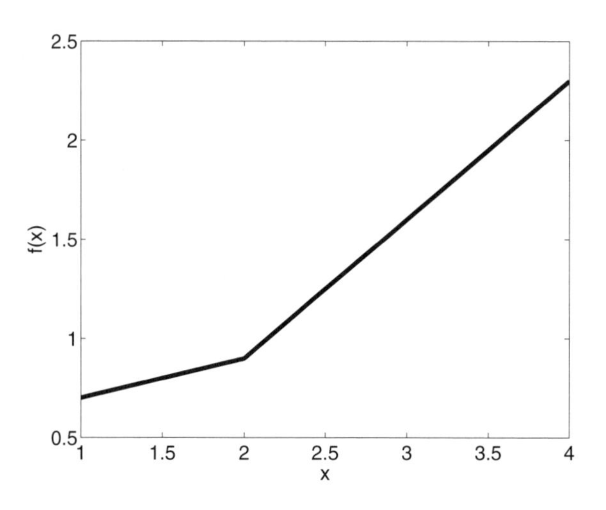

Fig. 13.8. *Example of a linear spline with a knot at 2.*

13.3.2 Linear splines with many knots

Plus functions are very convenient when defining linear splines with more than one knot because plus functions automatically join the component linear functions together so that the spline is continuous. For example, suppose we want a linear spline to have K knots, $t_1 < \cdots < t_K$, for the spline to equal $s(x) = \beta_0 + \beta_1 x$ for $x < t_1$, and for the first derivative of the spline to jump by the amount b_k at knot t_k, for $k = 1, \ldots, K$. Then the spline can be constructed from linear plus functions, one for each knot:

$$s(x) = \beta_0 + \beta_1 x + b_1(x-t_1)_+ + b_2(x-t_2)_+ + \cdots + b_K(x-t_K)_+.$$

Because the plus functions are continuous, the spline is the sum of continuous functions and is therefore continuous itself. The continuity of the spline is automatically inherited from the plus functions.

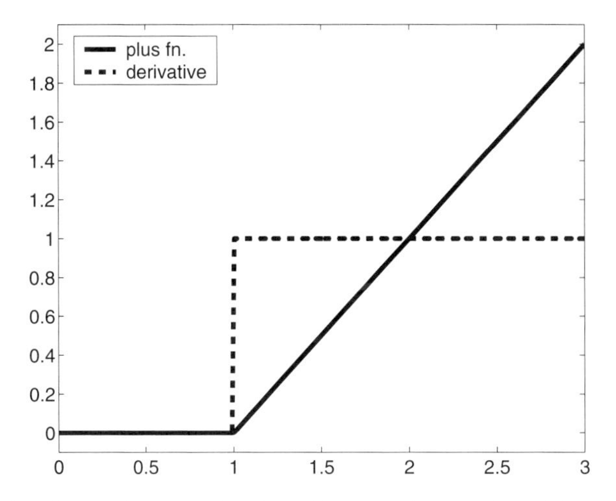

Fig. 13.9. *The linear plus function* $(x-1)_+$ *with knot at* 1 *and its first derivative.*

13.4 Other Degree Splines

13.4.1 Quadratic splines

A linear spline is continuous but has "kinks" at its knots where its first derivative jumps. If we want a function without these kinks, we cannot use a linear spline. A quadratic spline is a function obtained by piecing together quadratic polynomials. More precisely, $s(x)$ is a quadratic spline with knots $t_1 < \cdots < t_K$ if $s(x)$ equals one quadratic polynomial to the left of t_1 and equals a second quadratic polynomial between t_1 and t_2, and so on. The quadratic polynomials are pieced together so that the spline is continuous and, to guarantee no kinks, its first derivative is also continuous.

As with linear splines, continuity can be enforced by using plus functions. Define the quadratic plus function

$$\begin{aligned}(x-t)_+^2 &= 0, & x < t \\ &= (x-t)^2, & x \geq t.\end{aligned}$$

Notice that $(x-t)_+^2$ equals $\{(x-t)_+\}^2$, not $\{(x-t)^2\}_+ = (x-t)^2$.

Figure 13.10 shows a quadratic plus function and its first and second derivative. One can see that

$$\frac{d}{dx}(x-t)_+^2 = 2(x-t)_+$$

and

$$\frac{d^2}{dx^2}(x-t)_+^2 = 2(x-t)_+^0,$$

where $(x-t)_+^0 = \{(x-t)_+\}^0$ so that $(x-t)_+^0$ is the 0th degree plus function

$$(x-t)_+^0 = 0, \quad x < t,$$
$$\qquad\quad = 1, \quad x \geq t.$$

Therefore, the second derivative of $(x-t)_+^2$ jumps from 0 to 2 at the knot t.

A quadratic spline with knots $t_1 < \cdots < t_K$ can be written as

$$s(x) = \beta_0 + \beta_1 x + \beta_2 x^2 + b_1(x-t_1)_+^2 + b_2(x-t_2)_+^2 + \cdots + b_K(x-t_K)_+^2.$$

The second derivative of s jumps by the amount $2b_k$ at knot t_k for $k = 1, \ldots, K$.

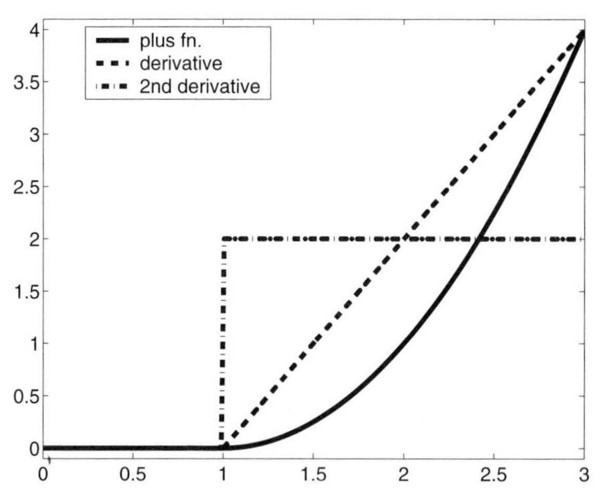

Fig. 13.10. *The quadratic plus function $(x-1)_+^2$ with knot at 1 and its first and second derivatives.*

13.4.2 pth degree splines

The way to define a general pth degree spline with knots $t_1 < \cdots < t_K$ should now be obvious:

$$s(x) = \beta_0 + \beta_1 x + \cdots + \beta_p x^p + b_1(x-t_1)_+^p + \cdots + b_K(x-t_K)_+^p, \tag{13.4}$$

where as we have seen for the specific case of $p = 2$, $(x-t)_+^p$ equals $\{(x-t)_+\}^p$. The first $p-1$ derivatives of s are continuous while the pth derivative takes a jump equal to $p!\, b_k$ at the kth knot.

It has been my experience that linear and quadratic splines work well in practice and I do not see a need to use cubic or higher-degree splines except for some special purposes.[4] One application where higher-order splines are useful is when one wants to estimate not the regression function itself, but rather one of the derivatives of the regression function. The derivative of a pth degree spline is a spline of degree $p-1$. So, for example, if one starts with a linear spline and differentiates it, then one obtains a 0-degree spline, which takes jumps that may not be desirable.[5]

13.5 Least Squares Estimation

A pth degree spline with knots $t_1 < \cdots < t_K$ can be easily fit to data by least squares. One simply sets up a multiple regression model with predictor variables $x, \ldots, x^p, (x-t_1)_+^p, \ldots, (x-t_K)_+^p$. Suppose that Y_i is the response variable and X_i is the predictor. The regression model is

$$Y_i = \beta_0 + \beta_1 X_i + \cdots + \beta_p X_i^p + b_1(X_i - t_1)_+^p + \cdots + b_K(X_i - t_K)_+^P + \epsilon_i \quad (13.5)$$

so that the regression function is the function s given by equation (13.4).

Here is a SAS program to fit a linear spline to the data shown in the top of Figure 13.2. The first section of the program is the data step that reads in three variables, eu01, eu03, and eu06. These variables are one-, three-, and six-month Euro dollar interest rates. Only eu01 is used in the program and it is divided by 100 to convert from a percentage. In the data step eu01 is differenced to create the variable diff and lagged to create rate, so that diff is the variable Δr_t in equation (13.1) and rate is the variable r_{t-1} in that equation. The variables plus1, plus2, plus3, and plus4 are linear plus functions with knots at 0.08, 0.12, 0.16, and 0.2, respectively.

```
data EuroRate ;
infile 'c:\book\SAS\euro_rates.dat' ;
input month day year eu01 eu03 eu06;
eu01 = eu01/100 ;
diff=dif(eu01) ;
rate = lag(eu01) ;
plus1 = (rate - 0.08) * (rate > 0.08) ;
plus2 = (rate - 0.12) * (rate > 0.12) ;
plus3 = (rate - 0.16) * (rate > 0.16) ;
plus3 = (rate - 0.2)) * (rate > 0.2) ;
run ;
```

In the remaining section of the program, a linear spline is fit using PROC REG by regressing diff on rate and the four plus functions. The fitted values are saved in the output data set. The fitted spline is plotted using

[4] This is not to say that there is anything wrong with using cubic or higher-degree splines. In fact, cubic splines are used routinely. However, I see no reason why they are better than quadratic splines for most purposes.

[5] A 0-degree spline is a constant (0-degree polynomial) between its knots and takes jumps at the knots. 0-degree splines are often called step functions or piecewise constant functions.

PROC GPLOT by plotting the fitted values against `rate`; see Figure 13.11. Although the spline was very easy to plot in this manner, there is a drawback in that the spline is not shown as the continuous function that it is but rather as an asterisk (the plotting symbol) above each of the observed values of `rate`.

```
title 'One Month Euro dollar deposit rates' ;
title1 'Linear spline estimate of drift' :
proc reg ;
model diff = rate plus1 plus2 plus3 plus4 ;
output out=EuroOut p=yhat ;
run ;
proc gplot ;
plot yhat*rate ;
run ;
```

Notice that the units of the vertical axis are 10^{-3}, so the estimated drift function is very close to zero. The decrease at the right boundary is due to only a few observations. The kinks in the fitted function at the knots are evident.

Here is a modification of the SAS program to fit a quadratic spline. Notice that the variable `rate2`, which is `rate` squared, is added to the data set and the plus functions have been changed to quadratic plus functions. The plot of fitted values is not shown, being similar to that in Figure 13.11 but without the kinks seen in the linear spline estimate.

```
data EuroRate ;
infile 'c:\book\SAS\euro_rates.dat' ;
input month day year eu01 eu03 eu06;
eu01 = eu01/100 ;
diff=dif(eu01) ;
rate = lag(eu01) ;
rate2 = rate**2 ;
plus1 = ((rate - 0.08)**2) * (rate > 0.08) ;
plus2 = ((rate - 0.12)**2) * (rate > 0.12) ;
plus3 = ((rate - 0.16)**2) * (rate > 0.16) ;
plus4 = ((rate - 0.2)**2) * (rate > 0.2) ;
run ;
title 'One Month Euro dollar deposit rates' ;
title2 'Quadratic spline estimate of drift' ;
proc reg ;
model diff = rate rate2 plus1 plus2 plus3 plus4 ;
output out=EuroRate p=yhat ;
run ;
proc gplot ;
plot yhat*rate ;
run ;
```

13.6 Selecting the Spline Parameters

When using a spline model there are several choices to be made: What degree should be used? How many knots should be used? Where should the knots be put?

My suggestion for the first choice is to use either linear or quadratic splines. Linear splines are somewhat simpler to understand than quadratic,

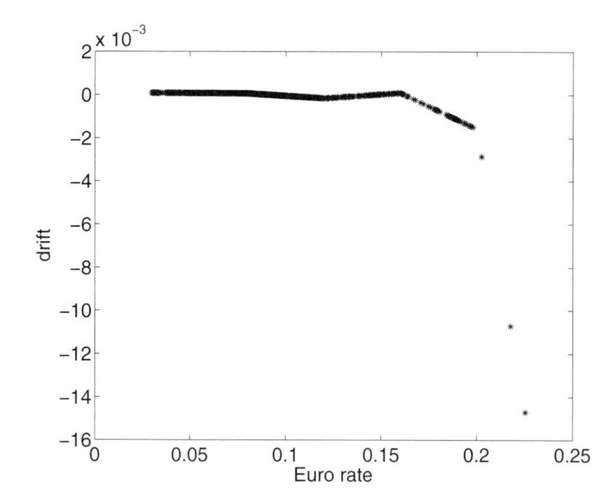

Fig. 13.11. *Euro interest rate example. Linear spline estimate of the drift function.*

but quadratic have the advantage of being smoother with no kinks. For the third question, I recommend that one start with approximately an equal number of data points between the knots. For example, one could place a knot at every Lth x-value where L would depend on how many knots one wanted to use.

The hard question, the second one, is how many knots to use. If we fit by least squares estimation, then the choice of how many knots is crucial. This fact can be appreciated by looking at Figure 13.12 where quadratic spline estimates of the drift function are shown for the Euro interest rates. Estimates with 2, 10, and 20 knots are plotted. The three splines are rather different. The spline with 20 knots seems rather wiggly and is undoubtedly overfitting the data. The 10-knot spline is also wiggly and an overfit, though not quite as much as the 20-knot spline. The spline with 2 knots seems like a reasonable estimate, though it is difficult to know for sure. The 2-knot spline doesn't bend as much on the right side as the 10- and 20-knot estimates, and that might be an underfit.

One can select the number of knots by using some model selection criteria such as C_p, AIC, or SBC. To do this one can fit models with 1, 2, 3, ... knots and choose the model that minimizes C_p, say.

Another possibility is to define a number of *potential* knots and to select from among these by model selection software. Here are some SAS programs to compute a quadratic spline estimate with *potential* knots at 0.08, 0.12, 0.16, and 0.2. The first program uses model selection by C_p to select the actual knots from the potential knots. PROC REG is SAS's linear regression program. There are several "methods" available for the regression in PROC REG. In the program below, "method" is "cp."

```
data EuroRate ;
infile 'c:\book\SAS\euro_rates.dat' ;
input month day year eu01 eu03 eu06;
eu01 = eu01/100 ;
diff=dif(eu01) ;
rate = lag(eu01) ;
rate2 = rate**2 ;
plus1 = ((rate - 0.08)**2) * (rate > 0.08) ;
plus2 = ((rate - 0.12)**2) * (rate > 0.12) ;
plus3 = ((rate - 0.16)**2) * (rate > 0.16) ;
plus3 = ((rate - 0.2)**2) * (rate > 0.2) ;
run ;
title 'One Month Euro dollar deposit rates' ;
proc reg ;
model diff = rate rate2 plus1 plus2 plus3 plus4 / method=cp;
run ;
```

Here is the part of the output. Models with higher values of C_p are not included.

```
                          Dependent Variable: diff
                           C(p) Selection Method

Number in
  Model         C(p)  R-Square  Variables in Model

      2      -0.5962    0.0172  plus3 plus4
      2      -0.1048    0.0168  plus2 plus4
      1       0.0526    0.0150  plus3
      2       0.3649    0.0164  plus1 plus4
      1       0.6221    0.0145  plus4
      2       0.7035    0.0161  rate2 plus4
      2       1.0446    0.0158  plus2 plus3
      2       1.1528    0.0157  rate plus4
      3       1.2585    0.0173  rate plus3 plus4
      3       1.2855    0.0173  rate2 plus3 plus4
      3       1.3833    0.0172  plus1 plus3 plus4
      3       1.4036    0.0172  plus2 plus3 plus4
      2       1.6279    0.0153  plus1 plus3
      3       1.8535    0.0168  plus1 plus2 plus4
      3       1.8671    0.0168  rate plus2 plus4
      3       1.8827    0.0168  rate2 plus2 plus4
```

Here is a SAS program to rerun regression with the model suggested by C_P that has only the quadratic plus functions with knots at 0.16 and 0.2, i.e., with the variables `plus3` and `plus4`.

```
(data step omitted but same as before)
proc reg ;
model diff = plus3 plus4;
output out=EuroOut p = yhat ;
run ;
proc gplot ;
plot yhat*rate ;
run ;
```

In this program, PROC REG was requested to produce an output data set containing the predicted variable as well as all the original variables, which are included by default. As in a previous program, PROC GPLOT plots the predicted values to show us the spline. If one wanted to be finicky about knot placement, one could have started with more potential knots, say at 0.06, 0.07, 0.08, ..., 0.19, 0.20, and 0.21. Then a few of these knots could be selected in the same way as we selected 0.16 and 0.2 from among 0.08, 0.12, 0.16, and 0.2.

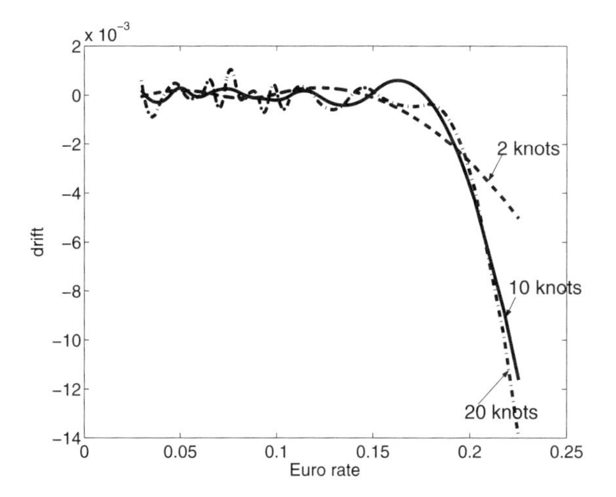

Fig. 13.12. *Euro bond interest rate example. Quadratic spline estimates of the drift with* 2, 10, *and* 20 *knots fit by ordinary least squares (OLS).*

13.6.1 Estimating the volatility function

The volatility function $\sigma(\cdot)$ in equation (13.1) can be estimated by regression. One starts with the residuals from estimation of the drift function $\mu(\cdot)$ and regresses the squared residuals against `rate` to estimate $\sigma^2(\cdot)$.

The next SAS program estimates the volatility function. First the drift function is estimated and an output data set is created that contains the residuals, which are named "resid." In a second data step the residuals are squared and the squared residuals are named "resid2." The statement "data EuroRate" starts the second data step and the statement "set EuroRate" tells SAS to use the data set "EuroRate" as input rather than reading in data from a file. Thus, "EuroRate" is used as input, updated with a new variable, and then overwritten. The net effect of both statements is to add the variable "resid2" to the data set "EuroRate."[6] Next, the squared residuals are used to estimate the volatility function. PROC REG is called again, this time with

[6] In some advanced SAS programming, there is a need for having several data sets in memory. If we had used the statement "data EuroRate2" in place of "data EuroRate," then the data set "EuroRate" would have been unchanged and a new data set called "EuroRate2" would have been created containing all of the variables in "EuroRate" plus "resid2."

"resid2" as the response and with "method=cp" to find a good model for the volatility function.

```
(data step omitted by as before)
title 'One Month Euro dollar deposit rates' ;
title2 'Preliminary fit to estimation drift function' ;
proc reg ;
model diff = plus3 plus4 ;
output out=EuroRate r=resid ;
run ;
comment The next step creates the squared residuals ;
data EuroRate ;     comment Second data step ;
set EuroRate ;
resid2 = resid*resid ;
title2 'Estimation of volatility using squared residuals' ;
proc reg  ;
model resid2 = rate rate2 plus1 plus2 plus3 plus4 / method=cp ;
run ;
```

Some of the output is shown below. The model with the smallest value of C_p uses just the first two plus functions.

```
                    Dependent Variable: resid2
                       C(p) Selection Method

Number in
  Model         C(p)  R-Square  Variables in Model

      2       2.4838    0.1220  plus1 plus2
      3       3.1000    0.1230  rate plus1 plus2
      3       3.3554    0.1228  rate2 plus1 plus2
      3       4.1612    0.1222  plus1 plus2 plus3
      3       4.4328    0.1220  plus1 plus2 plus4
      4       4.4790    0.1234  rate rate2 plus1 plus2
      3       4.8465    0.1217  rate plus1 plus3
      2       4.8578    0.1202  rate2 plus1
      4       4.8907    0.1231  plus1 plus2 plus3 plus4
      3       4.9479    0.1216  rate2 plus1 plus3
      4       4.9629    0.1231  rate plus1 plus2 plus4
      4       5.0109    0.1231  rate plus1 plus2 plus3
      2       5.0932    0.1201  rate plus1
      2       5.1177    0.1200  rate rate2
```

The following program fits the model just selected by C_p.

```
(data step omitted byut is the same as above)
proc reg ;
model resid2 = plus1 plus2 ;
output out=EuroRate p=yhat_vol ;
run ;
proc gplot ;
plot yhat_vol*rate ;
run ;
```

The fitted values are called "yhat_vol" in the output data set. The fitted values are plotted against "rate" to graph the estimated squared volatility function; see Figure 13.13. Notice that the estimated squared volatility function is a constant (the estimated intercept) until the first knot at 0.08, then assumes a positive slope (the coefficient of `plus1`) at 0.08, and changes slope at 0.12, the knot of `plus2`.

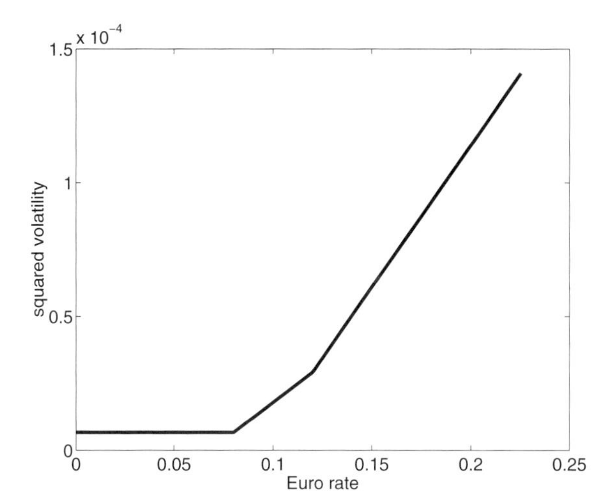

Fig. 13.13. *Euro interest rate example. Linear spline estimate of the squared volatility function.*

13.7 Additive Models*

So far we have only considered spline modeling with a single predictor variable. Now let us assume that for the ith case there are two predictors, $X_{1,i}$ and $X_{2,i}$. The most general nonparametric model is

$$Y_i = \mu(X_{1,i}, X_{2,i}) + \epsilon_i,$$

where $\mu(X_{1,i}, X_{2,i})$ is a completely arbitrary function of $X_{1,i}$ and $X_{2,i}$. However, developing spline models with this level of generality is beyond the scope of this book and, besides, requires a large amount of data. We instead use a simpler type of model called an additive model:

$$Y_i = \mu_1(X_{1,i}) + \mu_2(X_{2,i}) + \epsilon_i.$$

Here $\mu_1(X_{1,i})$ is a function of $X_{1,i}$ only and similarly for $\mu_2(X_{2,i})$. An additive spline model assumes that μ_1 and μ_2 are splines. Additive spline models are easy to fit, for example, in SAS. One just constructs monomials and plus functions in $X_{1,i}$ and monomials and plus functions in $X_{2,i}$.

The following SAS program fits an additive model to the Euro dollar one-month rates. The predictors are `rate` and `lagdiff` which is the lagged value of the response `diff`. For `rate` there are knots at 0.08, 0.12, 0.16, and 0.2. For `lagdiff` there is a single knot at 0. The variable `plusLD1` is the plus function in `lagdiff` with a knot at 0. The first program searches for a good model using C_p.

```
data EuroRate ;
infile 'c:\book\SAS\euro_rates.dat' ;
```

```
input month day year eu01 eu03 eu06;
eu01 = eu01/100 ;
diff=dif(eu01) ;
rate = lag(eu01) ;
lagdiff = lag(diff) ;
rate2 = rate**2 ;
lagdiff2 = lagdiff**2 ;
plus1 = ((rate - 0.08)**2) * (rate > 0.08) ;
plus2 = ((rate - 0.12)**2) * (rate > 0.12) ;
plus3 = ((rate - 0.16)**2) * (rate > 0.16) ;
plus4 = ((rate - 0.2)**2) * (rate > 0.2) ;
plusLD1 = ((lagdiff - 0)**2) * (lagdiff > 0) ;
run ;
title 'One Month Euro dollar deposit rates' ;
title2 'Additive Model' ;
proc reg  ;
model diff = rate rate2 plus1 plus2 plus3 plus4
lagdiff lagdiff2 plusLD1 / method=cp ;
run ;
```

Here is the output.

```
                         The REG Procedure
                      Dependent Variable: diff
                       C(p) Selection Method

Number in
  Model        C(p)  R-Square  Variables in Model

      4      0.8096    0.0623  plus2 plus3 lagdiff plusLD1
      3      1.1730    0.0604  plus4 lagdiff plusLD1
      4      1.1833    0.0620  plus3 plus4 lagdiff plusLD1
      3      1.3698    0.0603  plus3 lagdiff plusLD1
      4      1.4818    0.0617  plus1 plus3 lagdiff plusLD1
      4      2.2736    0.0611  plus2 plus4 lagdiff plusLD1
      5      2.4572    0.0626  plus2 plus3 plus4 lagdiff plusLD1
      4      2.5082    0.0609  rate2 plus3 lagdiff plusLD1
      5      2.6407    0.0624  rate2 plus2 plus3 lagdiff plusLD1
      5      2.6446    0.0624  plus1 plus3 plus4 lagdiff plusLD1
      5      2.6669    0.0624  rate plus2 plus3 lagdiff plusLD1
      5      2.6790    0.0624  plus2 plus3 lagdiff lagdiff2
                               plusLD1
(further output omitted)
```

As an illustration, the following program fits the model with the three variables `plus4`, `lagdiff`, and `plusLD1` that has the second lowest C_p value.[7]

```
(data read in and manipulated as in previous program)
proc reg ;
model diff = plus4 lagdiff plusLD1;
output out=EuroOut p = yhat ;
run ;
```

Here is the output. Notice that the plus function `plus4` in `rate`, `lagdiff`, and the plus function `plusLD1` in `lagdiff` are significant, but R^2 is very small. Thus, both predictors are statistically significant and have nonlinear effects but the practical significance of the predictive model seems small.

```
                         The REG Procedure
                      Dependent Variable: diff
                        Analysis of Variance
```

[7] One might, of course, prefer the model with the lowest C_p values, which is a four-variable model, but the difference in C_p between the two models is small.

```
                                  Sum of          Mean
Source                 DF        Squares        Square    F Value

Model                   3        0.00125    0.00041818      25.59
Error                1194        0.01951    0.00001634
Corrected Total      1197        0.02077

                           Analysis of Variance
                         Source            Pr > F
                         Model             <.0001
                         Error
                         Corrected Total

        Root MSE                0.00404    R-Square      0.0604
        Dependent Mean    -0.00002346    Adj R-Sq      0.0580
        Coeff Var              -17235

                          Parameter      Standard
Variable      DF         Estimate         Error    t Value    Pr > |t|
Intercept      1       0.00013987    0.00012053       1.16      0.2461
plus4          1        -19.96184       5.96808      -3.34      0.0008
lagdiff        1          0.27061       0.03544       7.64      <.0001
plusLD1        1        -16.79625       3.63164      -4.62      <.0001
```

The following SAS program uses the parameter estimates from this listing to compute and plot the estimates of μ_1 and μ_2.

```
data EuroRate ;
infile 'c:\book\SAS\euro_rates.dat' ;
input month day year eu01 eu03 eu06;
eu01 = eu01/100 ;
diff=dif(eu01) ;
rate = lag(eu01) ;
lagdiff = lag(diff) ;
plus4 = ((rate - 0.2)**2) * (rate > 0.2) ;
plusLD1 = ((lagdiff - 0)**2) * (lagdiff > 0) ;
mu1= -19.96184 * plus4 ;
mu2 = 0.27061*lagdiff - 16.79625 * plusLD1 ;
run ;
title 'One Month Euro dollar deposit rates' ;
proc gplot ;
plot mu1*rate ;
plot mu2*lagdiff ;
run ;
```

MATLAB plots of μ_1 and μ_2 are shown in Figure 13.14. The vertical scale (10^{-3}) shows again that the effects of the predictor variables are small and perhaps of little practical relevance despite their statistical significance; see Section 2.20.2 for a discussion of statistical and practical significance. To appreciate the small size of the effects, it is helpful to look at Figure 13.15 which has the same functions as Figure 13.14 but includes the data. Notice that the effects are rather small compared to the scatter in the data.

Figure 13.15 also shows that $\widehat{\mu}_2$ is increasing over a range that includes the bulk of the data (in fact, over a range including approximately 96% of the data). Thus, there is a small but positive association between the differences and the lagged differences of the interest rates, except that for the largest values of the lagged differences the association becomes negative. In other words, an increase in the interest rate tends to be followed by another increase, except that a very large increase is followed by a decrease. Unexpected phenomena such as this behavior of changes in interest rate are best detected by nonparametric regression, because nonparametric regression has the capacity to adapt to unanticipated features of the data.

Given the wide scatter of the data, it is natural to wonder whether μ_2 really does switch from being increasing to being decreasing. Might the estimate exhibit this behavior solely because of estimation error? There is a simple method to address this question. We can estimate μ_2', the derivative of μ_2, and provide confidence bounds. This is done in Figure 13.16. Notice that to the left the confidence bounds include only positive values showing that the derivative is positive (with 95% confidence). Similarly, the confidence bounds on μ_2' only include negative values at larger values of `rate`, so μ_2 is decreasing in this region (with 95% confidence).

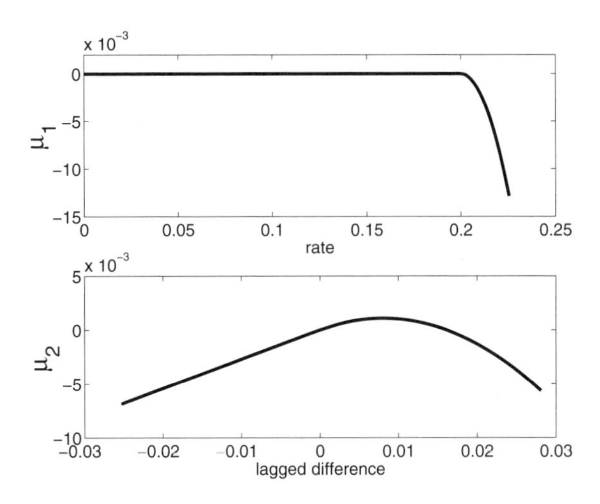

Fig. 13.14. *Estimates of* μ_1 *and* μ_2.

13.8 Penalized Splines*

So far, we have only considered estimation by least squares, which is also called ordinary least squares or OLS. An alternative to using model selection to choose the number of knots or to choose which plus functions to use in the final model is to use a large, though somewhat arbitrary, number of knots and to replace least squares estimation with an estimation method that

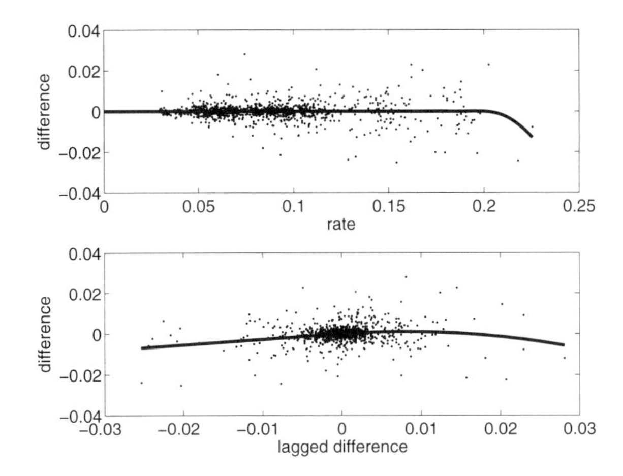

Fig. 13.15. *Estimates of μ_1 (top) and μ_2 (bottom) with data. The estimates are the same as in Figure* 13.14.

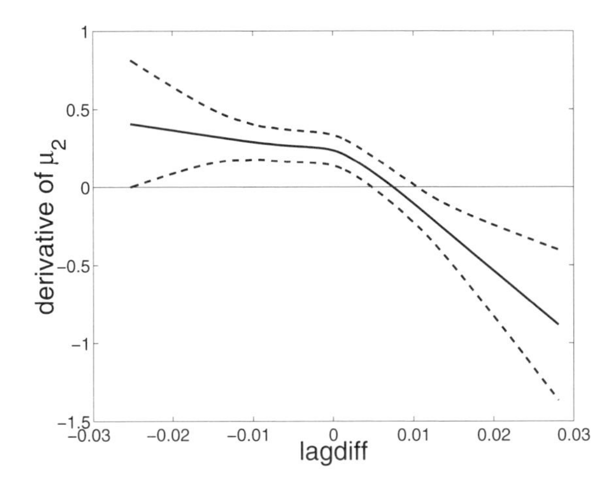

Fig. 13.16. *Estimates of $\mu_2^{(1)}$ (solid) with 95% confidence bounds (dashed).*

prevents overfitting. This is done in this section. The advantage of penalized least squares estimation is that the penalty prevents overfitting automatically without the need to choose the number of knots carefully.

Figure 13.17 shows 2-, 10-, and 20-knot spline estimates of the drift function by penalized least squares. The estimates from penalized least squares are called penalized splines, or simply *P*-splines. Notice that the 10- and 20-knot estimates are very similar, but the 2-knot estimate does not seem quite flexible enough to fit the data.

In this example, a spline with about 5 knots is flexible enough to fit the data. All that happens as we increase the number of knots beyond 5 is that we increase the *potential* for overfitting. However, the penalty that is being applied here prevents overfitting. Therefore, the estimated regression function does not depend greatly upon the number of knots as long as there are enough, which in this example is about 5 or more. This means we do not need to concern ourselves with the choice of number of knots. We can simply use a rather large number of knots, say 20, so that we have enough flexibility and let the penalty prevent overfitting.

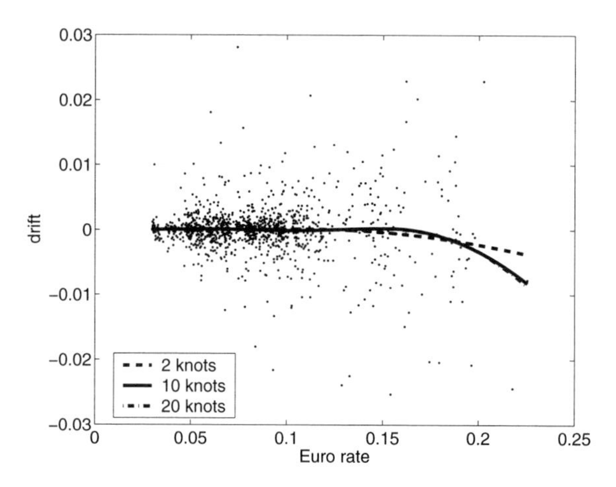

Fig. 13.17. *Euro bond interest rate example. Quadratic splines estimates of the drift with* 2, 10, *and* 20 *knots by penalized least squares (PLS). The* 10- *and* 20-*knot splines are very similar and difficult to separate visually. The penalty parameter was selected by GCV.*

13.8.1 Penalizing the jumps at the knots

We know that if we are using quadratic splines, then the spline's second derivative jumps by an amount $2b_k$ at the kth knot. Overfitting occurs when there are many jumps (many knots) and the jumps are unconstrained. Overfitting can be prevented by restricting either the number of jumps or the size of the jumps. Using model selection to choose the number of knots does the former. Penalization does the later.

Here's how penalization works. The sum of the squared jumps is four times

$$\sum_{k=1}^{K} b_k^2. \tag{13.6}$$

We wish to make small both the sum of squared residuals and the sum of the squared jumps, so we add these two together with a weight λ attached to the latter.[8] This gives us the estimation criterion

$$\sum_{i=1}^{n} \Big\{ Y_i - \Big(\beta_0 + \beta_1 x + \beta_2 x^2 + b_1 (x - t_1)_+^2 + \cdots$$
$$+ b_K (x - t_K)_+^2 \Big) \Big\}^2 + \lambda \sum_{k=1}^{K} b_k^2. \qquad (13.7)$$

The penalty parameter can be any value between 0 and ∞, inclusive. If $\lambda = \infty$, then $b_1, \ldots, b_K$ are constrained to be zero since any positive value of these parameters makes the penalty $\lambda \sum_{k=1}^{K} b_k^2$ infinite.[9] The value of λ is critical and so must be chosen carefully. To do that, we need to understand the role that λ plays in estimation. Since λ multiplies the term (13.6) in (13.7), λ determines how much we penalize jumps in the pth derivative. If $\lambda = 0$ then there is no penalty and we are using OLS. On the other hand, as $\lambda \to \infty$, the jumps in the second derivative are forced to 0. This makes the quadratic spline equal to a single quadratic polynomial. Therefore, using $\lambda = \infty$ gives the OLS fit with a quadratic *polynomial*, not a quadratic spline.

Figure 13.18 shows penalized spline fits with 25 knots and with λ equal to 0, 5, and 10^{10} (essentially ∞). We can see that the fit with $\lambda = \infty$ has little flexibility. It is a quadratic polynomial fit. The fit with $\lambda = 0$ is the OLS to a 25-knot spline already seen in Figure 13.12 to be an overfit. Using $\lambda = 5$ is a good compromise between these two extremes.

Although using 5 as the value of λ seems better than either 0 or infinity, it is natural to ask whether 5 is the best choice. In Figure 13.19 we compare the fits with λ equal to 0.5, 5, and 50. The three estimates of the regression function are clearly different, but it is not certain which is most appropriate. What is desirable is an automatic data-driven choice of λ. We turn to this topic next.

13.8.2 Cross-validation

We use the following principle for selection of λ.

> The optimal choice of λ is the value that give us the most accurate predictions of new data.

But can we determine how well we can predict new and as yet unobserved data? A technique called cross-validation does precisely this. Cross-validation,

[8] Only the relative weights matter so we can fix the weight given to the sum of squared residuals equal to 1.

[9] We interpret $(\infty)(0) = 0$ so the penalty is zero if all of $b_1, \ldots, b_K$ are zero.

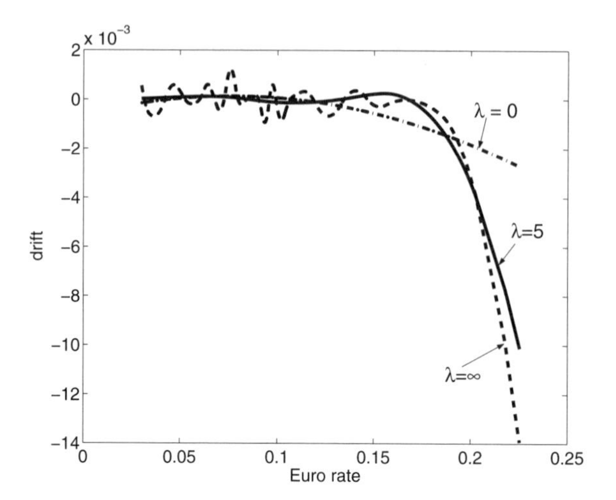

Fig. 13.18. *Euro bond interest rate example. Quadratic spline estimates of the drift function, each with* 25 *knots, Estimation is by penalized least squares (PLS) with* $\lambda = 0$, 5, *and* ∞. $\lambda = 0$ *corresponds to OLS.*

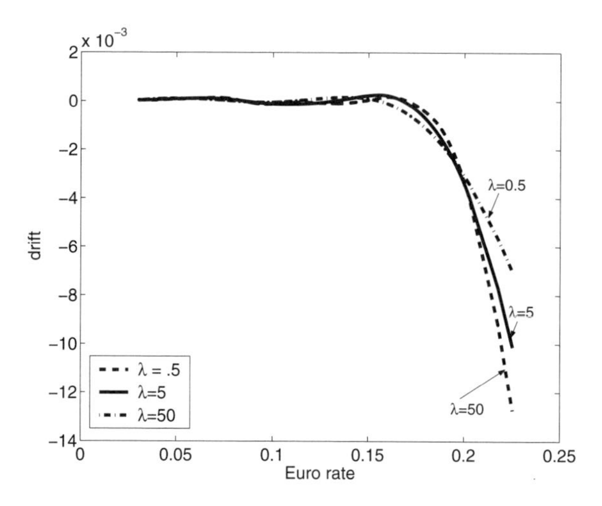

Fig. 13.19. *Euro bond interest rate example. Quadratic spline estimates of the drift, each with* 25 *knots. Estimation is by penalized least squares (PLS) with* $\lambda = 0.5$, 5, *and* 50.

also called CV, is conceptually simple: to see how well using a particular trial value of λ works for prediction, we can delete one data point, estimate the regression function with the other data, and then use the estimated regression function to predict the observation that was deleted. We can repeat this procedure n times using each data point in the sample as the deleted point. This gives us n separate estimates of the expected squared prediction error using this λ.

More precisely, CV goes through the following steps for any trial value of λ.

1. For $j = 1, \ldots, n$, let $\widehat{s}(\cdot; \lambda, -j)$ be the regression function estimated using this λ and with the jth data point deleted.[10] In other words, $\widehat{s}(\cdot; \lambda, -j)$ is the spline with coefficients equal to the parameter values that minimizes

$$\sum_{i \neq j} \left\{ Y_i - \left(\beta_0 + \beta_1 x + \beta_2 x^2 + b_1 (x - t_1)_+^2 + \cdots \right.\right.$$
$$\left.\left. + b_K (x - t_K)_+^2 \right) \right\}^2 + \lambda \sum_{k=1}^{K} b_k^2. \tag{13.8}$$

2. Let $s(X_j; \lambda, -j)$ be the prediction of the value of Y_j using the other observations. Define

$$\mathrm{CV}(\lambda) = n^{-1} \sum_{j=1}^{n} \{ Y_j - s(X_j; \lambda, -j) \}^2$$

to be the average squared error from these n predictions.

Since $\mathrm{CV}(\lambda)$ estimates the expected squared prediction error, the value of λ that minimizes $\mathrm{CV}(\lambda)$ is considered best. The final estimate of the regression function uses all the data and the value of λ that minimizes CV.

13.8.3 The effective number of parameters

If we use a pth degree spline with K knots, then there are $1 + p + K$ parameters, the intercept, the p coefficients of the powers x to x^p, and the K coefficients of the plus functions. In one extreme case, if we use $\lambda = 0$, then all of these parameters are free to vary. However, a positive value of λ constrains the size of these last K parameters. In the other extreme case where $\lambda = +\infty$, the estimated coefficients of the plus function are all be constrained to equal 0, so there are only $1 + p$ free parameters.

When λ is positive but not $+\infty$, then the "effective number of parameters" should be somewhere between $1 + p$ and $1 + p + K$. How can we measure the

[10] The notation "$-j$" is intended to suggest that the jth data point has been removed or "subtracted" from the data set.

effective number of parameters? The theory is somewhat complex and is not be introduced here. Only the end result is given. Let $\boldsymbol{X}$ be the matrix

$$\boldsymbol{X} = \begin{pmatrix} 1 & X_1 & \cdots & X_1^p & (X_1 - t_1)^p & \cdots & (X_1 - t_K)^p \\ 1 & X_2 & \cdots & X_2^p & (X_2 - t_1)^p & \cdots & (X_2 - t_K)^p \\ \vdots & \vdots & \ddots & \vdots & \vdots & \ddots & \vdots \\ 1 & X_n & \cdots & X_n^p & (X_n - t_1)^p & \cdots & (X_n - t_K)^p \end{pmatrix}. \tag{13.9}$$

Then let $\boldsymbol{D}$ be a $(1+p+K) \times (1+p+K)$ square matrix with all off-diagonal elements equal to zero and with its diagonal elements equal to $1+p$ zeros followed by K ones. Then the effective number of parameters, also called the effective degrees of freedom (DF), is

$$\mathrm{DF}(\lambda) = \mathrm{trace}\{(\boldsymbol{X}^{\mathsf{T}}\boldsymbol{X})(\boldsymbol{X}^{\mathsf{T}}\boldsymbol{X} + \lambda\boldsymbol{D})^{-1}\}.$$

Here trace($\boldsymbol{A}$) is the trace of the matrix $\boldsymbol{A}$ and is defined to be the sum of its diagonal elements.

You should realize that $\mathrm{DF}(\lambda)$ is generally not an integer so it is possible, for example, to have 2.5 effective parameters. To understand how this can happen, consider a linear spline with one knot which, of course, has 3 parameters. If the jump in the first derivative is constrained to be 0 (penalty is infinite), then there are only 2 effective parameters. If there is no penalty at all, then there are 3 effective parameters. If the jump is penalized but not completely constrained, then the number of effective parameters is somewhere strictly between 2 and 3.

Figure 13.20 is a semi-log plot of $\mathrm{DF}(\lambda)$ versus λ for a 25-knot quadratic spline ($p = 2$ and $K = 25$). We can see that $\mathrm{DF}(\lambda)$ decreases from $28 = 1+p+K$ to $3 = 1+p$ as λ increases from 0 to ∞. $\mathrm{DF}(\lambda)$ is essentially equal to 28 if $\lambda < 10^{-8}$ and $\mathrm{DF}(\lambda)$ is essentially equal to 3 if $\lambda > 10^4$.

If $\lambda = 0$, so that we are using ordinary least squares, then DF gives us precisely the number of parameters in the model. To appreciate this, note that

$$\mathrm{DF}(0) = \mathrm{trace}\{(\boldsymbol{X}^{\mathsf{T}}\boldsymbol{X})(\boldsymbol{X}^{\mathsf{T}}\boldsymbol{X})^{-1}\} = \mathrm{trace}(\boldsymbol{I}_{1+p+K}) = 1 + p + K,$$

where $\boldsymbol{I}_{1+p+K}$ is the $(1+p+K) \times (1+p+K)$ identity matrix.

DF can be used to get a "corrected" and nearly unbiased estimate of σ^2. Let $\widehat{s}(\cdot;\lambda)$ be the estimated regression function using *all* the data. Then

$$\widehat{\sigma}^2(\lambda) = \frac{\sum\{Y_i - \widehat{s}(X_i;\lambda)\}^2}{n - \mathrm{DF}(\lambda)}. \tag{13.10}$$

Notice that $\widehat{\sigma}^2(\lambda)$ is a spline analogue of the residual error MS defined by equation (6.7), so $\widehat{\sigma}^2(\lambda)$ is analogous to the usual estimator of σ^2 in linear regression.

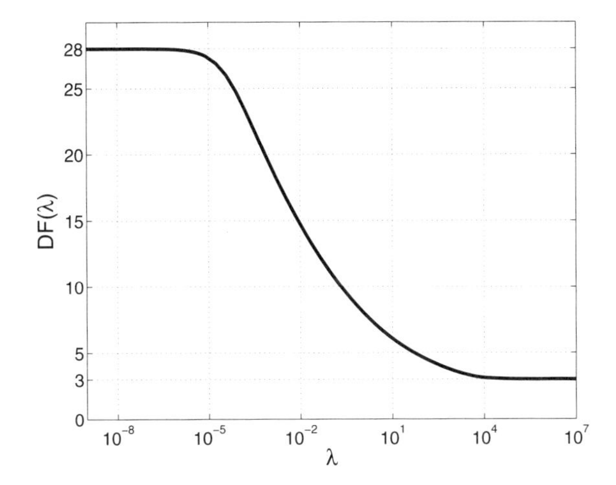

Fig. 13.20. *Euro bond interest rate example. Estimation of the drift function with 25-knot quadratic splines. Plot of DF as a function of* λ.

13.8.4 Generalized cross-validation

Computing $\mathrm{CV}(\lambda)$ is somewhat time consuming, though there are some tricks that are used to make the computations easier than they would seem.[11] However, there is an approximation to CV that is very quick to compute, faster than CV. The generalized cross-validation statistic (GCV) is

$$\mathrm{GCV}(\lambda) = \frac{n^{-1}\sum\{Y_i - \widehat{s}(X_i;\lambda)\}^2}{\left\{1 - \frac{\mathrm{DF}(\lambda)}{n}\right\}^2}.$$

The point is that GCV only uses the estimate $\widehat{s}$ computed from all the data.

Figure 13.21 is a semi-log plot of $\mathrm{GCV}(\lambda)$ versus λ for the Euro bond example. We can see that GCV is minimized by λ somewhere near 10. In fact, the minimum occurs at $\lambda = 13.3$. Figure 13.22 is a plot of $\mathrm{GCV}(\lambda)$ versus $\mathrm{DF}(\lambda)$. This is essentially the same function as in the curve of Figure 13.21 except that the horizontal scale is $\mathrm{DF}(\lambda)$ rather than λ. Notice that GCV is minimized by approximately 6 effective parameters. In fact, the minimum is at 5.8 DF.

[11] In particular, one does not really need to compute n separate regressions. There are formulas that show the effect of deleting an observation without actually needing to recalculate the estimate with the observation deleted.

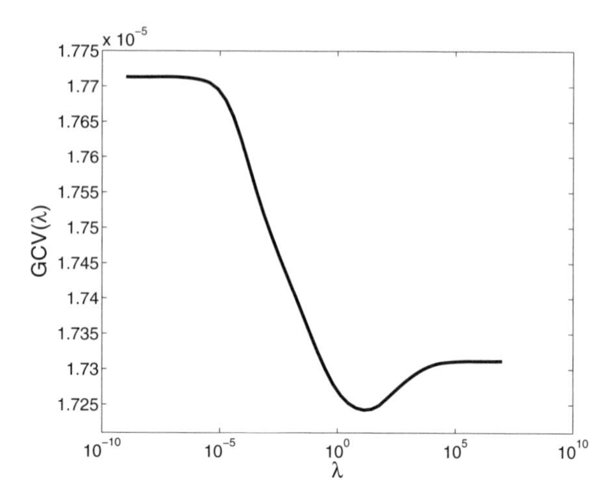

Fig. 13.21. *Euro bond interest rate example. Estimation of the drift function with 25-knot quadratic splines. Semilog plot of GCV versus* λ.

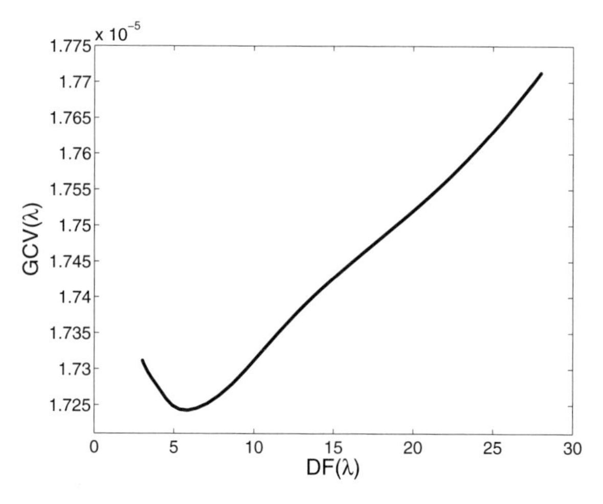

Fig. 13.22. *Euro bond interest rate example. Estimation of the drift function with 25-knot quadratic splines. Plot of GCV versus DF. GCV selects approximately* 6 *degrees of freedom.*

13.8.5 AIC

For linear regression models, AIC is

$$\text{AIC} = n \log(\widehat{\sigma}^2) + 2(1 + p),$$

where $1+p$ is the number of parameters in a model with p predictor variables; the intercept gives us the final parameter. Since $\text{DF}(\lambda)$ is the effective number of parameters of a P-spline, we can define AIC for P-splines as

$$\text{AIC}(\lambda) = n \log\{\widehat{\sigma}^2(\lambda)\} + 2\text{DF}(\lambda).$$

We can then select λ by minimizing AIC. Using AIC to select λ usually gives a similar estimate as using CV or GCV. In fact, it has been shown theoretically that all three criteria should give similar estimates. Figure 13.23 is a plot of AIC versus DF(λ). Notice that the curve has a similar shape to the one in Figure 13.22 and the minimum in both figures occurs around DF = 6.

Why do we have so many criteria for selecting the penalty parameter? The answer is that this is mostly a historical accident. AIC was developed by time series analysts, CV arose in parametric statistics, and GCV was introduced in nonparametric regression, specifically in spline estimation. Three different groups of researchers were attacking the same general problem and produced three similar, though not identical, answers. In fact, there were four similar answers, since C_p is similar to CV, GCV, and AIC.

AIC and GCV can both be computed very quickly and usually give essentially the same answers, and then it really does not matter which is used. CV gives similar answers to AIC and GCV but is a bit more work to compute, so I do not recommend CV. I use GCV because I have a lot of experience using GCV and have found it to be reliable.

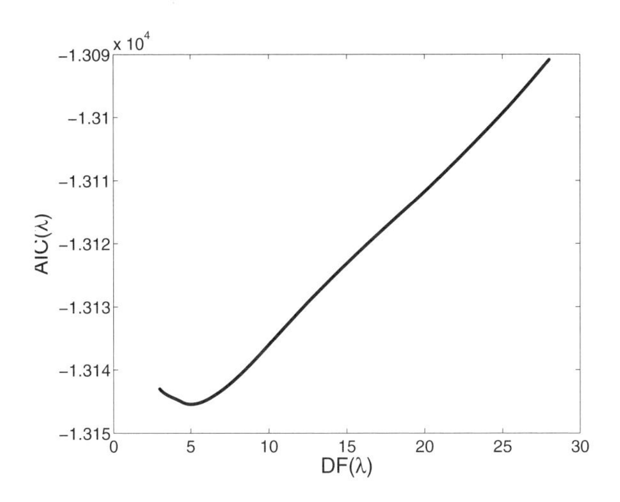

Fig. 13.23. *Euro bond interest rate example. Estimation of the drift function with 25-knot quadratic splines. Plot of AIC versus DF(λ). AIC selects approximately 6 degrees of freedom, the same as GCV.*

13.8.6 Penalized splines in MATLAB

The book website has three MATLAB programs for computing P-spline estimates:

- Pspline01.m The main program. Computes P-spline estimates with the penalty parameter chosen by GCV.
- powerbasis01.m Given the knots and the degree of the spline, computes the power functions and plus functions.
- quantileknots01.m Finds knot locations that put, as nearly as possible, equal numbers of x values between knots.

The "01" in "Pspline01" is the version number and the website may contain higher versions in the future. "Pspline01.m" calls "powerbasis01.m" which calls "quantileknots01.m." For applications, you only need to understand how to use "Pspline01.m." The other two programs work in the background. The call to "Pspline01.m" is simply

```
outputname = Pspline01(x,y,param)
```

where "param" is a MATLAB structure containing *optional* input such as the degree of the spline, the number of knots, and "penwt" which is a grid of λ values on which GCV is computed. For most applications, you can use the default values in which case "param" can be omitted so that the call is

```
outputname = Pspline01(x,y)
```

A list of all optional input that can be specified by "param" is given in the listing of the program. In the calling program, "outputname" can be any name you wish and names a structure containing all of the output including the following.

- degree = degree of the spline;
- knots = knots of the spline;
- penwt = values of the penalty parameter used to search for the minimum of GCV;
- imin = index of value of penwt that minimizes GCV, so, for example, if imin is 14 then the 14th value of penwt is where the minimum occurs;
- yhat = fitted values = regression function estimate at each x value;
- beta = regression coefficients in order: intercept, monomial coefficients, plus function coefficients;
- ulimit = upper 95% confidence limits of regression function at each x value;
- llimit = lower 95% confidence limits of regression function at each x value;
- yhatder = estimate of the derivative of the regression function at each x value;
- ulimitder = upper 95% confidence limits of the derivative of the regression function at each x value;
- llimitder = lower 95% confidence limits of the derivative of the regression function at each x value;
- xgrid = fine grid of x values (not the values in the data);
- mhat = estimate of regression function evaluated on xgrid.

The easiest way to use "Pspline01.m" is to put "Pspline01.m," "powerbasis01.m," and "quantileknots01.m" in a directory along with the data and your program that loads the data and calls "Pspline01.m."

The following MATLAB code is part of a program that estimates the drift and volatility functions and their first derivatives for the Euro deposit rates. The complete program is called "euro_interest.m" and is on the book website. The data are read by the first line of code from the file "Euro_hw.dat" which is also on the book website. After some data manipulation which is omitted here, `rate` is the interest rate while `diff` is the first differences of `rate`.

```
load Euro_hw.dat ;
(code to manipulate data omitted)
fit = Pspline01(rate,diff,struct('nknots',15,'degree',3)) ;
subplot(2,1,1) ;
p = plot(rate,fit.yhat,rate,fit.ulimit,'--',rate,fit.llimit,'--',[0 0.25],[0 0]) ;
(code omitted that enhances the plot)
subplot(2,1,2) ;
p=plot(rate,fit.yhatder,rate,fit.ulimitder,'--',rate,fit.llimitder,'--',[0 0.25],[0 0]) ;
(code omitted that enhances the plot)
```

Here "outputname" is "fit." The number of knots is specified as 15 rather than the default value, which is the smaller of 20 and $\lfloor 0.3n \rfloor$ where n is the sample size.[12] Also, the degree of the spline is 3, the default value of which is 2.

In the program "fit.yhat" contains the fitted values, "fit.ulimit" contains the upper confidence limits, and "fit.llimit" contains lower confidence limits. Also, "fit.yhatder" is the derivative of the estimated curve and "fit.ulimitder" and "fit.llimit" are the confidence limits. The plots of the drift function and its derivative are in Figure 13.24.

Further in the program, the code

```
fit_vol=Pspline01(rate,(diff-fit.yhat).^2,struct('nknots',15,'degree',3));
```

estimates the volatility function by regressing the squared residuals on `rate`. Notice the use of the name "fit_vol" to distinguish the output of this regression from the output of the earlier regression. The plots of the volatility function and its derivative are in Figure 13.25.

The drift function is very close to 0 for `rate` less than 0.16. For larger values of `rate` the estimated drift function is negative. However, there are very little data with `rate` greater than 0.16, so a natural question is whether the negative estimate of the drift function for `rate` greater than 0.16 is something real or instead due to estimation error. This question can be answered by the confidence intervals. Notice that on the far right the confidence intervals lie entirely below zero, showing that with 95% confidence the drift function is negative.

---

[12] The notation $\lfloor x \rfloor$ means the greatest integer less than or equal to x. In other words, $\lfloor x \rfloor$ equals x if x is an integer and otherwise is x rounded down to an integer.

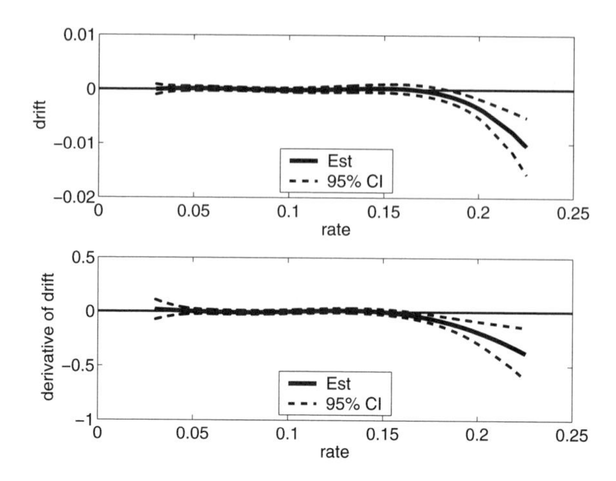

Fig. 13.24. Top: *Estimate of the drift function with 95% confidence intervals. The estimate is a 15-knot cubic P-spline with the penalty parameter estimated by GCV.* **Bottom:** *First derivative of the estimated drift function with 95% confidence intervals.*

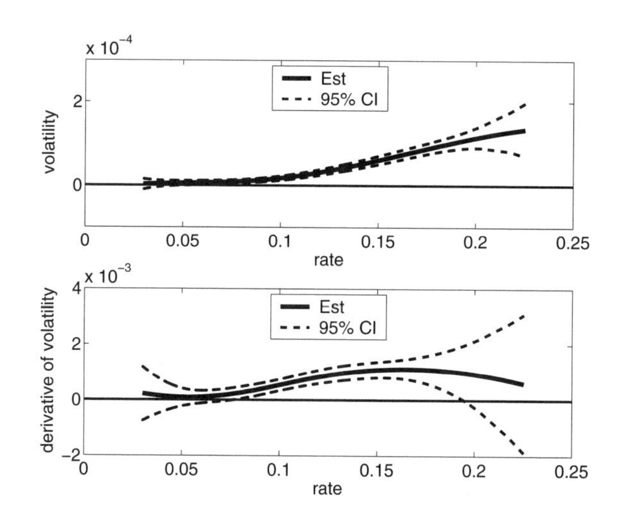

Fig. 13.25. Top: *Estimate of the volatility function with 95% confidence intervals. The estimate is a 15-knot cubic P-spline with the penalty parameter estimated by GCV.* **Bottom:** *First derivative of the estimated volatility function with 95% confidence intervals.*

13.9 Summary

Nonparametric regression can be used when one knows that a regression function is nonlinear but does not have a model for that function. There are many methods for nonparametric regression, with local polynomial regression and splines being the most commonly used.

A spline is a function that equals a fixed polynomial in each interval between two consecutive knots. The polynomials can change at the knots, but if the polynomials are of the pth degree, then their coefficients are constrained so that the first $p-1$ derivatives of the spline are continuous at the knots. Splines can be represented as the sum of a polynomial and plus functions, one plus function for each knot.

A spline can be fit to data by least squares, but the number and location of the knots should be chosen carefully. One method to accomplish this is to start with a large number of knots and to choose the knots by some model selection method.

Spline regression can be extended easily from one to many predictor variables by using an additive model. An additive model in p predictors is the sum of functions μ_1 to μ_p, where μ_j is a function only of the jth predictor.

Penalized splines are splines that are fit by penalized least squares. Penalized least squares minimizes the sum of squared residuals plus a penalty on the roughness of the spline. One commonly used roughness penalty is a nonnegative penalty parameter, denoted here by λ, times the sum of the squared coefficients of the plus functions. Choosing λ properly is crucial, and generalized cross-validation (GCV) is an effective method for making this choice.

13.10 Bibliographic Notes

Ruppert, Wand, and Carroll (2003) offers a comprehensive introduction to nonparametric and semiparametric modeling and their applications. Wand and Jones (1995) and Fan and Gijbels (1996) are introductions to local polynomial regression.

13.11 References

Chan, K. C., Karolyi, G. A., Longstaff, F. A., and Sanders, A. B. (1992) An empirical comparison of alternative models of the short-term interest rate, *Journal of Finance*, **47**, 1209–1227.

Cox, J. C., Ingersoll, J. E., and Ross, S. A. (1985) A theory of the term structure of interest rates, *Econometrica*, **53**, 385–407.

Fan, J. and Gijbels, I. (1996) *Local Polynomial Modelling and Its Applications*, Chapman & Hall, London.

Marchall, J. F. (2000) *Dictionary of Financial Engineering*, John Wiley & Sons, New York.

Merton, R. C. (1973) Theory of rational option pricing, *Bell Journal of Economics and Management Science*, **4** , 141–183.

Ruppert, D., Wand, M. P., and Carroll, R. J. (2003) *Semiparametric Regression*, Cambridge University Press, Cambridge, UK.

Vasicek, O. A. (1977) An equilibrium characterization of the term structure, *Journal of Financial Economics*, **5**, 177–188.

Wand, M. P. and Jones, M. C. (1995) *Kernel Smoothing*, Chapman & Hall, London.

Yau, P. and Kohn, R. (2001) Estimation and variable selection in nonparametric heteroscedastic regression, unpublished manuscript.

13.12 Problems

1. A linear spline $s(t)$ has knots at 1, 2, and 3. Also, $s(0) = 1$, $s(1) = 1.5$, $s(2) = 5$, $s(4) = 6$, and $s(5) = 6$.
   (a) What is $s(0.5)$?
   (b) What is $s(3)$?
   (c) What is $\int_2^4 s(t)\,dt$?
2. Suppose that (13.1) holds with $\mu(r) = 0.1(0.03 - r)$ and $\sigma(r) = 2.1r$.
   (a) What is the expected value of r_t given that $r_{t-1} = 0.04$?
   (b) What is the variance of r_t given that $r_{t-1} = 0.02$?
3. Let the spline $s(x)$ be defined as

$$s(x) = (x)_+ - 2(x-1)_+ + (x-2)_+.$$

   (a) Is $s(x)$ either a probability density function or a cumulative distribution (cdf) function? Could a function be both a pdf and a cdf? Explain your answers.
   (b) If X is a random variable and s is its pdf or cdf (whichever is the correct answer in (a)), then what is the 90th percentile of X?
4. Let s be the spline

$$s(x) = 1 + 0.5x + x^2 + (x-1)_+^2 + 0.6(x-2)_+^2.$$

   (a) What are $s(1.5)$ and $s'(1.5)$?
   (b) What is $s''(2.5)$?
5. Section 7.9 contains an analysis of whether beta might assume one value when the return R_M on the market portfolio is negative and another value when R_M is positive. A more general question is whether the regression of the return R_j of the jth asset on R_M is nonlinear. This problem revisits that question. You should obtain at least two years of daily returns on

the S&P 500 and on the Ford Motor Company or any other stock that interests you.

(a) Using SAS or other regression software, regress the daily returns of the stock on the S&P 500's returns using a three-knot quadratic spline. Plot the regression function and its derivative. The derivative can be thought of as a "beta curve". Why?

(b) Using the MATLAB program Pspline01 available at this book's web site, fit a quadratic Pspline to the same data as in part a). Plot the regression function and its derivative. How do these estimates compare with those in part a)?

(c) Plot the derivative in part b) along with 95% confidence bounds which are part of the output of Pspline01. Do you believe that the derivative is constant, that is, that regression function is linear? Why or why not?

(d) How might you use the estimated second derivative and its confidence bounds to answer the question in part c)?

14

Behavioral Finance

14.1 Introduction

Behavioral finance is the application of cognitive psychology to the study of the participants in financial markets. The question being investigated in this field is how humans actually perceive risk and make investment decisions. Economists have long assumed that people act so as to maximize the utility of their wealth, but we are learning that human behavior is more complex than this. People use rules-of-thumb, called *heuristics*, as short-cuts to reasoning. Moreover, how a person makes a financial decision depends on how the decision problem is stated, a phenomenon called *frame-dependence*. For example, people have a strong aversion to losses and a decision might depend on whether a decision problem is posed in a way that mentions the word "loss." Shefrin (2000) mentions the example of a stock broker who realized that clients were extremely reluctant to sell stocks at a loss in order to buy other stocks, even if this were the best investment decision for them. However, this broker found that clients were willing to sell at a loss when told that they were "transferring assets" rather than selling losers. Whether selling losers is a good thing for an investor could be debated, but the point is that the decision made by clients depends on how the problem is framed by the broker.

One of the most hotly debated issues in finance is the efficiency of financial markets. The results of behavioral finance are often in disagreement with the efficient market hypothesis (EMH). The EMH has been the central tenet of finance for almost 30 years and the power of the EMH assumption has had a remarkable influence on economic theory. The EMH started in the 1960s and was considered an immediate success, both in theory and empirically. Early empirical work gave overwhelming support to EMH. To a some extent, the EMH was invented at the University of Chicago and Chicago became a world center of research in finance. Jensen (1978) stated that "no other proposition in economics ... has more solid empirical support."

The initial enthusiasm for the EMH is changing. One reason is that the efficiency of arbitrage is now believed to be much weaker than expected. True arbitrage possibilities where a guaranteed profit can be locked in are rare. Near arbitrage, where two *nearly* equivalent assets are selling at different prices, is riskier than expected. If the prices of two nearly equivalent, but differently priced, assets were guaranteed to converge to a common value, then one could lock in a certain profit. For example, if two stocks were essentially equivalent, but one was priced higher than the other, then one could sell the higher-priced stock short and buy the lower-priced one. The beta of a short position in a stock is minus the beta of the stock. Therefore, a portfolio with a suitable choice of weights for the one stock short and the other stock long would have a beta of zero. This would eliminate market risk on the zero-beta portfolio, creating a near arbitrage opportunity. As the stocks converged to a common value, this portfolio would obtain a profit since the higher-price stock which was sold short would decline in price relative to the lower-priced stock. Unfortunately, mispricing can persist and even grow worse. According to Roger Lowenstein in his book, *When Genius Failed: The Rise and Fall of Long-Term Capital Management*, this type of irrational market behavior was a major reason for the highly publicized failure of the firm Long-Term Capital Management.[1] Lowenstein gives an incisive quote from John Maynard Keynes: "Markets can remain irrational longer than you can remain solvent."

Behavioral finance is important even if markets are efficient, because individual investors can make serious mistakes due to irrational investing. Indeed, a recent popular book, *Why Smart People Make Big Money Mistakes and How to Correct Them*, by Belsky and Gilovich (2000) is intended to remedy this problem. Belsky, a journalist specializing in business and finance, and Gilovich, a psychologist, represent the two fields brought together in behavioral finance.

14.2 Defense of the EMH

There are three lines of defense of the EMH, each serving to back up the previous ones:

- Investors are rational, so market prices equal the net present values (NPV's) of assets;
- Even if some investors are not rational, the trading of irrational investors is random and their trades cancel each other. Thus, market prices will still equal net present values even if some investors are willing to pay more than net present values for assets;
- Even if a "herd" of irrational investors trades in similar ways, rational arbitrageurs will eliminate their influence on market price.

As discussed in this chapter, each of these defenses is weaker than had been thought. As just mentioned, the discovery that arbitrage has less influence on market prices than expected has weakened the third line of defense, which perhaps was the one felt to be most secure.

[1] Other reasons were certainly hubris and greed.

Rational investing means valuing a security by its fundamental value, where "fundamental value" is the net present worth of all future cash flows, that is, the sum of all future cash flows discounted at a discount rate r; see Section 14.6.

Initial tests of the semi-strong form of efficiency supported that theory. **Event studies** look at the prices of a stock both before and after a new announcement about that stock. These studies showed that the market did react immediately to the news announcements and then stopping reacting, as the EMH predicts. Moreover, Scholes (1972) found little reaction to "nonnews." For example, the sale of a large block of shares of a stock might lead irrational "noise traders" to also sell, causing a decline in price. However, a block sale is considered to be "nonnews" since it contains no information about the fundamental value of the asset, so block sales should not affect stock prices if investors are rational as the EMH assumes. Scholes found that block sales of stocks had little effect on prices as the EMH predicts.

14.3 Challenges to the EMH

It can be proved that under certain assumptions, rational investing implies that prices are (geometric) random walks, but the converse is not true. Prices being random walks (or nearly so) does *not* imply rational investing, because random walk behavior could also be due to random irrational behavior of investors. Therefore, evidence that prices behave as do random walks should not necessarily be interpreted as favoring the EMH. Moreover, the random walk hypothesis, one of the foundations of the EMH, is now subject to serious doubt due to empirical evidence that is discussed in this chapter. For example, although prices appear to be geometric random walks over short horizons, there are signs of correlation over long horizons.

There is good evidence that irrational trading occurs, for example, during the internet stock bubble during the late 1990s, where the prices of Internet and other technology stocks rose to greatly inflated prices before crashing. One explanation of irrational trading is the overreaction hypothesis discussed in Section 14.5.2.

Many investors do react to irrelevant information. Black (1986) calls them noise traders. Investors act irrationally when they fail to diversify, purchase actively and expensively managed mutual funds, and churn their portfolios, i.e., buy and sell so excessively that transaction fees overwhelm any potential profits.

Investors are irrational in other ways as well. They do not look only at final levels of wealth when assessing risky situations but rather make situations based on the way a decision problem is framed. This behavior has led to the prospect theory of Kahneman and Tversky (1979). There is serious "loss aversion" in the market; investors behave in ways that avoid large losses on some assets even if their final wealth is thereby decreased. It is believed that the emotional impact of a loss is twice as great as that of an equal-sized gain. People appear not to follow Bayes' law for evaluating new information, but rather to pay too much attention to recent history.[2] Overreaction, meaning placing too much importance on recent news and not enough emphasis on historical data, is commonplace. These deviations from fully rational behavior are *not independent*. Noise traders will follow each other's mistakes, so noise trading is correlated across investors.

It is often argued that noise traders are small investors and have little effect upon the market. However, managers of large funds are themselves human and appear to make irrational trades too. Since these managers control a large portion of the total assets in the market, their behavior has a large influence. Managers also have their own types of irrationalities not found among small investors such as buying portfolios excessively close to a benchmark, buying the same stocks as other fund managers so as not to look bad, and window dressing, i.e., adding stocks to the portfolio that have been performing well recently so that these stocks can be listed in the fund's report. On average, pension and mutual fund managers underperform passive investment strategies such as indexing,[3] which is further evidence that these managers might be noise traders too.

14.4 Can Arbitrageurs Save the Day?

The last defense of the EMH depends on arbitrage. The argument is that even if investor sentiment is correlated and noise traders create incorrectly priced assets, arbitrageurs are expected to take the other side of their trades and drive prices back to fundamental values.

However, real-world arbitrage is risky and limited. Arbitrage depends on the existence of "close substitutes" for assets whose prices have been driven to incorrect levels by noise traders. Many securities do not have true substitutes; i.e., often there are no risk-free hedges for arbitrageurs. For example, it has been observed that stocks rise if their companies are put on the S&P 500 index. This is reaction to "nonnews" since being placed on this index does not affect fundamental value. The problem is that investors irrationally view the S&P 500 as *the* market index, rather than just an approximation to the market index, so many index funds must replicate the S&P 500 *exactly*. So when a stock is added to the S&P 500, many index funds are forced to buy that stock, which creates new demand and drives up the price. In one case, America Online rose 18% when included on the S&P 500. This might appear to have been an arbitrage opportunity. If AOL was overpriced at the time it entered the S&P 500, then an arbitrageur would have considered selling AOL

[2] See Kahneman and Tversky (1973).

[3] Indexing means designing one's portfolio to track some standard index such as the S&P 500, in essence creating an index fund.

short. However, there would have been the risk that AOL's price would rise, despite being overpriced, because of an overall rise in the market or because of economic developments favorable to all Internet access providers. One could have hedged the short sale of AOL by buying a company identical to AOL except not recently added to the S&P 500 and so not overpriced. The problem with this strategy should be obvious — there are no other companies exactly comparable to AOL. Moreover, even if there were, it might take a very long time for the mispricing of AOL to be corrected.

Mispricing can even get worse, as the managers of Long-Term Capital Management learned. This problem is called noise trader risk.

14.5 What Do the Data Say?

14.5.1 Excess price volatility

Shiller in his 1981 paper, "Do stock prices move too much to be justified by subsequent changes in dividends?" found that asset prices were too volatile to be consistent with a model where prices equal expected net present values. However, this work has been criticized by Merton (1987) who believes that Shiller did not correctly specify fundamental value. We discuss Shiller's work in Section 14.6.

14.5.2 The overreaction hypothesis

In 1985, De Bondt and Thaler published the paper "Does the stock market overreact?" which was based on De Bondt's Cornell PhD thesis. This has been a very influential paper and is frequently cited and reprinted. De Bondt and Thaler compare extreme winners and losers and find strong evidence of overreaction. More explicitly, every third year starting in 1933 they formed portfolios of the best performing stocks over the previous three years which were called "winner portfolios." They also formed portfolios of the worse performing stocks called "loser portfolios." They then examined returns on these portfolios over the next three years. Their surprising finding was that three years past portfolio formation the loser portfolios consistently outperformed the winner portfolios. De Bondt and Thaler found the better performance of loser to winner portfolios to be statistically significant by using a independent-sample t-test.

De Bondt and Thaler believe that this is evidence of overreaction. Their argument is that investors overreact to both good and bad news so that stocks that have risen over the last three years because of good news are priced too high and similarly that stocks that have fallen over recent years due to bad news are priced too low. De Bondt and Thaler see the poor relative performance of winner portfolios as corrections of these mispricings. Losers are too cheap and bounce back, while winners are too high and must slump. These adjustments of prices are a sign that markets work eventually, but according to the EMH none of these stocks should have been mispriced to begin with — overreaction does not occur in the perfect world of the EMH. Researchers who believe in the efficiency of markets do not accept the overreaction hypothesis but argue instead that the difference in performance between winner and loser portfolios is due to differences in risk. We know from Chapters 5 and 7 that riskier assets should have higher expected returns. So the higher returns on the loser portfolios could be interpreted as evidence that they are riskier, not evidence of mispricing, at least according to those who believe in efficient markets. The differences between loser and winner portfolios are difficult to explain as due to differences in risk, at least according to standard models such as CAPM. De Bondt and Thaler did try adjusting returns for risk, but still found that loser portfolios outperformed winner portfolios. However, as discussed shortly, using Fama and French's (1995, 1996) three-factor one can explain De Bondt and Thaler's findings as due to risk factors.

There are other empirical findings that are used to support the theory of inefficient markets. Historically, small stocks have earned higher returns than large stocks, and there is little evidence that the difference is due to higher risk. The superior returns of small stocks have been concentrated in January. The small firm effect and January effect seem to have disappeared over the last 15 years. This disappearance may be evidence of some market efficiency, but as with the correction of mispricing due to overreaction, some will argue that according to EMH the small firm and January effects should never have occurred in the first place.

14.5.3 Reactions to earnings announcements

There is a sizeable academic literature that studies how securities analysts react to the earnings announcements that companies make each quarter. Recall that the job of these analysts is to forecast future earnings that drive prices. When a quarterly earnings is announced it could either bring "good news" that the earning is above the forecast value or "bad news" that it is below. The difference between the actual and the forecast earnings is called the *earnings surprise*. The earnings surprise can be divided by the standard deviation of all earnings surprises of that quarter to obtain the *standardized unexpected earnings* or *SUE*.

If security analysts' forecasts are the best predictors of future earnings, then the quarterly SUE values of any stock are uncorrelated as a consequence of the following fact.

Result 14.5.1 *Let $X_1, X_2, \ldots$ be a sequence of random variables at times $t = 1, 2, \ldots$ and let $\widehat{X}_{t+1}$ be either the best linear predictor or the best predictor of X_{t+1} based on all information available at time t including the values of $X_1, \ldots, X_t$ and forecasts $\widehat{X}_1, \ldots, \widehat{X}_t$. Then the sequence of forecast errors $(X_2 - \widehat{X}_2), (X_3 - \widehat{X}_3), \ldots$ is uncorrelated.*

The result is not proven here, but Problem 6 verifies the result in a special case. The intuition behind this result is that if the next forecast error were correlated with the known value of the present forecast error, then this information could be used to predict the next forecast error. Then we could adjust that forecast leading to an improved forecast, which would be inconsistent with the assumption that the forecasts are already best or best linear.

When researchers have looked at SUE values for a stock, they have found autocorrelation. When a company reports good news, so that the SUE value is positive, the SUE of that company tends to be positive for the next two or three quarters, and similarly bad news means not only a negative SUE that quarter but a tendency for the SUE values to be negative for two or three additional quarters. The behavioral finance literature interprets this finding as due to *underreaction*. Apparently, analysts do not revise their beliefs about a company enough following good or bad news. Also, share prices underreact to news, perhaps because of the underreaction of analysts' forecast. The result is that after an announcement of good news, the price of that stock *drifts* upward causing abnormally high returns for several quarters rather than reacting instantaneously as the efficient market hypothesis predicts. Similarly, after bad news a stock has abnormally low returns for several quarters.

Suppose that at the one quarter stocks are divided into deciles according to SUE for that quarter and a portfolio goes long the highest decile and short the lowest decile. What happens? According to Shefrin (2000) this portfolio would have an abnormal return of 4.2% with much of this return occurring after subsequent earnings announcements which would tend to show more good news for the highest decile and more bad news for the lowest decile.

The result of underreaction to earnings announcements is short-term momentum in stock prices, which is the opposite of the longer-term reversals found by De Bondt and Thaler, though not inconsistent with these reversals.

Shefrin (2000) reports that the firm Fuller and Thaler Asset Management manages the "Behavioral Growth" Fund which holds stock in companies with high SUE. This fund has substantially outperformed both the Russell 2500 Growth index and the S&P 500 index over the period from January 1992 to January 1999.

14.5.4 Counter-arguments to pricing anomalies

Proponents of the EMH have countered these arguments against the EMH with the theory that the small firm and other anomalous effects reveal heretofore unknown risk factors that the market *is* correctly pricing. Market-to-book value is a measure of "cheapness." High market-to-book value firms are considered "growth" stocks while low market-to-book stocks are "value" stocks. However, in contradiction to the CAPM, these stocks tend to *both* underperform the market *and* to be riskier, especially in severe down markets. Does this finding support the EMH or provide evidence against the EMH? Economists have argued both ways.

Fama and French (1996) use book equity to market equity (BE/ME); that is, the reciprocal of market-to-book value, as one of the factors in their three-factor model discussed in Section 7.8.2. In their Table II, they look at returns on portfolios formed from the deciles of BE/ME, that is, the first portfolio is the bottom 10% based on BE/ME, and so forth. The returns are from a period *after* portfolio formation, since the portfolios are formed by looking at BE/ME at some time point and then returns are measured during a subsequent period. Mean returns increase steadily from the 1st decile (smallest BE/ME values and therefore growth stocks) to the 10th decile (highest BE/ME and therefore "value" stocks). The standard deviations of the returns are lowest in deciles 5 to 8, i.e., extreme deciles (deciles 1 to 4 and 9 to 10) have the highest risk.

Fama and French (1996) claim that their three -factor model, which was described in Section 7.8.2, "explains" the findings of De Bondt and Thaler. Fama and French note that losers tend to load more than winners on the factors SMB and HML of their model; i.e., losers behave as distressed stocks. Since SML and HML have high average risk premiums, portfolios that load highly on them have higher average returns. Whether losers are really distressed companies and whether their high returns are rational or due to overreaction is still a matter of considerable debate.

Fama and French (1996) discuss one empirical finding that they claim is unexplained by CAPM, their three-factor model, and the overreaction hypothesis. That finding, which is due to Jegadeesh and Titman (1993), is the continuation of return trends when portfolios are formed over *short* periods — short-term losers continue to lose and short-term winners to win, the opposite of what De Bondt and Thaler found with portfolios formed over *longer* periods. This momentum is consistent with the underreaction to earnings announcements discussed in Section 14.5.3.

There appears as yet to be no firm understanding of why we see underreaction in some situations such as to earnings announcements and overreaction in other cases. Fama (1998) argues that the existence in financial markets of both underreaction and overreaction is consistent with the EMH, because the EMH predicts the expected values of abnormal returns to be zero with pricing anomalies random split between overreaction and underreaction.

14.5.5 Reaction to non-news

If markets are efficient, then large price changes should only occur immediately after major news items, but the evidence appears to contradict this hypothesis. On October 19, 1987, the Dow Jones index dropped 22.6% although there was no apparent news that day other than the market crash itself. Cutler, Poterba, and Summers (1991) looked at the 50 largest one-day market changes. Many of these came on days with no major news announcements.

Roll (1988) tried to predict the proportion of return variation that could be explained by economic influences, returns on other stocks in the same industry, and public firm-specific news. Roll's findings were that $R^2 = 0.35$

for monthly data and $R^2 = 0.2$ for daily data. These numbers are quite small. If the EMH is true, they should be closer to 1.0. Roll's study also shows that there are no "close substitutes" for stocks and this lack of close substitutes limits arbitrage, a point mentioned before in connection with AOL.

14.6 Market Volatility and Irrational Exuberance

Robert Shiller has made market volatility a major part of his research and has written two books on the subject. The first, *Market Volatility*, published in 1989, is a collection of papers. The second, *Irrational Exuberance*, published in 2000, was written for a popular audience and has become a best-seller.

What is a stock worth? Let V_t be the stock's intrinsic value (fundamental value) at time t. By definition

$$V_t = \frac{C_{t+1}}{1+r} + \frac{C_{t+2}}{(1+r)^2} + \cdots = \sum_{i=1}^{\infty} \frac{C_{t+i}}{(1+r)^i}.$$

Here C_i is the cash flow at time i, and r is the discount rate. A little bit of algebra shows that

$$V_t = \sum_{i=1}^{T-t} \frac{C_{t+i}}{(1+r)^i} + \frac{V_T}{(1+r)^{T-t}}. \tag{14.1}$$

Shiller (1981) used formula (14.1) with T equal to present time and with t some time in the past, i.e., with $t < T$. Therefore, he could use past and present data to estimate V_t by the right-hand side of (14.1). The past and present cash flows, i.e., the C_{t+i}, in the right-hand side of (14.1) are known. The only unknown quantity in the right-hand side of (14.1) is the current fundamental value V_T, and as an approximation Shiller used the known stock price P_T in place of V_T.[4] With this approximation, all quantities in the right-hand side of (14.1) are known at time T which gives an estimate of V_t. Once Shiller had this estimate of the fundamental value V_t, he could then compare it with P_t, the *actual* stock price.

The EMH says that P_t is the optimal forecast (best predictor) at time t of V_t. How do V_t and P_t compare? The market data show that P_t is *much more* volatile than V_t. Does this agree with the hypothesis that P_t is the optimal forecast of V_t? The answer is *no!* An optimal forecast of V_t *must* be *less* volatile than V_t itself. To appreciate this, we need to look more closely at the theory of best prediction.

[4] If the discount factor $1/(1+r)^{T-t}$ is reasonably small, then the replacement of the unknown V_T by the known P_T should cause only a small approximation error.

14.6.1 Best prediction

If $\widehat{X}$ is the best predictor of X, then

$$\widehat{X} \text{ and } X - \widehat{X} \text{ are uncorrelated,} \tag{14.2}$$

$$\text{Var}(X) = \text{Var}(\widehat{X}) + E(X - \widehat{X})^2, \tag{14.3}$$

and

$$\text{Var}(\widehat{X}) \leq \text{Var}(X). \tag{14.4}$$

Equation (14.4) follows immediately from (14.3) and has an important interpretation.

Result 14.6.1 *An optimal predictor is less variable than the quantity being predicted.*

Example 14.1. (Random Walk) Suppose that $W_t = \epsilon_1 + \cdots + \epsilon_t$, where $\epsilon_1, \ldots$ are i.i.d. $N(\mu, \sigma^2)$. At time t, $0 \leq t < T$, the best predictor of $W_T = W_t + \epsilon_{t+1} + \cdots + \epsilon_T$ is $\widehat{W_T} = W_t + (T-t)\mu$. The reason for this is that, first, at time t, W_t is "predicted" by itself since it is known. Second, the sum of the future steps, $\epsilon_{t+1} + \cdots \epsilon_T$, is predicted by its expectation $(T-t)\mu$ since these terms are independent of all information available at time t. Note that $\text{Var}(W_T) = T\sigma^2$, $\text{Var}(\widehat{W_T}) = \text{Var}(W_t) = t\sigma^2$, $W_T - \widehat{W_T} = (\epsilon_{t+1} - \mu) + \cdots + (\epsilon_T - \mu)$, and $\text{Var}(W_T - \widehat{W_T}) = (T-t)\sigma^2$. In agreement with (14.3),

$$\text{Var}(W_T) = \text{Var}(\widehat{W_T}) + \text{Var}(W_T - \widehat{W_T})$$

and therefore

$$\text{Var}(\widehat{W_T}) \leq \text{Var}(W_T).$$

This verifies that, in accordance with the theory of best prediction, the predictor $\widehat{W_T}$ is less variable than the quantity being predicted W_T.

As t increases to T we cumulate more information about W_T and

- $\text{Var}(\widehat{W_T}) = \text{Var}(W_t) = t\sigma^2$ increases;
- $\text{Var}(W_T - \widehat{W_T}) = (T-t)\sigma^2$ decreases; and
- $\text{Var}(W_T) = T\sigma^2$ stays the same, of course, since it does not depend on t.

Equation (14.3) shows precisely how much smaller the variance of optimal forecast $\widehat{X}$ is compared to the variance of X, but the main point is simply that an optimal forecast is less variable than what is being forecast. But Shiller's (2000) analysis of the market data shows that stock prices are *much more* variable than the present values of future discounted dividends. Therefore, prices *cannot* be optimal forecasts of the present value of discounted future dividends.

Irrational Exuberance has a very interesting discussion of market psychology, bubbles, "naturally occurring Ponzi schemes," and other possible explanations of why the market seemed overpriced in 2000. Shiller presents fascinating evidence that periods where the market is either over- or under-priced have occurred, often several times, in many countries.

14.7 The Current Status of Classical Finance

By classical finance, I mean finance theory based on the assumptions of completely rational behavior including the EMH, the CAPM, and portfolio theory. There is now a considerable amount of empirical evidence and theoretical arguments that contradict the EMH. This evidence was not found during early testing of the EMH. Rather, researchers needed considerable time to learn what to look for. Nonetheless, proponents of the EMH are countering these arguments with theories about risk factors in addition to the market and unique risks in the CAPM. Their arguments sacrifice the CAPM to save the EMH. The empirical evidence against the EMH is not conclusive. It is certain, however, that two cornerstones of classic finance, the EMH and CAPM, are powerful null hypotheses and testing them is leading to much interesting research and a better understanding of how financial markets actually behave. Some economists such as Fama still believe firmly in the EMH, but the CAPM seems to have no strong adherents in academia despite its apparent widespread use in practice.

Classical finance is still useful, regardless of its shortcomings. In physics, there is no doubt that relativity is a more accurate description of nature than Newtonian mechanics, yet the latter is quite useful and frequently used. Classical finance has a similar role in finance to that of Newton mechanics in physics. Both are approximations to reality that are simple enough to be very useful. There is of yet no theory of behavioral finance that can replace the EMH and CAPM, so in finance there is no analogue to relativity theory. Work has begun, however, on economic models based upon concepts from behavioral finance; see, for example, models for noise trader risk and investor sentiment in Shleifer (2000).

14.8 Bibliographic Notes

Good overviews of the subject are available in the books by Shleifer (2000) and Shefrin (2000), and my understanding of behavioral finance owes much to their work. Shefrin is an excellent place to start for further study. Shleifer is more technical and more suitable for a reader with some background in economics. It is a good introduction to the economic models being developed in behavioral finance. An excellent shorter introduction to behavioral finance is De Bondt and Thaler (1995). Event studies are discussed in Campbell, Lo, and MacKinlay (1997, Chapter 4) and Bodie, Kane, and Marcus (1999, Section 12.3). Shiller (2000) is highly recommended. It is a very engaging exposition of the irrational behavior of markets. The effects of noise trading are analyzed in Black (1986) and De Long, Shleifer, Summers, and Waldman (1990) develop a model for noise trader risk; see also Shleifer (2000). The underreaction to earnings announcements is discussed by Bernard (1993) and Shefrin (2000).

14.9 References

Belsky, G. and Gilovich, T. (2000) *Why Smart People Make Big Money Mistakes And How To Correct Them*, Fireside, New York.

Bernard, V. (1993) Stock price reactions to earnings announcements: A summary of recent anomalous evidence and possible explanations, In *Advances in Behavioral Finance*, edited by Thaler, R., pp. 303–340, Russell Sage Foundation, New York.

Black, F. (1986) Noise, *Journal of Finance*, **41**, 529–543. (Reprinted in Thaler (1993).)

Bodie, Z., Kane, A., and Marcus, A. (1999) *Investments, 4th Ed.*, Irwin/McGraw-Hill, Boston.

Campbell, J., Lo, A., and MacKinlay, A. (1997) *The Econometrics of Financial Markets*, Princeton University Press, Princeton, NJ.

Cutler, D., Poterba, J., and Summers, L. (1991) Speculative dynamics, *Review of Economic Studies*, **53**, 1839–1885.

De Bondt, W. and Thaler, R. (1985) Does the stock market overreact?, *Journal of Finance*, **40**, 793–805. (Reprinted in Thaler (1993).)

De Bondt, W. and Thaler, R. (1995) Financial decision-making in markets and firms: A behavioral perspective, In *Handbook of Operations Research and Industrial Engineering, Vol. 9, Finance*, edited by Jarrow, R., Maksimovic, V., and Ziemba, W. T., pp. 385–410, Elsevier, Amsterdam.

De Long, J. B., Shleifer, A., Summers, L. H., and Waldman, R. J. (1990) Noise trader risk in financial markets, *Journal of Political Economy*, **98**, 703–738. (Reprinted in Thaler (1993).)

Fama, E. F. (1998) Market efficiency, long-term returns, and behavioral finance, *Journal of Financial Economics*, **49**, 283–306

Fama, E. F. and French, K. R. (1995) Size and book-to-market factors in earnings and returns, *Journal of Finance*, **50**, 131–155.

Fama, E. F. and French, K. R. (1996) Multifactor explanations of asset pricing anomalies, *Journal of Finance*, **51**, 55–84.

Jegadeesh, N. and Titman, S. (1993) Returns to buying winners and selling losers: Implications for stock market efficiency. *Journal of Finance*, **48**, 65–91.

Jensen, M. (1978) Some anomalous evidence regarding market efficiency, *Journal of Financial Economics*, **6**, 95–101.
Kahneman, D. and Tversky, A. (1973) On the psychology of prediction. *Psychological Review*, **80**, 237–251.
Kahneman, D. and Tversky, A. (1979) Prospect theory: An analysis of decision under risk, *Economerica*, **47**, 263–291.
Lowenstein, R. (2000) *When Genius Failed: The Rise and Fall of Long-Term Capital Management*, Random House, New York.
Merton, R. (1987) On the current state of the stock market rationality hypothesis, In *Macroeconomics and Finance: Essays in Honor of Franco Modigliani*, edited by Dornbusch, R., Fischer, S., and Bossons, J., MIT Press, Cambridge, MA.
Roll, R. (1988) R^2, *Journal of Finance*, **43**, 541–566.
Scholes, M. (1972) The market for securities: Substitution versus price pressure and effects of information on share prices, *Journal of Finance*, **45**, 179–211.
Shefrin, H. (2000) *Beyond Greed and Fear: Understanding Behavioral Finance and the Psychology of Investing*, Harvard Business School Press, Boston.
Shiller, R. (1981) Do stock prices move too much to be justified by subsequent changes in dividends?, *American Economic Review*, **71**, 421–436. (Reprinted in Thaler (1993).)
Shiller, R. (1992) *Market Volatility, Reprint Ed.*, MIT Press, Cambridge, MA.
Shiller, R. (2000) *Irrational Exuberance*, Broadway, New York.
Shleifer, A. (2000) *Inefficient Markets: An Introduction to Behavioral Finance*, Oxford University Press, Oxford.
Thaler, R. H. (1993) *Advances in Behavioral Finance*, Russell Sage Foundation, New York.

14.10 Problems

1. In Section 14.6.1 it was stated that if $\widehat{X}$ is the best predictor of X, then
   (a) $\widehat{X}$ and $X - \widehat{X}$ are uncorrelated.
   (b) $\text{Var}(X) = \text{Var}(\widehat{X}) + E(X - \widehat{X})^2$.
   (c) $\text{Var}(\widehat{X}) \leq \text{Var}(X)$.
   Verify these results when $\widehat{X}$ is the best *linear* predictor of X given W so that by (2.45)
$$\widehat{X} = E(X) + \frac{\sigma_{XW}}{\sigma_W^2}\{W - E(W)\}.$$
2. De Bondt and Thaler (1985) formed winner and loser portfolios and then in each year after portfolio formation they compared the returns on the two portfolios. To test for a difference in expected returns, they used the

independent-samples t-test. Do you agree that this was the correct test to use? Their finding was that the expected return is higher for loser than for winner portfolios. If you believe that they used an incorrect test, then do you feel that their conclusion is mistaken? Discuss.

3. The following decision problem is due to Kahneman and Tversky and is discussed by Shefrin (2000). Suppose you must simultaneously make two decisions:

   Decision 1*:* Choose between
   (A) a gain of $2400 with probability 1.
   (B) a gain of $10,000 with probability 0.25 or no gain with probability 0.75.

   Decision 2*:* Choose between
   (C) a loss of $7500 with probability 1.
   (D) a loss of $10,000 with probability 0.75 or no loss with probability 0.25.

   Since the decisions are made concurrently, the choices should depend only on the probability distribution of the total gain/loss for the combined decisions and one's tolerance for risk (which could be quantified by specifying a utility function). Many people choose (A) in Decision 1 and (D) in Decision 2, and we denote this joint decision by [A,D].
   (a) Discuss why [A,D] might be favored.
   (b) There are four possible joint decisions, [A,C], [A,D], [B,C], and [B,D]. For each of the four, list the possible values of the total gain/loss and their probabilities. Is there a decision that is better than [A,D] regardless of one's tolerance for risk? If so, which one and why?

4. As mentioned in Section 14.5.3, the Behavioral Growth Fund invests in companies with high SUE and has substantially outperformed both the Russell 2500 Growth index and the S&P 500 index over the period from January 1992 to January 1999. Does this disprove the EMH? Discuss.

5. As mentioned in Section 14.5.4, Fama (1998) believes that the existence of both underreaction and overreaction is consistent with the EMH, which predicts the expected values of abnormal returns to be zero with pricing anomalies randomly split between overreaction and underreaction. Do you agree with Fama's argument? Also, is Fama's argument in agreement with the excess volatility of stock prices found by Shiller? Discuss.

6. Verify Result 14.5.1 in the special case of an AR(1) process where $X_{t+1} = \phi X_t + \epsilon_t$ where ϵ_t is WhiteNoise$(0, \sigma^2)$. Assume that ϕ is known.

Glossary

AIC Akaike's Information Criterion. A statistic used to choose models that neither overfit nor underfit the data. "Good" models have a small value of this criterion. Often a moderate to large number of rather different models will all have small values of AIC, and if possible one should use subject matter knowledge to choose between them. *See also* SBC.

Alternative hypothesis H_1. In hypothesis testing, the alternative hypothesis is true if the null hypothesis is false.

American option An option that can be exercised on or before the exercise date. *Compare with* European option.

Arbitrage Making a risk-free profit without investing capital.

Arbitrage price The price of a security that is determined by the requirement that the market has no arbitrage opportunities.

AR(1) Autoregressive process where the present observation depends only on the most recent previous observation.

ARMA process A process that combines AR and MA terms. An ARMA(1,1) process is a combination of AR(1) and MA(1). Similarly ARMA(p, q) combines AR(p) and MA(q).

AR(p) Autoregressive process where the present observations depends on the previous p observations.

Ask price Price at which a market maker will sell a security.

At the money An option is at the money if the underlying security's price is on the boundary between where the option is in the money and out of the money.

Autocorrelation function Correlation between observations of a stationary process as function of the time interval between them (lag).

Autocovariance function Covariance between observations of a stationary process as function of the time interval between them (lag).

Autoregressive process A statistical model for a time series where the dependencies are modeled by expressing the present observation as a weighted average of past observations.

Bayesian statistics A branch of statistics that treats unknown parameters as random variables. A Bayesian analysis starts with a prior distribution that specifies information about the parameters that is known before data are collected. The posterior distribution, which is the conditional distribution of the parameters given the data, is the basis for estimation, testing, and inference.

Bayes' law Method of calculating conditional probabilities. Used in Bayesian statistics for updating the prior probability distribution of the parameters to the posterior distribution.

Behavioral finance The application of psychology to finance.

Best linear prediction Prediction of an unobserved random variable using a linear function of observed random variablse chosen in such a way as to minimize the expected squared prediction error.

Beta A measure of aggressiveness of a stock. Beta is the covariance between a security's return and the market return divided by the variance of the market return. Beta is also the slope of the regression of the stock's excess return on the market's excess return. By definition, the beta of the market portfolio is one. Stocks with beta greater (less) than one are considered aggressive (not aggressive).

Bid price Price at which a market maker will buy a security.

Binomial distribution Probability distribution of the number of "successes" in a sequence of i.i.d. binary trials. Binomial(n, p) is the the binomial distribution with n trials and the probability of success on any trial equal to p.

Book value The value of a firm according to its accounting books.

Bootstrap A method of computing standard errors, confidence limits, and other measures of sampling error by taking samples with replacement from the original sample. Also known as resampling.

Brownian motion The continuous time limit of a random walk. Brownian motion takes its name from the movement of microscopic particles in a water due to bombardment by water molecules.

Call A derivative security giving the owner the right but not the obligation to buy the underlying security.

CAPM The capital asset pricing model. One of the fundamental theories of classical finance but now often replaced by more comprehensive factor models.

Central limit theorem The central limit theorem states that the distribution of the sample mean will converge to the normal distribution as the sample size increases to ∞. Sometimes misunderstood to say that large populations are approximately normally distributed, which is not true in general. Also called the CLT.

CML Capital market line. Relates the excess expected return on an efficient portfolio to its risk (standard deviation). Also called CML.

Conditional distribution The probability distribution of a random variable or set of random variables given known values of other random variables.

Conditional heteroscedasticity *See* heteroscedasticity.

Confidence interval A range of values likely to contain an unknown parameter. If a large number of confidence intervals are constructed from independent samples, then the proportion of them that contain the parameter is approximately equal to the confidence coefficient. The confidence coefficient is denoted by

$1 - \alpha$ and is typically equal to 90%, 95%, or 99% though other values are also used.

Continuous compounding The limit of discrete compounding as the time interval from one compounding to the next shrinks to zero. Under continuous compounding at rate r, one dollar increases to e^{rt} dollars over a time interval of length t.

Coupon bond A bond making periodic payments of interest until maturity, when a payment of principal is also made. Generally coupon payments are semiannual.

Coupon rate The interest rate of the coupon payments of a bond as a percentage of the par value. If a bond is selling at its par value, then the coupon rate, current yield, and yield to maturity are all equal.

Correlation A quantity used to measure how well one random variable can linearly predict another. Also, a measure of association between random variables. Correlation, also called the correlation coefficient, takes values between -1 and 1.

Covariance A quantity that is used to calculate the correlation between two random variables. Used also to calculate the variance of a linear combination (weighted average) of two random variables.

Covariance matrix A matrix containing the variances and covariances of a collection of random variables. The covariance matrix can be used to find the variance of any weighted average of these random variables.

Credit risk The risk that a counterparty does not meet contractual obligations, for example, that interest or principal on a bond is not paid.

Cumulative distribution function Also called CDF. The CDF of a random variables is the function whose value at x is the probability the random variable is at most x.

Current yield The interest rate of the coupon payments as a percentage of the current price. Current yield differs from the coupon rate in that the latter is the coupon rate as a percentage of the par value. Unlike yield to maturity, the current yield does not take into account the loss or gain of principal if the current price differs from the par value.

Delta The derivative of a option's price with respect to the price of the underlying security. Also called the hedge ratio.

Delta hedging Hedging a stock with an option (or vice versa) using the option's delta to calculate the appropriate amount of the option (or stock) to purchase.

Discrete compounding Compounding at fixed time intervals, e.g., daily, monthly, quarterly, and so forth.

Drift The drift of a random walk or Brownian motion determines the expected change in that process. A positive (negative) drift means that the process is increasing (decreasing) on average.

Duration The duration of a coupon bond, denoted here by DUR, is the weighted average of the maturities of future cash flows with weights proportional to the net present value of the cash flows. For a zero-coupon bond, DUR equals time to maturity. Duration is a measure of interest rate risk.

Duration analysis Risk analysis using duration.

Efficient frontier Locus of points formed by the set of mean-variance efficient portfolios when standard deviations of returns are plotted against expected returns.

Efficient portfolio *See* mean-variance efficient portfolio.

Efficient market hypothesis EMH. Hypothesis that market prices fully reflect all available information. No one believes that the market is completely efficient, but the extent to which it is inefficient is a matter of much controversy.

Empirical distribution A probability distribution generated by a sample by assigning probability $1/n$ to each of the n observations in the sample. Bootstrapping uses resamples which are samples drawn from the empirical distribution.

Estimator A statistic used to estimate an unknown parameter.

Event studies Tests of the efficient market hypothesis that look at the price of a stock both before and after a news announcement about that stock.

Excess return The difference between the return on an asset and the risk-free return.

Exercise date *See* strike date.

Exercise price *See* strike price.

Expectation The expectation of a random variable is its average value with weighting according to its probability distribution. Expectation is the population analogue of the sample mean, and the sample mean is expectation with respect to the empirical distribution.

Expected shortfall The average (or expectation) of all losses exceeding a certain limit or threshold. Expected shortfall has recently been advocated as an alternative to VaR, because expected shortfall takes into account both the probability of a large loss (larger than the threshold) and the expected loss given that the loss exceeds the threshold.

Exponential random walk *See* geometric random walk.

European option An option that can only be exercised on the exercise date. *Compare with* American option.

Face value *See* par value.

Factor model A regression model that expresses the return on an asset as a function of factors such as the market return, the difference between returns on two portfolios, and macroeconomic variables. The noise term in the regression model represents the asset's unique risk. The security characteristic line is a simple factor with only the market return as a factor.

Fitted value The prediction of an observation made by a statistical model. The residual is the difference between the observed value and the fitted value. Least squares estimation minimizes the sum of the squared deviations between the observed and fitted values.

Financial engineering The construction of financial products such as stock options, interest rate derivatives, and credit derivatives.

Forward contract A contract to purchase something in the future at a prearranged price.

Forward interest rate The rate for future borrowing that one can lock in at the present time.

Fundamental analysis Selection of stocks by forecasting future earnings using accounting information and other data about the business prospects of firms.

Fundamental value The value of a firm based on future earnings. The fundamental value is the net present value of all future cash flows from the stock.

GARCH model A time series model with conditional heteroscedasticity. "GARCH" is an acronym for "generalized autoregressive conditional heteroscedasticity." GARCH models are becoming widely used in finance because financial market data often exhibit nonconstant volatility.

Geometric random walk The exponential of a random walk, that is, a process whose logarithm is a random walk. Also called exponential random walk. Geometric random walks are always nonnegative and therefore are better models for stock prices that random walks. If a stock price process is a geometric random walk, then its logarithm is a random walk and its log returns are white noise.

Gross return The ratio of the initial to the final price of an asset. Equal to 1 plus the net return. Also equal to the exponential of the log return.

Heavy tails A probability distribution has heavy tails if outliers are more likely than for a normal distribution. Heavy-tailed return distributions indicate that extremely large losses are more likely than for normally distributed returns.

Hedge ratio *See* Delta.

Hedging Assuming a securities position to offset risk due to another position.

Homoscedasticity Having a constant variance. Conditional homoscedasticity means having a constant conditional variance. ARMA models are conditionally homoscedastic.

Heteroscedasticity Having a nonconstant variance. Conditional heteroscedasticity means having a nonconstant conditional variance. GARCH models are conditionally heteroscedastic.

Holding period The period of time over which a return is calculated.

Implied volatility The value of the volatility parameter (standard deviation) at which the market price equals the price given by the Black-Scholes formula.

I.i.d. Independent and identically distributed. A sequence of random variables is i.i.d. if the random variables are mutually independent and they have the same probability distribution. An i.i.d. sequence from a probability distribution is called a random sample from that distribution.

In the money An option is in the money if the underlying security's price is such that the option would be worth a positive amount if the option could be exercised immediately. A call (put) is in the money if the the share price of the underlying stock is above (below) the strike price.

Interest rate risk Risk due to changes in bond prices due to fluctuations in interest rates.

Knot A location where the functional form of a spline changes from one polynomial to another.

Kurtosis Kurtosis is used to measure tail weight of a probability distribution. Large kurtosis means heavy tails.

Law of large numbers The law of large numbers states that the sample mean will converge to the population mean as the sample size increases.

Law of one price The law that states that two portfolios (or self-financing trading strategies) with exactly the same future cash flows should have the same price. Any violation of this law is an arbitrage opportunity, because one can buy the cheaper portfolio and sell the more one expensive short. Doing this secures an immediate profit after which all future cash flows cancel.

Least squares Estimation by minimizing the sum of squared residuals.

Leverage In finance, leverage refers to the sensitivity of an option's price to the price of the underlying. In this text, leverage is denoted by L and quantified as the ratio of the return on an option to the return on the underlying security. Leverage is related to but not the same as Delta. In regression, the leverage of an observation is the sensitivity of the fitted value of that observation to changes in the regression coefficients.

Leverage effect Predicts that an asset's returns may become more volatile when its price decreases.

Liquidity risk The risk due to the possible extra cost of liquidating a position because buyers may be difficult to locate.

Log return The change in the (natural) logarithm of the price of an asset. Also the logarithm of the gross return. Sometimes called the continuously compounded return.

Quantile The q-quantile of a random variable is a value x such that the probability that the random variable is less than or equal to x equals q. Quantiles are the same as percentiles, e.g., the 0.5-quantile is the 50-percentile.

Independence Property of a set of random variables that the conditional distribution of any subset of them does not depend upon the observed values of the others.

Likelihood function The probability of the observed data as a function of unknown parameters. Used to estimate parameters and test hypotheses.

Likelihood ratio test A test of a statistical hypothesis that uses the likelihood function. Also called the LRT. The LRT is widely used because of its simplicity and efficient use of the data.

Market portfolio The portfolio of all securities held in proportion to their market values.

Market return Return on the market portfolio.

Market risk The systematic component of risk due to correlation between an asset's return and the market return. A second meaning is risk due to changes in prices.

Market value The value of a firm according to the market. It is the share price times the number of outstanding shares.

Martingale A mathematical model of a fair game. A stochastic process is a martingale if any expected future value of the process, conditional given the present and past values of the process, equals the present value.

Martingale measure Also known as the risk-neutral measure or pricing measure. The measure under which the discounted price process is a martingale.

Maximum likelihood estimator MLE. The value of the parameter at which the likelihood function is largest. The MLE is widely used because it is generally very accurate and can be computed using standard optimization software.

Mean-variance efficient portfolio A portfolio whose mean return is as large as the mean return of any other portfolio with the same or smaller risk (measured by the standard deviation).

Measure The set of path probabilities of a stochastic process.

MA(1) A moving average process that is a weighted average of the present and the most recent past observation of a white noise process.

MA(q) A moving average process that is a weighted average of the present and the q most recent past observations of a white noise process.

Minimum variance portfolio The portfolio with the smallest possible standard deviation. The minimum variance portfolio is the leftmost and lowest point on the efficient frontier.

Model selection Selection of the predictor variables in a model and the degree of complexity of the model. The usual goal of model selection is to find a model that generalizes well in that it gives accurate predictions of new data. Model selection is used to achieve a balance between underfitting and overfitting. Often, different predictor variables contain nearly the same information so that many models give similar predictions. Therefore, there is generally no single "correct" or "true" model, and model selection is used to avoid a poor model giving inaccurate predictions. Cross-validation, AIC , C_p, SBC, and adjusted R^2 are often used to select a good model. R^2 (unadjusted) is not used for selection of a regression model, because R^2 is biased in favor of complex models and maximization of R^2 always selects the most complex model. *See also* overfitting and underfitting.

Moving average process A time series model where the present observation is a weighted average of past and present values of a white noise process. *See* MA(1) and MA(q). *Also see* ARMA process.

Net present value NPV. The sum of all discounted future cash flows. Also called present value, and discounted value.

Net return The change in price of an asset divided by the initial price. Equal to the gross return minus 1.

Nonlinear regression Regression where the regression function is nonlinear in the parameters, e.g., has unknown parameters in an exponential or is a sine function of unknown frequency. Nonlinear regression is better called nonlinear parametric regression to distinguish it from nonparametric regression.

Nonmarket risk The unique component of risk. This is the variation in an asset's return that cannot be predicted by the market return.

Nonparametric regression Regression when the shape of the regression function is determined by the data, not the model. Used when this shape is not known prior to collection of the data.

Normal distribution The bell-curve. $N(\mu, \sigma^2)$ is the normal distribution with mean μ and variance σ^2. The central limit theorem states that the distribution of the sample mean will converge to the normal distribution as the sample size increases.

Normal probability plot An informal graphical method for checking whether a sample came from a normal distribution. Normality tests such as the Shapiro-Wilk, Kolmogoroff-Smirnov, and Cramér-von Mises tests are formal methods of testing the null hypothesis of normality.

Null hypothesis H_0. One of the two hypotheses in statistical testing. It is called "null" because frequently it is created with the intention of attempting to disproving it. Generally, the null hypothesis is the simpler of the two hypotheses.

Opacity Risks are said to be opaque when they are not easily understood. Risks due to holding or selling derivatives are often considered to be opaque.

Out of the money An option is out of the money if the underlying security's price is such that the option would be worth nothing if exercised immediately. A call (put) is out of the money if the the share price of the underlying stock is below (above) the strike price.

Overfitting Using a statistical model that is overly complex so that the model fits random features of the data and therefore predicts future observations poorly. *See also* underfitting and model selection.

Par value The face value of a bond. This is the amount of principal that is returned at maturity. Generally, a coupon bond will sell initially at par and will stay at par if the term structure of interest rates does not change.

Population In probability theory, "population" is the probability distribution from which a sample is taken.

Probability distribution The set of possible values of a random variable and their probabilities.

Premium The initial cost of an option.

Probability density function Function giving the probability distribution of a continuous random variable. Also called PDF, density function, or density.

Put A derivative security giving the owner the right but not the obligation to sell the underlying security.

***p*-value** In hypothesis testing the *p*-value indicates the amount of evidence against the hull hypothesis. Small *p*-values lead us to reject the null hypothesis. A type I error occurs if the null hypothesis is incorrectly rejected. The probability of a type I error is called the level of the test and denoted by α, with typical values of α being 0.01, 0.05, and 0.1. The null hypothesis is rejected in the *p*-value is less than α. Statistical software routinely includes *p*-values of various tests in its output.

Quadratic programming A method for minimizing a quadratic function subject to linear equality and inequality constraints. Quadratic programming is used to find mean-variance efficient portfolios and can accommodate constraints such as a prohibition against selling short.

Random sample A random sample from a probability distribution is a collection of independent random variables each with that probability distribution.

Random variable A quantity that can take on many possible values of which exactly one will occur according to known probabilities.

Random walk A stochastic process that changes by steps that are a white noise process. This means that future changes in a random walk are uncorrelated with the present and past changes and so future changes are not predictable. The efficient market hypothesis predicts that stock prices (or their logarithms) will be random walks, and random walks are often used as models for stock (log) prices. *See also* geometric random walk.

Regression Modeling the expected value of a response variable as a function of predictor variables. Straight line regression models the response as a linear function of one predictor. In multiple linear regression, the response is a linear function of several predictors. In linear regression, the intercept and slopes are unknown parameters that are often estimated by least squares. In nonlinear regression, the response is a *known* function of the predictors and unknown parameters. Nonparametric regression is used when the response is an *unknown* function of the predictors.

Resampling *See* bootstrap.

Residual The difference between an observation and its prediction from a statistical model. Residuals are used to check models and to estimate expected prediction error. Least squares estimation minimizes the sum of the squared residuals.

Return Revenue during a holding period from an investment expressed as a fraction of the initial investment. There are several types of returns; *see* net return, gross return, and log return. Single period returns are for one holding period and multiperiod returns over several consecutive holding periods.

Risk Uncertainty in the future return from an investment. Risk is often understood to imply that the uncertainty is measurable, meaning that the return has a *known* probability distribution.

Risk factor A predictor variable in a factor model for an asset's return. Variation in a risk factor causes variation in the asset's return, that is, risk. Also called factor. The market return is the most commonly used risk factor. The difference between the returns on two portfolios is another common type of risk factor.

Risk-free asset An asset whose return is known in advance and therefore has a standard deviation of zero, for example, a U.S. Treasury bill that expires at the end of the holding period.

Risk premium The extra expected return that investors demand for bearing risk.

Risk-free rate The interest rate of risk-free assets.

Sample correlation The correlation calculated from the sample and used to estimate the population correlation.

Sample mean The average of the values in the sample.

Sample standard deviation A quantity that describes the dispersion of the sample.

SBC Schwarz's Bayesian Criterion. A statistic used for model selection. Often called SBC or BIC. SBC is similar to AIC, but, compared to AIC, models with a smaller number of parameters tend to be selected by SBC.

Security characteristic line A factor model with only one factor, the market return.

Security market line Relates an asset's risk premium (expected excess return of the asset) to its beta (the slope of the regression of the stock's excess return on the excess return of the market portfolio). Also called SML. The SML differs from the CML in that the SML applies to all assets while the CML applies only to efficient portfolios. The SML is obtained by replacing all random variables in the security characteristic line's equation by their expected values.

Self-financing A trading strategy is self-financing if after the initial investment no further capital is either invested or withdrawn.

Selling short Borrowing a stock to sell it, and then purchasing the stock at the end of a holding period to return it to the lender. Selling short is used to earn a profit when a stock's price decreases. Holding a short position is an asset means that the weight of that asset in one's portfolio is negative.

Skewness Skewness is used to measure the asymmetry of a probability distribution.

Sharpe's ratio The ratio of the excess expected return of an asset to its standard deviation.

Spline A function that is a polynomial on each interval between adjacent knots. Splines are frequently used as nonparametric regression models.

Spot rate The n-year spot rate is the yield to maturity of an zero-coupon bond maturing in n years.

Standard normal distribution The bell-curve with mean equal to 0 and standard deviation equal to 1.

Standard deviation A quantity that describes the dispersion of a random variable. The standard deviation of a random variable is the square root of its variance. The standard deviation of the return describes an asset's volatility and risk.

Standard error The standard error is an indication of the precision of an estimator. It is the standard deviation of the estimator.

Statistic A statistic is a quantity that is determined solely by the data, not the unknown values of parameters.

Stationary A stochastic processes whose probability distribution is unchanged by shifts in time is said to be stationary.

Stochastic process A sequence of random variables. Stochastic processes are used as models of time series.

Strike date The date at which a European option can be exercised. An American option can be exercised at or prior to the strike date.

Strike price The price at which an option can be exercised.

Tangency portfolio The portfolio with the largest Sharpe's ratio. The tangency portfolio is optimal for combining with the risk-free asset; that is, when a risk-free asset is available then the mean-variance efficient portfolios contain only the tangency portfolio and the risk-free asset.

Technical analysis Prediction of stock prices using only market data such as past returns and trading volume.

Term structure The dependence of forward rates and yields on time to maturity.

Time series A sequence of observations taken in time, for example, the daily prices of a stock.

Type I error The error when the null hypothesis is true and it is falsely rejected. The probability of a type I error is denoted by α.

Type II error The error when the null hypothesis is not true and it is accepted.

Underfitting Using a statistical model that is overly simple so that the model cannot fit all systematic features of the data and therefore gives predictions that are biased. *See also* overfitting and model selection.

VaR. Value at risk. A threshold such that the probability that a loss exceeds this threshold is equal to the confidence coefficient. VaR has become the industry standard for risk management though it has been criticized recently for ignoring the expected loss given that the VaR is exceeded. Expected shortfall is now being suggested as an alternative to VaR.

Variance The average squared deviation of a random variable from its expectation. The square root of the variance is called the standard deviation.

Volatility The variability of returns, typically measured by their standard deviation.

Volatility smile The curve seen when implied volatility is plotted against either strike price or strike date. According to the Black-Scholes theory, this plot should be a horizontal line. Thus, the existence of volatility smiles indicates that some assumptions of the Black-Scholes model are not met. More complex models where the underlying stock price follows a GARCH model can "explain" the volatility smile.

White noise WhiteNoise(μ, σ^2). An uncorrelated stochastic process with a constant mean μ and a constant conditional (and unconditional) variance σ^2. Gaussian white noise is an i.i.d. sequence of normal random variables.

Yield to maturity The average rate of return of a bond, including the loss (or gain) of capital because the bond was purchased above (or below) par. Often called simply "yield."

Zero coupon bond A bond making no payments of interest or principal until maturity.

Index

Springer Texts in Statistics *(continued from p.ii)*

Lehmann: Elements of Large-Sample Theory
Lehmann and Romano: Testing Statistical Hypotheses, Third Edition
Lehmann and Casella: Theory of Point Estimation, Second Edition
Lindman: Analysis of Variance in Experimental Design
Lindsey: Applying Generalized Linear Models
Madansky: Prescriptions for Working Statisticians
McPherson: Applying and Interpreting Statistics: A Comprehensive Guide, Second Edition
Mueller: Basic Principles of Structural Equation Modeling: An Introduction to LISREL and EQS
Nguyen and Rogers: Fundamentals of Mathematical Statistics, Volume I: Probability for Statistics
Nguyen and Rogers: Fundamentals of Mathematical Statistics, Volume II: Statistical Inference
Noether: Introduction to Statistics: The Nonparametric Way
Nolan and Speed: Stat Labs: Mathematical Statistics Through Applications
Peters: Counting for Something: Statistical Principles and Personalities
Pfeiffer: Probability for Applications
Pitman: Probability
Rawlings, Pantula, and Dickey: Applied Regression Analysis
Robert: The Bayesian Choice: From Decision-Theoretic Foundations to Computational Implementation, Second Edition
Robert and Casella: Monte Carlo Statistical Methods, Second Edition
Rose and Smith: Mathematical Statistics with *Mathematica*
Ruppert: Statistics and Finance: An Introduction
Santner and Duffy: The Statistical Analysis of Discrete Data
Saville and Wood: Statistical Methods: The Geometric Approach
Sen and Srivastava: Regressions Analysis: Theory, Methods, and Applications
Shao: Mathematical Statistics, Second Edition
Shorack: Probability for Statisticians
Shumway and Stoffer: Time Series Analysis and Its Applications
Simonoff: Analyzing Categorical Data
Terrell: Mathematical Statistics: A Unified Introduction
Timm: Applied Multivariate Analysis
Toutenburg: Statistical Analysis of Designed Experiments, Second Edition
Wasserman: All of Nonparametric Statistics
Wasserman: All of Statistics: A Concise Course in Statistical Inference
Weiss: Modeling Longitudinal Data
Whittle: Probability via Expectation, Fourth Edition
Zacks: Introduction to Reliability Analysis: Probability Models and Statistical Methods